T0215654

# Communications in Computer and Information Science 1102

*Commenced Publication in 2007*
Founding and Former Series Editors:
Phoebe Chen, Alfredo Cuzzocrea, Xiaoyong Du, Orhun Kara, Ting Liu,
Krishna M. Sivalingam, Dominik Ślęzak, Takashi Washio, Xiaokang Yang,
and Junsong Yuan

More information about this series at http://www.springer.com/series/7899

Quan-Lin Li · Jinting Wang · Hai-Bo Yu (Eds.)

# Stochastic Models in Reliability, Network Security and System Safety

Essays Dedicated to Professor Jinhua Cao
on the Occasion of His 80th Birthday

Springer

*Editors*
Quan-Lin Li
School of Economics and Management
Beijing University of Technology
Beijing, China

Hai-Bo Yu
School of Economics and Management
Beijing University of Technology
Beijing, China

Jinting Wang
School of Management Science
and Engineering
Central University of Finance
and Economics
Beijing, China

ISSN 1865-0929          ISSN 1865-0937   (electronic)
Communications  in Computer and Information Science
ISBN 978-981-15-0863-9          ISBN 978-981-15-0864-6   (eBook)
https://doi.org/10.1007/978-981-15-0864-6

This Springer imprint is published by the registered company Springer Nature Singapore Pte Ltd.
The registered company address is: 152 Beach Road, #21-01/04 Gateway East, Singapore 189721, Singapore

Jinhua Cao (a picture in the Xiangshan Park, Beijing, 2008)

Jinhua Cao's happy family (a picture in Lake Louise, Canada, 2018)

# Preface

This book is dedicated to Jinhua Cao on the occasion of his 80th birthday. The title "Stochastic Models in Reliability, Network Security and System Safety" reflects the research interests and achievements of this outstanding scientist. Jinhua is an ordinary Chinese scientist with a complete historical experience: from the victory in World War II to the rapid economic growth of China; from the behindhand environment of science and technology to their fast development in China. During such a critical development period of China, Jinhua has been able to combine high-level innovation ability in scientific work with remarkable successes and opportunities in high administrative and organizational positions. In the research field of reliability in China, he has played a key role in not only theoretical research and engineering applications but also in cultivating more and more students and engineering technicians, many of whom have become outstanding talents and excellent leaders in a lot of enterprises and industries related to reliability in China. In addition, we also need to emphasize that one of Jinhua's main contributions to China is that through offering many early classes and courses of reliability, he cultivated a lot of reliability talents who successfully fulfilled the urgent need of many practical areas in the process of China's economic development, such as machinery, electronics, aerospace, weaponry, and so forth. Therefore, the high admiration that Jinhua enjoys in the scientific community in China and even all over the world was witnessed by the enthusiastic response from the contributors of this book.

Jinhua Cao was born in Shanghai in December 1939, and received his education in the same city. Jinhua's middle school was affiliated with the Shanghai Hujiang University (i.e. a church university), whose name was changed to Shanghai Beijiao Middle School one year after his enrolment. When studying at the junior middle school, the first thing stimulating his interest in mathematics was the planar geometry class taught by an excellent teacher, whose talent in the planar geometry inspired Jinhua to look for extracurricular math books to read independently. Moreover, the education at senior middle school further promoted Jinhua's great interest in mathematics. In 1957, Jinhua's second year of senior middle school, he participated in the Shanghai Mathematical Competition and won first prize as well as the Shanghai Excellence Award, becoming the only winner from his school. Shortly thereafter, Chinese mathematicians living in Shanghai, such as Buqing Su, Chaohao Guo, and Daoxing Xia, made a series of mathematical reports for those winners (including Jinhua) of the Shanghai Mathematical Competition. All of these strengthened Jinhua's excellent grades in mathematics so that he happily completed his studies at the senior middle school. Of course, Jinhua's early success and learning confidence immeasurably made him believe that mathematics was his biggest interest and even the most ideal career to pursue in his life.

Jinhua Cao entered the University of Science and Technology of China (Beijing) with excellent scores and marks. He started his undergraduate studies in 1958 and finished in 1963. This was his favorite university with many outstanding professors in the field of mathematics from China, who had all undertaken the basic and professional courses, for example, Loo-keng Hua, Yuan Wang, Minyi Yue, Zhexian Wan, Xiaqi Ding, Zhongci Shi, Xiru Chen, Guozhi Xu, and so on. In the first half of 1963, Jinhua luckily followed Minyi Yue, and began his research of queuing theory as his under-graduate thesis. Based on this, Jinhua published his first paper (1963): "Some Problems of M/G/1 Queueing System in Which the Probability That a Customer Joins in Queue Depends on the Queue Size" – achieving an indelible memory.

After graduating with a much better understanding of queueing research, Jinhua started his first job at the Institute of Mathematics, Chinese Academy of Sciences, Beijing. Since then, his academic titles are listed as follows: 1963–1978, Research Assistant; 1978–1981, Research Associate; 1981–1987, Associate Professor; 1987–present, Professor; and he has been retired happily since 2005. During his stable career, it is worthwhile to mention that the Institute of Mathematics was changed to Institute of Applied Mathematics from 1978 to 1999, and further to the Academy of Mathematics and Systems Sciences in 1999.

Following his undergraduate thesis, Jinhua Cao joined the research team of Minyi Yue as his first job at the end of 1963, and continued his research in queuing theory. In the process of an initial research work, which got his entire career started, there were two things that Jinhua remembers deeply. One is studying *Markov chains with stationary transition probabilities* by Chung (1960), and the other is discussing frontier directions of queuing theory. In fact, in the early 1960s, the international frontier of queuing research included the embedding Markov chain, the supplementary variable method, the random walk, the differential integral equation, and so forth. Although the theoretical research and technological developments were simple by today's standards, they were the most advanced research topics in the field of international queuing theory in the 1960s. The queueing research group (including Jinhua) at the Institute of Mathematics, Chinese Academy of Sciences, completed high-quality research works on queuing systems during that period.

Jinhua's career as a researcher and teacher is outstanding. He is the author of eight scientific books or book chapters as well as over 100 papers in various journals and conference proceedings. He has supervised a number of MSc and PhD theses with excellent marks. He has received nearly ten research awards from the Chinese Academy of Sciences and the government of China. On the other hand, he also has a rich background and experiences in international cooperation and exchange with peers in many countries and districts, such as the USA, Canada, Japan, Europe, Taiwan (China), Hong Kong (China), and so forth. In this book, an internationally famous reliability expert (the president of the City University of Hong Kong), Way Kuo, wrote a Chinese couplet dedicated to Jinhua's 80th birthday celebrations, which was truly inspiring and exciting, with a lot of appreciation for Jinhua's valuable contributions.

One of the main contributions by Jinhua Cao is on how to set up some basic relations between queueing theory and reliability. See "Analysis of M/G/1 Queueing System with Repairable Service Station" by Cao and Cheng (1982) for more details, which is always interesting but difficult in the study of stochastic models (or systems). In fact, Jinhua's early work quickly became very important and useful in much sub-sequent research in emerging engineering fields, for example, manufacturing systems in the 1980s to 2000s (see the books: *Stochastic Models of Manufacturing Systems* by Buzacott and Shanthikumar (1993), and *Manufacturing Systems Engineering* by Gershwin (1994)); information and network security (see the books: *Information Security Management Handbook* by Tipton and Nozaki (2007) and *Computer Security: Principles and Practice* by Stallings et al. (2012)); and so forth. The research of repairable queues by Cao and Cheng (1982) is so fundamental and significant that it has promoted a long-term research boom of queueing theory as well as reliability engineering in China. Also, some of their excellent papers were published in leading journals that receive much attention. In addition, the extension and generalization of repairable queues were further developed to either block structure or network architecture.

Although the *embedding Markov chain* by Kendall (1953), together with the semi-Markov process, the Markov renewal process, and the regeneration point process, had obtained many theoretically critical advances in the 1960s to 1980s, it is still very difficult and challenging to deal with complicated stochastic models in practice when using these mathematical methods. Within such a field, Jinhua Cao completed more works to discuss various reliability (repairable) models which provided better reference examples for how to apply the Markov renewal process, the embedding Markov chain, and the regeneration point process to deal with practical and real stochastic systems. Based on this, the book *Introduction to Reliability Mathematics* by Cao and Cheng (1986) provided a systematic summary for the reliability (repairable) models analyzed by using the Markov renewal process, the embedding Markov chain, the regeneration point process, and the supplementary variable method. In addition, the book also conducted some analysis for other interesting topics of reliability theory, for instance, life distribution classes, coherent systems, fault trees, maintenance policies, reliability statistics, and so on. So far, Cao and Cheng (1986) have played an important role in and generated a high impact on not only reliability courses of undergraduate seniors and graduate students but also academic research and engineering applications in China. To master the basic methods of stochastic modeling and analysis, the key is to cultivate and develop the ability of students to solve various practical stochastic problems by applying Markov processes, Markov renewal processes, Markov reward processes, Markov decision processes, and stochastic game theory. Therefore, from the perspective of cultivating students, during the last four decades, Jinhua's research and his book have played a key role in reliability education, academic research, and engineering applications in China. Finally, we also mention that Jinhua's book is the most frequently cited one among many Chinese reliability books.

Over the last four decades (1963 to 2005), Jinhua's research group carried out a number of outstanding research works including life distribution classes, repairable systems, repairable queueing systems, multi-server vacation queues, retrial queues, production-inventory systems, computer integrated manufacturing systems, and the matrix-analytic methods. Here, we refer to one of Jinhua's theoretical works, which gives a basic understanding of general Markov processes. Since the late 1970s, Marcel F. Neuts proposed and developed two seminal works: The phase-type (PH) distribution and the Markovian arrival process (MAP), see the book: *Matrix-Geometric Solutions in Stochastic Models: An Algorithmic Approach* by Neuts (1981) for more details. Based on these works, he further developed the matrix-geometric solutions in stochastic models, which is regarded as an important breakthrough of theoretical research in the field of queuing theory. In light of this research direction, the paper "Two Types of RG-factorizations of Quasi-birth-and-death Processes" by Li and Cao (2004) proposed and developed two types of RG-factorizations in a general quasi-birth-and-death (QBD) process. The RG-factorizations play an important role in the study of stochastic models. First, the RG-factorizations construct a new theoretical system of Markov processes by means of the Wiener-Hopf equations as well as the infinite-dimensional Gaussian elimination, thus they are a unified common property of general Markov processes. Second, by using the RG-factorizations, some difficult problems of Markov processes have been solved recently, for example, quasi-stationary distribution, Poisson's equation, perturbed Markov chain, and so on. Third, the RG-factorizations extend and generalize the matrix-geometric solution by Neuts (1981) to a new version such that the stationary performance, the transient solution, the first passage time, and the sojourn time can be discussed in a unified computational framework. In addition, the RG-factorizations by Li and Cao (2004) were further discussed and practically applied by some famous scholars. Crucially, by using various useful relations between the random walk and the Markov chain, the paper "LU-factorization Versus Wiener-Hopf Factorization for Markov Chains" by Vigon (2013) systematically proved that the two types of RG-factorizations offer a complete factorization framework of general Markov processes.

Starting from 1978, China has been strengthening its economic construction and development. In such an economic climate, reliability theory and engineering applications become increasingly important. There is an especially high demand for reliability in management departments of Chinese government and in many kinds of enterprises and industries. Therefore, Jinhua held many courses of reliability theory and engineering applications in China from 1981 to 2005, in which those learners came not only from faculty and graduate students in Chinese universities and research institutes, but also from reliability engineering technicians in enterprises as well as managers in Chinese government departments. Later, many of these people became outstanding talents and excellent leaders in a lot of enterprises and industries related to reliability in China, so that they were able to meet China's actual need and demand with respect to reliability theory, technologies, and engineering applications. Until today, many major scientific and engineering projects in China still insist on the one-vote veto system of reliability assessment. From these levels and perspectives, Jinhua's reliability work is a long-term contribution to the economic development of China.

It is no wonder that Jinhua Cao is one of the most active and influential reliability experts in China. It goes back to 1975 when Jinhua began to organize some reliability seminars for Chinese scholars and students in Beijing. Based on such early, solid preparation for both scholars and reliability architecture, in 1981 Jinhua pushed to set up the Reliability Society of China, which belongs to the Operations Research Society of China. Under Jinhua's great patience and efforts, the first seven Chinese conferences of reliability mathematics were held from 1982 to 2005 in mainland China: 1982 (Wuxi), 1985 (Shanghai), 1989 (Xi'an), 1992 (Guilin), 1995 (Chengde), 1998 (Taian), and 2005 (Beijing). After Jinhua retired in 2005, as an honorary president of the Reliability Society of China, he continued to provide active enthusiastic guidance and help for the subsequent three conferences: 2009 (Nanjing), 2013 (Changsha), and 2017 (Beijng). In addition, from 1975 to 2005, Jinhua still established other interesting conferences, for example, Sino-Japanese reliability academic conference, Shanghai, September 13–16, 1987, in which *Reliability Theory and Applications* with 42 selected papers, edited by S. Osaki and J. Cao, was published by World Scientific. Moreover, Jinhua's academic activities and organization efforts also include other research directions, such as software reliability, risk management, network and information security, system maintenance and safety, etc. Therefore, we cannot thank Jinhua enough for not only his selfless and generous but also hard work of academic organization during such a long period. This really drives the whole development of reliability theory, technologies, and applications in China, and also builds a strong bridge of exchanges and cooperation among the reliability organizations from China to other countries.

Since 2005, Jinhua has been living a very happily retired life with his family in China and Canada. Jinhua's son, Ye Cao, is excellent, with a happy family of three, the most important one of whom is Jinhua's lovely small granddaughter, who likes physical exercise, loves science, has the best imagination, and so much more. Jinhua loves his son's family, but his son's family loves him best. Because of this, Jinhua travels between China and Canada, living a life surrounded by beauty, happiness, and endless joy. In addition, Jinhua loves his country for trees, mountains, rivers, and blue skies; and he also likes Beijing very much for imperial palaces, historic parks, and grand and spacious city layouts. From here and from there, really and truly, Jinhua has a very healthy body, and he always briskly walks in either beautiful Beijing streets or interesting parks every day. Such a beautiful environment surely sustains his good mood and healthy mindset which are very important for him at all times.

For this book, we received 31 submissions from mainland China, Hong Kong, Japan, the UK, and the USA, most of which were invited from famous research groups related to reliability, stochastic operations research, and artificial intelligence. All papers were peer-reviewed and evaluated on the quality, originality, soundness, and significance of their contributions by many high-level reviewers invited from different countries. Finally, 25 papers were accepted as full papers appearing in this CCIS proceedings published by Springer, 4 congratulatory messages within 2 pages were accepted as congratulations on Jinhua's 80th birthday, and an appendix lists Jinhua's past publications. We are very grateful to each author of this book and all the invited reviewers. Our special thanks also go to professors Way Kuo, Xiaodong Hu, and Wei Li for their great help and support in publishing this book. We thank the National Natural Science Foundation of China, the Operations Research Society of China, and the Reliability Society of China for their great support and encouragement.

Each of the three editors of this book has a warm and close friendship with Jinhua, which developed throughout the past many years. In particular, Quan-Lin and Jinting are grateful to him as their doctoral supervisor; while Haibo is also grateful for many years of great friendship. We wish Jinhua continuing success, happiness, satisfaction, and health in science, life, and travels in the years to come.

With the help of this book, all of us, new and old friends, warmly congratulate Jinhua on his 80th birthday, and wish him a long and healthy life.

September 2019

Quan-Lin Li
Jinting Wang
Hai-Bo Yu

# A Sincere Congratulations on Jinhua Cao's 80th Birthday

Way Kuo

President and University Distinguished Professor
City University of Hong Kong, Kowloon, Hong Kong
office.president@cityu.edu.hk

I am pleased to have known Professor Jinhua Cao for over 30 years. When I first visited the Institute of Applied Mathematics, Chinese Academy of Sciences in 1988, Professor Cao was the host of my visit. Since then, we met socially in various professional

Way Kuo offered congratulations in September 2019

meetings and international conferences. I was so impressed by Professor Cao's accomplishments and personality. He is renowned for his work in reliability theory, stochastic operations research, and stochastic models of manufacturing systems. He has always been an ardent supporter of high-quality engineering education and is a pioneer of reliability mathematics in the mainland. I hold him in high esteem.

# A Tribute to Professor Jinhua Cao on His 80th Birthday

Wei Li

Department of Computer Science
Texas Southern University
Houston, TX 77004, USA
LiW@tsu.edu

Talented. Jolly. Determined. Just a few words that describe my wonderful professor, Jinhua Cao, an emeritus fellow from the Chinese Academy of Science (CAS). As the first PhD student under Professor Cao's supervision, currently a Professor in the Department of Computer Science and the Director of Center for Research on Complex Networks supported by the National Science Foundation (NSF) at Texas Southern University, I sincerely wish you a very happy 80th birthday, Professor Cao! May all your wishes come true.

I could purchase almost anything I want with money, but the manners, morals, and integrity you taught me, I cannot. Thank you, Sir, for teaching me how to be a good human being. I still clearly remember the first time I visited you in your home in 1990 – it was an important event in my life. This visit, augmented by everything I learned from you later, completely changed my career. You have always fully supported me, from pursuing my PhD degree under your co-supervision in 1991 to the excitement of starting my employment in 1996 at the CAS, and even to assuming challenging positions as a tenure-track and eventually tenured Professor and Administrator at several research universities in the United States. The beautiful memories of interacting with you for almost over 30 years are always joyous.

At this moment of celebrating your 80th birthday, I find myself deeply reflecting the impact of the manners, morals, and integrity I learnt from you. I first witnessed your serious and thoughtful manner of research whenever I joined your research group in 1991, although at that time I already knew that you had been recognized as a well-known researcher who published numerous research papers in the reliability theory and related subjects and also a highly cited book with the title of *Introduction to Reliability Mathematics*. Your various corrections and modifications on my first research paper in 1992, among the 15 published papers I accomplished under your supervision for prestigious publications, were very impressive and made me realize how much I could learn from you and how extensive and perceptive your knowledge was. Participating in the 4th National Conference of Reliability Society of the Operations Research Society of China (RS-ORSC) in Guilin in 1992 with you gave me the opportunity to perceive your wisdom and acumen as the President of RS-ORSC. These observations have had a significant impact on me in all my academic jobs both in China and in the United States. Thank you, Professor Cao, for inspiring me not only to read professional journals but also to look beyond classroom walls and reach heights that were beyond my expectations. Your guidance during my co-organization for the

First Youth Reliability Conference of the RS-ORSC in Qinhuangdao in 1994 taught me the first lesson of your moral and ethical standards in dealing with various partnerships, which was the key to making the conference a great success. Your exemplary daily working style in professional research has been the consummate model for my humble efforts. Your amiable countenance in dealing with various societies, no matter how difficult or contentious they were, has influenced me to respond in a similar manner. Your example has taught me, as well as your other PhD Students and Postdoctoral Fellows I know, such as Bin Liu, Dequan Yue, Quan-Lin Li, Jinting Wang, and Chenghong Wang, et al., how to maintain thoughtful manners, good morals, and high integrity throughout our entire careers.

You opened my mind to a thousand new things. Thank you for being a terrific professor and giving me wings. Have a wonderful 80th birthday, Prof. Jinhua Cao!

# Dedicated to Prof. Jinhua Cao's 80th Birthday

Xiaodong Hu

Institute of Applied Mathematics
Academy of Mathematics and Systems Science
Chinese Academy of Sciences
Beijing 100190, China

I was a graduate student at the Operations Research Department (ORD) of Institute of Applied Mathematics (IAM), Chinese Academy of Sciences (CAS), during 1985–1989, where I came to know and got acquainted with Professor Jinhua Cao. I joined IAM after I received my PhD in 1989. Prof. Cao was the director of ORD during 1987–1997. As one of the graduate students and young researchers at ORD, I learned quite a lot from Prof. Cao. Time flies, 30 years have passed since I graduated and joined IAM. On the occasion of celebrating Prof. Cao's 80th birthday, I'm very pleased to write an essay about Prof. Cao.

Prof. Cao graduated from the University of Science and Technology of China (Beijing) in 1963. Because of his excellent performance at the university, he was recommended to the Institute of Mathematics of CAS as an Intern Research Fellow after his graduation. He joined the Operations Research Group in the institute and began his academic career under the supervision of Prof. Minyi Yue, who was the leader of the group and now is considered one of the pioneers of operations research in China. With the strong promotion from the world-famous mathematician Prof. Loo-keng Hua, CAS established the Institute of Applied Mathematics (IAM) in 1979 and appointed Prof. Hua as the founding director of IAM. As Prof. Cao worked on reliability and queueing theory, he was assigned to the new institute, together with some colleagues including Prof. Yue. Since then, Prof. Cao has always been working in ORD of IAM until he retired in 2005.

In nearly 50 years of his academic career, Prof. Cao's work covers many key topics of operations research. In particular, he has made great contributions to reliability, queuing theory, stochastic operations research, and stochastic optimization. He has published more than 100 research articles, some of them have had a great impact worldwide. In addition, he has supervised a lot of graduate students including Ph.D. students, Master student as well as Postdoctoral fellows during his whole academic life at IAM, some of them have become famous scholars not only in China but also in the world. As one of the major contributors, Prof. Cao received the National Science Congress Award of China in 1978.

Furthermore, I would like to mention a few words on Prof. Cao's contribution to the development of Operations Research Society of China (ORSC), which was founded in 1980. In 1981 ORSC established the Reliability Society, which is the earliest established one among all 15 Societies of ORSC. Prof. Cao played an important rule in the creation of this Reliability Society and then paid great attention to its development and innovation until his retirement in 2005. As the executive manager and also once the

President of Reliability Society of ORSC since 1981, it is no doubt that Prof. Cao will be recognized forever as a key founder in the history of the Reliability Society of ORSC.

At the end, as the president of ORSC, I would like to express my sincere thanks to Prof. Cao's contributions to the operations research development in China, and I personally feel very grateful to his invaluable support towards myself in the past 30 years.

# Foreword in Honor of Professor Jinhua Cao's Birthday

Jinting Wang
School of Management Science and Engineering
Central University of Finance and Economics
Beijing 100081, China
jtwang@cufe.edu.cn

While celebrating Professor Jinhua Cao's 80th birthday, as President and on behalf of the Reliability Society of the Operations Research Society of China (RS-ORSC), I would like to express our hearty gratitude to him for his lifetime contribution to the exploration and development of Reliability Theory research in China and particularly to the creation and advancement of the RS-ORSC.

Professor Cao devoted himself entirely to the education, training, and research of various nationwide reliability theory activities when he was first employed as a research assistant in the Chinese Academy of Science (CAS) in 1963. Having continued on the first reliability mathematics workshop, which was organized by him in Beijing in 1980, Professor Cao organized over 30 different training classes on reliability theory in various universities, research departments, and enterprises, among others, to cultivate graduate students and young teachers in his subsequent 25-year academic career in the CAS. His contribution through the lectures and research examples has established a solid foundation in the research of reliability for all attendees. Several hundred academic researchers and engineering technicians benefited greatly from him, most of whom later on became outstanding professors and excellent leaders in the reliability area of China. In addition to his more than 100 prestigious research publications, it is worth noting that the book "Introduction to Reliability Mathematics," authored by Jinhua Cao and Kan Cheng in 1986, has played a principal part in the history of reliability theory research in China. As of today, this book has been cited over 2,000 times, according to CNKI citation, by scholars from all over the world. Now, it has become the first book for beginners in the reliability theory area to read because it condensed many years of the authors' experience and wisdom. From my observations over the last 20 years, it is clear that several generations of graduate students and young teachers in both operations research and applied statistics have learned their reliability theory from this book. Without a doubt, it will continue to be used and to be highly cited in the literature. Nowadays, many reliability talents working in diverse areas such as machinery, electronics, aerospace, and so forth, are influenced by Professor Cao's research through this book. And truthfully, as one of three editors of this book, when I edited the collection of essays in honor of Professor Cao's 80th birthday, I witnessed and felt in my heart the enthusiastic responses and respects from contributors around the world who benefited from Professor Cao's book and knowledge.

In addition to his devotion to Chinese reliability theory education and research, Prof. Cao's life was also dedicated to the initiation and the innovation of Reliability Society of the Operations Research Society of China (ORSC) and his tremendous effort finally resulted in the official establishment of RS-ORSC in 1981. Prof. Cao was an active executive, as well as President, of RS-ORSC for 25 years, from 1981 to 2005. He planned and designed each of the RS-ORSC annual meetings together with other senior leaders; attended with guidance each of national meetings within RS-ORSC, including the first National Youth Reliability Conference in 1994; and originated almost every one of the subcommittees within RS-ORSC. With his leadership, the RS-ORSC established a clear vision and mission for the promotion and innovation of Reliability Theory research in China. Over the last about 40 years under the proposed direction, the RS-ORSC has grown rapidly and has been widely recognized as the premier established reliability society in the nation and beyond, with a large number of registered members who are working in diverse research directions of the reliability field. As of now, the RS-ORSC has organized over 50 national academic conferences and organized/sponsored over 20 international academic events since 1981. I cannot omit mentioning the first RS-ORSC international conference – the Sino-Japan joint conference in 1987 – organized with Professor Cao's key leadership. This conference was a tremendous success and had a far-reaching impact internationally. As Professor Cao directed, the RS-ORSC will continuously strive to achieve its long-term goal, that is, to unite the vast number of reliability researchers, to create a relaxed, harmonious and united academic atmosphere, and to contribute to the development of China's society and economy. Professor Cao served once as our former President and has been serving as an Honorary President of the RS-ORSC since he started enjoying his retirement life in 2005, and he will continue to play an active role in the promotion and development of reliability theory research in China. We do believe, that if we continue to advance with the innovation guidance established by Prof. Cao and all other former senior leaders, the RS-ORSC will certainly be moving at a vigorous pace towards the international scientific and technological arena.

Other students are probably jealous of those who are fortunate to have Professor Cao as a supervisor. As I know, many of his students have achieved great success later in various positions, e.g., as the president of a professional society, as an established professor in the nation and overseas, or as a successful manager in the business world, etc. Personally, the commemoration of his 80th birthday has a dual significance for me, not only because I acquired a rich knowledge of reliability theory from him but also because I eventually chose reliability research as my academic career. All due to his impact on me. Finally, I would like to point out that Professor Cao's example in research and education has influenced me very much during the past 20 years and particularly has taught me how to maintain integrity and honesty during my whole career. Thank you very much, Professor Cao! Wishing you a long and healthy life!

# Contents

# Review Papers

# Consecutive $k$ and Related Models—A Survey

Lirong Cui[1($\boxtimes$)] and Qinglai Dong[1,2]

[1] School of Management and Economics, Beijing Institute of Technology,
Beijing, China
Lirongcui@bit.edu.cn
[2] School of Mathematics and Computer Science, Yan'an University,
Yan'an, Shaanxi, China

**Abstract.** As one of the most popular reliability models, the previous several decades have witnessed remarkable developments and extensive applications of consecutive $k$ systems, and a number of related models have been developed. In the paper, a summary of the state of the arts in the field is provided. After a brief introduction of conventional consecutive $k$ systems, we focus on variants of the consecutive $k$ systems by considering failure criteria (single failure criterion and multiple failure criteria), geometric structure of the system, states of components and the system, weight of each component, dependency of components. Finally, several future challenges deserving further studies are highlighted.

**Keywords:** Consecutive $k$-out-of-$n$ systems · System reliability · Component importance · System signature

## 1 Introduction

Consecutive $k$ systems were first introduced by Kontoleon (1978), and the name consecutive $k$-out-of-$n$: F system was first coined by Chiang and Niu (1981). In practice, consecutive $k$ systems can be used to model the spatial distributed systems, such as the telecommunication systems, oil pipeline systems, photography of a nuclear accelerator, vacuum systems in an electron accelerator, computer ring networks, supervision systems, pattern detection systems, rows of street lights, and so on (Chang et al. 2000). Based on the geometric structure of the system, there are two basic models for consecutive $k$ systems, i.e., the linear consecutive $k$-out-of-$n$: F systems and the circular consecutive $k$-out-of-$n$: F systems, whose closed reliability formulae were given by Lambiris and Papastavridis (1985). The linear (or circular) consecutive $k$-out-of-$n$: F system consists of $n$ components ordered in a line (or a circle), and the system fails if and only if there exist at least consecutive $k$ failed components. For the circular consecutive $k$-out-of-$n$: F system, it is assumed that the first component is adjacent to (and follows) the $n$th component in the system (Derman et al. 1982). If the condition for continuous arrangement of $n$ failed components is weakened, i.e., the system fails if there are at least $k$ failed components, it will become the $k$-out-of-$n$: F system. When $k = 1$, the $k$-out-of-$n$: F system and the consecutive $k$-out-of-$n$: F system reduce to the series system, and when $k = n$, the $k$-out-of-$n$: F system and the consecutive $k$-out-of-$n$: F system reduce to the parallel system. Comparing with series systems (the reliability is low) and parallel systems (the reliability is high but they tend to be very expensive),

© Springer Nature Singapore Pte Ltd. 2019
Q.-L. Li et al. (Eds.): Cao Festschrift 2019, CCIS 1102, pp. 3–18, 2019.
https://doi.org/10.1007/978-981-15-0864-6_1

consecutive $k$ systems have two main advantages: much higher reliability than series systems and less expensive than parallel systems (Chao et al. 1995). Based on the description of the system from the perspective of working or failed, consecutive $k$-out-of-$n$: F systems and consecutive $k$-out-of-$n$: G systems are two basic models of consecutive $k$ systems. The linear (or circular) consecutive $k$-out-of-$n$: G system is defined as a system that consists of $n$ components ordered in a line (or a circle), and the system works if and only if there exist at least consecutive $k$ working components. From the definition of a dual structure (Barlow and Proschan 1965), the consecutive $k$-out-of-$n$: F system and the consecutive $k$-out-of-$n$: G system are dual systems, and duality also holds between series and parallel systems, $k$-out-of-$n$: F and $k$-out-of-n: G systems (Kuo et al. 1990; Cui et al. 2006).

Due to the motivation of theory development and practical application in reliability field, consecutive $k$ systems have caught the attention of many engineers and researchers and have been widely studied. So far, there is a concise monograph written by Chang et al. (2000). Extensive reviews can be found in Ge (1993), Chao et al. (1995), Kuo and Zuo (2003), Eryilmaz (2010), Daus and Beiu (2014), Sen et al. (2015). To facilitate the research on reliability, this paper devotes to the review of consecutive $k$ and related models. In the following, a number of references will be mentioned, but not all published research on consecutive $k$ and related systems will be covered. We apologize that some interesting articles might be omitted from this short survey.

At present, in order to accommodate more flexible operation principles, a lot of modifications or generalizations of the consecutive $k$-out-of-$n$: F (G) system have been proposed in the following directions: failure criteria (single failure criterion and multiple failure criteria), geometric structure of the system, states of components and the system, weight of each component, dependency of components, and so on. The taxonomy of consecutive $k$ and related models is shown in Fig. 1.

## 2  Extended Systems Based on the Failure Criteria

The system failure criteria may be single or multiple; therefore, there are two classes of extended systems based on the failure criteria.

(1) Extended systems based on the single failure criterion. In the case of single failure criterion, there are three basic models of consecutive $k$ systems: consecutive $k$-out-of-$n$: F (G) systems, $m$-consecutive-$k$-out-of-$n$: F (G) systems and consecutive $k$-within-$m$-out-of-$n$: F (G) systems, where the latter two systems are also called window systems, which were first introduced by Griffith (1986). In the literature, there are some alternative names, such as consecutive $k$-out-of-$m$-from-$n$: F (G) and $k$-within-consecutive $m$-out-of-$n$: F (G) which were all used for the consecutive $k$-within-$m$-out-of-$n$: F (G) system. The $m$-consecutive-$k$-out-of-$n$: F (G) system fails (or works) if and only if there exist at least $m$ non-overlapping runs of consecutive $k$ failed (or working) components, and if $m = 1$, it reduces to the consecutive $k$-out-of-$n$: F (G) system. Agarwal et al. (2007b) extended the Griffith's model to the $m$-consecutive-atleast-$k$-out-of-$n$: F system. The consecutive $k$-within-$m$-out-of-$n$: F (G) system fails (or works) if there exists a window

consisting of $m$ non-overlapping consecutive components in which at least $k$ components fail (or work). If $k = m$, it reduces to the consecutive $k$-out-of-$n$: F (G) system, and if $k = n$, it reduces to the $k$-out-of-$n$: F (G) system. In fact, except of non-overlapping, there exists another way to count the runs: overlapping. $M$-consecutive-$k$-out-of $n$: F system with overlapping runs was proposed and studied by Agarwal and Mohan (2008), in which "the system fails if and only if there exist at least $m$ overlapping runs of $k$ consecutive failures". Considering that the system fails if and only if at least $m$ times $l$-overlapping runs of consecutive failed (working) components, Eryilmaz and Mahmoud (2012) proposed the linear $m$-consecutive-$k$, $l$-out-of-$n$ system and gave the number of path sets, reliability and signature by using of a combinatorial method. Cui et al. (2015) gave the reliability formulae for the linear and circular $m$-consecutive-$k$, $l$-out-of-$n$ systems by use of the finite Markov chain imbedding approach (FMCIA), which was invented by Fu and Koutras (1994), and a survey paper on the developments and applications of FMCIA in reliability was given by Cui et al. (2010). In order to gain more information on both ways to count the runs, we can refer the reader to Levitin (2005), Zhao et al. (2007), Levitin and Dai (2011), Gera (2011), Eryilmaz (2012), Eryilmaz and Bayramoglu (2012), Zhu et al. (2017), and so on.

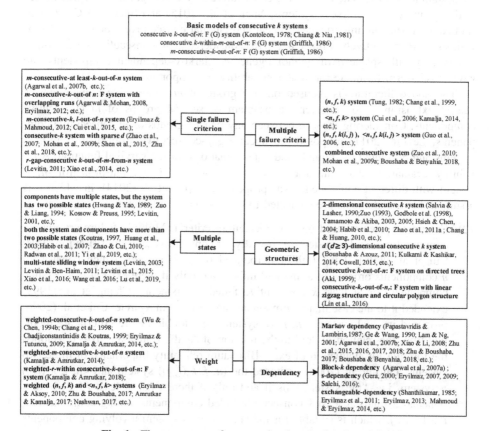

**Fig. 1.** The taxonomy of consecutive $k$ and related models

The research contents of three basic models of consecutive $k$ systems involve reliability analysis (e.g., Bollinger 1982; Sfakianakis et al. 1992), importance analysis (e.g., Chang et al. 2002; Kamalja 2012, 2014; Eryilmaz 2013a; Shen et al. 2015; Zhu et al. 2017, 2018); signature (e.g., Eryilmaz et al. 2011; Triantafyllou and Koutras 2011; Kulkarni and Kashikar 2014), the bounds for the system reliability (e.g., Cai 1994; Daus and Beiu 2014), algorithms to evaluate the system reliability (e.g., Papastavridis and Sfakianakis 1991; Psillakis 1995), stochastic orderings (e.g., Boland and Samaniego 2004; Eryilmaz 2011; Salehi et al. 2012; Salehi 2016), optimal arrangement of components (e.g., Malon 1985; Sfakianakis 1993; Cui and Hawkes 2008; Shingyochi et al. 2010, 2015, 2016), repairable consecutive $k$ systems (e.g., Papastavridis and Koutras 1992; Zhang and Lam 1998; Lam and Ng 2001; Xiao and Li 2008; Tang and Hou 2012; Lam and Zhang 2015), and so on.

Based on three basic models, many other extended models have also been proposed. Considering the concept of sparse $d$ (i.e., "if the number of working components between two failed components is less than $d$, the two failed components can be viewed as consecutive"), Zhao et al. (2007) proposed the consecutive $k$ systems with sparse $d$ and gave formulae for the system reliability. Mohan et al. (2009b) applied the Graphical Evaluation and Review Technique (GERT) in the same systems and studied the system reliability. When $d = 0$, the consecutive-$k$ systems with sparse $d$ will reduce to the ordinary consecutive $k$ systems. Shen et al. (2015) and Shen and Cui (2015) studied the Birnbaum importance of linear and circular consecutive-$k$-out-of-$n$ systems with sparse $d$, respectively. Zhu et al. (2018) considered an $m$-consecutive-$k$-out-of-$n$: F (G) system with sparse $d$ of nonhomogeneous Markov-dependent components and gave closed-form formulae for system reliability and importance.

The above literature is focused on one group of $m$ consecutive components, a natural extension is to consider the problem of several groups of $m$ consecutive components. Considering the gap between any pair of groups of $m$ consecutive components, Levitin (2011) proposed the linear $r$-gap-consecutive $k$-out-of-$m$-from-$n$: F systems and presented the bounds and approximations for the survival function. A reliability evaluation algorithm for a linear $r$-gap-consecutive $k$-out-of-$m$-from-$n$: F system with different components was proposed by Xiao et al. (2014), in which it is assumed that the elements are subjected to the common cause failures.

(2) Extended systems based on multiple failure criteria. Multiple failure criteria are common for complex systems, especially for consecutive $k$ systems (Levitin 2004; Cui et al. 2006). Tung (1982) introduced $(n, f, k)$ system, which consists of $n$ components ordered in a line and fails if and only if there exist at least $f$ failed components or at least $k$ consecutive failed components. Chang et al. (1999) extended it to the case that $n$ components are ordered in a circle. Cui et al. (2006) proposed its dual system: $<n, f, k>$ system, i.e., double failure criteria are modified from the relation of "or" to the relation of "and". Kamalja (2014) developed the formula for evaluation of exact Birnbaum reliability importance of these two systems. Guo et al. (2006) proposed the $(n, f, k (i, j))$ system and $<n, f, k (i, j) >$ system, i.e., "the system fails if and only if there exist at least $f$ failed components or (and) at least $k$ consecutive failed components among components $i$, $i + 1, \ldots, j$, which is suitable for the analysis of a system involving components

requiring special attention". In fact, $(n, f, k)$, $<n, f, k>$, $(n, f, k\ (i, j))$ and $<n, f, k\ (i, j)>$ systems can be viewed as the combination of failure criteria of $k$-out-of-$n$ system and consecutive-$k$-out-of-$n$ system. Based on combined failure criteria of consecutive $k$ systems, Zuo et al. (2000) introduced three combined consecutive systems including combined $k$-out-of-$n$: F & consecutive-$k$-out-of-$n$: F, combined $k$-out-of-$mn$: F & linear connected-$(r, s)$-out-of-$(m, n)$: F system, combined $k$-out-of-$mn$: F, consecutive-$k_c$-out-of-$n$: F & linear $(r, s)$-out-of-$(m, n)$: F system. Considering two failures criteria, i.e., if and only if there exist at least non-overlapping runs of $k$ consecutive failed components, or at least $k_b$ consecutive failed components, where $k_c < mk$, Mohan et al. (2009a) developed a combined-consecutive-$k$-out-of-$n$: F & consecutive-$k_b$-out-of-$n$: F system. Boushaba and Benyahia (2018) proposed a combined $m$-consecutive-$k$-out-of-$n$: F & consecutive-$k_b$-out-of-$n$: F systems.

# 3 Multi-state Consecutive $k$ Systems

In practice, a system may operate in degraded states, i.e., there may exist many intermediate states between full working state and failed state (Zhao and Cui 2010). Therefore, it is much more flexible to consider the multi-state consecutive $k$ systems. Hwang and Yao (1989) proposed the first multi-state generalization of the consecutive $k$-out-of-$n$ system. Thereafter, Zuo and Liang (1994) and Kossow and Preuss (1995) studied the algorithms of the system reliability. For these systems, their components have more than two possible states, but the system has only two possible states. Koutras (1997) proposed a three-state consecutive-$k$-out-of-$n$: F system, in which the system and each component have one working state and two different failure states. For more general multi-state systems, there are more than two possible states for both the system and components. Huang et al. (2003) generalized the consecutive $k$-out-$n$: F system to the multi-state case and gave the exact reliability of decreasing multi-state system and bounds of increasing multi-state system, in which the system state is below $j$ if at least $k_l$ consecutive components are in states below $l$ for all $j \leq l \leq M$. Similarly, Radwan et al. (2011) generalized the consecutive $k$-out-of-$r$-from-$n$: F system to multi-state case and gave bounds of increasing multi-state system. Assuming that if at least $k_j$ components out of $r$ consecutive components are in state $j$ or above the system is in state $j$ or above, Habib et al. (2007) proposed a multistate consecutive $k$-out-of-$r$-from-$n$: G system.

Another generalization of the linear consecutive $k$-out-of-$r$-from-$n$: G system was provided by Levitin (2003) in which the system is named as "linear multi-state sliding window system", which fails "if the sum of the performance rates of any $r$ consecutive multi-state elements is lower than a minimum allowable level". The linear multi-state sliding window system was generalized to the multi-state sliding window system with $m$ consecutive overlapping windows by Levitin and Ben-Haim (2011). Xiao et al. (2016) proposed the multi-state $k$-within-$m$-from-$r/n$ sliding window system, which fails "if at least $k$ groups out of $m$ consecutive groups of $r$ consecutive multi-state elements have cumulative performance lower than the demand $W$", and the system covers the models of Levitin (2003), Levitin and Dai (2011), Levitin and Ben-Haim (2011) as special cases. Lu et al. (2019) considered a sliding window system with two failure modes: "the total number of failed groups of $r$ consecutive elements reaches $k$ or the gap between any two adjacent failed groups of $r$ consecutive elements is less than $m$ working groups" and analyzed the reliability, importance and the optimal component allocation. In practice, the sliding window system can also be regarded as an effective tool to describe the consecutive connected systems with multi-phase missions, for example, see Levitin et al. (2015), Wang et al. (2016).

In order to obtain the system reliability evaluation of the multi-state systems, several methods were also investigated in the literature. The universal $z$-transform technique is adopted by Levitin (2003), Levitin and Dai (2011), Levitin and Ben-Haim (2011), and so on. Huang et al. (2003) used a minimal path vector method to obtain the state distributions or state distribution bounds. However, the methods used in the above papers can only be applied for monotonic multistate consecutive-$k$-out-of-$n$: F (G) systems. Combinational analysis method was used by Belaloui and Ksir (2007) to compute the reliability of monotonic multistate consecutive-$k$-out-of-$n$: G systems. Yamamoto et al. (2006) provided an algorithm for state distributions of general multistate consecutive $k$-out-of-$n$: G systems. By using the FMCIA, Zhao and Cui (2010) presented a unified formula for evaluating the system state distribution for generalized multi-state $k$-out-of-$n$: F systems which cover many multi-state systems. Zhao et al. (2012) obtained a unified formula for evaluating the system state distribution of two kinds of multistate consecutive-$k$ systems including multi-state consecutive-$k$-out-of-$n$: G systems and multi-state consecutive-$k$-out-of-$r$-from-$n$ systems. Yi et al. (2019) gave the reliability formulae of four multistate consecutive-$k$ systems including a multistate linear $m$-consecutive-$k$-out-of-$n$: G system, a multistate linear consecutive-$k$-out-of-$n$: G system with sparse $d$, a multistate linear $m$-consecutive-$k$-out-of-$n$: G system with sparse $d$, and a multistate linear $<n, f, k>$: G system.

# 4 Consecutive $k$ Systems Based on the Geometric Structure of Systems

A line and a circle are the most widely studied geometric structures of consecutive $k$ systems (Preuss and Boehme 1994). However, Aki (1999) pointed out it is much more practical to assume that the components of a system are not necessarily allocated in a line or a circle. A natural generalization is the topology of a system being extended to a plane or space, i.e., the consecutive $k$ systems can be extended to multi-dimensional cases (Yamamoto and Akiba 2003). Salvia and Lasher (1990) first proposed a two-dimensional consecutive-$k$-out-of-$n$: F system which is denoted as $k^2/n^2$: F system. The $k^2/n^2$: F system is a square grid of side $n$ (containing $n^2$ components), which fails if and only if there is at least one square of side $k$ ($2 \leq k \leq n\text{-}1$) that contains all failed components. Boehme et al. (1992) extended the above model to the general connected-$X$-out-of-$(m, n)$: F lattice system, in which $X$ is the figure consisting of the failed components, such as the connected-$(r, s)$-out-of-$(m, n)$: F system, which fails "whenever there exists a grid of $r$ rows and $s$ columns that consists of all failed components". Since then, many authors have devoted to the studies of these systems, for example, see Zuo (1993), Godbole et al. (1998), Yamamoto and Akiba (2005), Habib et al. (2010), Zhao et al. (2011a). By using artificial perfect components, Hsieh and Chen (2004) converted the two-dimensional consecutive-$k$-out-of-$n$: F system into a general one-dimensional consecutive-$k$-out-of-$n$: F system. Yamamoto and Akiba (2005) proposed a two-dimensional $k$-within consecutive-$r \times s$-out-of-$m \times n$: F system and gave upper and lower bounds of system reliability, and Chang and Huang (2010) provided an evaluation method of the reliability by use of FMCIA. A two-dimensional linear connected-$k$ system was proposed by Zhao et al. (2011b), in which the system and components have three states: full working, degrading working, and failure. Boushaba and Azouz (2011) considered a 3-dimensional consecutive $k$-out-of-$n$: F system and gave the lower bound of the reliability. Akiba et al. (2005) introduced the consecutive-$(r_1, r_2, r_3)$-out-of-$(n_1, n_2, n_3)$: F system, which fails "only when failed components in the system form a cuboid of size $(r_1, r_2, r_3)$" and can be viewed as a generalization of the two-dimensional consecutive-$(r, s)$-out-of-$(m, n)$: F system. Combining the 3-dimensional consecutive-$(s, s, s)$-out-of-$(s, s, m)$: F system and 2 $s^3$-out-of-$ms^2$: F system, Kulkarni and Kashikar (2014) introduced a conditional 3-dimensional consecutive $(s, s, s)$-out-of-$(s, s, m)$: F system and obtained the expressions of the signature and reliability. Cowell (2015) provided a novel formula for the reliability of a general d-dimensional consecutive-k-out-of-n: F system, as an exact polynomial in $q$.

Except of the above extensions, Aki (1999) proposed a consecutive $k$-out-of-$n$: F system on directed trees. It is assumed that $n$ components are placed at the vertices one by one, and the system fails if and only if there exist at least consecutive $k$ failed components along the direction. Aki's model includes linear consecutive $k$ systems as special cases, because "the tree will reduces to a line system if there is only one child for every vertex of the tree except for a leaf". Considering the components' spatial distribution, Lin et al. (2016) introduced and studied the consecutive-$k_r$-out-of-$n_r$: F system with linear zigzag structure and circular polygon structure. The consecutive-$k_r$-out-of-$n_r$: F system with linear zigzag structure is composed of $m$ lines, in which two adjacent lines overlap with one shared node, and each line is regarded as a consecutive-$k_r$-out-of-$n_r$: F subsystem ($r = 1, 2, \ldots, m$). If and only if all $m$ lines are reliable, the system is reliable. Moreover, if the $m$ lines form a circle, it is called the consecutive-$k_r$-out-of-$n_r$: F system with circular polygon structure. Obviously, if $r = 1$, it reduces to the linear (or circular) consecutive $k$-out-of-$n$: F system.

## 5 Weighted Consecutive $k$ Systems

In practice, it is common that the components have different weights and reliabilities (Eryilmaz and Sarikaya 2014), i.e., the components are not always identical. Considering unequal weights for the components, Wu and Chen (1994a) introduced the weighted $k$-out-of-$n$: F (G) system which has $n$ components, each with its own positive integer weight and "the system fails (or works) if and only if the total weight of failed (or working) components is at least $k$". Since then, the weighted $k$-out-of-$n$ system has been discussed by many authors, for example, Chen and Yang (2005), Li and Zuo (2008), Ding et al. (2010), Faghih-Roohi et al. (2014), Zhuang et al. (2018), and so on.

In the above research, the weight of each component is assumed to be positive integer. Li et al. (2016) considered a more general weighted $k$-out-of-$n$ system, in which the weight of each component can take any positive value. Eryilmaz (2013b) proposed a $k$-out-of-$n$ systems with components having random weights and gave a recursive formula to compute the system state probabilities, and Meshkat and Mahmoudi (2017) gave the joint importance of two components.

A consecutive-weighted-$k$-out-of-$n$ system was proposed by Wu and Chen (1994b), in which the evaluation algorithms of reliability were given. Chang et al. (1998) provided a fast reliability-algorithm, and Chadjiconstantinidis and Koutras (1999) gave recursive reliability-formula by use of FMCIA. When each component has weight 1, the usual $k$-out-of-$n$: system and the consecutive-$k$-out-of-$n$ system are special case of the weighted consecutive $k$ systems. Eryilmaz and Tutuncu (2009) provided the exact formulae for the system reliability of linear systems with independent and non-identical components and the approximate formulae in the case that the components are non-homogeneous Markov dependent. Kamalja and Amrutkar (2014) developed a formula for the evaluation of reliability and importance measures of the weighted-consecutive-$k$-out-of-$n$ and weighted-$m$-consecutive-$k$-out-of-$n$ systems. A $r$-within consecutive-$k$-out-of-$n$: F system with weighted components was proposed by Kamalja and Amrutkar (2018), in which a binomial-type weighted scan statistic was introduced to study the system reliability and importance. The linear weighted $(n, f, k)$ system was introduced by Eryilmaz and Aksoy (2010), and the recursive formula of the system reliability was obtained, where the system fails if and only if the total weight of failed components is at least $f$, or the total weight of failed consecutive components is at least $k$. Amrutkar and Kamalja (2017) studied the system reliability and importance measures of linear weighted $(n, f, k)$ and $<n, f, k>$ systems. Nashwan (2017) provided the reliability and the failure functions of linear and circular weighted $(n, f, k)$ and $<n, f, k>$ systems. Zhu and Boushaba (2017) extended the results of the weighted consecutive-$k$-out-of-$n$ systems and the weighted $k$-out-of-$n$ systems to the case of a linear weighted $(n, f, k)$ system with non-homogeneous Markov-dependent components.

# 6    Consecutive $k$ Systems Based on Dependency

In most of the above literature, it is assumed that consecutive $k$ systems consist of independent components. However, in reality it is inevitable for components to be dependent with each other because they are influenced by common production and operating environment. Considering the Markov dependency, Papastavridis and Lambiris (1987) assumed that the probability of failure of a component depends on the preceding one and proposed the system reliability of a consecutive $k$-out-of-$n$: F system by use of a recurrence relation. Fu and Hu (1987) developed a consecutive $k$-out-of-$n$: F system with $(k$-1)-step Markov dependence, i.e., "each component-failure probability depends on the number of consecutive failures immediately preceding the component". Lam and Ng (2001) proposed a consecutive-$k$-out-of-$n$: F repairable system with exponential distribution and $(k$-1)-step Markov dependence. Xiao and Li (2008) gave direct simulation and conditional expectation estimation for some reliability indices of consecutive-$k$-out-of-$n$: F repairable systems with $(k$-1)-step Markov dependence.

Many other consecutive $k$ systems with Markov-dependent components have also been proposed, for example, $m$ consecutive-$k$-out-of-$n$: F systems with $(k$-1)-step Markov dependence components (Agarwal et al. 2007a), consecutive-$k$-out-of-$n$: F systems and m consecutive-$k$-out-of-$n$: F systems with homogeneous Markov-dependent components (Zhu et al. 2015), consecutive-$k$-within-$m$-out-of-$n$: F system with Markov-dependent components (Zhu et al. 2016), linear weighted $(n, f, k)$ systems with non-homogeneous Markov-dependent components (Zhu and Boushaba 2017), $m$-consecutive-$k$, $l$-out-of-$n$ system with non-homogeneous Markov-dependent components (Zhu et al. 2017), combined $m$-consecutive-$k$-out-of-$n$: F and consecutive-$k_b$-out-of-$n$: F systems with non-homogeneous Markov-dependent components (Boushaba and Benyahia 2018), linear $m$-consecutive-$k$-out-of-$n$ systems with sparse $d$ of nonhomogeneous Markov-dependent components (Zhu et al. 2018), and so on.

Except of the Markov dependency, Block-$k$ dependency and $s$-dependency of components are also considered in the literature. Agarwal et al. (2007a) assumed each subsequent occurrence of a block of $k$-consecutive failures increases the failure probability of the remaining components and proposed $m$-consecutive-$k$-out-of-$n$: F system with Block-$k$ dependence. By use of a matrix formulation, Gera (2000) obtained the reliability of consecutive $k$-out-of $n$: G system. Eryilmaz (2007) considered a consecutive $k$-out-of-$n$: F system with $s$-dependent components and showed that the mean residual life ordering was not preserved. Erylmaz (2009) discussed the reliability properties of consecutive $k$-out-of-$n$ systems consisting of arbitrarily dependent components. Some stochastic ordering properties of residual lifetime, inactivity time of the component of consecutive $k$-out-of-$n$ systems consisting of arbitrarily dependent components was presented by Salehi (2016). In fact, when the dependency is involved, it is much more difficult to analyze the reliability properties of consecutive $k$ systems. Therefore, many authors have considered the flexible cases, such as the consecutive $k$ systems with exchangeable components (Shanthikumar 1985; Eryilmaz et al. 2011; Eryilmaz 2013a, b; Mahmoud and Eryilmaz 2014).

## 7 Conclusion and Future Challenges

In the paper, we have presented a comprehensive review on consecutive $k$ and related models. The review is focused on conventional consecutive $k$ models and the associated variants of the consecutive $k$ systems considering failure criteria, geometric structure of the system, state of components and the system, weight of each component, dependency of components, and so on.

Although significant progress has been made to extend the flexibility of consecutive $k$ systems, there are still several challenges deserving further research. There are theoretical and practical implications for these challenges and problems, but they are not fully dealt with in the current literature. Several directions are listed as follows.

(1) The development of new consecutive $k$ systems, for example, high-dimensional consecutive $k$ systems, consecutive $k$ systems with non-Markov dependency, multistate consecutive $k$ systems, consecutive $k$ systems with non-integer weights, and so on.

(2) Consecutive $k$ systems considering the geometric structure of the system, for example consecutive $k$ systems on graph.
(3) The efficient and effective computation method of related reliability measurement including reliability, importance, the lifetime distribution, inactivity time, and so on.

**Acknowledgments.** The work was supported by the National Natural Science Foundation of China under Grant (71631001) and Scientific Research Program Funded by Shaanxi Provincial Education Department (18JK0877).

# References

Agarwal, M., Mohan, P.: GERT analysis of $m$-consecutive-$k$-out-of-$n$: F system with overlapping runs and $(k-1)$-step Markov dependence. Int. J. Oper. Res. **3**(1–2), 36–51 (2008)

Agarwal, M., Mohan, P., Sen, K.: GERT analysis of $m$-consecutive-$k$-out-of-$n$: F systems with dependence. Econ. Qual. Control **22**(1), 141–157 (2007a)

Agarwal, M., Sen, K., Mohan, P.: GERT analysis of $m$-consecutive-$k$-out-of-$n$ systems. IEEE Trans. Reliab. **56**(1), 26–34 (2007b)

Aki, S.: Distributions of runs and consecutive systems on directed trees. Ann. Inst. Stat. Math. **51**(1), 1–15 (1999)

Akiba, T., Yamamoto, H., Tsujimura, Y.: Evaluating methods for the reliability of a three-dimensional $k$-within system. J. Qual. Maintenance Eng. **11**(3), 254–266 (2005)

Amrutkar, K.P., Kamalja, K.K.: Efficient algorithm for reliability and importance measures of linear weighted-$(n, f, k)$ and $<n, f, k>$ systems. Comput. Ind. Eng. **107**, 85–99 (2017)

Barlow, R.E., Proschan, F.: Mathematical Theory of Reliability. Wiley, New York (1965)

Belaloui, S., Ksir, B.: Reliability of a multi-state consecutive $k$-out-of-$n$: G system. Int. J. Reliab. Qual. Saf. Eng. **14**(4), 361–377 (2007)

Boehme, T.K., Kossow, A., Preuss, W.: A generalization of consecutive-$k$-out-of-$n$: F systems. IEEE Trans. Reliab. **41**(3), 451–457 (1992)

Boland, P.J., Samaniego, F.J.: Stochastic ordering results for consecutive $k$-out-of-$n$: F systems. IEEE Trans. Reliab. **53**(1), 7–10 (2004)

Bollinger, R.: Direct computations for consecutive-$k$-out-of-$n$: F systems. IEEE Trans. Reliab. **31**, 444–446 (1982)

Boushaba, M., Azouz, Z.: Reliability bounds of a 3-dimensional consecutive-$k$-out-of-$n$: F system. Int. J. Reliab. Qual. Saf. Eng. **18**(1), 51–59 (2011)

Boushaba, M., Benyahia, A.: Reliability and importance measures for combined $m$-consecutive-$k$-out-of-$n$: F and consecutive-$k_b$-out-of-n: F systems with non-homogeneous Markov-dependent components. Int. J. Reliab. Qual. Saf. Eng. **25**(5), 1850022 (2018)

Cai, J.: Reliability of a large consecutive-$k$-out-of-$r$-from-$n$: F system with unequal component-reliability. IEEE Trans. Reliab. **43**(1), 107–111 (1994)

Chadjiconstantinidis, S., Koutras, M.V.: Measures of component importance for Markov chain imbeddable reliability structures. Naval Res. Logistics **46**(6), 613–639 (1999)

Chang, J.C., Chen, R.J., Hwang, F.K.: A fast reliability-algorithm for the circular consecutive-weighted-$k$-out-of-$n$: F system. IEEE Trans. Reliab. **47**(4), 472–474 (1998)

Chang, H.W., Chen, R.J., Hwang, F.K.: The structural Birnbaum importance of consecutive-$k$ systems. J. Comb. Optim. **6**(2), 183–197 (2002)

Chang, G.J., Cui, L.R., Hwang, F.K.: Reliabilities for $(n, f, k)$ systems. Stat. Probab. Lett. **43**(3), 237–242 (1999)

Chang, G.J., Cui, L.R., Hwang, F.K.: Reliabilities of Consecutive-$k$-Systems. Kluwer, Dordrecht (2000)

Chang, Y.M., Huang, T.H.: Reliability of a 2-dimensional $k$-within consecutive-$r \times s$-out-of-$m \times n$: F system using finite Markov chains. IEEE Trans. Reliab. **59**(4), 725–733 (2010)

Chao, M.T., Fu, J.C., Koutras, M.V.: Survey of reliability studies of consecutive-$k$-out-of-n: F and related systems. IEEE Trans. Reliab. **44**(1), 120–127 (1995)

Chen, Y., Yang, Q.: Reliability of two-stage weighted-$k$-out-of-$n$ systems with components in common. IEEE Trans. Reliab. **54**(3), 431–440 (2005)

Chiang, D.T., Niu, S.C.: Reliability of consecutive-$k$-out-of-n: F System. IEEE Trans. Reliab. **30**(1), 87–89 (1981)

Cowell, S.: A formula for the reliability of a $d$-dimensional consecutive-$k$-out-of-$n$: F system. Mathematics **2015**, 1–5 (2015). 140909

Cui, L.R., Hawkes, A.G.: A note on the proof for the optimal consecutive $k$-out-of-$n$: G line for $n \leq 2k$. J. Stat. Plann. Infer. **138**, 1516–1520 (2008)

Cui, L.R., Kuo, W., Li, J.L., Xie, M.: On the dual reliability systems of $(n, f, k)$ and $<n, f, k>$. Stat. Probab. Lett. **76**(11), 1081–1088 (2006)

Cui, L.R., Lin, C., Du, S.J.: $m$-consecutive-$k$, $l$-out-of-$n$ systems. IEEE Trans. Reliab. **64**(1), 386–393 (2015)

Cui, L.R., Xu, Y., Zhao, X.: Developments and applications of the finite Markov chain imbedding approach in reliability. IEEE Trans. Reliab. **59**(4), 685–690 (2010)

Daus, L., Beiu, V.: Review of reliability bounds for consecutive-$k$-out-of-$n$ systems. In: IEEE 14th International Conference on Nanotechnology (IEEE-NANO), pp. 302–307. IEEE (2014)

Derman, C., Lieberman, G.J., Ross, S.M.: On the consecutive-$k$-out-of-n: F system. IEEE Trans. Reliab. **31**, 57–63 (1982)

Ding, Y., Zuo, M.J., Lisnianski, A., Li, W.: A framework for reliability approximation of multi-state weighted $k$-out-of-$n$ systems. IEEE Trans. Reliab. **59**(2), 297–308 (2010)

Eryilmaz, S.: On the lifetime distribution of consecutive $k$-out-of-$n$: F system. IEEE Trans. Reliab. **56**(1), 35–39 (2007)

Erylmaz, S.: Reliability properties of consecutive $k$-out-of-$n$ systems of arbitrarily dependent components. Reliab. Eng. Syst. Saf. **94**(2), 350–356 (2009)

Eryilmaz, S.: Review of recent advances in reliability of consecutive $k$-out-of-$n$ and related systems. Proc. IMechE Part O: J. Risk Reliab. **224**, 225–237 (2010)

Eryilmaz, S.: Circular consecutive $k$-out-of-$n$ systems with exchangeable dependent components. J. Stat. Plann. Infer. **141**(2), 725–733 (2011)

Eryilmaz, S.: $m$-consecutive-$k$-out-of-$n$: F system with overlapping runs: signature-based reliability analysis. Int. J. Oper. Res. **15**(1), 64–73 (2012)

Eryilmaz, S.: Component importance for linear consecutive-$k$-out-of-$n$ and $m$-consecutive-$k$-out-of-$n$ systems with exchangeable components. Naval Res. Logistics **60**(4), 313–320 (2013a)

Eryilmaz, S.: On reliability analysis of a $k$-out-of-$n$ system with components having random weights. Reliab. Eng. Syst. Saf. **109**(1), 41–44 (2013b)

Eryilmaz, S., Aksoy, T.: Reliability of linear $(n, f, k)$ systems with weighted components. J. Syst. Sci. Syst. Eng. **19**(3), 277–284 (2010)

Eryilmaz, S., Bayramoglu, K.: Residual lifetime of consecutive $k$-out-of-$n$ systems under double monitoring. IEEE Trans. Reliab. **61**(3), 792–797 (2012)

Eryilmaz, S., Koutras, M.V., Triantafyllou, I.S.: Signature based analysis of $m$-Consecutive-$k$-out-of-$n$: F systems with exchangeable components. Naval Res. Logistics **58**(4), 344–354 (2011)

Eryilmaz, S., Mahmoud, B.: Linear $m$-Consecutive-$k$, $l$-out-of-$n$: F system. IEEE Trans. Reliab. **61**(3), 787–791 (2012)

Eryilmaz, S., Sarikaya, K.: Modeling and analysis of weighted-$k$-out-of-$n$: G system consisting of two different types of components. Proc. IMechE Part O: J. Risk Reliab. **228**(3), 265–271 (2014)

Eryilmaz, S., Tutuncu, G.Y.: Reliability evaluation of linear consecutive-weighted-$k$-out-of-$n$: F system. Asia-Pac. J. Oper. Res. **26**(6), 805–816 (2009)

Faghih-Roohi, S., Xie, M., Ng, K.M., et al.: Dynamic availability assessment and optimal component design of multi-state weighted $k$-out-of-$n$ systems. Reliab. Eng. Syst. Saf. **123**(4), 57–62 (2014)

Fu, J.C., Hu, B.: On reliability of a large consecutive-$k$-out-of-$n$: F system with $(k\text{-}1)$-step Markov dependence. IEEE Trans. Reliab. **36**(1), 75–77 (1987)

Fu, J.C., Koutras, M.V.: Distribution theory of runs: a Markov chain approach. Publ. Am. Stat. Assoc. **89**, 1050–1058 (1994)

Ge, G.P.: On consecutive $k$-out-of-$n$: F systems. Adv. Math. **22**(4), 306–311 (1993). (in Chinese)

Gera, A.E.: A consecutive $k$-out-of-$n$: G system with dependent elements—a matrix formulation and solution. Reliab. Eng. Syst. Saf. **68**(1), 61–67 (2000)

Gera, A.E.: Combined $m_1$-consecutive-$k$-out-of-$n$ and $m_2$-consecutive-$k$-out-of-$n$ systems. IEEE Trans. Reliab. **60**, 493–497 (2011)

Godbole, A.P., Potter, L.K., Sklar, J.K.: Improved upper bounds for the reliability of $d$ dimensional consecutive $k$-out-$f$-$n$: F systems. Naval Res. Logistics **45**(2), 219–230 (1998)

Griffith, W.S.: On consecutive $k$-out-of-$n$ failure systems and their generalizations. In: Basu, A. P. (ed.) Reliability and Quality Control, pp.157–165. Elsevier, North Holland (1986)

Guo, Y.L., Cui, L.R., Li, J.L., Gao, S.: Reliabilities for $(n, f, k\,(i, j))$ and $< n, f, k\,(i, j) >$ systems. Commun. Stat. Theory Methods **35**(10), 1779–1789 (2006)

Habib, A., Al-Seedy, R.O., Radwan, T.: Reliability evaluation of multi-state consecutive $k$-out-of-$r$-from-$n$: G system. Appl. Math. Model. **31**(11), 2412–2423 (2007)

Habib, A.S., Yuge, T., Al-Seedy, R.O., Ammar, S.I.: Reliability of a consecutive $(r, s)$-out-of-$(m, n)$: F lattice system with conditions on the number of failed components in the system. Appl. Math. Model. **34**(3), 531–538 (2010)

Hsieh, Y.C., Chen, T.C.: Reliability lower bounds for two-dimensional consecutive-$k$-out-of-$n$: F systems. Comput. Oper. Res. **31**(8), 1259–1272 (2004)

Huang, J.S., Zuo, M.J., Fang, Z.D.: Multi-state consecutive-$k$-out-of-$n$ systems. IIE Trans. **35**(6), 527–534 (2003)

Hwang, F., Yao, Y.: Multistate consecutively-connected systems. IEEE Trans. Reliab. **38**, 472–474 (1989)

Kamalja, K.K.: Birnbaum importance for consecutive-$k$ systems. Int. J. Reliab. Qual. Saf. Eng. **19**(4), 1250016 (2012)

Kamalja, K.K.: Birnbaum reliability importance for $(n, f, k)$ and $<n, f, k>$ system. Commun. Stat. Theory Methods **43**(10–12), 2406–2418 (2014)

Kamalja, K.K., Amrutkar, K.P.: Computational methods for reliability and importance measures of weighted-consecutive-system. IEEE Trans. Reliab. **63**(1), 94–104 (2014)

Kamalja, K.K., Amrutkar, K.P.: Reliability and reliability importance of weighted-$r$-within-consecutive-$k$-out-of-$n$: F system. IEEE Trans. Reliab. **67**(3), 951–969 (2018)

Kontoleon, J.M.: Optimum allocation of components in a special 2-port network. IEEE Trans. Reliab. **27**(2), 112–115 (1978)

Kossow, A., Preuss, W.: Reliability of linear consecutively-connected systems with multistate components. IEEE Trans. Reliab. **44**(3), 518–522 (1995)

Koutras, M.V.: Consecutive-$k$, $r$-out-of-$n$: DFM systems. Microelectron. Reliab. **37**(4), 597–603 (1997)

Kulkarni, M.G., Kashikar, A.S.: Signature and reliability of conditional three-dimensional consecutive-$(s, s, s)$-out-of-$(s, s, m)$: F system. Int. J. Reliab. Qual. Saf. Eng. **21**(2), 1450009 (2014)

Kuo, W., Zuo, M.J.: Optimal Reliability Modelling-Principles and Applications. Wiley, New Jersey (2003)

Kuo, W., Zhang, W., Zuo, M.J.: A consecutive-$k$-out-of-$n$: G system: the mirror image of a consecutive-$k$-out-of-$n$: F system. IEEE Trans. Reliab. **39**(2), 244–253 (1990)

Lam, Y., Ng, H.K.: A general model for consecutive-$k$-out-of-$n$: F repairable system with exponential distribution and $(k$-$1)$-step Markov dependence. Eur. J. Oper. Res. **129**, 663–682 (2001)

Lam, Y., Zhang, Y.L.: Repairable consecutive-$k$-out-f-$n$: F system with Markov dependence. Naval Res. Logistics **47**(1), 18–39 (2015)

Lambiris, M., Papastavridis, S.: Exact reliability formulas for linear & circular consecutive-$k$-out-of-$n$: F systems. IEEE Trans. Reliab. **34**(2), 124–126 (1985)

Levitin, G.: Linear multi-state sliding-window systems. IEEE Trans. Reliab. **52**(2), 263–269 (2003)

Levitin, G.: Consecutive $k$-out-of-$r$-from-$n$ system with multiple failure criteria. IEEE Trans. Reliab. **53**(3), 394–400 (2004)

Levitin, G.: The Universal Generating Function in Reliability Analysis and Optimization. Springer, London (2005)

Levitin, G.: Linear $m$-gap-consecutive $k$-out-of-$r$-from-$n$: F systems. Reliab. Eng. Syst. Saf. **96**(2), 292–298 (2011)

Levitin, G., Ben-Haim, H.: Consecutive sliding window systems. Reliab. Eng. Syst. Saf. **96**(10), 1367–1374 (2011)

Levitin, G., Dai, Y.: Linear $m$-consecutive-$k$-out-of-$r$-from-$n$: F systems. IEEE Trans. Reliab. **60**, 640–646 (2011)

Levitin, G., Xing, L., Ben-Haim, H., Dai, Y.: $M/n$ CCS: Linear consecutively connected systems subject to combined gap constraints. Int. J. Gen Syst **44**(7), 1–16 (2015)

Li, W., Zuo, M.J.: Optimal design of multi-state weighted $k$-out-of-$n$ systems based on component design. Reliab. Eng. Syst. Saf. **93**, 1673–1681 (2008)

Li, X.H., You, Y.P., Fang, R.: On weighted $k$-out-of-$n$ systems with statistically dependent component lifetimes. Probab. Eng. Inf. Sci. **30**(04), 533–546 (2016)

Lin, C., Cui, L.R., Coit, D.W., Lv, M.: Reliability modeling on consecutive-$k_r$-out-of-$n_r$: F linear zigzag structure and circular polygon structure. IEEE Trans. Reliab. **65**(3), 1–13 (2016)

Lu, S.Q., Shi, D.M., Xiao, H.: Reliability of sliding window systems with two failure modes. Reliab. Eng. Syst. Saf. **188**, 366–376 (2019)

Mahmoud, B., Eryilmaz, S.: Joint reliability importance in a binary $k$-out-of-$n$: G system with exchangeable dependent components. Qual. Technol. Quant. Manag. **11**(4), 453–460 (2014)

Malon, D.M.: Optimal consecutive-$k$-out-of-$n$: F component sequencing. IEEE Trans. Reliab. **34**(1), 46–49 (1985)

Meshkat, R.S., Mahmoudi, E.: Joint reliability and weighted importance measures of a $k$-out-of-$n$ system with random weights for components. J. Comput. Appl. Math. **326**, 273–283 (2017)

Mohan, P., Agarwal, M., Sen, K.: Combined $m$-consecutive-$k$-out-of-$n$: F & consecutive $k_c$-out-of-n: F system. IEEE Trans. Reliab. **58**(2), 328–337 (2009)

Mohan, P., Agarwal, M., Sen, K.: Reliability analysis of sparsely connected consecutive-$k$ systems: GERT approach. In: International Conference on Reliability, pp. 213–218 (2009b)

Nashwan, I.I.H.: Reliability and failure functions of some weighted systems. Int. J. Appl. Math. Res. **6**(1), 7–13 (2017)

Papastavridis, S.T., Koutras, V.: Consecutive-$k$-out-of-$n$ systems with maintenance. Ann. Inst. Stat. Math. **44**, 605–612 (1992)

Papastavridis, S., Lambiris, M.: Reliability of a consecutive-*k*-out-of-*n*: F system for Markov-dependent components. IEEE Trans. Reliab. **36**(1), 78–79 (1987)

Papastavridis, S.T., Sfakianakis, M.: Optimal-arrangement & importance of the components in a consecutive-k-out-of-r-from-n: F system. IEEE Trans. Reliab. **40**, 277–279 (1991)

Preuss, W.W., Boehme, T.K.: On reliability analysis of consecutive-*k*-out-of-*n*: F systems and their generalizations - a survey. In: Anastassiou, G., Rachev, S.T. (eds.) Approximation, Probability, and Related Fields, pp. 401–411. Springer, Boston (1994). https://doi.org/10.1007/978-1-4615-2494-6_31

Psillakis, Z.M.: A simulation algorithm for computing failure probability of a consecutive-*k*-out-of-*r*-from-*n*: F system. IEEE Trans. Reliab. **44**(3), 523–531 (1995)

Radwan, T., Habib, A., Alseedy, R., Elsherbeny, A.: Bounds for increasing multi-state consecutive *k*-out-of-*r*-from-*n*: F, system with equal components probabilities. Appl. Math. Model. **35**(5), 2366–2373 (2011)

Salehi, E.T.: On reliability analysis of consecutive *k*-out-of-*n* systems with arbitrarily dependent components. Appl. Math. **61**(5), 565–584 (2016)

Salehi, E.T., Asadi, M., Eryılmaz, S.: On the mean residual lifetime of consecutive *k*-out-of-*n* systems. Test **21**, 93–115 (2012)

Salvia, A.A., Lasher, W.C.: Two-dimensional consecutive-*k*-out-of-*n*: F models. IEEE Trans. Reliab. **39**(3), 382–385 (1990)

Sen, K., Agarwal, M., Mohan, P.: GERT analysis of consecutive-*k* systems: an overview. Oikos **94**(1), 101–117 (2015)

Sfakianakis, M., Kounias, S., Hillaris, A.: Reliability of consecutive *k*-out-of-*r*-from-*n*: F systems. IEEE Trans. Reliab. **41**, 442–447 (1992)

Sfakianakis, M.: Optimal arrangement of components in a consecutive *k*-out-of-*r*-from *n*: F system. Microelectron. Reliab. **33**(10), 1573–1578 (1993)

Shanthikumar, J.G.: Lifetime distribution of consecutive-*k*-out-of-*n*: F systems with exchangeable lifetimes. IEEE Trans. Reliab. **34**(5), 480–483 (1985)

Shen, J.Y., Cui, L.R.: Reliability and Birnbaum importance for sparsely connected circular consecutive-*k* systems. IEEE Trans. Reliab. **64**(4), 1140–1157 (2015)

Shen, J.Y., Cui, L.R., Du, S.J.: Birnbaum importance for linear consecutive-*k*-out-of-*n* systems with sparse *d*. IEEE Trans. Reliab. **64**(1), 359–375 (2015)

Shingyochi, K., Yamamoto, H., Tsujimura, Y., Akiba, T.: Proposal of Simulated annealing algorithms for optimal arrangement in a circular consecutive-*k*-out-of-*n*: F system. Qual. Technol. Quant. Manag. **7**(4), 395–405 (2010)

Shingyochi, K., Yamamoto, H., Tsujimura, Y.: Genetic algorithm for solving optimal component arrangement problem of circular consecutive-*k*-out-of-*n*: F system. Ieice Tech. Rep. **105**(480), 13–18 (2015)

Shingyochi, K., Yamamoto, H., Yamachi, H.: Comparative study of several simulated annealing algorithms for optimal arrangement problems in a circular consecutive-*k*-out-of-*n*: F system. Qual. Technol. Quant. Manag. **9**(3), 295–303 (2016)

Tang, S.D., Hou, W.G.: A repairable linear *m*-consecutive-*k*-out-of-*n*: F system. Chin. Phys. Lett. **29**(9), 098401 (2012)

Triantafyllou, I.S., Koutras, M.V.: Signature and IFR preservation of 2-within-consecutive *k*-out-of-*n*: F systems. IEEE Trans. Reliab. **60**(1), 315–322 (2011)

Tung, S.S.: Combinatorial analysis in determining reliability. In: Annual Reliability and Maintainability Symposium, Los Angeles, CA, pp. 262–266 (1982)

Wang, M. Q., Yang, J., Yu, H.: Reliability of phase mission linear consecutively-connected systems with constrained number of consecutive gaps. In: 2016 International Conference on System Reliability & Science, pp. 148–151. IEEE (2016)

Wu, J.S., Chen, R.J.: An algorithm for computing the reliability of weighted-$k$-out-of-$n$ systems. IEEE Trans. Reliab. **43**(2), 327–328 (1994a)

Wu, J.S., Chen, R.J.: Efficient algorithms for $k$-out-of-$n$ and consecutive-weighted-$k$-out-of-$n$: F system. IEEE Trans. Reliab. **43**(4), 650–655 (1994b)

Xiao, G., Li, Z.: Estimation of dependability measures and parameter sensitivities of a consecutive-$k$-out-of-$n$: F repairable system with ($k$-1)-step Markov dependence by simulation. IEEE Trans. Reliab. **57**(1), 71–83 (2008)

Xiao, H., Peng, R., Levitin, G.: Optimal replacement and allocation of multi-state elements in $k$-within-$m$-from-$r/n$ sliding window systems. Appl. Stochast. Models Bus. Ind. **32**(2), 184–198 (2016)

Xiao, H., Peng, R., Wang, W. B., Zhao, F.: Linear $m$-gap-consecutive $k$-out-of-$r$-from-$n$ system with common supply failures. In: 2014 International Conference on Reliability, Maintainability and Safety (ICRMS). IEEE (2014). https://doi.org/10.1109/icrms.2014.7107201

Yamamoto, H., Akiba, T.: Survey of reliability studies of multi-dimensional consecutive-$k$-out-of-$n$: F systems. Reliab. Eng. Assoc. Japan **25**(8), 783–796 (2003)

Yamamoto, H., Akiba, T.: Evaluating methods for the reliability of a large 2-dimensional rectangular $k$-within-consecutive-($r$, $s$)-out-of-($m$, $n$): F system. Naval Res. Logistics **52**(3), 243–252 (2005)

Yamamoto, H., Zuo, M.J., Akiba, T., Tian, Z.: Recursive formulas for the reliability of multi-state consecutive-$k$-out-of-$n$: G systems. IEEE Trans. Reliab. **55**(1), 98–104 (2006)

Yi, H., Cui, L.R., Gao, H.D.: Reliabilities of some multistate consecutive-$k$ systems. IEEE Trans. Reliab. (2019). https://doi.org/10.1109/tr.2019.2897726

Zhang, Y.L., Lam, Y.: Reliability of consecutive-$k$-out-of-$n$: G repairable system. Int. J. Syst. Sci. **29**, 1375–1379 (1998)

Zhao, X., Cui, L.R.: Reliability evaluation of generalised multi-state $k$-out-of-$n$ systems based on FMCI approach. Int. J. Syst. Sci. **41**(12), 1437–1443 (2010)

Zhao, X., Cui, L.R., Kuo, W.: Reliability for sparsely connected consecutive-$k$ systems. IEEE Trans. Reliab. **56**(3), 516–524 (2007)

Zhao, X., Cui, L.R., Zhao, W., Liu, F.: Exact reliability of a linear connected-($r$, $s$)-out-of-($m$, $n$): F system. IEEE Trans. Reliab. **60**(3), 689–698 (2011a)

Zhao, X., Xu, Y., Liu, F.Y.: State distributions of multi-state consecutive-$k$ systems. IEEE Trans. Reliab. **61**(2), 274–281 (2012)

Zhao, X., Zhao, W., Xie, W.J.: Two-dimensional linear connected-$k$ system with trinary states and its reliability. J. Syst. Eng. Electron. **22**(5), 866–870 (2011b)

Zhu, X.Y., Boushaba, M.: A linear weighted ($n$, $f$, $k$) system for non-homogeneous Markov-dependent components. IISE Transactions **49**(7), 722–736 (2017)

Zhu, X.Y., Boushaba, M., Coit, D.W., Benyahia, A.: Reliability and importance measures for $m$-consecutive-$k$, $l$-out-of-$n$ system with non-homogeneous Markov-dependent components. Reliab. Eng. Syst. Saf. **167**, 1–9 (2017)

Zhu, X.Y., Boushaba, M., Boulahia, A., Zhao, X.: A linear $m$-consecutive-$k$-out-of-$n$ system with sparse $d$ of non-homogeneous Markov-dependent components. Proc. IMechE Part O: J. Risk Reliab. (2018). https://doi.org/10.1177/1748006x18776189

Zhu, X.Y., Boushaba, M., Reghioua, M.: Joint reliability importance in a consecutive-$k$-out-of-$n$: F system and an $m$-consecutive-$k$-out-of-$n$: F system for Markov-dependent components. IEEE Trans. Reliab. **64**(2), 784–798 (2015)

Zhu, X.Y., Boushaba, M., Reghioua, M.: Reliability and joint reliability importance in a consecutive-$k$-within-$m$-out-of-$n$: F system with Markov-dependent components. IEEE Trans. Reliab. **65**(2), 802–815 (2016)

Zhuang, X.C., Yu, T.X., Shen, L.J.: On capacity evaluation for multi-state weighted $k$-out-of-$n$ system. Commun. Stat. Simul. Comput. **3**, 1–16 (2018)

Zuo, M.J.: Reliability & design of 2-dimensional consecutive $k$-out-of-$n$: F systems. IEEE Trans. Reliab. **42**(3), 488–490 (1993)

Zuo, M.J., Liang, M.: Reliability of multistate consecutively-connected systems. Reliab. Eng. Syst. Saf. **44**(2), 173–176 (1994)

Zuo, M.J., Lin, D., Wu, Y.: Reliability evaluation of combined $k$-out-of-$n$: F, consecutive-$k$-out-of-$n$: F and linear connected-$(r, s)$-out-of-$(m, n)$: F system structures. IEEE Trans. Reliab. **49**(1), 99–104 (2000)

# Recent Advances on Reliability of Phased Mission Systems

Di Wu[1], Rui Peng[2(✉)], and Liudong Xing[3]

[1] School of Management, Xi'an Jiaotong University, Xi'an, China
wd_0824@stu.xjtu.edu.cn
[2] School of Economics and Management, Beijing University of Technology,
Beijing, China
pengruisubmit@163.com
[3] Department of Electrical and Computer Engineering,
University of Massachusetts, Dartmouth, MA, USA
lxing@umassd.edu

**Abstract.** Many practical systems need to fulfill their mission in multiple phases, where the system in different phases may have to perform different tasks undergoing different environmental conditions and success criteria. In the last few decades, reliability modeling and evaluation of phased mission systems (PMSs) and related optimization problems have attracted a lot of attention. This chapter is dedicated to a comprehensive review of recent developments on reliability evaluation and optimization of PMSs. Different evaluation methods are classified and their applicability to different types of PMSs (e.g., static versus dynamic; redundant versus non-redundant; perfect coverage versus imperfect coverage) are discussed. Traditional and recently-developed optimization problems related to PMSs are introduced. Some directions for future researches are suggested.

**Keywords:** Analytical techniques · Multi-valued decision diagram · Optimization · Phased mission systems · Reliability

## 1 Introduction

Complex systems, such as those in aerospace, chemical control, communication networks, electronics, transportation and nuclear, usually consist of several subsystems each aiming to accomplish a specific and different task. The process of mission executed by these systems can typically be divided into several consecutive phases typically involving different subsystems or components. For instance, an aircraft system needs to undergo take-off, ascent, level flight, descent, and landing phases. These systems are referred to as phased mission systems (PMSs). Compared to single-phase systems, the analysis of PMSs is more challenging due to the following aspects: (1) the system structure function (determined based on the reliability requirements) can change from phase to phase; (2) components may occupy different weights or importance values in different phases due to the changing system structure functions; (3) components may have different reliability parameters in different phases since they may be

© Springer Nature Singapore Pte Ltd. 2019
Q.-L. Li et al. (Eds.): Cao Festschrift 2019, CCIS 1102, pp. 19–43, 2019.
https://doi.org/10.1007/978-981-15-0864-6_2

exposed in diverse environment conditions in different phases; (4) the state of a component at the beginning of a phase depends on its state at the end of the previous phase.

Considerable research efforts have been expended in the reliability analysis of PMSs. Two classes of approaches have been developed: simulations and analytical modelling methods. Simulations can typically offer great generality in system representation, but can only offer approximate results and are often expensive in computational requirements. This chapter focuses on the analytical modelling methods that can typically offer accurate results with reasonable computational overhead. The analytical modeling methods can be further classified into four categories: combinatorial methods, state-space oriented methods, modular methods that combine the former two methods, and recursive methods (Sect. 3). Based on these evaluation methods, many researchers have formulated and solved different optimization problems to improve performance of PMSs.

The remaining of this chapter is organized as follows: Sect. 2 gives classifications of PMSs from different aspects. Section 3 reviews diverse analytical modeling methods for PMS reliability evaluation. Section 4 introduces optimization problems for PMSs and solution methods. Section 5 discusses potential future works on PMS research.

The abbreviations used in this chapter are summarized in Table 1.

**Table 1.** Abbreviations

| | |
|---|---|
| BDD | binary decision diagram |
| CCF | common-cause failure |
| CPR | combinatorial phase requirement |
| CTMC | continuous-time Markov chain |
| ELC | element-level coverage |
| FLC | fault-level coverage |
| ite | if-then-else |
| MDD | multi-valued decision diagram |
| PDO | phase dependent operation |
| PMF | probability mass function |
| PMS | phased mission system |
| PN | petri net |
| RAP | redundancy allocation problem |
| RBD | reliability block diagram |
| UGF | universal generating function |

## 2  Classification of PMSs

A component in a PMS may fail by itself (e.g., due to wear-out) or due to common-cause failures (CCFs) (Xing et al. 2009). CCFs are simultaneous failures of multiple components due to a shared root cause, which can be an *external* or *internal cause* (Xing 2007; Xing and Levitin 2013). It has been shown by many studies that CCFs can increase the joint failure probabilities of system components and lead to the augment of the overall system unreliability. Both individual failures and CCFs (when applicable) should be considered for reliability analysis of PMSs.

To facilitate the review of PMS reliability evaluation methods, some common categories of PMSs studied in literature are presented below.

*Repairable Versus Non-repairable PMS:* In a non-repairable PMS, the state of a component at the beginning of a phase should be identical to its state at the end of the previous phase, and the system state depends mainly on failure characteristics of its components. Nonetheless, in a repairable PMS, the state of the system depends not only on failure characteristics of its components but also on maintenances conducted during the mission. The maintenance policies can be divided into three categories: 1. *Failure-driven maintenance* is triggered by the occurrence of a component failure; 2. *Time-driven maintenance* is conducted based on a pre-determined schedule; 3. *Condition-driven maintenance* is conducted based on the monitored condition of the system (e.g., the maintenance is triggered when some components are out of work while the whole system remains functional) (Tinga 2013; Ding and Kamaruddin 2015). While the study on non-repairable PMSs is extensive (Peng et al. 2019), the study on repairable PMSs is still limited. Example works on repairable PMSs include Shrestha et al. (2011), in which decision diagrams and Markov models were integrated to analyze reliability of repairable PMSs with multiple component states. Lu and Wu (2014) employed continuous time Markov chains (CTMCs) to evaluate the reliability of a generalized PMS with repairable components. Lu et al. (2015) analyzed the reliability of large PMSs with repairable components based on success-state sampling (or the discretization method). Wu and Wu (2015) proposed an extended object-oriented Petri net model for mission reliability simulation of repairable PMSs with CCFs.

*Static Versus Dynamic PMS:* A PMS is considered to be static if the structure of the reliability model for any phase of the system is combinational. That is, the failure of the mission in any phase depends only on combinations of component failure events, not on the occurrence sequence of the input events. In contrast, if the structure of the reliability model is not only combinational but also involves certain dependencies among the system components, the PMS is considered to be dynamic. Particularly, dynamic PMSs can involve sequence dependence (the order in which the component failure events occur affects the system outcome or status), function dependence (the failure of certain component causes other components within the same system to become isolated, i.e., unusable or inaccessible) or spares management (Xing et al. 2019). For the analysis of static PMSs, combinatorial methods like binary decision diagrams (Sect. 3.1) can be employed; for the evaluation of dynamic PMSs, state-space methods like CTMCs (Sect. 3.2) are often applied.

*Binary-State Versus Multi-state PMS:* If components in a PMS only assume two states (*function* or *failure*), then the system is known as binary-state. Levitin et al. (2012) developed a recursive algorithm for reliability evaluation of PMSs with binary-state elements. In recent years, multi-state systems have been widely studied to describe multiple levels of working performance or the degradation process of components and systems. Decision diagram methods, stochastic process methods, universal generating function methods have been proposed to obtain the reliability of generic multi-state systems (Levitin 2005; Xing and Amari 2015). Example works on addressing multi-state PMSs include an integrated modeling approach proposed by Shrestha et al. (2011) for reliability analysis of repairable PMSs with multiple ordered or unordered component states. Li et al. (2018) employed a multi-state multi-valued decision diagram (MDD) algorithm for modeling reliability of PMSs with non-repairable, multi-state components. Besides addressing challenges in analyzing binary-state PMSs (described in Introduction), reliability analysis of multi-state PMSs also needs to consider dependencies among different states across different phases for each system component during both modeling and evaluation procedures.

*Coherent Versus Non-coherent PMS:* If at least one of each component's states contributes to the system state and any additional component failure may only make the system status worse or at least no improvement in the system performance, the PMS is considered to be coherent. In contrast, the state of a non-coherent PMS does not monotonically increase with an additional number of functioning components. For a non-coherent system, both component failures and repairs can contribute to the system failure. Example works of non-coherent systems can be found in Niu et al. (2012), Chu (2009), Matuzas and Contini (2015). This chapter concentrates on the analysis of coherent systems.

*Redundant Versus Non-redundant PMS:* In the case of no extra unit being provided, a PMS is non-redundant. If extra units are used to enhance system reliability, then the PMS is considered a redundant system. Three types of redundancy are widely employed: (1) *Standby redundancy*, also known as backup redundancy, involves one or multiple primary online units and some standby units that take over the mission when an online unit fails. There are three standby modes: cold, warm, and hot depending on the readiness of the standby unit. (2) *Active redundancy*, also known as parallel redundancy, involves multiple units functioning in parallel. (3) *1: N redundancy*, involves a single backup shared by multiple units. If the backup is used by one of the units, it is not available for the remaining units. Levitin et al. (2017) formulated the redundancy allocation problem (RAP) in PMSs and applied several optimization methods to solve the problem.

*PMS with Perfect Coverage Versus Imperfect Coverage:* If the system can adequately detect, locate, and recover from a component fault, then the system has perfect coverage, which is a common assumption for PMS studies. In contrast, the system's fault recovery mechanism is not perfect so that an undetected or uncovered component fault leads to the corruption of the entire system. This behavior is known as imperfect coverage. It introduces two types of failure modes: a covered failure that affects only a single component, and an uncovered failure that can propagate through the system and lead to the failure of the entire system. Example works on PMSs with imperfect coverage include Xing and Dugan (2002), where a separable binary decision diagram-based method was proposed to address imperfect coverage in reliability analysis of PMSs with combinatorial phase requirements. Xing (2007) addressed imperfect coverage in reliability analysis of PMSs subject to CCFs. Xing et al. (2012) modeled reliability of $k$-out-of-$n$ PMSs with identical components and imperfect coverage using a recursive method; Wang et al. (2018a) modeled the same problem using a record value-based method. The above-mentioned works assume the element-level coverage (ELC) model for addressing imperfect coverage, where the fault coverage probability of an element does not depend on states of other system elements (Myers and Rauzy 2008; Levitin and Amari 2009). Another common model for imperfect coverage is fault-level coverage (FLC), where the fault coverage probability depends on the number of faults happening to a group within a certain recovery window. Example works of considering PMSs with FLC include Peng et al. (2014), (2016).

*PMS with Fixed Versus Dynamic Duration:* The phase duration can be fixed and remain unchanged during the whole mission; it can also be stochastic and follow certain distribution, e.g., the log-normal distribution. Literature on PMSs with dynamic durations is limited. For example, Li and Peng (2014) considered stochastic phase durations in multi-state series-parallel PMSs and employed the combination of Markov chains and universal generating functions to calculate the system availability and operation cost. Si et al. (2015) employed stochastic filtering to model dynamic phase durations in the reliability analysis of PMSs subject to phase-dependent degradation processes.

*PMS with Series Versus Combinatorial Phase Requirement.* If the entire mission fails as long as the system fails in any period of its life span, then the PMS is classified as a series PMS or a PMS with phase-OR requirements. As a generalization, Xing and Dugan (2002) introduced PMSs with combinatorial phase requirements (CPRs), where a phase failure does not necessarily result in the entire mission failure. Specifically, the failure criteria of a PMS with CPRs can be illustrated as a logical AND/OR combination of phase failures.

*PMS with Fixed Versus Dynamic Phase Sequence:* For some PMSs, the sequence of phases traversed by the system to accomplish its mission goal is always composed of a single path from the first phase to the last phase. For other PMSs, at the end of each phase, the next phase to perform is dynamically determined based on the state of the system, i.e., the sequence of phases traversed by the mission can be dynamic. While the existing works have mostly focused on sequential PMSs, little studies were dedicated to modeling PMSs with dynamic phase sequences. For example, Mura and Bondavalli (2002), Bondavalli et al. (2004) employed Markov regenerative stochastic Petri nets to model PMSs with dynamic phase sequences.

# 3 Evaluation Techniques of PMSs

Four different types of analytical modeling methods for reliability analysis of PMSs are reviewed, including the combinatorial methods in Sect. 3.1, the state space-based approaches in Sect. 3.2, the modular methods in Sect. 3.3, and the recursive methods in Sect. 3.4. In Sect. 3.5, we summarize the applicability of the different methods reviewed in this section.

## 3.1   Combinatorial Methods

Four different combinatorial methods are discussed, including the mini-component technique, Boolean algebraic-based binary decision diagrams, multi-valued decision diagrams, and the universal generating function-based method.

### The Mini-component Technique

Esary and Ziehms (1975) first introduced the mini-component technique that deals with s-dependence across phases by replacing a component in each phase with a series of independent mini-components. It is proved that the reliability of the modified system (after the replacement) is same as that of the original PMS. Figure 1 illustrates the replacement procedure in two different system reliability models. Specifically, if the reliability block diagram (RBD) is used as the reliability model, a component $X$ in phase $j$ of a PMS is replaced by a set of s-independent mini-component $\{x_i\}, i = 1, \ldots, j$ connected in a series structure and the relation between a component and its mini-component can be denoted as: $X_j = x_1 \cdot x_2 \cdot \ldots \cdot x_j$. This equation represents that the component is operational ($X_j = 1$ or $\overline{X}_j = 1$) in phase $j$ if and only if it has functioned in all phases from phase 1 to phase $j$. If the fault tree is used as the system reliability model, then the event representing component $X$ failing in phase $j$ is replaced by an OR gate with $j$ inputs, each being the failure of one mini-component $\{x_i\}, i = 1, \ldots, j$, as illustrated in Fig. 1. This replacement process represents that component $X$ is failed in phase $j$ if one of the mini-components $\{x_i\}, i = 1, \ldots, j$ is failed.

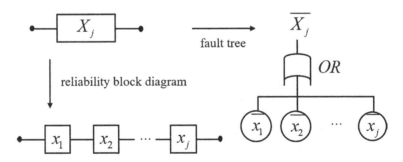

**Fig. 1.**   Illustration of the mini-component technique.

**Illustrative Example:** The example PMS has three components employed in two non-overlapping and consecutive phases. The system demand in each phase is denoted as $d_j, j = 1, 2$; the capacity of three independent components in each phase is denoted as $q_{ij}, i = A, B, C, j = 1, 2$. In phase one, $d_1 = 6, q_{A1} = 6, q_{B1} = 4, q_{C1} = 2$; in phase two, $d_2 = 8, q_{A2} = 5, q_{B2} = 4, q_{C2} = 3$. The system fails in a phase if the summation of components' capacity is smaller than the desired demand in that phase. Figure 2 presents the fault tree model of each phase.

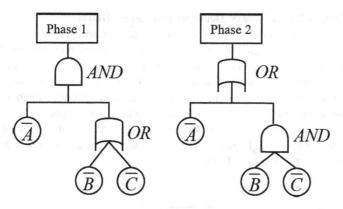

**Fig. 2.** Fault tree of the example PMS.

Applying the mini-component technique, Fig. 3 shows the fault tree of the example PMS after the replacement, which can be analyzed without considering s-dependencies across phases for each component. However, the disadvantage of this method is the size of the system model after the replacement can become very large as the number of components increases and the number of phases increases. Thus, significant computational power and space are needed for obtaining the solution.

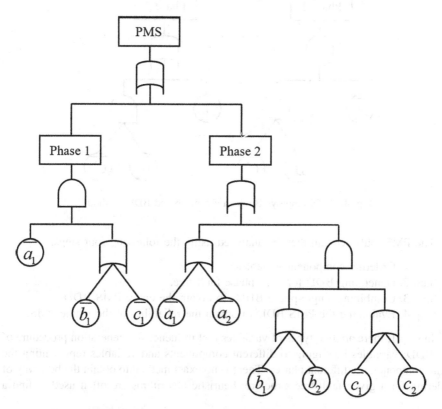

**Fig. 3.** Equivalent mini-component system model.

**Boolean Algebra-Based Binary Decision Diagrams (BDDs)**

A BDD is a directed acyclic graph, where all the paths start at the root node, traverse some non-sink nodes, and end at one of two terminal nodes representing the system being failed (labeled by 1) or operational (labeled by 0) (Xing and Amari 2015; Mo 2009), respectively. Each non-sink node (corresponding to a system component) has two outgoing branches or edges, representing the failure (right edge or then edge) and operation (left edge or else edge) of the corresponding component.

In the Boolean algebra-based BDD method, the PMS fault tree is obtained by simply connecting each single-phase fault tree model using a logic OR gate, as illustrated in Fig. 4 for the example PMS.

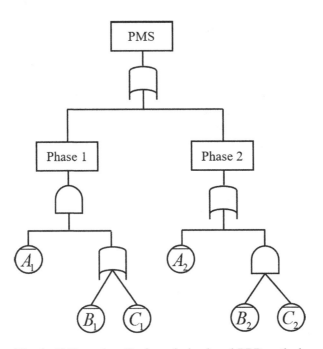

**Fig. 4.** PMS employs Boolean algebra based BDD method.

The PMS fault tree can then be analyzed using the following four steps.

Step 1. Ordering component variables.
Step 2. Generating BDD for each phase fault tree.
Step 3. Combining single-phase BDDs to obtain the entire PMS BDD.
Step 4. Evaluating the PMS BDD to obtain the reliability of the whole PMS.

In step 1, there are two types of variables that influence the generation procedure of the BDD: variables belonging to different components and variables representing the same component in different phases. There is no exact method to obtain the best way of ordering for a given fault tree structure; heuristic algorithms are often used to find a reasonable ordering (or indexes) of variables belonging to different components. For variables representing the same component in different phases, forward or backward ordering can be used. In the forward phase dependent operation (PDO), the variables belonging to the same component in different phases stay together and the order of these variables is consistent with the phase order. In contrast, in the backward PDO, though the variables belonging to the same component in different phases still stay together, the order of these variables is opposite to the phase order.

In step 2, BDD is generated for each single phase fault tree based on the manipulation rules in Zang et al. (1999) and Xing and Amari (2015). Specifically, Boolean logic expressions $g$ and $h$ encoding two BDD models can be represented using the if-then-else (ite) format as:

$$g = ite(x, g_{x=1}, g_{x=0}) = ite(x, G_1, G_2), \tag{1}$$

$$h = ite(y, h_{y=1}, h_{y=0}) = ite(y, H_1, H_2). \tag{2}$$

$x$ and $y$ are root nodes of the BDD models to be combined. A logic operation (denoted by "$\circ$") between $g$ and $h$ can be represented using the following manipulations rules:

$$ite(x, G_1, G_2) \circ ite(y, H_1, H_2) = \begin{cases} ite(x, G_1 \circ H_1, G_2 \circ H_2) & index(x) = index(y) \\ ite(x, G_1 \circ h, G_2 \circ h) & index(x) < index(y) \\ ite(y, g \circ H_1, g \circ H_2) & index(x) > index(y) \end{cases} . \tag{3}$$

For example, Fig. 5 illustrates the BDD model for each phase of the example PMS.

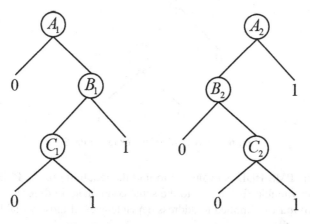

**Fig. 5.** BDD model for each phase.

In step 3, the rules of phase algebra are introduced for combining terms containing multiple variables belonging to the same component but different phases. Specifically, if we use $A_i$ and $A_j, j < i$, to respectively represent that component $A$ works in phase $i$ and in phase $j$, then the following phase algebra rules hold:

$$A_j \cdot A_i \rightarrow A_i, \tag{4}$$

$$\overline{A_j} \cdot \overline{A_i} \rightarrow \overline{A_j}, \tag{5}$$

$$\overline{A_j} \cdot A_i \rightarrow 0, \tag{6}$$

$$\overline{A_j} + \overline{A_i} \rightarrow \overline{A_i}, \tag{7}$$

$$A_j + A_i \rightarrow A_j, \tag{8}$$

$$A_j + \overline{A_i} \rightarrow 1. \tag{9}$$

The proof of these rules can be found in Zang et al. (1999) by using the relation between a component and its mini-components. The phase algebra rules are used in combining single phase BDDs into the PMS BDD. Figure 6 illustrates the PMS BDD of the example PMS.

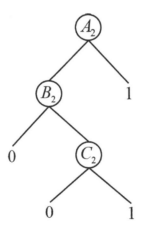

**Fig. 6.** PMS BDD of the example PMS.

In step 4, the PMS BDD is evaluated to find the reliability of the PMS. In the case of edges linking variables belonging to the same component in different phases, phase algebra rules (6) must be applied to address dependencies of those variables during the PMS BDD evaluation.

In Wang et al. (2012), the PMS BDD method is utilized in a combinatorial procedure for addressing competing failure propagation and failure isolation effects in the reliability analysis of PMSs subject to function dependence in one of the mission phases.

## Multivalued Decision Diagrams (MDDs)

MDDs are extensions of BDDs to the multi-valued logic (Mo et al. 2014b). Xing and Dai (2009) first adapted MDDs for the reliability analysis of generic multi-state systems (MSSs), where the two sink nodes represent the system being in (labeled 1) or not in (labeled 0) a particular state. A non-sink node in the MDD has multiple outgoing edges each corresponding to a different state of the multi-state component represented by the node. Later on, Mo et al. (2014a) employed MDDs for analyzing reliability of PMSs, where each PMS component is modeled as a multi-state non-sink node with multiple edges including edge 0 and one edge for each phase.

The MDD-based method for PMS reliability analysis can be summarized as the following four-step procedure:

Step 1. Ordering all component variables using a heuristic method.
Step 2. Generating MDD for each phase fault tree.
Step 3. Combining single-phase MDDs to obtain the entire PMS MDD.
Step 4. Evaluating the PMS MDD to obtain the reliability of the whole PMS.

Different from the BDD-based method in previous subsection, no dependencies exist among variables in the PMS MDD method, thus no phase algebra rules are needed during the model generation and evaluation.

Figure 7 gives the MDD model for the example PMS generated in step 3 using ordering of $A < B < C$.

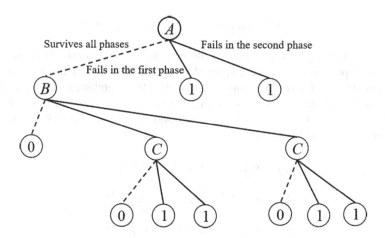

**Fig. 7.** MDD of the example PMS.

In Wang et al. (2018b), the MDD-based method is utilized in a combinatorial procedure for addressing competing failure propagation and probabilistic failure isolation effects in the reliability analysis of PMSs subject to probabilistic function dependence (e.g., body sensor networks).

**Universal Generating Function**

Universal generating function (UGF) is a flexible and powerful technique to depict different states of multi-state systems since it can represent the performance distributions of system elements using algebraic procedures (Levitin 2005). Specifically, the UGF represents the probability mass function (PMF) $\alpha_{j,h} = \Pr(Y_j = y_{j,h})$ of a discrete random variable $Y_j$ with $k_j$ possible values in the form of

$$u_j(z) = \sum_{h=1}^{k_j} \alpha_{j,h} z^{y_{j,h}}. \tag{10}$$

Moreover, the operation $\varphi(Y_1, Y_2, \ldots, Y_n)$ performed on $n$ independent random variables can be denoted as

$$U(z) = \underset{\varphi}{\otimes}(u_1(z), \ldots, u_n(z)) = \underset{\varphi}{\otimes}(\sum_{h_1=1}^{k_1} \alpha_{1h_1} z^{y_{1,h_1}}, \ldots, \sum_{h_n=1}^{k_n} \alpha_{nh_n} z^{y_{n,h_n}})$$

$$= \sum_{h_1=1}^{k_1} \sum_{h_2=1}^{k_2} \cdots \sum_{h_n=1}^{k_n} (\prod_{i=1}^{n} \alpha_{ih_i} z^{\varphi(y_{1,h_1}, \ldots, y_{n,h_n})}). \tag{11}$$

For the reliability evaluation of the example PMS, the failure probability of each component $i$ in phase $j$, denoted by $\mu_{ij}$ should be given or can be derived from input parameters. The UGF of component $i$ in two phases can thus be denoted as

$$U_i = \mu_{i1} z^{(0,0)} + \mu_{i2} z^{(q_{i1},0)} + (1 - \mu_{i1} - \mu_{i2}) z^{(q_{i1},q_{i2})}. \tag{12}$$

where $q_{ij}, i = A, B, C, j = 1, 2$ represents the capacity of component $i$ in phase $j$.

Recursive procedures can be employed to obtain the UGF of the entire system. Specifically, $U_A$ and $U_B$ are first combined, which is then combined with $U_C$ to obtain the UGF of the whole system. Assume that the benchmark takes the value of $\mu_{A1} = 0.1, \mu_{A2} = 0.1, \mu_{B1} = 0.2, \mu_{B2} = 0.1, \mu_{C1} = 0.1$ and $\mu_{C2} = 0.2$. The UGFs of the three components in the example PMS are

$$U_A = 0.1z^{(0,0)} + 0.1z^{(6,0)} + 0.8z^{(6,5)},$$

$$U_B = 0.1z^{(0,0)} + 0.2z^{(4,0)} + 0.7z^{(4,4)},$$

$$U_C = 0.2z^{(0,0)} + 0.1z^{(2,0)} + 0.7z^{(2,3)}.$$

The UGF of the combination of components $A$ and $B$ is obtained as

$$U_{A-B} = U_A \underset{\times}{\otimes} U_B = 0.02z^{(0,0)} + 0.01z^{(4,0)} + 0.07z^{(4,4)} + 0.02z^{(6,0)} + 0.01z^{(10,0)}$$

$$+ 0.07z^{(10,4)} + 0.16z^{(6,5)} + 0.08z^{(10,5)} + 0.56z^{(10,9)}.$$

Note that the composition operator $\otimes_{\times}$ implements the multiplication of coefficients and the summation of exponents for every pair of terms of the two UGFs at both sides of the operator. The UGF of the system can be further obtained by combining $U_{A-B}$ with $U_C$ as

$$U_x = U_{A-B} \otimes_{\times} U_C = 0.002z^{(0,0)} + 0.004z^{(2,0)} + 0.014z^{(2,3)} + 0.001z^{(4,0)} + 0.002z^{(6,0)}$$
$$+ 0.007z^{(6,3)} + 0.007z^{(4,4)} + 0.014z^{(6,4)} + 0.049z^{(6,7)} + 0.002z^{(6,0)} + 0.004z^{(8,0)}$$
$$+ 0.014z^{(8,3)} + 0.001z^{(10,0)} + 0.002z^{(12,0)} + 0.007z^{(12,3)} + 0.007z^{(10,4)} + 0.014z^{(12,4)}$$
$$+ 0.049z^{(12,7)} + 0.016z^{(6,5)} + 0.032z^{(8,5)} + 0.112z^{(8,8)} + 0.008z^{(10,5)} + 0.016z^{(12,5)}$$
$$+ 0.056z^{(12,8)} + 0.0056z^{(10,9)} + 0.112z^{(12,9)} + 0.392z^{(12,12)}.$$

Recall that the system fails when the summation of capacity in any phase cannot satisfy the demand. Therefore, the system fails when the first exponent of $z$ is lower than 6 or the second exponent of $z$ is lower than 8. In other words, the possible cases for the survival of the system are $(8,8)$, $(12,8)$, $(10,9)$, $(12,9)$ and $(12,12)$. Therefore, the reliability of the example PMS can be calculated as the summation of coefficients of those cases.

$$R = 0.112 + 0.056 + 0.056 + 0.112 + 0.392 = 0.728.$$

Recently, Peng et al. (2016) employed the UGF technique to analyze the reliability of PMSs subject to fault-level coverage. The multivariable UGFs are introduced to link states of a certain component throughout the mission with its corresponding state probabilities.

## 3.2 State Space Based Approaches

Conventionally, if the failure criteria in any one phase of the PMS are dynamic, then a state space-based approach should be employed. We first discuss the CTMC solution and then introduce the application of Petri nets to the PMS analysis in this subsection.

### Continuous-Time Markov Chain (CTMC)

The basic idea is to construct a CTMC to represent the failure behavior of the entire PMS, or construct several Markov chains each representing the failure behavior of each phase of the PMS. These Markov models account for dependence among components within a phase as well as dependence across phases for a given component. The probability of the system in each state can be obtained through solving the CTMC models constructed. The unreliability of the whole PMS can thus be obtained through summing all the failure state probabilities.

For a specific example, Smotherman and Zemoudeh (1989) used a single non-homogeneous Markov chain model to perform the reliability analysis of a PMS. The method is known as the SZ approach. Consider the example PMS in Fig. 2. Assume the failure rate of component $i \in \{A, B, C\}$ in each phase $j \in \{1, 2\}$ is denoted as $\mu_{ij}$. Figure 8 illustrates the Markov chain model under the SZ approach, where the system state is represented by a 3-tuple indicating statuses of the three components ("1": operational, "0": failed). For instance, state (101) implies that $A$ and $C$ are operational and B has failed. Note that "*F*" in Fig. 8 represents the system failure. Transition function $h(t)$ represents the failure rate from one phase to another phase, associated with the time at which the phase change occurs.

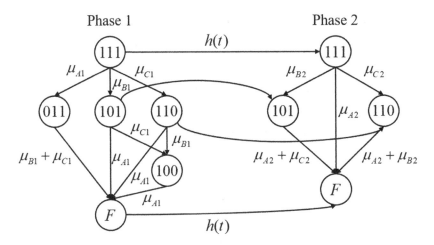

**Fig. 8.** Markov chain model under the SZ approach.

For another example, Somani et al. (1992) proposed to generate and solve separate Markov chains for individual phases of a PMS. While analyzing a phase, only states relevant to that phase are considered. Figure 9 illustrates the Markov model in each phase of the example PMS. The initial state occupation probability vector consists of a "1" for the initial state and "0" for all other states when solving the Markov chain of the first phase. State occupation probabilities at the end of any phase become the initial state occupation probabilities of the corresponding states when solving the Markov chain of the next phase. The summation of operational state probabilities obtained by solving the Markov chain of the last mission phase gives the reliability of the whole PMS.

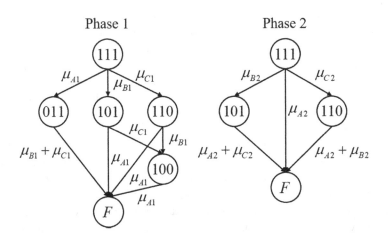

**Fig. 9.** Markov model for each phase of the example PMS.

Wang et al. (2017) extended the separated Markov method of Somani et al. (1992) for the reliability analysis of PMSs subject to competitions between failure propagation and failure isolation caused by the function dependence happening in multiple phases.

Lu and Wu (2014) proposed a CTMC-based analytical approach to evaluate the reliability of PMSs considering both CPR and repairable components. In their work, the CPR is analyzed by decomposing the overall mission success into the corresponding system behavior in each phase. The method can exclude a large number of redundant states from the model, instead of incorporating all states like the traditional CTMC-based approaches.

### Petri Nets

Integrating benefits from both analytical modeling and simulations, Petri nets (PNs) have been widely employed for dynamic system modelling. We explain the typical features of a Petri net using an example in Fig. 10.

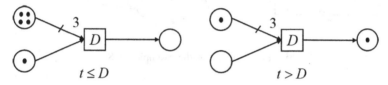

**Fig. 10.** An example of Petri nets.

In Fig. 10, the circles and squares respectively represent places and transitions. Solid dots in circles represent tokens. A PN containing tokens is referred to as a marked PN. The location of tokens indicates the marking or state of the system at the moment. As tokens move through the system, the system state changes, making the dynamic process representation possible. The movement of tokens is facilitated by the firing of a transition. To enable a transition firing, each input place should be populated with at least as many tokens as the weight of the respective directed edge input to the transition. Upon firing a transition, tokens are removed from the input places in a quantity that matches the weight of the respective input directed edge, new tokens are created in the output place(s) in a quantity that matches the weight of the respective output directed edges. Figure 10(b) illustrates the PN after firing the transition in Fig. 10(a).

Figure 11 illustrates the PN of the example PMS, where "*up*" means that the component state is working, and "*dn*" means that the component state is failed. In this PN model, the master part is used to describe the phase sequence, the components part is used to describe the components state change, and the logic part is used to present phase failure logic.

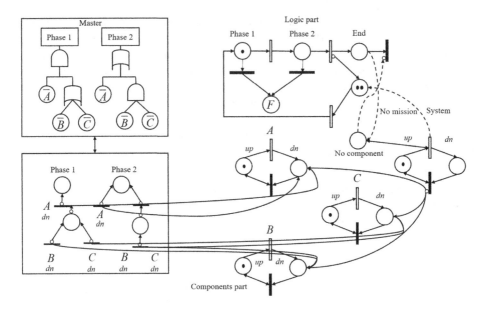

**Fig. 11.** Petri net of the example PMS.

Chew et al. (2008) proposed a simulation model to assess the overall reliability, which can be summarized as the following procedure:

Step 1. Sampling switching times for each newly enabled timed transition in each subnet from the switching time distribution randomly.
Step 2. Locating the transition with the earliest switching time and firing it.
Step 3. Searching through each of the immediate transitions and if there are any enabled, firing it.
Step 4. Repeating step 3 until no more immediate transitions are enabled.
Step 5. Testing and logging the conditions based on several rules:

(a) Begin next simulation if the system has failed;
(b) Begin next operating period if the system has been abandoned;
(c) Begin next mission if the mission has completed;

Step 6. If simulations complete before the maximal running time, go to step 1, otherwise end.

Another method to evaluate PN is to convert the reachability graph of the PN to the isomorphic CTMC model for analysis (Bouali et al. 2012).

Recently, Remenyte-Prescott and Andrews (2011) considered fault propagation modeling using Petri nets. Michael et al. (2014) employed a Petri net approach to fault verification in PMSs using a standard deviation technique. Wu and Wu (2015) proposed an extended object-oriented Petri net model for mission reliability simulation of a repairable PMS with CCFs.

### 3.3   Modular Methods

Combinatorial methods can analyze PMSs efficiently, but can only handle systems with s-independent components. The state space-based approaches are flexible and powerful in modelling PMSs with complex dynamic behaviors. However, they suffer from the state space explosion problem. Modular methods have been developed to analyze PMSs with a large number of components, which merge advantages of both combinatorial methods and state space-based approaches (Xing and Amari 2008; Yang and Wu 2014; Zhai et al. 2018).

The phase-modular approach involves the following major steps (Meshkat 2000; Meshkat et al. 2003; Xing and Amari 2008):

Step 1. Representing each mission phase with a fault tree and linking the phase fault trees with gates representing the CPRs.

Step 2. Dividing each phase fault tree into independent subtrees/modules.

Step 3. Characterizing each phase module as static or dynamic.

Step 4. Identifying each phase module as bottom-level or upper-level.

Step 5. Finding the system-level independent modules and finding the unions of components in all the phase modules that overlap in at least one component.

Step 6. Identifying each system-level module as static or dynamic across the phases (identifying a component as dynamic in at least one mission phase is sufficient for the identification of the corresponding system-level module as dynamic).

Step 7. Grouping the phase modules according to the corresponding system-level module.

Step 8. Finding the joint phase module probabilities for all system-level modules. The BDD method can be employed for modules that are static across all phases and the combined Markov chain can be employed for dynamic modules.

Step 9. Considering each module as a basic event of a static fault tree of the entire system and solving the corresponding fault tree using BDD to find the overall system reliability based on the reliability of the modules.

### 3.4   Recursive Methods

The recursive method involves two major steps: generating combinations of element failures and employing the backward recursion algorithm.

Consider a random vector $X_j = (x_1(j), \ldots, x_n(j))$, representing the system state (composition of states of $n$ elements) at the end of phase $j$. Each $x_i(j), 1 \leq i \leq n$ is a Boolean variable assuming value of either "1" (representing the working state of component) or "0" (representing the failed state of the component). It is further assumed that a realization $Y = (y_1(j), \ldots, y_n(j))$ of vector $X_j$ consists of $s$ zeros. In other words, $s$ out of $n$ elements fail before the end of phase $j$. Specifically, zero elements have numbers $c(k), 1 \leq k \leq s$. $x_i(j), 1 \leq i \leq n$ are non-increasing functions of $j$, and any $y_i(j) = 1$ implies that $x_i(m) = 1$ for $m = 1, \ldots, j - 1$. Therefore, different system states that can precede state $Y$ can be obtained only by replacing zeros with ones in $Y$. This relationship represents that the corresponding elements work at the end of phase $j - 1$ and fail during phase $j$. For the sake of obtaining all possible realization of state $X_{j-1}$ preceding state $Y$, one has to enumerate all the possible combinations of vector elements that take the value of one in $X_{j-1}$ and the value of 0 in $Y$. Further, to obtain any combination of elements failed during phase $j$, one can run the integer index $\sigma$ from $\sigma = 0$ to $\sigma = 2^s - 1$ given that the system state at the end of this phase is $Y$. If $\text{mod}_2 \lfloor \frac{\sigma}{2^{i-1}} \rfloor = 1$, element $c(i)$ remains functioning; if $\text{mod}_2 \lfloor \frac{\sigma}{2^{i-1}} \rfloor = 0$, the element fails during phase $j$. Thus, one can obtain the possible $\sigma_{th}$ realizations of system state vector $X_{j-1}$ preceding the realization of system state vector $X_j$ using the operator $X_{j-1} = \pi(Y, \sigma)$, where $\pi(Y, \sigma)$ represents a binary vector $Y$ with zero elements $y_{c(1)}(j), \ldots, y_{c(s)}(j)$ and integer number $\sigma$. Specifically, the following recursive procedure can be used to obtain $\pi(Y, \sigma)$.

Step 1. $d = 1$
Step 2. For $i = 1$ to $s$,
2.1 If $\text{mod}_2 \lfloor \frac{\sigma}{d} \rfloor = 1$, assign $y_{c(i)}(h) = 1$;
2.2 Assign $d = 2d$.

The probability of the combination of failures causing the system transition from state $\pi(Y, \sigma)$ in the beginning of phase $j$ to state $Y$ in the end of phase $j$ can be denoted through substituting the value of index $\sigma$ and the conditional failure probabilities $q_{c(i)}$ of elements $c(i)$. Specifically, the probability can be represented as

$$Q_j(\sigma) = \prod_{i=0}^{s} (q_{c(i)}(h))^{\text{mod}_2 \lfloor \frac{\sigma}{2^{i-1}} \rfloor}. \tag{13}$$

Further, we let $Z_{j,Y}$ represent the probability of the event when $X_j = Y$ and $\varphi_l(X_l) = 1$ for all $l < j$. Among, $\varphi_l(X_l), 1 \leq l \leq j$ represents system state acceptability function in phase $j$. In other words, we have

$$Z_{j,Y} = \text{Pr}\{X_j = Y; \varphi_{j-1}(X_{j-1}) = 1; , \ldots, \varphi_1(X_1) = 1\}. \tag{14}$$

Because of the Markov property of $X_j$, the conditional distribution of $X_j$ given the entire sequence $(X_1, \ldots, X_{j-1})$ depends only on the most recent value $X_{j-1}$. Thus, $Z_{j,Y}$ can be calculated through

$$Z_{j,Y} = \prod_{i=0}^{n} p_i(j)^{y_k} \times \sum_{\sigma=0}^{2^s-1} \varphi_{j-1}(\pi(Y, \sigma)) Z_{j-1, \pi(Y, \sigma)} Q_j(\sigma). \tag{15}$$

where $p_i(j)$ represents the conditional reliability of element $i$ at phase $j$. It is easy to derive $Z_{1,Y}$ as

$$Z_{1,Y} = \prod_{i=0}^{2^n-1} (q_i(1))^{1-y_i} (p_i(1))^{y_i}. \tag{16}$$

Thus, the reliability of the whole system in $j$ phases can be denoted as

$$R = \sum_{\sigma=0}^{2^n-1} \varphi_j(\pi(0,\sigma))Z_{j,\pi(0,\sigma)}. \tag{17}$$

As an illustration, we employ the recursive algorithm to evaluate the reliability of the example PMS in Fig. 2. Since there are three components in the example system, there are eight different combinations in any phase $j, j = 1, 2$: $A_jB_jC_j, A_jB_j\overline{C_j}, A_j\overline{B_j}C_j, \overline{A_j}B_jC_j, \overline{A_j}\overline{B_j}C_j, \overline{A_j}B_j\overline{C_j}, A_j\overline{B_j}\overline{C_j}$ and $\overline{A_j}\overline{B_j}C_j$. Among these combinations, $A_2B_2C_2, A_2B_2\overline{C_2}$ and $A_2\overline{B_2}C_2$ make the system survive in the second phase. We first consider the case for $A_2B_2C_2$, the only possible combination that can occur in phase one is $A_1B_1C_1$. Therefore, we have

$$R_{A_1B_1C_1-A_2B_2C_2} = 0.8 \times 0.8 \times 0.7 = 0.392.$$

When the combination in phase two is $A_2B_2\overline{C_2}$, the possible combinations in phase one are $A_1B_1C_1$ and $A_1B_1\overline{C_1}$. Thus, we have

$$R_{A_1B_1C_1-A_2B_2\overline{C_2}} + R_{A_1B_1\overline{C_1}-A_2B_2\overline{C_2}} = 0.8 \times 0.7 \times 0.2 + 0.8 \times 0.7 \times 0.1 = 0.168.$$

When the combination in phase two is $A_2\overline{B_2}C_2$, the possible combinations in phase one are $A_1B_1C_1$ and $A_1\overline{B_1}C_1$. Under this case, we have

$$R_{A_1B_1C_1-A_2\overline{B_2}C_2} + R_{A_1\overline{B_1}C_1-A_2\overline{B_2}C_2} = 0.8 \times 0.7 \times 0.1 + 0.8 \times 0.7 \times 0.2 = 0.168..$$

Therefore, the reliability of the whole system evaluated by the recursive method can be obtained as

$$R = 0.392 + 0.168 + 0.168 = 0.728,$$

which matches the results obtained using the UGF method.

The main advantage of the recursive method is that it does not require the composition of any graph models. Moreover, both its computational time and memory requirements are linear in terms of the system size.

The recursive algorithm has been applied or extended in several directions. For example, Amari and Xing (2011) proposed an efficient recursive method for exact reliability evaluation of $k$-out-of-$n$ PMS with identical components. Levitin et al. (2012) employed the recursive algorithm to evaluate the reliability of arbitrary binary or multi-state PMS consisting of non-identical, binary and non-repairable elements. Levitin et al. (2013b) proposed a recursive and exact method for reliability evaluation of PMSs with failures originating from some system elements that can propagate causing common cause failures of groups of elements. Levitin et al. (2013a) extended the recursive method for the exact reliability evaluation of PMS consisting of non-identical, non-repairable, multistate elements.

### 3.5  Summary of PMS Analysis Methods

Table 2 summarizes the PMS evaluation methods reviewed in this section and their applicability based on main characteristics used for the PMS classification in Sect. 2. Note that these characteristics are not necessarily exclusive; but different perspectives to characterize a PMS. With this check table, the specific kind of PMSs each method is applicable to can be easily identified. For example, BDD is applicable to "Coherent", but not "Multi-State". As a multistate system may also be coherent, by integrating the properties checked in the table, it can be judged that BDD is applicable to coherent binary systems, but not coherent multistate systems.

**Table 2.** Applicability of Methods to PMSs with different characteristics.

| | Mini-Comp. | BDD | MDD | UGF | MC | PN | Modular | Recursive |
|---|---|---|---|---|---|---|---|---|
| Repairable | | | | ✓ | ✓ | ✓ | ✓ | ✓ |
| Non-repairable | ✓ | ✓ | ✓ | ✓ | ✓ | ✓ | ✓ | ✓ |
| Static structure | ✓ | ✓ | ✓ | ✓ | ✓ | ✓ | ✓ | ✓ |
| Dynamic structure | ✓ | | | | ✓ | ✓ | ✓ | ✓ |
| Binary-state | ✓ | ✓ | ✓ | ✓ | ✓ | ✓ | ✓ | ✓ |
| Multi-State | | ✓ | ✓ | ✓ | ✓ | ✓ | ✓ | ✓ |
| Coherent | ✓ | ✓ | ✓ | ✓ | ✓ | ✓ | ✓ | ✓ |
| Non-coherent | ✓ | ✓ | ✓ | ✓ | ✓ | ✓ | ✓ | ✓ |
| Redundant | ✓ | ✓ | ✓ | ✓ | ✓ | ✓ | ✓ | ✓ |
| Non-redundant | ✓ | ✓ | ✓ | ✓ | ✓ | ✓ | ✓ | ✓ |
| Perfect Coverage | ✓ | ✓ | ✓ | ✓ | ✓ | ✓ | ✓ | ✓ |
| Imperfect Coverage | | ✓ | ✓ | ✓ | ✓ | ✓ | ✓ | ✓ |
| Fixed Duration | ✓ | ✓ | ✓ | ✓ | ✓ | ✓ | ✓ | ✓ |
| Dynamic Duration | | ✓ | ✓ | ✓ | ✓ | ✓ | ✓ | ✓ |
| Series PMS | ✓ | ✓ | ✓ | ✓ | ✓ | ✓ | ✓ | ✓ |
| PMS with CPR | ✓ | ✓ | ✓ | ✓ | ✓ | ✓ | ✓ | ✓ |
| Fixed phase sequence | ✓ | ✓ | ✓ | ✓ | ✓ | ✓ | ✓ | ✓ |
| Dynamic phase sequence | | ✓ | ✓ | | ✓ | ✓ | ✓ | ✓ |

## 4  Optimization of PMSs

Several optimization issues of PMSs have been formulated and solved: system structure optimization, component separation and combination optimization, redundancy optimization, standby element sequencing optimization, and component test and maintenance optimization.

System structure optimization is a well-studied problem in the field of reliability engineering, aiming at balancing reliability and cost for the system design. Traditionally, this problem was only solved for systems that do not change their tasks and configurations during the mission. In the past decade, the system structure optimization problem was solved for PMSs with dynamic system configuration, success criteria and element behavior. In particular, Dai et al. (2013) employed the recursive algorithm to evaluate the reliability of a non-repairable PMS and then applied the genetic algorithm to determine the optimal system structure. Peng et al. (2016) employed the UGF to evaluate reliability of a PMS consisting of subsystems connected in series, where each subsystem contains components with different capacities. The components within the same subsystem are divided into several disjoint work-sharing groups. Since different partitions of the working groups may lead to different reliability, optimizing the system structure is relevant and significant. Based on the UGF evaluation, the genetic algorithm method was applied to find the optimal structure of PMS maximizing the entire system reliability. Yu et al. (2017) considered the reliability of a phased-mission common bus system with CCFs using a recursive algorithm. The genetic algorithm was employed to search the optimal allocation strategies of the service elements.

The improvement in the survivability of a system can result from separating its elements. Levitin et al. (2013c), (2014a) formulated the optimal element separation problem in non-repairable PMSs. Levitin et al. (2014c) studied the problem in linear consecutively-connected systems subject to multiple phases of mission and CCFs.

The traditional redundancy allocation problem (RAP) has also been solved for PMSs. Yu et al. (2010) considered the RAP for PMSs by employing the particle swarm optimization. Levitin et al. (2017) formulated a constrained RAP and solved it through the brute force approach.

Levitin et al. (2014b), (2016) considered the optimization on standby element sequencing problem. Specifically, Levitin et al. (2014b) analyzed the optimal cold standby element sequencing problem. The initiation sequence of the system elements was investigated to minimize the expected mission cost while providing a desired level of system reliability. In Levitin et al. (2016), they considered a warm standby system to obtain the optimal activation sequence that maximizes system reliability, or minimizes expected mission cost, or minimizes expected uncompleted work.

Feyzioğlu et al. (2008) considered the optimal component test plans for PMSs. They formulated the optimal component testing problem as a semi-infinite linear program and employed an algorithmic procedure to compute optimal test times based on the column generation technique.

Jia et al. (2018) proposed a selective maintenance model for PMSs subject to random CCFs and optimally identified a subset of maintenance activities to be performed on some elements of the system to maximize the reliability of the whole system.

# 5   Future Research

In this chapter, we extend the review in Xing and Amari (2008) by introducing the PMS analysis methods and optimization problems developed in the last decade. One of the future research directions is to analyze cascading failures in multi-state PMSs, especially those that can influence function of subsequent phases. In other words, the failure of some component in one phase may lead to failures of other system components in subsequent phases. Moreover, the analysis of PMSs with storage components is also interesting, where the redundant capacity in one phase can be stored for use in later phases (Qiu et al. 2018). Another direction is to study the mission abortion strategy for PMSs used in life-critical applications where accomplishing a specified mission and aborting mission objectives in the case of certain condition being met to survive the system are both crucial (Levitin et al. 2018, 2019).

## References

Amari, S.V., Xing, L.: Reliability analysis of $k$-out-of-$n$ systems with phased-mission requirements. Int. J. Performability Eng. **7**(6), 604–609 (2011)

Bondavalli, A., Chiaradonna, S., Di Giandomenico, F., Mura, I.: Dependability modeling and evaluation of multiple-phased systems using deem. IEEE Trans. Reliab. **53**(4), 509–522 (2004)

Bouali, M., Barger, P., Schon, W.: Backward reachability of colored petri nets for systems diagnosis. Reliab. Eng. Syst. Saf. **99**, 1–14 (2012)

Chew, S.P., Dunnett, S.J., Andrews, J.D.: Phased mission modelling of systems with maintenance-free operating periods using simulated petri nets. Reliab. Eng. Syst. Saf. **93**(7), 980–994 (2008)

Chu, T.L.: Methods for probabilistic analysis of non-coherent fault trees. IEEE Trans. Reliab. **29**(5), 354–360 (2009)

Dai, Y., Levitin, G., Xing, L.: Structure optimization of non-repairable phased mission systems. IEEE Trans. Syst. Man Cybern. Syst. **44**(1), 121–129 (2013)

Ding, S.H., Kamaruddin, S.: Maintenance policy optimization—literature review and directions. Int. J. Adv. Manuf. Technol. **76**(5–8), 1263–1283 (2015)

Esary, J.D., Ziehms, H.: Reliability and Fault Tree Analysis: Theoretical and Applied Aspects of System Reliability and Safety Assessment. SIAM, Philadelphia (1975)

Feyzioğlu, O., Altınel, İ.K., Özekici, S.: Optimum component test plans for phased-mission systems. Eur. J. Oper. Res. **185**(1), 255–265 (2008)

Jia, X.S., Cao, W.B., Hu, Q.W.: Selective maintenance optimization for random phased-mission systems subject to random common cause failures. In: of the Institution of Mechanical Engineers, Part O: Journal of Risk and Reliability (2018). https://doi.org/10.1177/1748006X18791724

Levitin, G.: Universal Generating Function in Reliability Analysis and Optimization. Springer, London (2005)

Levitin, G., Amari, S.V.: Three types of fault coverage in multi-state systems. In: International Conference on Reliability. IEEE (2009)

Levitin, G., Amari, S.V., Xing, L.: Algorithm for reliability evaluation of non-repairable phased-mission systems consisting of gradually deteriorating multistate elements. IEEE Trans. Syst. Man Cybern. Syst. **43**(1), 63–73 (2013a)

Levitin, G., Finkelstein, M., Dai, Y.: Redundancy optimization for series-parallel phased mission systems exposed to random shocks. Reliab. Eng. Syst. Saf. **167**, 554–560 (2017)

Levitin, G., Xing, L., Amari, S.V.: Recursive algorithm for reliability evaluation of non-repairable phased mission systems with binary elements. IEEE Trans. Reliab. **61**(2), 533–542 (2012)

Levitin, G., Xing, L., Amari, S.V., Dai, Y.: Reliability of non-repairable phased-mission systems with propagated failures. Reliab. Eng. Syst. Saf. **119**, 218–228 (2013b)

Levitin, G., Xing, L., Amari, S.V., Dai, Y.: Optimal elements separation in non-repairable phased-mission systems. Int. J. Gen Syst **43**(8), 864–879 (2014a)

Levitin, G., Xing, L., Dai, Y.: Optimal allocation of connecting elements in phase mission linear consecutively-connected systems. IEEE Trans. Reliab. **62**(3), 618–627 (2013c)

Levitin, G., Xing, L., Dai, Y.: Minimum mission cost cold-standby sequencing in non-repairable multi-phase systems. IEEE Trans. Reliab. **63**(1), 251–258 (2014b)

Levitin, G., Xing, L., Dai, Y.: Reliability versus expected mission cost and uncompleted work in heterogeneous warm standby multiphase systems. IEEE Trans. Syst. Man Cybern. Syst. **47**(3), 462–473 (2016)

Levitin, G., Xing, L., Dai, Y.: Co-optimization of state dependent loading and mission abort policy in heterogeneous warm standby systems. Reliab. Eng. Syst. Saf. **172**, 151–158 (2018)

Levitin, G., Xing, L., Luo, L.: Influence of failure propagation on mission abort policy in heterogeneous warm standby systems. Reliab. Eng. Syst. Saf. **183**, 29–38 (2019)

Levitin, G., Xing, L., Yu, S.: Optimal connecting elements allocation in linear consecutively-connected systems with phased mission and common cause failures. Reliab. Eng. Syst. Saf. **130**, 85–94 (2014c)

Li, X.Y., Huang, H.Z., Li, Y.F., Zio, E.: Reliability assessment of multi-state phased mission system with non-repairable multi-state components. Appl. Math. Model. **61**, 181–199 (2018)

Li, Y.F., Peng, R.: Availability modeling and optimization of dynamic multi-state series–parallel systems with random reconfiguration. Reliab. Eng. Syst. Saf. **127**, 47–57 (2014)

Lu, J.M., Wu, X.Y.: Reliability evaluation of generalized phased-mission systems with repairable components. Reliab. Eng. Syst. Saf. **121**, 136–145 (2014)

Lu, J.M., Wu, X.Y., Liu, Y., Lundteigen, M.A.: Reliability analysis of large phased-mission systems with repairable components based on success-state sampling. Reliab. Eng. Syst. Saf. **142**, 123–133 (2015)

Matuzas, V., Contini, S.: Dynamic labelling of BDD and ZBDD for efficient non-coherent fault tree analysis. Reliab. Eng. Syst. Saf. **144**, 183–192 (2015)

Meshkat, L. Dependency Modeling and Phase Analysis for Embedded Computer Based Systems. University of Virginia, Charlottesville (2000)

Meshkat, L., Xing, L., Donohue, S.: An overview of the phase-modular fault tree approach to phased-mission system analysis. In: Proceedings of the 1st International Conference on Space Mission Challenges for Information Technology (SMC-IT), Pasadena, CA, pp. 393–398 (2003)

Michael, D.L., John, D.A., Rasa, R.P., John, T.P., Peter, H.: A petri net approach to fault verification in phased mission systems using the standard deviation technique. Qual. Reliab. Eng. Int. **30**(1), 83–95 (2014)

Mo, Y.: New insights into the BDD-based reliability analysis of phased-mission systems. IEEE Trans. Reliab. **58**(4), 667–678 (2009)

Mo, Y., Xing, L., Amari, S.V.: A multiple-valued decision diagram based method for efficient reliability analysis of non-repairable phased-mission systems. IEEE Trans. Reliab. **63**(1), 320–330 (2014a)

Mo, Y., Xing, L., Dugan, J.B.: MDD-based method for efficient analysis on phased-mission systems with multimode failures. IEEE Trans. Syst. Man Cybern. Syst. **44**(6), 757–769 (2014b)

Mura, I., Bondavalli, A.: Markov regenerative stochastic petri nets to model and evaluate phased mission systems dependability. IEEE Trans. Comput. **50**(12), 1337–1351 (2002)

Myers, A., Rauzy, A.: Efficient reliability assessment of redundant systems subject to imperfect fault coverage using binary decision diagrams. IEEE Trans. Reliab. **57**(2), 336–348 (2008)

Niu, R., Blum, R.S., Varshney, P.K., Drozd, A.L.: Target localization and tracking in non-coherent multiple-input multiple-output radar systems. IEEE Trans. Aerosp. Electron. Syst. **48**(2), 1466–1489 (2012)

Peng, R., Wu, D., Xiao, H., Xing, L.D., Gao, K.Y.: Redundancy versus protection for a non-reparable phased-mission system subject to external impacts. Reliab. Eng. Syst. Saf. **191**, 106556 (2019)

Peng, R., Zhai, Q., Xing, L., Yang, J.: Reliability of demand-based phased-mission systems subject to fault level coverage. Reliab. Eng. Syst. Saf. **121**, 18–25 (2014)

Peng, R., Zhai, Q., Xing, L., Yang, J.: Reliability analysis and optimal structure of series-parallel phased-mission systems subject to fault level coverage. IIE Trans. **48**(8), 736–746 (2016)

Qiu, H., Yan, X.B., Ma, X.Y., Peng, R.: Reliability of a phased-mission system with a storage component. Syst. Sci. Control Eng. **6**(1), 279–292 (2018)

Remenyte-Prescott, R., Andrews, J.D.: Modeling fault propagation in phased mission systems using Petri nets. In: Reliability & Maintainability Symposium. IEEE (2011)

Shrestha, A., Xing, L., Dai, Y.: Reliability analysis of multistate phased-mission systems with unordered and ordered states. IEEE Trans. Syst. Man Cybern. Part A Syst. Hum. **41**(4), 625–636 (2011)

Si, X.S., Hu, C.H., Zhang, Q., Li, T.: An integrated reliability estimation approach with stochastic filtering and degradation modeling for phased-mission systems. IEEE Trans. Cybern. **47**(1), 67–80 (2015)

Smotherman, M., Zemoudeh, K.: A non-homogeneous Markov model for phased-mission reliability analysis. IEEE Trans. Reliab. **38**(5), 585–590 (1989)

Somani, A.K., Ritcey, J.A., Au, S.H.L.: Computationally-efficient phased-mission reliability analysis for systems with variable configurations. IEEE Trans. Reliab. **41**(4), 504–511 (1992)

Tinga, T.: Towards a usage driven maintenance concept: improving maintenance value. In: Concurrent Engineering Approaches for Sustainable Product Development in a Multi-Disciplinary Environment. Springer, London (2013)

Wang, C., Xing, L., Levitin, G.: Competing failure analysis in phased-mission systems with functional dependence in one of phases. Reliab. Eng. Syst. Saf. **108**, 90–99 (2012)

Wang, C., Xing, L., Peng, R., Pan, Z.: Competing failure analysis in phased-mission systems with multiple functional dependence groups. Reliab. Eng. Syst. Saf. **164**, 24–33 (2017)

Wang, G., Peng, R., Xing, L.: Reliability evaluation of unrepairable $k$-out-of-$n$: G systems with phased-mission requirements based on record values. Reliab. Eng. Syst. Saf. **178**, 191–197 (2018a)

Wang, Y., Xing, L., Levitin, G., Huang, N.: Probabilistic competing failure analysis in phased-mission systems. Reliab. Eng. Syst. Saf. **176**, 31–51 (2018b)

Wu, X., Wu, Y.: Extended object-oriented petri net model for mission reliability simulation of repairable PMS with common cause failures. Reliab. Eng. Syst. Saf. **136**, 109–119 (2015)

Xing, L.: Reliability evaluation of phased-mission systems with imperfect fault coverage and common-cause failures. IEEE Trans. Reliab. **56**(1), 58–68 (2007)

Xing, L., Amari, S.V.: Reliability of phased-mission systems. In: Misra, K.B. (eds.) Handbook of Performability Engineering. Springer, London (2018). https://doi.org/10.1007/978-1-84800-131-2_23

Xing, L., Amari, S.V.: Binary Decision Diagrams and Extensions for System Reliability Analysis. Wiley-Scrivener, MA (2015)

Xing, L., Dugan, J.B.: Analysis of generalized phased-mission system reliability, performance, and sensitivity. IEEE Trans. Reliab. **51**(2), 199–211 (2002)

Xing, L., Amari, S.V., Wang, C.: Reliability of $k$-out-of-$n$ systems with phased-mission requirements and imperfect fault coverage. Reliab. Eng. Syst. Saf. **103**, 45–50 (2012)

Xing, L., Dai, Y.: A new decision-diagram-based method for efficient analysis on multistate systems. IEEE Trans. Dependable Secure Comput. **6**(3), 161–174 (2009)

Xing, L., Levitin, G.: BDD-based reliability evaluation of phased-mission systems with internal/external common-cause failures. Reliab. Eng. Syst. Saf. **112**, 145–153 (2013)

Xing, L., Levitin, G., Wang C.: Dynamic System Reliability: Modeling and Analysis of Dynamic and Dependent Behaviors, Wiley (2019)

Xing, L., Shrestha, A., Meshkat, L., Wang, W.: Incorporating common-cause failures into the modular hierarchical systems analysis. IEEE Trans. Reliab. **58**(1), 10–19 (2009)

Yang, X., Wu, X.: Mission reliability assessment of space TT&C system by discrete event system simulation. Qual. Reliab. Eng. Int. **30**(8), 1263–1273 (2014)

Yu, H., Yang, J., Lin, J., Zhao, Y.: Reliability evaluation of non-repairable phased-mission common bus systems with common cause failures. Comput. Ind. Eng. **111**, 445–457 (2017)

Yu, J., Hu, T., Yang, J.J., Zhao, L.P.: Redundancy optimization of standby phased-mission systems. In: International Conference on Intelligent Computing & Integrated Systems, pp. 395–398 (2010)

Zang, X., Sun, H., Trivedi, K.S.: A BDD-based algorithm for reliability analysis of phased mission systems. IEEE Trans. Reliab. **48**(1), 50–60 (1999)

Zhai, Q., Xing, L., Peng, R., Yang, J.: Aggregated Combinatorial Reliability Model for Non-Repairable Parallel Phased-Mission Systems. Reliab. Eng. Syst. Saf. **176**, 242–250 (2018)

# An Overview for Markov Decision Processes in Queues and Networks

Quan-Lin Li[1], Jing-Yu Ma[2(✉)], Rui-Na Fan[2], and Li Xia[3]

[1] School of Economics and Management, Beijing University of Technology,
Beijing 100124, China
`liquanlin@tsinghua.edu.cn`
[2] School of Economics and Management, Yanshan University,
Qinhuangdao 066004, China
`mjy0501@126.com, fanruina@stumail.ysu.edu.cn`
[3] Bussiness School, Sun Yat-sen University, Guangzhou 510275, China
`xiali5@sysu.edu.cn, xial@tsinghua.edu.cn`

**Abstract.** Markov decision processes (MDPs) in queues and networks
have been an interesting topic in many practical areas since the 1960s.
This paper Provides a detailed overview on this topic and tracks the evo-
lution of many basic results. Also, this paper summarizes several interesting directions in the future research. We hope that this overview can
shed light to MDPs in queues and networks, and also to their extensive
applications in various practical areas.

**Keywords:** Queueing systems · Queueing networks · Markov Decision
processes · Sensitivity-based optimization · Event-based optimization

## 1 Introduction

One main purpose of this paper is to provide an overview for research on MDPs in
queues and networks in the last six decades. Also, such a survey is first related to
several other basic studies, such as, Markov processes, queueing systems, queueing networks, Markov decision processes, sensitivity-based optimization, stochastic optimization, fluid and diffusion control. Therefore, our analysis begins from
three simple introductions: Markov processes and Markov decision processes,
queues and queueing networks, and queueing dynamic control.

Quan-Lin Li was supported by the National Natural Science Foundation of China
under grants No. 71671158 and 71932002, and by the Natural Science Foundation of
Hebei province under grant No. G2017203277. Li Xia was supported by the National
Natural Science Foundation of China under grant No. 61573206. The authors thank
X.R. Cao and E.A. Feinberg for their valuable comments and suggestions to improve
the presentation of this paper.

Q.-L. Li et al. (Eds.): Cao Festschrift 2019, CCIS 1102, pp. 44–71, 2019.
https://doi.org/10.1007/978-981-15-0864-6_3

## (a) Markov processes and Markov decision processes

The Markov processes, together with the Markov property, were first introduced by a Russian mathematician: Andrei Andreevich Markov (1856–1922) in 1906. See Markov [238] for more details. From then on, as a basically mathematical tool, the Markov processes have extensively been discussed by many authors, e.g., see some excellent books by Doob [99], Karlin [175], Karlin and Taylor [176], Chung [80], Anderson [21], Kemeny et al. [181], Meyn and Tweedie [241], Chen [77], Ethier and Kurtz [110] and so on.

In 1960, Howard [165] is the first to propose and discuss the MDP (or stochastic dynamic programming) in terms of his Ph.D thesis, which opened up a new and important field through an interesting intersection between Markov processes and dynamic programming (e.g., see Bellman and Kalaba [32]). From then on, not only are the MDPs an important branch in the area of Markov processes, but also it is a basic method in modern dynamic control theory. Crucially, the MDPs have been greatly motivated and widely applied in many practical areas in the past 60 years. Readers may refer to some excellent books, for example, the discrete-time MDPs by Puterman [261], Glasserman and Yao [143], Bertsekas [33], Bertsekas and Tsitsiklis [34], Hernádez-Lerma and Lasserre [155,156], Altman [10], Koole [193] and Hu and Yue [166]; the continuous-time MDPs by Guo and Hernández-Lerma [145]; the partially observable MDPs by Cassandra [67] and Krishnamurthy [196]; the competitive MDPs (i.e., stochastic game) by Filar and Vrieze [127]; the sensitivity-based optimization by Cao [58]; some applications of MDPs by Feinberg and Shwartz (Eds.) [122] and Boucherie and Van Dijk (eds.) [44]; and so on.

## (b) Queues and queueing networks

In the early 20th century, a Danmark mathematician: Agner Krarup Erlang, published a pioneering work [109] of queueing theory in 1909, which started the study of queueing theory and traffic engineering. Over the past 100 years, queueing theory has been regarded as a key mathematical tool not only for analyzing practical stochastic systems but also for promoting theory of stochastic processes (such as Markov processes, semi-Markov processes, Markov renew processes, random walks, martingale theory, fluid and diffusion approximation, and stochastic differential equations). On the other hand, the theory of stochastic processes can support and carry forward advances in queueing theory and applications (for example single-server queues, multi-server queues, tandem queues, parallel queues, fork-join queues, and queueing networks). It is worthwhile to note that so far queueing theory has been widely applied in many practical areas, such as manufacturing systems, computer and communication networks, transportation networks, service management, supply chain management, sharing economics, healthcare and so forth.

*The Single-Server Queues and the Multi-server Queues:* In the early development of queueing theory (1910s to 1970s), the single-server queues were a main topic with key results including Khintchine formula, Little's law, birth-death processes of Markovian queues, the embedded Markov chain, the supplementary variable method, the complex function method and so on. In 1969, Professor J.W. Cohen published a wonderful summative book [81] with respect to theoretical progress of single-server queues.

It is a key advance that Professor M.F. Neuts proposed and developed the phase-type (PH) distributions, Markovian arrival processes (MAPs), and the matrix-geometric solution, which were developed as the matrix-analytic method in the later study, e.g., see Neuts [245,246] and Latouche and Ramaswami [210] for more details. Further, Li [218] proposed and developed the RG-factorizations for any generally irreducible block-structured Markov processes. Crucially, the RG-factorizations promote the matrix-analytic method to a unified matrix framework both for the steady-state solution and for the transient solution (for instance the first passage time and the sojourn time). In addition, the matrix-analytic method and the RG-factorizations can effectively deal with small-scale stochastic models with several nodes.

In the study of queueing systems, some excellent books include Kleinrock [184,185], Tijms [304] and Asmussen [23]. Also, an excellent survey on key queueing advances was given in Syski [302]; and some overview papers on different research directions were reported by top queueing experts in two interesting books by Dshalalow [104,105].

*The Queueing Networks:*In 1957, J.R. Jackson published a seminal paper [168] which started research on queueing networks. Subsequent interesting results include Jackson [169], Baskett et al. [29], Kelly [178,180], Disney and König [97], Dobrushin *et al.* [98], Harrison [152], Dai [86] and so on. For the queueing networks, the well-known examples contain Jackson networks, BCMP networks, parallel networks, tandem networks, open networks, closed networks, polling queues, fork-join networks and distributed networks. Also, the product-form solution, the quasi-reversibility and some approximation algorithms are the basic results in the study of queueing networks.

For the queueing networks, we refer readers to some excellent books such as Kelly [179], Van Dijk [310], Gelenbe *et al.* [138], Chao *et al.* [72], Serfozo [284], Chen and Yao [76], Balsamo *et al.* [27], Daduna [85], Bolch *et al.* [41], Bramson [46] and Boucherie and Van Dijk (Eds.) [43].

For applications of queueing networks, readers may refer to some excellent books, for example, manufacturing systems by Buzacott and Shanthikumar [52], communication networks by Chang [71], traffic networks by Garavello and Piccoli [133], healthcare by Lakshmi and Iyer [206], service management by Demirkan et al. [92] and others.

### (c) Queueing dynamic control

In 1967, Miller [242] and Ryokov and Lembert [277] are the first to apply the MDPs to consider dynamic control of queues and networks. Those two works opened a novel interesting research direction: MDPs in queues and networks.

For MDPs of queues and networks, we refer readers to three excellent books by Kitaev and Rykov [182], Sennott [282] and Stidham [298].

In MDPs of queues and networks, so far there have been some best survey papers, for instance, Crabill *et al.* [83,84], Sobel [289], Stidham and Prabhu

[299], Rykov [272,274], Kumar [198], Stidham and Weber [300], Stidham [296] and Brouns [50].

For some Ph.D thesises by using MDPs of queues and networks, reader may refer to, such as, Farrell [114], Abdel-Gawad [1], Bartroli [28], Farrar [112], Veatch [314], Altman [8], Atan [25] and Efrosinin [107].

Now, MDPs of queues and networks play an important role in dynamic control of many practical stochastic networks, for example, inventory control [54,116,117], supply chain management [111], maintenance and quality [95,200], manufacturing systems [52,172], production lines [323], communication networks [6,12,251], wireless and mobile networks [4,96], cloud service [301], healthcare [254,308], airport management [211,271], energy-efficient management [250,262] and artificial intelligence [188,287]. With rapid development of Internet of Things (IoT), big data, cloud computing, blockchain and artificial intelligence, it is necessary to discuss MDPs of queues and networks under an intelligent environment.

From the detailed survey on MDPs of queues and networks, this paper suggests a future research under an intelligent environment from three different levels as follows:

1. *Networks with several nodes:* Analyzing MDPs of policy-based Markov processes with block structure, for example, QBD processes, Markov processes of GI/M/1 type, and Markov processes of M/G/1 type, and specifically, discussing their sensitivity-based optimization.
2. *Networks with a lot of nodes:* discussing MDPs of practical big networks, such as blockchain systems, sharing economics, intelligence healthcare and so forth.
3. *Networks with a lot of clusters:* studying MDPs of practical big networks by means of the mean-field theory, e.g., see Gast and Gaujal [134], Gast et al. [135] and Li [219].

The remainder of this paper is organized as follows. Sections 2 to 5 provide an overview for MDPs of single-server queues, multi-server queues, queueing networks, and queueing networks with special structures, respectively. Section 6 sets up specific objectives to provide an overview for key objectives in queueing dynamic control. Section 7 introduce the sensitivity-based optimization and the event-based optimization, both of which are applied to analyze MDPs of queues and networks. Finally, we give some concluding remarks in Sect. 8.

## 2   MDPs of Single-Server Queues

In this section, we provide an overview for MDPs of single-server queues, including the M/M/1 queues, the M/G/1 queues, the GI/M/1 queues and others. In the early research on MDPs of queues and networks, the single-server queues have been an active topic for many years.

**(1) MDPs of M/M/1 queues**
Kofman and Lippman [187], Rue and Rosenshine [268,269], Yeh and Thomas [338], Lu and Serfozo [231], Plum [258], Altman [18], Kitaev and Serfozo [183], Savaşaneril et al. [279] and Dimitrakopoulos and Burnetas [94].

**(2) MDPs of M/G/1 queues**
Mitchell [243], Doshi [100,101], Gallisch [130], Rue and Rosenshine [270], Jo and Stidham [173], Mandelbaum and Yechiali [236], Kella [177], Wakuta [317], Altman and Nain [7], Feinberg and Kim [120], Feinberg and Kella [119] and Sanajian et al. [278].

**(3) MDPs of GI/M/1 queues**
Stidham [293] and Mendelson and Yechiali [239].

**(4) MDPs of more genernal single-server queues**
Stidham [292], Crabill [82], Lippman [227], Schassberger [280], Stidham [294], Hordijk and Spieksma [164], Federgruen and So [115], Lamond [207], Towsley et al. [307], Koole [190], Haviv and Puterman [153], Lewis et al. [216], George and Harrison [139], Johansen and Larsen [174], Piunovskiy [256], Stidham [297], Adusumilli and Hasenbein [3], Kumar et al. [199] and Yan et al. [335].

**(5) MDPs of single-server batch queues**
Deb and Serfozo [90], Deb [89] and Powell and Humblet [259] with batch services; and Nobel and Tijms [248] with batch arrivals.

**(6) MDPs of single-server queues with either balking, reneging or abandonments**
Blackburn [39] with balking, Down et al. [102] with reneging, and Legros [215] with abandonments.

**(7) MDPs of single-server priority queues**
Robinson [264], Browne and Yechiali [47], Groenevelt et al. [144] and Brouns and Van Der Wal [51].

**(8) MDPs of single-server processor-sharing queues**
De Waal [93], Altman et al. [16], Van der Weij et al. [309] and Bhulai et al. [38].

**(9) MDPs of single-server retrial queues**
Liang and Kulkarni [225], Winkler [322] and Giovanidis et al. [141].

**(10) MDPs of single-server information-based queues**
Kuri and Kumar [201,202], Altman and Stidham [19] and Honhon and Seshadri [159].

**(11) MDPs of single-server queues with multiple classes of customers**
Harrison [151], Chen [73], Browne and Yechiali [49], De Serres [87,88], Ata [24], Feinberg and Yang [123] and Larrañaga et al. [209].

**(12) MDPs of single-server queues with optimal pricing**
Low [229], Chen [73], Yoon and Lewis [340], Çelik and Maglaras [69], Economou and Kanta [106] and Yildirim and Hasenbein [339].

**(13) MDPs of single-server manufacturing queues**

(a) *The make-to-stock queues:* Savaşaneril et al. [279], Sanajian et al. [278], Perez and Zipkin [255], Jain [170] and Cao and Xie [54].

(b) *The make-to-order queues:* Besbes and Maglaras [36] and Çelik and Maglaras [69].

(c) *The assemble-type queues:* Argon and Tsai [22] and Nadar et al. [244].

(d) *The inventory control queues:* Veatch [314], Savaşaneril et al. [279], Federgruen and Zipkin [117], Federgruen and Zheng [116], Feinberg [118], Feinberg and Liang [121].

**(14) MDPs of inventory rationing across multiple demand classes**
Ha [146–148], Gayon et al. [137] and Li et al. [222].

# 3   MDPs of Multi-server Queues

In this section, we provide an overview for MDPs of multi-server queues, which are another important research direction.

**(1) MDPs of M/M/c queues**
Low [230], Anderson [20], Printezis and Burnetas [260] and Feinberg and Yang [123,124].

**(2) MDPs of GI/M/c queues**
Yechiali [337], Van Nunen and Puterman [312] and Feinberg and Yang [125].

**(3) MDPs of two-server queues**
Larsen and Agrawala [208], Lin and Kumar [226], Hajek [149], Varma [313], Chen et al. [75] and Xu and Zhao [333].

**(4) MDPs of multi-server queues**
Emmons [108], Helm and Waldmann [154], Blanc et al. [40], Bradford [45], Koçağa and Ward [186] and Lee and Kulkarni [212].

**(5) MDPs of heterogeneous server queues**
Rosberg and Kermani [265], Nobel and Tijms [249], Rykov [273], Rykov and Efrosinin [275] and Tirdad et al. [306].

# 4   MDPs of Queueing Networks

In this section, we provide an overview for MDPs of queueing networks. Note that the MDPs of queueing networks have been an interesting research direction for many years, and they have also established key applications in many practical areas.

**(1) MDPs of more general queueing networks**
Ross [267], Weber and Stidham [320], Stidham [295], Shanthikumar and Yao [285], Veatch and Wein [315], Tassiulas and Ephremides [303], Papadimitriou and Tsitsiklis [252], Bäuerle [30], Bäuerle [31] and Solodyannikov [290].

**(2) MDPs of queueing networks with multiple classes of customers**
Shioyama [286], Bertsimas et al. [35], Maglaras [235], Chen and Meyn [78] and Cao and Xie [55].

**(3) Queueing applications of Markov decision processes**
Serfozo [283] studied the MDPs of birth-death processes and random walks, and then discussed dynamic control queueing networks. White [321] focused on the MDPs of QBD processes, which were used to deal with dynamic control of queueing networks. Robinson [263] and Hordijk et al. [163] studied the MDP which were applied to the study of queueing networks. Sennott [281] analyzed the semi-MDP and applied the obtained results to discuss the queueing networks.

Other key research includes Van Dijk and Puterman [311], Liu et al. [228], Altman et al. [13] and Adlakha et al. [2].

# 5   MDPs of Queueing Networks with Special Structure

In this section, we provide an overview for MDPs of queueing networks with special Structure, for example, multi-station tandem queues, multi-station parallel queues, polling queues, fork-join queues, distributed queueing networks, heavy traffic analysis for queueing control and so on.

**(1) MDPs of two-station tandem queues**
Ghoneim and Stidham [140], Nishimura [247], Farrar [113], Iravani et al. [167], Ahn et al. [5] and Zayas-Cabán et al. [341].

**(2) MDPs of multi-station tandem queues**
Rosberg et al. [266], Hordijk and Koole [160], Hariharan et al. [150], Gajrat et al. [129], Koole [192], Zhang and Ayhan [344] and Leeuwen and Núnez-Queija [213].

**(3) MDPs of parallel queues**
Weber [319], Bonomi [42], Menich and Serfozo [240], Xu et al. [332], Hordijk and Koole [161], Chen et al. [70], Sparaggis et al. [291], Koole [189], Ku and Jordan [197], Down and Lewis [103], Delasay et al. [91] and Feinberg and Zhang [126].

**(4) MDPs of polling queues**
Browne and Yechiali [48], Gandhi and Cassandras [131], Koole and Nain [195] and Gaujal et al. [136].

**(5) MDPs of fork-Join queueing networks**
Pascual et al. [253], Zeng et al. [342], Marin and Rossi [237] and Zeng et al. [343].

**(6) MDPs of Call Centers**
Koole [194], Bhulai [37], Legros et al. [214], Gans et al. [132] and Koole and Mandelbaum [191].

**(7) MDPs of distributed queueing networks**
Chou and Abraham [79], e Silva and Gerla [288], Franken and Haverkort [128], Li and Kameda [217], Nadar et al. [244] and Vercraene et al. [316].

**(8) Competitive MDPs of distributed queueing networks**
The competitive MDPs are called stochastic games. Altman and Hordijk [14] studied the zero-sum Markov game and applied the obtained results to the worst-case optimal control of queueing networks. Altman [9] studied non-zero stochastic games and applied their results to admission, service and routing control in queueing networks. Altman [11] proposed a Markov game approach for analyzing the optimal routing of a queueing network. Hordijk et al. [162] studied a multi-chain stochastic game which was applied to the worst case admission control in a queueing network. Xu and Hajek [334] studied the game problem of supermarket models. Xia [324] applied the stochastic games to analyzing the service rate control of a closed queueing network.

**(9) Heavy traffic analysis for controlled queues and networks**
Heavy traffic analysis can be used to deal with a class of important problems of controlled queues and networks by means of fluid and diffusion approximation. Readers may refer to, for example, Kushner [203], Kushner and Ramachandran [205], Kushner and Martins [204]; Harrison [152], Plambeck et al. [257]; Chen and Yao [76], Atar et al. [26].

# 6 Key Objectives in Queueing Dynamic Control

In this section, we introduce some key objectives to classify the literature of queueing dynamic control, for example, input control, service control, dynamic control under different service mechanisms, dynamic control with pricing, threshold control and so forth.

**Objective One: Input Control**
The input control is to apply the MDPs to dynamically control the input process of customers in the queues and networks, including the input rate control, the interval time control, and the admission access control (e.g., probability that an arriving customer chooses entering the system or some servers).

*(a) The input rate control:* Kitaev and Rykov [182], Sennott [282], Crabill et al. [84], Stidham and Weber [300], Crabill [82] and Lee and Kulkarni [212].

*(b) The input process control:* Kitaev and Rykov [182], Sennott [282], Crabill et al. [84], Stidham and Weber [300], Abdel-Gawad [1], Doshi [100], Stidham [293], Piunovskiy [256], Kuri and Kumar [201], Kuri and Kumar [202], Van Nunen and Puterman [312], Helm and Waldmann [154], Ghoneim and Stidham [140] and Nishimura [247].

*(c) The admission access control:* Crabill et al. [83,84], Stidham and Weber [300], Brouns [50], Rue and Rosenshine [268–270], Dimitrakopoulos and Burnetas [94], Mandelbaum and Yechiali [236], Mendelson and Yechiali [239], Stidham [294], Hordijk and Spieksma [164], Lamond [207], Lewis et al. [216], Adusumilli and Hasenbein [3], Altman et al. [16], Honhon and Seshadri [159], Yoon and Lewis [340], Yildirim and Hasenbein [339], Anderson [20], Emmons [108], Blanc et al. [40], Koçağa and Ward [186], Zhang and Ayhan [344], Altman [9], Hordijk et al. [162] and Xia [324].

**Objective Two: Service Control**

The service control is to use the MDPs to dynamically control the service process in queues and networks, including the service rate control, the service time control, and the service process control.

*(a) The service rate control:* Kitaev and Rykov [182], Sennott [282], Stidham [296,298], Crabill et al. [83,84], Stidham and Weber [300], Yao and Schechner [336], Dimitrakopoulos and Burnetas [94], Mitchell [243], Doshi [101], Jo and Stidham [173], Adusumilli and Hasenbein [3], Kumar et al. [199], Anderson [20], Lee and Kulkarni [212], Weber and Stidham [320], Ma and Cao [232], Xia [324], Xia and Shihada [331] and Xia and Jia [329].

*(b) The service time control:* Gallisch [130].

*(c) The service process control:* Kitaev and Rykov [182], Sennott [282], Stidham [298], Crabill et al. [83,84], Stidham and Weber [300], Schassberger [280], Johansen and Larsen [174], Stidham [297], Nishimura [247], Rosberg et al. [266], Altman [9] and Hordijk et al. [162].

**Objective Three: Dynamic Control Under Different Service Mechanisms**

Many practical and real problems lead to introduction of different service mechanisms which make some interesting queueing systems, for example, priority queues, processor-sharing queues, retrial queues, vacation queues, repairable queues, fluid queues and so on.

*(a) The priority queueing control:* The priority is an important service mechanism, and it is a precondition that sets up useful relations among key customers, segmenting market and adhering to long-term cooperation. Note that the priority makes dynamic control of queues with multi-class customers. Readers may refer to Rykov and Lembert [277], Crabill et al. [83,84], Stidham and Weber [300], Kofman and Lippman [187], Robinson [264], Browne and Yechiali [47], Groenevelt et al. [144], Brouns and Van Der Wal [51], Jain [170], Printezis and Burnetas [260] and Koole and Nain [195].

*(b) The processor-sharing queueing control:* Crabill et al. [83,84], Stidham and Weber [300], De Waal [93], Altman et al. [16], Van der Weij et al. [309], Bhulai et al. [38] and Bonomi [42].

*(c) The retrial queueing control:* Bhulai et al. [38], Liang and Kulkarni [225], Winkler [322] and Giovanidis et al. [141].

*(d) The vacation queueing control:* Li et al. [220], Altman and Nain [7,18], Kella [177] and Federgruen and So [115].

*(e) The repairable queueing control:* Dimitrakos and Kyriakidis [95], Rykov and Efrosinin [276], Tijms and van der Duyn Schouten [305].

*(f) The removable server control:* For dynamic control of working servers, it is necessary to real-time response to the peak period or an emergency phenomenon through increasing or decreasing the number of working servers according to either customer number or system workload. We refer the readers to Feinberg and Kim [120], Feinberg and Kella [119] and Iravani et al. [167].

*(g) The dynamic control of queueing behavior:* blocking by Blackburn [39] and Economou and Kanta [106]; reneging and impatience by Li et al. [220] and Anderson [20]; and abandonment by Down et al. [102], Legros et al. [215], Larrañaga et al. [209] and Zayas-Cabán et al. [341].

## Objective Four: Threshold Control

In dynamic control of queues and networks, the threshold-type policy is a simple and effective mode, including single-threshold and dual-threshold.

*(a) The single-threshold policy:* Brouns [50], Altman and Nain [18], Federgruen and So [115], Brouns and Van Der Wal [51];

*(b) The dual-threshold policy:* Lu and Serfozo [231], Plum [258] and Kitaev and Serfozo [183].

## Objective Five: Optimal Routing Control

*(a) The entering parallel-server policy:* Rosberg et al. [266], Weber [319], Bonomi [42], Menich and Serfozo [240], Xu et al. [332], Hordijk and Koole [161], Chang et al. [70], Sparaggis et al. [291], Koole [189], Ku and Jordan [197], Down and Lewis [103], Delasay et al. [91] and Li and Kameda [217].

*(b) The routing policy:* Abdel-Gawad [1], Altman [8,10], Towsley et al. [307], Liang and Kulkarni [225], Xu and Zhao [333], Bradford [45], Rosberg and Kermani [265], Ross [267], Stidham [295], Tassiulas and Ephremides [303], Menich and Serfozo [240], Koole [189], Browne and Yechiali [47], Altman and Nain [18] and Ho and Cao [157].

*(c) The assignment policy:* Weber [319], Bonomi [42] and Xu et al. [332].

## Objective Six: Controlled Queues and Networks with Useful Information

In the queueing networks, the useful information plays a key role in dynamic control of queueing networks. Readers may refer to Kuri and Kumar [201], Altman and Stidham [19], Honhon and Seshadri [159], Altman et al. [17], Altman and Jiménez [15] and Rosberg and Kermani [265].

Load balancing is an interesting research direction in queueing networks with simply observable information, e.g., see Down and Lewis [103], Chou and Abraham [79], e Silva and Gerla [288], Li and Kameda [217], Li et al. [220,221], Li [219] and Li and Lui [224].

## Objective Seven: Controlled Queues and Networks with Pricing

The optimal pricing policy is an important research direction in dynamic control of queues and networks, e.g., see Low [229], Chen and Frank [74], Yoon and Lewis [340], Çelik and Maglaras [69], Economou and Kanta [106], Yildirim and Hasenbein [339], Feinberg and Yang [125], Bradford [45], Xia and Chen [327] and Federgruen and Zheng [116].

# 7    Sensitivity-Based Optimization for MDPs of Queueing Networks

In this section, we simple introduce the sensitivity-based optimization in the MDPs, and then provide an overview on how to apply the sensitivity-based optimization in dynamic control of queues and networks.

In the late 1980s, to study dynamic control of queueing systems, Professors Yu-Chi Ho and Xi-Ren Cao proposed and developed the infinitesimal perturbation method for discrete event dynamic systems (DEDS), which is a new research direction for online simulation optimization of the DEDS. See Ho and Cao [158] for more interpretation. Further excellent books include Glasserman [142], Cao [56] and Cassandras and Lafortune [68].

**Sensitivity-Based Optimization:** Cao et al. [66] and Cao and Chen [65] published a pioneer work that transforms the infinitesimal perturbation of DEDS, together with the MDPs, into the so-called sensitivity-based optimization by means of the policy-based Markov processes and the policy-based Poisson equations, in which they also developed new concepts, for example, performance potential, and performance difference equation. On this research line, Cao [58] summarized many basic results of the sensitivity-based optimization. In addition, Li and Liu [223] and Chapter 11 in Li [218] extended and generalized the sensitivity-based optimization to a more general perturbed Markov process with infinite states by means of the RG-factorizations.

So far some work has applied the sensitivity-based optimization to deal with MDPs of queues and networks, e.g., see Xia and Cao [326], Xia and Shihada [331], Xia [324], Xia and Jia [329], Xia et al. [328] and Xia and Chen [327]; Ma et al. [233, 234] for data centers; and Li et al. [222] for inventory rationing control. It is worthwhile to note that the sensitivity-based optimization of queues and networks can be effectively supported and developed by means of the matrix-analytic method by Neuts [245, 246] and the RG-factorizations by Li [218]. Also see Ma et al. [233, 234] and Li et al. [222] for more details.

Recently, Xi-Ren Cao further extended and generalized the sensitivity-based optimization to the more general case of diffusion processes, called *relative optimization of continuous-time and continuous-state stochastic systems* (see Cao [64] with a complete draft). Important examples include Cao [59–63] and references therein.

**Event-Based Optimization Approach:** In many practical systems, an event usually has a specific physical meaning and can mathematically correspond to a set of state transitions with the same characteristics. In general, the number of events from change of system states is much smaller than the state number of the system. Therefore, such an event can be used to describe an approximate MDP, hence this sets up a new optimal framework, called event-based optimization. The event-based optimization can directly capture the future information and the structure nature of the system, which are reflected in the event to aggregate performance potential. Note only can the event-based optimization greatly save the calculation, but also it alleviates the dimensional disaster of a network decision process.

For the event-based optimization, readers may refer to, for example, dynamic control of queueing systems by Koole [190] and Koole [195]; dynamic control of Markov systems by Cao [57], Cao [53], Xia [330] and Jia [171]; partially observable Markov decision processes by Wang and Cao [318]; and admission control of open queueing networks by Xia [325].

# 8 Concluding Remarks

In this survey, we provide an overview for the MDPs of queues and networks, including single-server queues, multi-server queues and queueing networks. At the same time, the overview is also related to some specific objectives, for example, input control, service control, dynamic control based on different service mechanisms, dynamic control based on pricing, threshold control and so on.

Along such a line, there are still a number of interesting directions for potential future research, for example:

- Developing effective and efficient algorithms to find the optimal polices and to compute the optimal performance measures, and also probably linking AI and learning algorithms;
- discussing structure properties of the optimal policy in the MDPs of queueing networks under intelligent environment (for example, IoT, big data, cloud service, blockchain and AI), and specifically, dealing with multi-dimensional queueing dynamic control;
- analyzing structure properties of the optimal policy in the MDPs with either QBD processes, Markov processes of GI/M/1 type or Markov processes of M/G/1 type, which are well related to various practical stochastic models;
- applying the sensitivity-based optimization and the event-based optimization to deal with dynamic control of practical stochastic networks, for example, production and inventory control, manufacturing control, transportation networks, healthcare, sharing economics, cloud service, blockchain, service management, energy-efficient management and so forth.

# References

1. Abdel-Gawad, E.F.: Optimal control of arrivals and routing in a network of queues. Ph.D. dissertation, North Carolina State University (1984)
2. Adlakha, S., Lall, S., Goldsmith, A.: Networked Markov decision processes with delays. IEEE Trans. Autom. Control **57**(4), 1013–1018 (2012)
3. Adusumilli, K.M., Hasenbein, J.J.: Dynamic admission and service rate control of a queue. Queueing Syst. **66**(2), 131–154 (2010)
4. Ahmed, M.H.: Call admission control in wireless networks: a comprehensive survey. IEEE Commun. Surv. Tutorials **7**(1), 49–68 (2005)
5. Ahn, H.S., Duenyas, I., Lewis, M.E.: Optimal control of a two-stage tandem queuing system with flexible servers. Probab. Eng. Inf. Sci. **16**(4), 453–469 (2002)

6. Alsheikh, M.A., Hoang, D.T., Niyato, D., Tan, H.P., Lin, S.: Markov decision processes with applications in wireless sensor networks: a survey, pp. 1–29. arXiv preprint arXiv:1501.00644 (2015)
7. Altman, E., Nain, P.: Optimal control of the M/G/1 queue with repeated vacations of the server. IEEE Trans. Autom. Control 38(12), 1766–1775 (1993)
8. Altman, E.: A Markov game approach for optimal routing into a queuing network. Ph.D. dissertation, INRIA (Institut National de Recherche en Informatique et en Automatique) (1994)
9. Altman, E.: Non zero-sum stochastic games in admission, service and routing control in queueing systems. Queueing Syst. 23(1–4), 259–279 (1996)
10. Altman, E.: Constrained Markov Decision Processes. CRC Press (1999)
11. Altman, E.: A Markov game approach for optimal routing into a queuing network. In: Bardi, M., Raghavan, T.E.S., Parthasarathy, T. (eds.) Stochastic and Differential Games, vol. 4, pp. 359–375. Birkhäuser, Boston (1999). https://doi.org/10.1007/978-1-4612-1592-9_9
12. Altman, E.: Applications of Markov decision processes in communication networks. In: Feinberg, E.A., Shwartz, A. (eds.) Handbook of Markov Decision Processes, vol. 40, pp. 489–536. Springer, Boston (2002). https://doi.org/10.1007/978-1-4615-0805-2_16
13. Altman, E., Gaujal, B., Hordijk, A.: Discrete-Event Control of Stochastic Networks: Multimodularity and Regularity. Springer, Heidelberg (2003). https://doi.org/10.1007/b93837
14. Altman, E., Hordijk, A.: Zero-sum Markov games and worst-case optimal control of queueing systems. Queueing Syst. 21(3–4), 415–447 (1995)
15. Altman, E., Jiménez, T.: Admission control to an M/M/1 queue with partial information. In: Dudin, A., De Turck, K. (eds.) ASMTA 2013. LNCS, vol. 7984, pp. 12–21. Springer, Heidelberg (2013). https://doi.org/10.1007/978-3-642-39408-9_2
16. Altman, E., Jiménez, T., Koole, G.: On optimal call admission control in resource-sharing system. IEEE Trans. Commun. 49(9), 1659–1668 (2001)
17. Altman, E., Jiménez, T., Núñez Queija, R., Yechiali, U.: Optimal routing among ·/M/1 queues with partial information. Stoch. Models 20(2), 149–171 (2004)
18. Altman, E., Nain, P.: Optimality of a threshold policy in the M/M/1 queue with repeated vacations. Math. Methods Oper. Res. 44(1), 75–96 (1996)
19. Altman, E., Stidham, S.: Optimality of monotonic policies for two-action Markovian decision processes, with applications to control of queues with delayed information. Queueing Syst. 21(3–4), 267–291 (1995)
20. Anderson, M.Q.: Optimal admission pricing and service rate control of an $M^X/M/s$ queue with reneging. Naval Res. Logistics 30(2), 261–270 (1983)
21. Anderson, W.J.: Continuous-Time Markov Chains: An Applications-Oriented Approach. Springer, New York (1991). https://doi.org/10.1007/978-1-4612-3038-0
22. Argon, N.T., Tsai, Y.C.: Dynamic control of a flexible server in an assembly-type queue with setup costs. Queueing Syst. 70(3), 233–268 (2012)
23. Asmussen, S.: Applied Probability and Queues. Springer, New York (2003). https://doi.org/10.1007/b97236
24. Ata, B.: Dynamic control of a multiclass queue with thin arrival streams. Oper. Res. 54(5), 876–892 (2006)
25. Atan, S.T.: Solution methods for controlled queueing networks. Ph.D. dissertation, Iowa State University (1997)
26. Atar, R., Mandelbaum, A., Reiman, M.I.: Scheduling a multi class queue with many exponential servers: asymptotic optimality in heavy traffic. Ann. Appl. Probab. 14(3), 1084–1134 (2004)

27. Balsamo, S., de Nitto Personé, V., Onvural, R.: Analysis of Queueing Networks with Blocking. Springer, Dordrecht (2001). https://doi.org/10.1007/978-1-4757-3345-7

28. Bartroli, M.: On the structure of optimal control policies for networks of queues. Ph.D. dissertation, University of North Carolina at Chapel Hill (1989)

29. Baskett, F., Chandy, K.M., Muntz, R.R., Palacios, F.G.: Open, closed, and mixed networks of queues with different classes of customers. J. ACM **22**(2), 248–260 (1975)

30. Bäuerle, N.: Asymptotic optimality of tracking policies in stochastic networks. Ann. Appl. Probab. **10**(4), 1065–1083 (2000)

31. Bäuerle, N.: Optimal control of queueing networks: an approach via fluid models. Adv. Appl. Probab. **34**(2), 313–328 (2002)

32. Bellman, R., Kalaba, R.E.: Dynamic Programming and Modern Control Theory. Academic Press, New York (1965)

33. Bertsekas, D.P.: Dynamic Programming and Optimal Control. Athena Scientific, Belmont (1995)

34. Bertsekas, D.P., Tsitsiklis, J.N.: Neuro-Dynamic Programming. Athena Scientific, Belmont (1996)

35. Bertsimas, D., Paschalidis, I.C., Tsitsiklis, J.N.: Optimization of multiclass queueing networks: polyhedral and nonlinear characterizations of achievable performance. Ann. Appl. Probab. **4**(1), 43–75 (1994)

36. Besbes, O., Maglaras, C.: Revenue optimization for a make-to-order queue in an uncertain market environment. Oper. Res. **57**(6), 1438–1450 (2009)

37. Bhulai, S.: Dynamic routing policies for multiskill call centers. Probab. Eng. Inf. Sci. **23**(1), 101–119 (2009)

38. Bhulai, S., Brooms, A.C., Spieksma, F.M.: On structural properties of the value function for an unbounded jump Markov process with an application to a processor sharing retrial queue. Queueing Syst. **76**(4), 425–446 (2014)

39. Blackburn, J.D.: Optimal control of a single-server queue with balking and reneging. Manage. Sci. **19**(3), 297–313 (1972)

40. Blanc, J.P.C., de Waal, P.R., Nain, P., Towsley, D.: Optimal control of admission to a multiserver queue with two arrival streams. IEEE Trans. Autom. Control **37**(6), 785–797 (1992)

41. Bolch, G., Greiner, S., De Meer, H., Trivedi, K.S.: Queueing Networks and Markov Chains: Modeling and Performance Evaluation with Computer Science Applications. Wiley, New York (2006)

42. Bonomi, F.: On job assignment for a parallel system of processor sharing queues. IEEE Trans. Comput. **39**(7), 858–869 (1990)

43. Boucherie, R.J., Van Dijk, N.M. (eds.): Queueing Networks: A Fundamental Approach. Springer, Boston (2011). https://doi.org/10.1007/978-1-4419-6472-4

44. Boucherie, R.J., Van Dijk, N.M. (eds.): Markov Decision Processes in Practice. Springer, Cham (2017). https://doi.org/10.1007/978-3-319-47766-4

45. Bradford, R.M.: Pricing, routing, and incentive compatibility in multiserver queues. Eur. J. Oper. Res. **89**(2), 226–236 (1996)

46. Bramson, M.: Stability of Queueing Networks. Springer, Heidelberg (2008). https://doi.org/10.1007/978-3-540-68896-9

47. Browne, S., Yechiali, U.: Dynamic priority rules for cyclic-type queues. Adv. Appl. Probab. **21**(2), 432–450 (1989)

48. Browne, S., Yechiali, U.: Dynamic routing in polling systems. Teletraffic Sci. ITC-12, 1455–1466 (1989)

49. Browne, S., Yechiali, U.: Dynamic scheduling in single-server multiclass service systems with unit buffers. Naval Res. Logistics **38**(3), 383–396 (1991)

50. Brouns, G.A.: Queueing models with admission andtermination control: monotonicity and threshold results. Technische Universiteit Eindhoven, pp. 1–198 (2003)

51. Brouns, G.A., Van Der Wal, J.: Optimal threshold policies in a two-class preemptive priority queue with admission and termination control. Queueing Syst. **54**(1), 21–33 (2006)

52. Buzacott, J.A., Shanthikumar, J.G.: Stochastic Models of Manufacturing Systems. Prentice Hall (1993)

53. Cao, F.: Event-based optimization for the continuous-time Markov systems. Doctoral dissertation, Hong Kong University of Science and Technology, Hong Kong (2008)

54. Cao, P., Xie, J.: Optimal control of an inventory system with joint production and pricing decisions. IEEE Trans. Autom. Control **61**(12), 4235–4240 (2016)

55. Cao, P., Xie, J.: Optimal control of a multiclass queueing system when customers can change types. Queueing Syst. **82**(3–4), 285–313 (2016)

56. Cao, X.R.: Realization Probabilities: The Dynamics of Queuing Systems. Springer, Heidelberg (1994). https://doi.org/10.1007/BFb0035250

57. Cao, X.R.: Basic ideas for event-based optimization of Markov systems. Discrete Event Dyn. Syst. **15**(2), 169–197 (2005)

58. Cao, X.R.: Stochastic Learning and Optimization: A Sensitivity-Based Approach. Springer, Boston (2007). https://doi.org/10.1007/978-0-387-69082-7

59. Cao, X.R.: Optimization of average rewards of time nonhomogeneous Markov chains. IEEE Trans. Autom. Control **60**(7), 1841–1856 (2015)

60. Cao, X.R.: Optimality conditions for long-run average rewards with underselectivity and nonsmooth features. IEEE Trans. Autom. Control **62**(9), 4318–4332 (2017)

61. Cao, X.R.: Semismooth potentials of stochastic systems with degenerate diffusions. IEEE Trans. Autom. Control **63**(10), 3566–3572 (2018)

62. Cao, X.R.: State classification and multi-class optimization of continuous-time and continuous-state Markov processes. IEEE Trans. Autom. Control, 1–14 (2019)

63. Cao, X.R.: Stochastic control of multi-dimensional systems with relative optimization. IEEE Tran Autom. Control, 1–15 (2019)

64. Cao, X.R.: Relative Optimization of Continuous-time and Continuous-state Stochastic Systems. A complete draft of Cao's new book by Springer (2019)

65. Cao, X.R., Chen, H.F.: Perturbation realization, potentials, and sensitivity analysis of Markov processes. IEEE Trans. Autom. Control **42**(10), 1382–1393 (1997)

66. Cao, X.R., Yuan, X.M., Qiu, L.: A single sample path-based performance sensitivity formula for Markov chains. IEEE Trans. Autom. Control **41**, 1814–1817 (1996)

67. Cassandra, A.R.: Exact and Approximate Algorithms for Partially Observable Markov Decision Processes. Doctoral Dissertation, Brown University Providence (1998)

68. Cassandras, C.G., Lafortune, S.: Introduction to Discrete Event Systems. Springer, Boston (2008). https://doi.org/10.1007/978-0-387-68612-7

69. Çelik, S., Maglaras, C.: Dynamic pricing and lead-time quotation for a multiclass make-to-order queue. Manag. Sci. **54**(6), 1132–1146 (2008)

70. Chang, C.S., Nelson, R., Yao, D.D.: Optimal task scheduling on distributed parallel processors. Perform. Eval. **20**(1–3), 207–221 (1994)

71. Chang, C.S.: Performance Guarantees in Communication Networks. Springer, London (2000). https://doi.org/10.1007/978-1-4471-0459-9
72. Chao, X., Miyazawa, M., Pinedo, M.: Queueing Networks: Customers. Signals and Product Form Solutions. Wiley, New York (1999)
73. Chen, H.: Optimal intensity control of a multi-class queue. Queueing Syst. **5**(4), 281–293 (1989)
74. Chen, H., Frank, M.Z.: State dependent pricing with a queue. IIE Trans. **33**(10), 847–860 (2001)
75. Chen, H., Yang, P., Yao, D.D.: Control and scheduling in a two-station queueing network: optimal policies and heuristics. Queueing Syst. **18**(3–4), 301–332 (1994)
76. Chen, H., Yao, D.D.: Fundamentals of Queueing Networks: Performance, Asymptotics, and Optimization. Springer, New York (2001). https://doi.org/10.1007/978-1-4757-5301-1
77. Chen, M.: From Markov Chains to Non-equilibrium Particle Systems. World Scientific, Singapore (2004)
78. Chen, R.R., Meyn, S.: Value iteration and optimization of multiclass queueing networks. Queueing Syst. **32**(1–3), 65–97 (1999)
79. Chou, T.C.K., Abraham, J.A.: Load balancing in distributed systems. IEEE Trans. Software Eng. **4**, 401–412 (1982)
80. Chung, K.L.: Markov Chains. Springer, Boston (1967). https://doi.org/10.1007/1-4020-0611-X
81. Cohen, J.W.: The Single Server Queue. North-Holland Publishing Company, Amsterdan (1969)
82. Crabill, T.B.: Optimal control of a service facility with variable exponential service times and constant arrival rate. Manage. Sci. **18**(9), 560–566 (1972)
83. Crabill, T.B., Gross, D., Magazine, M.J.: A survey of research on optimal design and control of queues. No. Serial T-280, Washington DC program in Logistics, George Washington University (1973)
84. Crabill, T.B., Gross, D., Magazine, M.J.: A classified bibliography of research on optimal design and control of queues. Oper. Res. **25**(2), 219–232 (1977)
85. Daduna, H.: Queueing Networks with Discrete Time Scale: Explicit Expressions for the Steady State Behavior of Discrete Time Stochastic Networks. Springer, Heidelberg (2001). https://doi.org/10.1007/3-540-44592-7
86. Dai, J.G.: On positive Harris recurrence of multiclass queueing networks: a unified approach via fluid limit models. Ann. Appl. Probab. **5**(1), 49–77 (1995)
87. De Serres, Y.: Simultaneous optimization of flow control and scheduling in a single server queue with two job classes. Oper. Res. Lett. **10**(2), 103–112 (1991)
88. De Serres, Y.: Simultaneous optimization of flow-control and scheduling in a single server queue with two job classes: numerical results and approximation. Comput. Oper. Res. **18**(4), 361–378 (1991)
89. Deb, R.K.: Optimal control of batch service queues with switching costs. Adv. Appl. Probab. **8**(1), 177–194 (1976)
90. Deb, R.K., Serfozo, R.F.: Optimal control of batch service queues. Adv. Appl. Probab. **5**(2), 340–361 (1973)
91. Delasay, M., Kolfal, B., Ingolfsson, A.: Maximizing throughput in finite-source parallel queue systems. Eur. J. Oper. Res. **217**(3), 554–559 (2012)
92. Demirkan, H., Spohrer, J.C., Krishna, V. (eds.): Service Systems Implementation. Springer, Boston (2011). https://doi.org/10.1007/978-1-4419-7904-9
93. De Waal, P.: A constrained optimization problem for a processor sharing queue. Naval Res. Logistics **40**(5), 719–731 (1993)

94. Dimitrakopoulos, Y., Burnetas, A.: The value of service rate flexibility in an M/M/1 queue with admission control. IISE Trans. **49**(6), 603–621 (2017)
95. Dimitrakos, T.D., Kyriakidis, E.G.: A semi-Markov decision algorithm for the maintenance of a production system with buffer capacity and continuous repair times. Int. J. Prod. Econ. **111**(2), 752–762 (2008)
96. Dinh, H.T., Lee, C., Niyato, D., Wang, P.: A survey of mobile cloud computing: architecture, applications, and approaches. Wireless Commun. Mob. Comput. **13**(18), 1587–1611 (2013)
97. Disney, R.L., König, D.: Queueing networks: a survey of their random processes. SIAM Rev. **27**(3), 335–403 (1985)
98. Dobrushin, R.L., Kelbert, M.Y., Rybko, A.N., Suhov, Y.M.: Qualitative methods of queueing network theory. In: Dobrushin, R.L., Kryukov, V.M., Toom, A.L. (eds.) Stochastic Cellular Systems: Ergodicity, Memory, Morphogenesis, pp. 183–224. University Press, Manchester (1990)
99. Doob, J.L.: Stochastic Processes. Wiley, New York (1953)
100. Doshi, B.T.: Continuous time control of the arrival process in an M/G/1 queue. Stoch. Process. Their Appl. **5**(3), 265–284 (1977)
101. Doshi, B.T.: Optimal control of the service rate in an M/G/1 queueing system. Adv. Appl. Probab. **10**(3), 682–701 (1978)
102. Down, D.G., Koole, G., Lewis, M.E.: Dynamic control of a single-server system with abandonments. Queueing Syst. **67**(1), 63–90 (2011)
103. Down, D.G., Lewis, M.E.: Dynamic load balancing in parallel queueing systems: stability and optimal control. Eur. J. Oper. Res. **168**(2), 509–519 (2006)
104. Dshalalow, J.H.: Advances in Queueing Theory, Methods, and Open Problems. CRC Press, Boca Raton (1995)
105. Dshalalow, J.H.: Frontiers in Queueing: Models and Applications in Science and Engineering. CRC Press, Boca Raton (1997)
106. Economou, A., Kanta, S.: Optimal balking strategies and pricing for the single server Markovian queue with compartmented waiting space. Queueing Syst. **59**(3–4), 237 (2008)
107. Efrosinin, D.: Controlled queueing systems with heterogeneous servers. Ph.D. dissertation, Universitätsbibliothek (University of Trier) (2004)
108. Emmons, H.: The optimal admission policy to a multiserver queue with finite horizon. J. Appl. Probab. **9**(1), 103–116 (1972)
109. Erlang, A.K.: The theory of probabilities and telephone conversations. Nyt Tidsskrift for Matematik **20**(B), 33–39 (1909)
110. Ethier, S.N., Kurtz, T.G.: Markov Processes: Characterization and Convergence. Wiley, New York (2005)
111. Ettl, M., Feigin, G.E., Lin, G.Y., Yao, D.D.: A supply network model with base-stock control and service requirements. Oper. Res. **48**(2), 216–232 (2000)
112. Farrar, T.M.: Resource allocation in systems of queues. Ph.D. dissertation, University of Cambridge (1992)
113. Farrar, T.M.: Optimal use of an extra server in a two station tandem queueing network. IEEE Trans. Autom. Control **38**(8), 1296–1299 (1993)
114. Farrell, W.: Optimal switching policies in a non-homogeneous exponential queueing system. Ph.D. dissertation, University of California at Los Angeles (1976)
115. Federgruen, A., So, K.C.: Optimality of threshold policies in single-server queueing systems with server vacations. Adv. Appl. Probab. **23**(2), 388–405 (1991)
116. Federgruen, A., Zheng, Y.S.: An efficient algorithm for computing an optimal $(r, Q)$ policy in continuous review stochastic inventory systems. Oper. Res. **40**(4), 808–813 (1992)

117. Federgruen, A., Zipkin, P.: An efficient algorithm for computing optimal $(s, S)$ policies. Oper. Res. **32**(6), 1268–1285 (1984)

118. Feinberg, E.A.: Optimality conditions for inventory control. In: Optimization Challenges in Complex, Networked and Risky Systems, pp. 14–45. INFORMS TutORials in Operations Research (2016)

119. Feinberg, E.A., Kella, O.: Optimality of D-policies for an M/G/1 queue with a removable server. Queueing Syst. **42**(4), 355–376 (2002)

120. Feinberg, E.A., Kim, D.J.: Bicriterion optimization of an M/G/1 queue with a removable server. Probab. Eng. Inf. Sci. **10**(1), 57–73 (1996)

121. Feinberg, E.A., Liang, Y.: Structure of optimal policies to periodic-review inventory models with convex costs and backorders for all values of discount factors. Ann. Oper. Res., 1–17 (2017)

122. Feinberg, E.A., Shwartz, A. (eds.): Handbook of Markov Decision Processes: Methods and Applications. Springer, New York (2002). https://doi.org/10.1007/978-1-4615-0805-2

123. Feinberg, E.A., Yang, F.: Optimality of trunk reservation for an M/M/k/N queue with several customer types and holding costs. Probab. Eng. Inf. Sci. **25**(4), 537–560 (2011)

124. Feinberg, E.A., Yang, F.: Dynamic price optimization for an M/M/k/N queue with several customer types. ACM SIGMETRICS Perform. Eval. Rev. **41**(3), 25–27 (2014)

125. Feinberg, E.A., Yang, F.: Optimal pricing for a GI/M/k/N queue with several customer types and holding costs. Queueing Syst. **82**(1–2), 103–120 (2016)

126. Feinberg, E.A., Zhang, X.: Optimal switching on and off the entire service capacity of a parallel queue. Probab. Eng. Inf. Sci. **29**(4), 483–506 (2015)

127. Filar, J., Vrieze, K.: Competitive Markov Decision Processes. Springer, New York (2012). https://doi.org/10.1007/978-1-4612-4054-9

128. Franken, L.J., Haverkort, B.R.: Reconfiguring distributed systems using Markov-decision models. In: Proceedings of the Workshop on Trends in Distributed Systems, pp. 219–228 (1996)

129. Gajrat, A., Hordijk, A., Ridder, A.: Large-deviations analysis of the fluid approximation for a controllable tandem queue. Ann. Appl. Probab. **13**(4), 1423–1448 (2003)

130. Gallisch, E.: On monotone optimal policies in a queueing model of M/G/1 type with controllable service time distribution. Ann. Appl. Probab. **11**(4), 870–887 (1979)

131. Gandhi, A.D., Cassandras, C.G.: Optimal control of polling models for transportation applications. Math. Comput. Modell. **23**(11–12), 1–23 (1996)

132. Gans, N., Koole, G., Mandelbaum, A.: Telephone call centers: tutorial, review, and research prospects. Manuf. Serv. Oper. Management. **5**(2), 79–141 (2003)

133. Garavello, M., Piccoli, B.: Traffic Flow on Networks. Springfield: American Institute of Mathematical Sciences (2006)

134. Gast, N., Gaujal, B.: A mean field approach for optimization in discrete time. Discrete Event Dyn. Syst. **21**(1), 63–101 (2011)

135. Gast, N., Gaujal, B., Le Boudec, J.Y.: Mean field for Markov decision processes: from discrete to continuous optimization. IEEE Trans. Autom. Control **57**(9), 2266–2280 (2012)

136. Gaujal, B., Hordijk, A., Van Der Laan, D.: On the optimal open-loop control policy for deterministic and exponential polling systems. Probab. Eng. Inf. Sci. **21**(2), 157–187 (2007)

137. Gayon, J.P., De Vericourt, F., Karaesmen, F.: Stock rationing in an $M/E_r/1$ multi-class make-to-stock queue with backorders. IIE Trans. **41**(12), 1096–1109 (2009)

138. Gelenbe, E., Pujolle, G., Gelenbe, E., Pujolle, G.: Introduction to Queueing Networks. Wiley, New York (1998)

139. George, J.M., Harrison, J.M.: Dynamic control of a queue with adjustable service rate. Oper. Res. **49**(5), 720–731 (2001)

140. Ghoneim, H.A., Stidham, S.: Control of arrivals to two queues in series. Eur. J. Oper. Res. **21**(3), 399–409 (1985)

141. Giovanidis, A., Wunder, G., Bühler, J.: Optimal control of a single queue with retransmissions: delay-dropping tradeoffs. IEEE Trans. Wireless Commun. **8**(7), 3736–3746 (2009)

142. Glasserman, P., Ho, Y.C.: Gradient Estimation via Perturbation Analysis. Springer, Boston (1991)

143. Glasserman, P., Yao, D.D.: Monotone Structure in Discrete-Event Systems. Wiley, New York (1994)

144. Groenevelt, R., Koole, G., Nain, P.: On the bias vector of a two-class preemptive priority queue. Math. Methods Oper. Res. **55**(1), 107–120 (2002)

145. Guo, X., Hernández-Lerma, O.: Continuous-Time Markov Decision Processes. Springer, Heidelberg (2009)

146. Ha, A.Y.: Inventory rationing in a make-to-stock production system with several demand classes and lost sales. Manage. Sci. **43**(8), 1093–1103 (1997)

147. Ha, A.Y.: Stock-rationing policy for a make-to-stock production system with two priority classes and backordering. Naval Res. Logistics **44**(5), 457–472 (1997)

148. Ha, A.Y.: Stock rationing in an $M/E_k/1$ make-to-stock queue. Manage. Sci. **46**(1), 77–87 (2000)

149. Hajek, B.: Optimal control of two interacting service stations. IEEE Trans. Autom. Control **29**(6), 491–499 (1984)

150. Hariharan, R., Moustafa, M.S., Stidham, S.: Scheduling in a multi-class series of queues with deterministic service times. Queueing Syst. **24**(1–4), 83–99 (1996)

151. Harrison, J.M.: Dynamic scheduling of a multiclass queue: discount optimality. Oper. Res. **23**(2), 270–282 (1975)

152. Harrison, J.M.: Brownian Motion and Stochastic Flow Systems. Wiley, New York (1985)

153. Haviv, M., Puterman, M.L.: Bias optimality in controlled queueing systems. J. Appl. Probab. **35**(1), 136–150 (1998)

154. Helm, W.E., Waldmann, K.H.: Optimal control of arrivals to multiserver queues in a random environment. J. Appl. Probab. **21**(3), 602–615 (1984)

155. Hernádez-Lerma, O., Lasserre, J.B.: Discrete-Time Markov Control Processes: Basic Optimality Criteria. Springer, New York (1996). https://doi.org/10.1007/978-1-4612-0729-0

156. Hernádez-Lerma, O., Lasserre, J.B.: Further Topics on Discrete-Time Markov Control Processes. Springer, New York (1999). https://doi.org/10.1007/978-1-4612-0561-6

157. Ho, Y.C., Cao, X.R.: Performance sensitivity to routing changes in queuing networks and flexible manufacturing systems using perturbation analysis. IEEE J. Rob. Autom. **1**(4), 165–172 (1985)

158. Ho, Y.C., Cao, X.R.: Perturbation Analysis of Discrete-Event Dynamic Systems. Kluwer Academic Publisher, Boston (1991)

159. Honhon, D., Seshadri, S.: Admission control with incomplete information to a finite buffer queue. Probab. Eng. Inf. Sci. **21**(1), 19–46 (2007)

160. Hordijk, A., Koole, G.: On the shortest queue policy for the tandem parallel queue. Probab. Eng. Inf. Sci. **6**(1), 63–79 (1992)
161. Hordijk, A., Koole, G.: On the optimality of LEPT and $\mu c$ rules for parallel processors and dependent arrival processes. Adv. Appl. Probab. **25**(4), 979–996 (1993)
162. Hordijk, A., Passchier, O., Spieksma, F.: Optimal service control against worst case admission policies: a multichained stochastic game. Math. Methods Oper. Res. **45**(2), 281–301 (1997)
163. Hordijk, A., Van Der Duyn Schouten, F.A.: Average optimal policies in Markov decision drift processes with applications to a queueing and a replacement model. Adv. Appl. Probab. **15**(2), 274–303 (1983)
164. Hordijk, A., Spieksma, F.: Constrained admission control to a queueing system. Ann. Appl. Probab. **21**(2), 409–431 (1989)
165. Howard, R.A.: Dynamic Programming and Markov Processes. MIT Press, Cambridge (1960)
166. Hu, Q.Y., Yue, W.Y.: Markov Decision Processes with Their Applications. Springer, Boston (2007). https://doi.org/10.1007/978-0-387-36951-8
167. Iravani, S.M., Posner, M.J.M., Buzacott, J.A.: A two-stage tandem queue attended by a moving server with holding and switching costs. Queueing Syst. **26**(3–4), 203–228 (1997)
168. Jackson, J.R.: Networks of waiting lines. Oper. Res. **5**(4), 518–521 (1957)
169. Jackson, J.R.: Jobshop-like queueing systems. Manage. Sci. **10**(1), 131–142 (1963)
170. Jain, A.: Priority and dynamic scheduling in a make-to-stock queue with hyperexponential demand. Naval Res. Logistics **53**(5), 363–382 (2006)
171. Jia, Q.S.: On solving event-based optimization with average reward over infinite stages. IEEE Trans. Autom. Control **56**(12), 2912–2917 (2011)
172. Jo, K.Y., Maimon, O.Z.: Optimal dynamic load distribution in a class of flow-type flexible manufacturing systems. Eur. J. Oper. Res. **55**(1), 71–81 (1991)
173. Jo, K.Y., Stidham, S.: Optimal service-rate control of M/G/1 queueing systems using phase methods. Ann. Appl. Probab. **15**(3), 616–637 (1983)
174. Johansen, S.G., Larsen, C.: Computation of a near-optimal service policy for a single-server queue with homogeneous jobs. Eur. J. Oper. Res. **134**(3), 648–663 (2001)
175. Karlin, S.: A First Course in Stochastic Processes. Academic Press, New York (1968)
176. Karlin, S., Taylor, H.E.: A Second Course in Stochastic Processes. Elsevier, New York (1981)
177. Kella, O.: Optimal control of the vacation scheme in an M/G/1 queue. Oper. Res. **38**(4), 724–728 (1990)
178. Kelly, F.P.: Networks of queues. Ann. Appl. Probab. **8**(2), 416–432 (1976)
179. Kelly, F.P.: Reversibility and Stochastic Networks. Cambridge University Press, Cambridge (1979)
180. Kelly, F.P.: Loss networks. Ann. Appl. Probab. **1**(3), 319–378 (1991)
181. Kemeny, J.G., Snell, J.L., Knapp, A.W.: Denumerable Markov Chains: with a Chapter of Markov Random Fields by David Griffeath. Springer, New York (1976). https://doi.org/10.1007/978-1-4684-9455-6
182. Kitaev, M.Y., Rykov, V.V.: Controlled Queueing Systems. CRC Press, Boca Raton (1995)
183. Kitaev, M.Y., Serfozo, R.F.: M/M/1 queues with switching costs and hysteretic optimal control. Oper. Res. **47**(2), 310–312 (1999)

184. Kleinrock, L.: Queueing Systems, Volume 1: Theory. Wiley Interscience, New York (1975)
185. Kleinrock, L.: Queueing Systems, Volume II: Computer Applications. Wiley Interscience, New York (1976)
186. Koçağa, Y.L., Ward, A.R.: Admission control for a multi-server queue with abandonment. Queueing Syst. **65**(3), 275–323 (2010)
187. Kofman, E., Lippman, S.A.: An M/M/1 dynamic priority queue with optional promotion. Oper. Res. **29**(1), 174–188 (1981)
188. Kolobov, A.: Planning with Markov decision processes: an AI perspective. Synthesis Lectures on Artificial Intelligence and Machine Learning **6**(1), 1–210 (2012)
189. Koole, G.: On the pathwise optimal Bernoulli routing policy for homogeneous parallel servers. Math. Oper. Res. **21**(2), 469–476 (1996)
190. Koole, G.: The deviation matrix of the M/M/1/∞ and M/M/1/N queue, with applications to controlled queueing models. In: Proceedings of the 37th IEEE Conference on Decision and Control, vol. 1, pp. 56–59 (1998)
191. Koole, G., Mandelbaum, A.: Queueing models of call centers: an introduction. Ann. Oper. Res. **113**(1–4), 41–59 (2002)
192. Koole, G.: Convexity in tandem queues. Probab. Eng. Inf. Sci. **18**(1), 13–31 (2004)
193. Koole, G.: Monotonicity in Markov reward and decision chains: theory and applications. Found. Trends Stoch. Syst. **1**(1), 1–76 (2007)
194. Koole, G.: Call Center Optimization (2013). Lulu.com
195. Koole, G., Nain, P.: On the value function of a priority queue with an application to a controlled polling model. Queueing Syst. **34**(1–4), 199–214 (2000)
196. Krishnamurthy, V.: Partially Observed Markov Decision Processes. Cambridge University Press, Cambridge (2016)
197. Ku, C.Y., Jordan, S.: Access control of parallel multiserver loss queues. Perform. Eval. **50**(4), 219–231 (2002)
198. Kumar, A.: Task allocation in multiserver systems–a survey of results. Sadhana **15**(4–5), 381–395 (1990)
199. Kumar, R., Lewis, M.E., Topaloglu, H.: Dynamic service rate control for a single-server queue with Markov-modulated arrivals. Naval Res. Logistics **60**(8), 661–677 (2013)
200. Kuo, Y.: Optimal adaptive control policy for joint machine maintenance and product quality control. Eur. J. Oper. Res. **171**(2), 586–597 (2006)
201. Kuri, J., Kumar, A.: Optimal control of arrivals to queues with delayed queue length information. IEEE Trans. Autom. Control **40**(8), 1444–1450 (1995)
202. Kuri, J., Kumar, A.: On the optimal control of arrivals to a single queue with arbitrary feedback delay. Queueing Syst. **27**(1–2), 1–16 (1997)
203. Kushner, H.J.: Heavy Traffic Analysis of Controlled Queueing and Communication Networks. Springer, New York (2001). https://doi.org/10.1007/978-1-4613-0005-2
204. Kushner, H.J., Martins, L.F.: Heavy traffic analysis of a controlled multiclass queueing network via weak convergence methods. SIAM J. Control Optim. **34**(5), 1781–1797 (1996)
205. Kushner, H.J., Ramachandran, K.M.: Optimal and approximately optimal control policies for queues in heavy traffic. SIAM J. Control Optim. **27**(6), 1293–1318 (1989)
206. Lakshmi, C., Iyer, S.A.: Application of queueing theory in health care: a literature review. Oper. Res. Health Care **2**(1–2), 25–39 (2013)
207. Lamond, B.F.: Optimal admission policies for a finite queue with bursty arrivals. Ann. Oper. Res. **28**(1), 243–260 (1991)

208. Larsen, R.L., Agrawala, A.K.: Control of a heterogeneous two-server exponential queueing system. IEEE Trans. Software Eng. **4**, 522–526 (1983)

209. Larrañaga, M., Ayesta, U., Verloop, I.M.: Index policies for a multi-class queue with convex holding cost and abandonments. ACM SIGMETRICS Perform. Eval. Rev. **42**(1), 125–137 (2014)

210. Latouche, G., Ramaswami, V.: Introduction to Matrix Analytic Methods in Stochastic Modeling. SIAM (1999)

211. Lautenbacher, C.J., Stidham, S.: The underlying Markov decision process in the single-leg airline yield-management problem. Transp. Sci. **33**(2), 136–146 (1999)

212. Lee, N., Kulkarni, V.G.: Optimal arrival rate and service rate control of multi-server queues. Queueing Syst. **76**(1), 37–50 (2014)

213. Leeuwen, D.V., Núñez-Queija, R.: Near-optimal switching strategies for a tandem queue. In: Boucherie, R., van Dijk, N. (eds.) Markov Decision Processes in Practice, vol. 248, pp. 439–459. Springer, Cham (2017). https://doi.org/10.1007/978-3-319-47766-4_17

214. Legros, B., Jouini, O., Koole, G.: Optimal scheduling in call centers with a callback option. Perform. Eval. **95**, 1–40 (2016)

215. Legros, B., Jouini, O., Koole, G.: A uniformization approach for the dynamic control of queueing systems with abandonments. Oper. Res. **66**(1), 200–209 (2018)

216. Lewis, M.E., Ayhan, H., Foley, R.D.: Bias optimality in a queue with admission control. Probab. Eng. Inf. Sci. **13**(3), 309–327 (1999)

217. Li, J., Kameda, H.: Load balancing problems for multiclass jobs in distributed/parallel computer systems. IEEE Trans. Comput. **47**(3), 322–332 (1998)

218. Li, Q.L.: Constructive Computation in Stochastic Models with Applications: The RG-Factorizations. Springer, Heidelberg and Tsinghua Press (2010). https://doi.org/10.1007/978-3-642-11492-2

219. Li, Q.L.: Nonlinear Markov processes in big networks. Spec. Matrices **4**(1), 202–217 (2016)

220. Li, Q.L., Dai, G., Lui, J.C.S., Wang, Y.: The mean-field computation in a supermarket model with server multiple vacations. Discrete Event Dyn. Syst. **24**(4), 473–522 (2014)

221. Li, Q.L., Du, Y., Dai, G., Wang, M.: On a doubly dynamically controlled supermarket model with impatient customers. Comput. Oper. Res. **55**, 76–87 (2015)

222. Li, Q.L., Li, Y.M., Ma, J.Y., Liu, H.L.: A complete algebraic transformational solution for the optimal dynamic policy in inventory rationing across two demand classes. Online Publication, pp. 1–62, arXiv:1908.09295 (2019)

223. Li, Q.L., Liu, L.M.: An algorithmic approach on sensitivity analysis of perturbed QBD processes. Queueing Syst. **48**(3–4), 365–397 (2004)

224. Li, Q.L., Lui, J.C.S.: Block-structured supermarket models. Discrete Event Dyn. Syst. **26**(2), 147C–182 (2016)

225. Liang, H.M., Kulkarni, V.G.: Optimal routing control in retrial queues. In: Shanthikumar, J.G., Sumita, U. (eds.) Applied Probability and Stochastic Processes, vol. 19, pp. 203–218. Springer, Boston (1999). https://doi.org/10.1007/978-1-4615-5191-1_14

226. Lin, W., Kumar, P.: Optimal control of a queueing system with two heterogeneous servers. IEEE Trans. Autom. Control **29**(8), 696–703 (1984)

227. Lippman, S.A.: Applying a new device in the optimization of exponential queuing systems. Oper. Res. **23**(4), 687–710 (1975)

228. Liu, Z., Nain, P., Towsley, D.: Sample path methods in the control of queues. Queueing Syst. **21**(3–4), 293–335 (1995)

229. Low, D.W.: Optimal pricing for an unbounded queue. IBM J. Res. Dev. **18**(4), 290–302 (1974)
230. Low, D.W.: Optimal dynamic pricing policies for an M/M/s queue. Oper. Res. **22**(3), 545–561 (1974)
231. Lu, F.V., Serfozo, R.F.: M/M/1 queueing decision processes with monotone hysteretic optimal policies. Oper. Res. **32**(5), 1116–1132 (1984)
232. Ma, D.J., Cao, X.R.: A direct approach to decentralized control of service rates in a closed Jackson network. IEEE Trans. Autom. Control **39**(7), 1460–1463 (1994)
233. Ma, J.Y., Xia, L., Li, Q.L.: Optimal energy-efficient policies for data centers through sensitivity-based optimization. Discrete Event Dyn. Syst., 1–40 (2019)
234. Ma, J.Y., Li, Q.L., Xia, L.: Optimal asynchronous dynamic policies in energy-efficient data centers. Online Publication, 1–63, arXiv:190103371 (2019)
235. Maglaras, C.: Dynamic scheduling in multiclass queueing networks: stability under discrete-review policies. Queueing Syst. **31**(3–4), 171–206 (1999)
236. Mandelbaum, A., Yechiali, U.: Optimal entering rules for a customer with wait option at an M/G/1 queue. Manage. Sci. **29**(2), 174–187 (1983)
237. Marin, A., Rossi, S.: Power control in saturated fork-join queueing systems. Perform. Eval. **116**, 101–118 (2017)
238. Markov, A.A.: Rasprostranenie zakona bol'shih chisel na velichiny, zavisyaschie drug ot druga. In: Izvestiya Fiziko-matematicheskogo obschestva pri Kazanskom universitete, 2-ya seriya, tom 15, pp. 135–156 (1906)
239. Mendelson, H., Yechiali, U.: Controlling the GI/M/1 queue by conditional acceptance of customers. Eur. J. Oper. Res. **7**(1), 77–85 (1981)
240. Menich, R., Serfozo, R.F.: Optimality of routing and servicing in dependent parallel processing systems. Queueing Syst. **9**(4), 403–418 (1991)
241. Meyn, S.P., Tweedie, R.L.: Markov Chains and Stochastic Stability. Springer, London (1996). https://doi.org/10.1007/978-1-4471-3267-7
242. Miller, B.L.: Finite state continuous time Markov decision processes with applications to a class of optimization problems in queueing theory. Ph.D. dissertation, Stanford University, California, USA (1967)
243. Mitchell, B.: Optimal service-rate selection in an M/G/1 Queue. SIAM J. Appl. Math. **24**(1), 19–35 (1973)
244. Nadar, E., Akan, M., Scheller-Wolf, A.: Technical note–optimal structural results for assemble-to-order generalized M-systems. Oper. Res. **62**(3), 571–579 (2014)
245. Neuts, M.F.: Matrix-Geometric Solutions in Stochastic Models: An Algorithmic Approach. The Johns Hopkins University Press, Baltimore (1981)
246. Neuts, M.F.: Structured Stochastic Matrices of M/G/1Type and Their Applications. Dekker (1989)
247. Nishimura, S.: Service mechanism control and arrival control of a two-station tandem queue. J. Oper. Res. Soc. Japan **29**(3), 191–205 (1986)
248. Nobel, R.D., Tijms, H.C.: Optimal control for an $M^X/G/1$ queue with two service modes. Eur. J. Oper. Res. **113**(3), 610–619 (1999)
249. Nobel, R.D., Tijms, H.C.: Optimal control of a queueing system with heterogeneous servers and setup costs. IEEE Trans. Autom. Control **45**(4), 780–784 (2000)
250. Okamura, H., Miyata, S., Dohi, T.: A Markov decision process approach to dynamic power management in a cluster system. IEEE Access **3**, 3039–3047 (2015)
251. Pajarinen, J., Hottinen, A., Peltonen, J.: Optimizing spatial and temporal reuse in wireless networks by decentralized partially observable Markov decision processes. IEEE Trans. Mob. Comput. **13**(4), 866–879 (2014)

252. Papadimitriou, C.H., Tsitsiklis, J.N.: The complexity of optimal queuing network control. Math. Oper. Res. **24**(2), 293–305 (1999)
253. Pascual, R., Martínez, A., Giesen, R.: Joint optimization of fleet size and maintenance capacity in a fork-join cyclical transportation system. J. Oper. Res. Soc. **64**(7), 982–994 (2013)
254. Patrick, J., Begen, M.A.: Markov decision processes and its applications in healthcare. In: Handbook of Healthcare Delivery Systems. CRC (2011)
255. Perez, A.P., Zipkin, P.: Dynamic scheduling rules for a multiproduct make-to-stock queue. Oper. Res. **45**(6), 919–930 (1997)
256. Piunovskiy, A.B.: Bicriteria optimization of a queue with a controlled input stream. Queueing Syst. **48**(1–2), 159–184 (2004)
257. Plambeck, E., Kumar, S., Harrison, J.M.: A multiclass queue in heavy traffic with throughput time constraints: asymptotically optimal dynamic controls. Queueing Syst. **39**(1), 23–54 (2001)
258. Plum, H.J.: Optimal monotone hysteretic Markov policies in an M/M/1 queueing model with switching costs and finite time horizon. Zeitschrift für Oper. Res. **35**(5), 377–399 (1991)
259. Powell, W.B., Humblet, P.: The bulk service queue with a general control strategy: theoretical analysis and a new computational procedure. Oper. Res. **34**(2), 267–275 (1986)
260. Printezis, A., Burnetas, A.: Priority option pricing in an M/M/m queue. Oper. Res. Lett. **36**(6), 700–704 (2008)
261. Puterman, M.L.: Markov Decision Processes: Discrete Stochastic Dynamic Programming. Wiley, New York (1994)
262. Qiu, Q., Pedram, M.: Dynamic power management based on continuous-time Markov decision processes. In: Proceedings of the 36th annual ACM/IEEE Design Automation Conference, pp. 555–561 (1999)
263. Robinson, D.R.: Markov decision chains with unbounded costs and applications to the control of queues. Adv. Appl. Probab. **8**(1), 159–176 (1976)
264. Robinson, D.R.: Optimization of priority queues-a semi-Markov decision chain approach. Manage. Sci. **24**(5), 545–553 (1978)
265. Rosberg, Z., Kermani, P.: Customer routing to different servers with complete information. Adv. Appl. Probab. **21**(4), 861–882 (1989)
266. Rosberg, Z., Varaiya, P., Walrand, J.: Optimal control of service in tandem queues. IEEE Trans. Autom. Control **27**(3), 600–610 (1982)
267. Ross, K.W.: Optimal dynamic routing in Markov queueing networks. Automatica **22**(3), 367–370 (1986)
268. Rue, R.C., Rosenshine, M.: Optimal control for entry of many classes of customers to an M/M/1 queue. Naval Res. Logistics **28**(3), 489–495 (1981)
269. Rue, R.C., Rosenshine, M.: Some properties of optimal control policies for entry to an M/M/1 queue. Naval Res. Logistics **28**(4), 525–532 (1981)
270. Rue, R.C., Rosenshine, M.: Optimal control of entry to an $M/E_k/1$ queue serving several classes of customers. Naval Res. Logistics **30**(2), 217–226 (1983)
271. Rue, R.C., Rosenshine, M.: The application of semi-Markov decision processes to queueing of aircraft for landing at an airport. Transp. Sci. **19**(2), 154–172 (1985)
272. Rykov, V.V.: Controllable queueing systems. Itogi Nauki i Tekhniki. Seriya "Teoriya Veroyatnostei. Matematicheskaya Statistika. Teoreticheskaya Kibernetika" 12, 45–152 (1975) (There is English translation in Journal of Soviet Mathematics). (In Russian)
273. Rykov, V.V.: Monotone control of queueing systems with heterogeneous servers. Queueing Syst. **37**(4), 391–403 (2001)

274. Rykov, V.V.: Controllable queueing systems: from the very beginning up to nowadays. Reliab. Theor. Appl. **12**(2(45)), 39–61 (2017)
275. Rykov, V.V., Efrosinin, D.: Optimal control of queueing systems with heterogeneous servers. Queueing Syst. **46**(3–4), 389–407 (2004)
276. Rykov, V.V., Efrosinin, D.: On optimal control of systems on their life time. In: Lisnianski, A., Frenkel, I. (eds.) Recent Advances in System Reliability. Springer Series in Reliability Engineering, pp. 307–319. Springer, London (2012). https://doi.org/10.1007/978-1-4471-2207-4_22
277. Rykov, V.V., Lembert, E.: Optimal dynamic priorities in single-line queueing systems. Eng. Cybern. **5**(1), 21–30 (1967)
278. Sanajian, N., Abouee-Mehrizi, H., Balcıoglu, B.: Scheduling policies in the M/G/1 make-to-stock queue. J. Oper. Res. Soc. **61**(1), 115–123 (2010)
279. Savaşaneril, S., Griffin, P.M., Keskinocak, P.: Dynamic lead-time quotation for an M/M/1 base-stock inventory queue. Oper. Res. **58**(2), 383–395 (2010)
280. Schassberger, R.: A note on optimal service selection in a single server queue. Manage. Sci. **21**(11), 1326–1331 (1975)
281. Sennott, L.I.: Average cost semi-Markov decision processes and the control of queueing systems. Probab. Eng. Inf. Sci. **3**(2), 247–272 (1989)
282. Sennott, L.I.: Stochastic Dynamic Programming and the Control of Queueing Systems. Wiley, New York (2009)
283. Serfozo, R.F.: Optimal control of random walks, birth and death processes, and queues. Adv. Appl. Probab. **13**(1), 61–83 (1981)
284. Serfozo, R.F.: Introduction to Stochastic Networks. Springer, New York (1999). https://doi.org/10.1007/978-1-4612-1482-3
285. Shanthikumar, J.G., Yao, D.D.: Stochastic monotonicity in general queueing networks. J. Appl. Probab. **26**(2), 413–417 (1989)
286. Shioyama, T.: Optimal control of a queuing network system with two types of customers. Eur. J. Oper. Res. **52**(3), 367–372 (1991)
287. Sigaud, O., Buffet, O. (eds.): Markov Decision Processes in Artificial Intelligence. Wiley (2013)
288. e Silva, E.D.S., Gerla, M.: Queueing network models for load balancing in distributed systems. J. Parallel Distrib. Comput. **12**(1), 24–38 (1991)
289. Sobel, M.J.: Optimal operation of queues. In: Clarke, A.B. (ed.) Mathematical Methods in Queueing Theory, vol. 98, pp. 231–261. Springer, Heidelberg (1974). https://doi.org/10.1007/978-3-642-80838-8_12
290. Solodyannikov, Y.V.: Control and observation for dynamical queueing networks I. Autom. Remote Control **75**(3), 422–446 (2014)
291. Sparaggis, P.D., Towsley, D., Cassandras, C.G.: Optimal control of multiclass parallel service systems. Discrete Event Dyn. Syst. **6**(2), 139–158 (1996)
292. Stidham, S.: On the optimality of single-server queuing systems. Oper. Res. **18**(4), 708–732 (1970)
293. Stidham, S.: Socially and individually optimal control of arrivals to a GI/M/1 queue. Manage. Sci. **24**(15), 1598–1610 (1978)
294. Stidham, S.: Optimal control of admission to a queueing system. IEEE Trans. Autom. Control **30**(8), 705–713 (1985)
295. Stidham, S.: Scheduling, routing, and flow control in stochastic networks. In: Fleming, W., Lions, P.L. (eds.) Stochastic Differential Systems, Stochastic Control Theory and Applications, vol. 10, pp. 529–561. Springer, New York (1988). https://doi.org/10.1007/978-1-4613-8762-6_31
296. Stidham, S.: Analysis, design, and control of queueing systems. Oper. Res. **50**(1), 197–216 (2002)

297. Stidham, S.: On the optimality of a full-service policy for a queueing system with discounted costs. Math. Methods Oper. Res. **62**(3), 485–497 (2005)

298. Stidham, S.: Optimal Design of Queueing Systems. Chapman and Hall/CRC, Boca Raton (2009)

299. Stidham, S., Prabhu, N.U.: Optimal control of queueing systems. In: Clarke, A.B. (ed.) Mathematical Methods in Queueing Theory, vol. 98, pp. 263–294. Springer, Heidelberg (1974). https://doi.org/10.1007/978-3-642-80838-8_13

300. Stidham, S., Weber, R.: A survey of Markov decision models for control of networks of queues. Queueing Syst. **13**(1–3), 291–314 (1993)

301. Sun, L., Dong, H., Hussain, F.K., Hussain, O.K., Chang, E.: Cloud service selection: state-of-the-art and future research directions. J. Netw. Comput. Appl. **45**, 134–150 (2014)

302. Syski, R.: A personal view of queueing theory. In: Frontiers in Queueing: Models and Applications in Science and Engineering, pp. 3–18. CRC Press (1997)

303. Tassiulas, L., Ephremides, A.: Throughput properties of a queueing network with distributed dynamic routing and flow control. Adv. Appl. Probab. **28**(1), 285–307 (1996)

304. Tijms, H.C.: Stochastic Models: An Algorithmic Approach. Wiley (1994)

305. Tijms, H.C., van der Duyn Schouten, F.A.: A Markov decision algorithm for optimal inspections and revisions in a maintenance system with partial information. Eur. J. Oper. Res. **21**(2), 245–253 (1985)

306. Tirdad, A., Grassmann, W.K., Tavakoli, J.: Optimal policies of $M(t)/M/c/c$ queues with two different levels of servers. Eur. J. Oper. Res. **249**(3), 1124–1130 (2016)

307. Towsley, D., Sparaggis, P.D., Cassandras, C.G.: Optimal routing and buffer allocation for a class of finite capacity queueing systems. IEEE Trans. Autom. Control **37**(9), 1446–1451 (1992)

308. Vanberkel, P.T., Boucherie, R.J., Hans, E.W., Hurink, J.L., Litvak, N.: A survey of health care models that encompass multiple departments. Int. J. Health Manag. Inf. **1**(1), 37–69 (2010)

309. Van der Weij, W., Bhulai, S., Van der Mei, R.: Optimal scheduling policies for the limited processor sharing queue. Technical report WS2008-5, Department of Mathematics, Vrije University (2008)

310. Van Dijk, N.M.: Queueing Networks and Product Forms: A Systems Approach. Wiley (1993)

311. Van Dijk, N.M., Puterman, M.L.: Perturbation theory for Markov reward processes with applications to queueing systems. Adv. Appl. Probab. **20**(1), 79–98 (1988)

312. Van Nunen, J.A.E.E., Puterman, M.L.: Computing optimal control limits for GI/M/s queuing systems with controlled arrivals. Manage. Sci. **29**(6), 725–734 (1983)

313. Varma, S.: Optimal allocation of customers in a two server queue with resequencing. IEEE Trans. Autom. Control **36**(11), 1288–1293 (1991)

314. Veatch, M.H.: Queueing control problems for production/inventory systems. Ph.D. dissertation, Massachusetts Institute of Technology (1992)

315. Veatch, M.H., Wein, L.M.: Monotone control of queueing networks. Queueing Syst. **12**(3–4), 391–408 (1992)

316. Vercraene, S., Gayon, J.P., Karaesmen, F.: Effects of system parameters on the optimal cost and policy in a class of multidimensional queueing control problems. Oper. Res. **66**(1), 150–162 (2018)

317. Wakuta, K.: Optimal control of an M/G/1 queue with imperfectly observed queue length when the input source is finite. J. Appl. Probab. **28**(1), 210–220 (1991)
318. Wang, D.X., Cao, X.R.: Event-based optimization for POMDPs and its application in portfolio management. In: Proceedings of the 18th IFAC World Congress, vol. 44, no. 1, pp. 3228–3233 (2011)
319. Weber, R.R.: On the optimal assignment of customers to parallel servers. J. Appl. Probab. **15**(2), 406–413 (1978)
320. Weber, R.R., Stidham, S.: Optimal control of service rates in networks of queues. Adv. Appl. Probab. **19**(1), 202–218 (1987)
321. White, L.B.: A new policy evaluation algorithm for Markov decision processes with quasi birth-death structure. Stoch. Models **21**(2–3), 785–797 (2005)
322. Winkler, A.: Dynamic scheduling of a single-server two-class queue with constant retrial policy. Ann. Oper. Res. **202**(1), 197–210 (2013)
323. Wu, C.H., Lin, J.T., Chien, W.C.: Dynamic production control in a serial line with process queue time constraint. Int. J. Prod. Res. **48**(13), 3823–3843 (2010)
324. Xia, L.: Service rate control of closed Jackson networks from game theoretic perspective. Eur. J. Oper. Res. **237**(2), 546–554 (2014)
325. Xia, L.: Event-based optimization of admission control in open queueing networks. Discrete Event Dyn. Syst. **24**(2), 133–151 (2014)
326. Xia, L., Cao, X.R.: Performance optimization of queueing systems with perturbation realization. Eur. J. Oper. Res. **218**(2), 293–304 (2012)
327. Xia, L., Chen, S.: Dynamic pricing control for open queueing networks. IEEE Trans. Autom. Control **63**(10), 3290–3300 (2018)
328. Xia, L., He, Q.M., Alfa, A.S.: Optimal control of state-dependent service rates in a MAP/M/1 queue. IEEE Trans. Autom. Control **62**(10), 4965–4979 (2017)
329. Xia, L., Jia, Q.S.: Parameterized Markov decision process and its application to service rate control. Automatica **54**, 29–35 (2015)
330. Xia, L., Jia, Q.S., Cao, X.R.: A tutorial on event-based optimization–a new optimization framework. Discrete Event Dyn. Syst. **24**(2), 103–132 (2014)
331. Xia, L., Shihada, B.: Max-Min optimality of service rate control in closed queueing networks. IEEE Trans. Autom. Control **58**(4), 1051–1056 (2013)
332. Xu, S.H., Righter, R., Shanthikumar, J.G.: Optimal dynamic assignment of customers to heterogeneous servers in parallel. Oper. Res. **40**(6), 1126–1138 (1992)
333. Xu, S.H., Zhao, Y.Q.: Dynamic routing and jockeying controls in a two-station queueing system. Adv. Appl. Probab. **28**(4), 1201–1226 (1996)
334. Xu, J., Hajek, B.: The supermarket game. Stoch. Syst. **3**(2), 405–441 (2013)
335. Yang, R., Bhulai, S., van der Mei, R.: Structural properties of the optimal resource allocation policy for single-queue systems. Ann. Oper. Res. **202**(1), 211–233 (2013)
336. Yao, D.D., Schechner, Z.: Decentralized control of service rates in a closed Jackson network. IEEE Trans. Autom. Control **34**(2), 236–240 (1989)
337. Yechiali, U.: Customers' optimal joining rules for the GI/M/s queue. Manage. Sci. **18**(7), 434–443 (1972)
338. Yeh, L., Thomas, L.C.: Adaptive control of M/M/1 queues–continuous-time Markov decision process approach. J. Appl. Probab. **20**(2), 368–379 (1983)
339. Yildirim, U., Hasenbein, J.J.: Admission control and pricing in a queue with batch arrivals. Oper. Res. Lett. **38**(5), 427–431 (2010)
340. Yoon, S., Lewis, M.E.: Optimal pricing and admission control in a queueing system with periodically varying parameters. Queueing Syst. **47**(3), 177–199 (2004)
341. Zayas-Cabán, G., Xie, J., Green, L.V., Lewis, M.E.: Dynamic control of a tandem system with abandonments. Queueing Syst. **84**(3–4), 279–293 (2016)

342. Zeng, Y., Chaintreau, A., Towsley, D., Xia, C.H.: A necessary and sufficient condition for throughput scalability of fork and join networks with blocking. ACM SIGMETRICS Perform. Eval. Rev. **44**(1), 25–36 (2016)
343. Zeng, Y., Chaintreau, A., Towsley, D., Xia, C.H.: Throughput scalability analysis of fork-join queueing networks. Oper. Res. **66**(6), 1728–1743 (2018)
344. Zhang, B., Ayhan, H.: Optimal admission control for tandem queues with loss. IEEE Trans. Autom. Control **58**(1), 163–167 (2013)

# Stochastic Monotonicity of the Mean-CVaRs and Their Applications to Inventory Systems with Stockout Cost: A Transformation Approach

Hai-Bo Yu[(✉)]

School of Economics and Management Sciences, Beijing University of Technology,
Beijing 100124, People's Republic of China
haibo_yu@126.com

**Abstract.** Motivated by various applications in inventory management, this article is devoted to the stochastic monotonicity and comparability of two special mean-risk models, called mean-conditional value-at-risk (abbreviated as MCVaR) measures. Firstly, we characterize the two MCVaR measures by the second quantile function of a random variable, and show that the two MCVaR measures are consistent with ascending stochastic dominance (abbreviated as ASD) or descending stochastic dominance (abbreviated as DSD) for risk lovers by using then relations between the second quantile function and two stochastic dominance. We also show that the two MCVaR measures have loss-aversion property for risk-aversion case, and have subadditivity (superadditivity) and convexity (convexity) for risk-seeking (risk-aversion), respectively. We obtain similar results by using a linear transformation of a random variable with a location and a scalar parameters, the transformation is corresponding to ASD and DSD as location parameter changes. The obtained results with respect to the stochastic monotonicity for the two MCVaR measures are used to solve a inventory decision problem with bi-objective maximization expected utility considering risk preference (including risk-aversion and risk-seeking) and stockout cost. We obtain the close solution and optimal expected utility for this problem. Due to the complexity of the solution, we provide several upper and lower bounds for the optimal order quantity, which are corresponding value in risk-neutral or without stockout cost cases. The obtained results in this paper has a certain insights and help for enterprises facing market uncertainty and considering decision maker's risk preference.

**Keywords:** Newsvendor model · Stochastic dominance · Levy's transformation · Mean-conditional value-at-risks · Stockout cost

## 1 Introduction

Consider a company, e.g. a retailer sourcing a product with short life cycle to stock using the framework of the newsvendor model, which is based upon risk

© Springer Nature Singapore Pte Ltd. 2019
Q.-L. Li et al. (Eds.): Cao Festschrift 2019, CCIS 1102, pp. 72–115, 2019.
https://doi.org/10.1007/978-981-15-0864-6_4

neutrality so that managers will place orders to maximize expected profits or minimize expected cost. However, in practice, there are many examples that imply managers' decisions do not always correspond to the expected profit-maximization order quantity (e.g., Kahn 1992; Fisher and Raman 1996; Patsuris 2001). Experimental findings show that for high-profit products the order quantity is less than the one of the risk-neutral decisions maker and for lower-profit products the order quantity is higher than the one of the risk-neutral decisions maker (e.g., Schweitzer and Cachon 2000; Bolton and Katok 2008). On the other hand, the optimal inventory policies may involve stockouts under stochastic demand. The effect of stockout in inventory models is usually taken into account by means of a stockout (penalty) cost, and often assumed to be proportional to the excess of demand over supply when the demand during stockout is met by a priority shipment or extra production run (see Schwartz 1966; Caine and Plaut 1976). For example, lost sales and production disruptions resulting from the stockouts prompted one quarter of the buyers to abandon their supplier by personal interviews and a mail survey of professional buyers in Canada and the U.S.A (see Banting et al. 2015). Hence, it is very important and interesting for the problems about the non-expected profit maximization (or cost minimization) and considering stockout cost in stochastic inventory management.

There are lots of approaches to analyze the non-expected profit maximization (or cost minimization) problems for risk averters: expected utility theory (abbreviated as EUT) (Eeckhoudt et al. 1995; Agrawal and Seshadri 2000a,b; Keren and Pliskin 2006); stochastic dominance (abbreviated as SD) (Quirk and Saposnik 1962; Fishburn 1964, 1974); mean-variance (abbreviated as MV) model (Markowitz 1952, 1970) etc. However, the above methods exist some limitations. The EUT is however, too conceptual to identify in practice, and it exists a limitation in the economics field, i.e., risk aversion within the EUT framework implies that people are approximately risk-neutral when economic risk are small (Arrow 1971), but it is unreasonable when economic risk are large (see the example presented by Rabin 2000). The stochastic dominance can model risk averters, but it is computationally very difficult, it is not quite suitable with some of the more complex inventory decision issues-it is a multiple criteria model with a continuum of criteria. The mean-variance valuation satisfies a class of decision makers with concave quadratic utility function to model risk averters, but it is inadequate in the sense that it equally quantifies desirable upside influence and undesirable downside outcomes.

In contrast to wide researches about profit maximization and risk control in newsvendor problem, the study of loss aversion and risk-seeking in newsvendor problem is relatively few. Motivated by the empirical supports and successful applications of loss aversion in other fields (see Rabin 1998 and Camerer 2001), we propose two special mean-risk models to model a newsvendor problem with stockout cost, the two mean-risk models can not only model the risk-aversion and risk-seeking, but also model loss aversion in newsvendor situation. That is, we analyze a complex newsvendor problem incorporating decision maker's risk preference or loss aversion and considering positive stockout cost.

The main contributions of the paper are:

- We introduce two class of special mean-risk models which the risk measures are the conditional expected values of low or high profit (i.e., mean-CVaR models). We show that the two mean-CVaR measures have loss aversion characterization when the risk preference coefficient is larger than one.
- We show that the two class of mean-CVaR measures are consistent with the ascending stochastic dominance (abbreviated as ASD) for risk averters or descending stochastic dominance (abbreviated as DSD) for risk lovers, it can also be characterized by a transformation of a random variable depending a two parameters representing the deterministic demand and demand variability.
- We provide a close solution of the optimal order quantity and expected utility for the newsvendor model with mean-CVaR constraint and stockout cost. We show that a risk-averse inventory manager holds more inventory level than that without stockout cost, a risk-taking inventory manager holds less inventory level than that without stockout cost.
- We provide a comparison result for the optimal inventory level of the newsvendor model in risk-averse, risk-taking and risk-neutral cases. We show that a risk-taking inventory manager holds more inventory level than in the classical newsvendor model (risk-neutral case), a risk-averse inventory manager holds less inventory level than in the classical newsvendor model.

Note that in this paper 'increasing' means 'nondecreasing' and 'decreasing' means 'nonincreasing'.

The rest of this paper is organized as follows. In Sect. 2 we briefly review some papers of related literature. In Sect. 3 we introduce two class of special mean-risk models, and characterize them by stochastic dominance and a transformation of a random variable with location and scalar parameters. In Sect. 4, from the perspective of application, we analyze a newsvendor problem considering stockout cost and newsvendor's preferences. In Sect. 5, we provide several numerical examples to illustrate our results. Finally, in Sect. 6, we offer concluding remarks and suggest opportunities for future research.

## 2    Literature Review

### 2.1    The Newsvendor Problem Considering Risk Averters

In modern operations management, there are lots of approaches to analyze the non-expected profit maximization (or cost minimization) problems for risk averters: expected utility theory (EUT); stochastic dominance; mean-variance and mean-risk model etc.

Many authors have studied the selection rules for risk averters by expected utility theory (e.g., Eeckhoudt et al. 1995; Agrawal and Seshadri 2000a; 2000b; Keren and Pliskin 2006). Readers may refer to Schoemaker (1982) for a comprehensive review of EUT.

Stochastic dominance is one of the fundamental concepts of decision theory (Whitmore and Findlay 1978; Levy 1992) and has been widely used in economics and finance (Bawa 1982; Levy 1992), risk theory (Denuit et al. 2005), reliability (Ross and Schechner 1984; Bhattacharjee 1991) and queueing model (Yu 2017). It originated in the majorization theory (Hardy, Littlewood and Polya 1934) for the discrete case and was later extended to general distributions (Hanoch and Levy 1969; Rothschild and Stiglitz 1969). Stochastic dominance is based on an axiomatic model of risk-averse preferences (Fishburn 1964), it introduces a partial order in the space of real random variables. The first degree relation was originated by Quirk and Saposnik (1962) which carries over to expectations of monotone utility functions, the second degree relation–to expectations of concave nondecreasing utility functions, and the increasing convex ordering relation–to expectations of convex nondecreasing utility functions. These increasing concave, linear and increasing convex utility functions corresponding three major types of persons: risk averters, risk neutrals and risk lovers. For equal mean distributions, a theorem proved by Karamata (1932) is similar to the second degree stochastic dominance (see also Lehmann 1955; Hardy et al. 1959; Lebreton 1987). More collections of such works include Marshall and Olkin (1979), Stoyan (1983), Shaked and Shanthikumar (1994) and Ross (1983). Song (1994) explored the effect of leadtime uncertainty in minimizing cost newsvendor model with two equal-mean demands by using more variable order. The auther finds optimal expected cost is increasing when demand variability increases. Yu (2014a) studied the effect of demand variability in maximizing profit newsvendor model with unequal-mean demands by using more variable order and generalized more variable order. The author finds optimal expected profit may be increasing when demand variability increases under a certain condition. Yu (2014) analyzed the effect of demand variability in inventory system with sales effort and replenishment decisions. Yu, Yang and Li (2018) discussed the effect of supply leadtime and demand uncertainty in inventory systems.

The mean-risk approach quantifies the problem in a expressive form of only two criteria: the mean, representing the expected outcome, and the risk, a scalar measure of the variability of outcomes (Markowitz 1952). It is appealing to decision makers and allows a simple trade-off analysis, analytical or geometrical. The portfolio optimization problem is modeled as a mean-risk bicriteria optimization problem. The classical Markowitz model uses the variance as the risk measure, and is used to analyze optimization problem for risk averters (Markowitz 1952, 1970; Tobin 1958, 1965). However, many authors have pointed out that the mean-variance model is, in general, not consistent with stochastic dominance rules.

The CVaR is known as risk measures which is coherent (Artzner et al. 1999), and consistent with the second (or higher) order stochastic dominance (Pflug 2000; Ogryczak and Ruszczyński 2002). These preferable properties are induced from some axiomatization of rational investors' behavior under uncertainty and, thus, they are meaningful also to a manager who faces uncertain profit/loss situation as in the newsvendor problem. In particular, the consistency with the stochastic dominance implies that minimizing the CVaR never conflicts with maximizing the expectation of any risk-averse utility function (Ogryczak and Ruszczyński 2002). Chen et al. (2003) proposed a model about CVaR of profit for newsvendor problem and obtained the optimal order quantity. Gotoh and Takano (2007) provided analytical solutions and linear programming formulation for minimizing CVaR in the newsvendor problem. Ma et al. (2012) considered a supply chain with a risk-neutral manufacturer and a risk-averse retailer who use the CVaR measure as his risk measure. Sun et al. (2013) proposed a stochastic programming model for the newsvendor problem by adopting the CVaR measure as the risk metric in the objective function. Wu et al. (2013) introduced the CVaR measure to study the effect of capacity uncertainty on the inventory decisions of a risk-averse newsvendor. Xu et al. (2013) introduced a tri-level model for the three-stage supply chain, in which the risk-averse retailer intends to maximize his CVaR about profit. Qiu et al. (2014) introduced a CVaR based profit maximization model to study the robust inventory decision-making problem faced by risk-averse newsvendors with incomplete demand information. Katariya et al. (2014) investigated the relationship between risk-neutral and risk-averse newsvendor problems under the CVaR measure. All these papers aim to control the potential risks and to reduce losses, either by maximizing the CVaR about profit or minimizing the CVaR about cost for the risk-averse newsvendor. Yu (2014c) discussed the stochastic monotonicity of mixture conditional value-at-risk and its applications to inventory system without considering stockout cost. Yu and Wang (2014) analyzed the effect of demand uncertainty in inventory systems with mixture CVaR constrain.

## 2.2 The Newsvendor Problem Considering Risk Averters and Risk Takers

From the perspective of decision-making, there are three major types of persons: risk averters, risk neutrals and risk lovers. Their corresponding utility functions are concave, linear and convex; all are increasing functions. Quirk and Saposnik (1962) and many others develop some univariate stochastic dominance (SD) rules for risk averters. On the other hand, Hammond (1974) and many others develop the univariate SD rules for risk seekers. Li and Wong (1999 Extension of...) developed some stochastic dominance theorems for the location and scale family of random variables and linear combinations of random variables and for risk lovers as well as risk averters that extend results in Hadar and Russell (1971). Li and Wong (1999) called stochastic dominance for risk lovers descending stochastic dominance (DSD), corresponding to the increasing convex ordering of Stoyan (1983). They called stochastic dominance for risk averters ascending stochastic dominance (ASD), corresponding to the increasing concave ordering (i.e., SSD) of Stoyan (1983). We note that stochastic dominance for risk neutrals is a special case in the theory of stochastic dominance for risk averters or risk lovers. Some authors analyzed the univariate ascending (descending) stochastic dominance applied to risk averters (seekers) (e.g., Sriboonchitta, et al. 2009). Guo and Wong (2016) first extends some well-known univariate stochastic dominance results to multivariate stochastic dominances for both risk averters and risk seekers, respectively. Jammernegg and Kischka (2007) proposed a newsvendor model where the inventory manager can control internal and customer-oriented performance measures. The objective function is a convex combination of conditional expected values of low and high profits, respectively. They showed that A risk-averse inventory manager cannot Pareto-dominate a risk-neutral or risk-taking inventory manager with respect to the expected profit and the level of product availability. Gotoh and Takano (2007) provided a optimal solution through a algorithm with three steps for the unconstrained mean-CVaR model with positive stockout cost, and showed that a closed form solution of the model did not exist. They only obtained a closed form solution of the model for zero stockout cost case. Yu (2014c) discussed the stochastic monotonicity of mixture conditional value-at-risk and its applications to inventory system without considering stockout cost. Yu and Wang (2014) analyzed the effect of demand uncertainty in inventory systems with mixture CVaR constrain.

## 2.3    The Newsvendor Problem by Mean-Preserving Transformation

The mean-preserving transformation (abbreviated as MPT) approach is used to study the effect of demand variability in stochastic modeles, it was introduced by Gerchak and Mossman (1992), which is used to study the magnitude of the optimal cost and order quantity change of newsboy with loss of sales when demand variability increases. They showed that higher variability will lead to higher costs. They also found that although the service level is not related to the optimal cost when demand variability increases, but the optimal order quantity increases at the low critical ratio when demand variability increases. Chen and Federgruen (2000) pointed out that compared with profit measure, cost measure is more complicated. From a corporate point of view, they analyszed the effect of reducing demand variability based on profits. Li and Atikins (2005) focused on the impact of coordination and information when market demand becomes more variable in the sense of specific MPT. Zhu and Wu (2014) provided a general framework for stochastic variability ordering under any mean-preserving transformation that can be parameterized by a single scalar, and apply it to a broad class of specific transformations, including the widely used mean-preserving affine transformation, truncation, and capping. Qin et al. (2011) analyzed the case in which marketing effort effects demand uncertainty with MPT. They found that demand mean increase due to marketing effort may leads to the increase about the optimal order quantity. But the effect about demand variability reduction is not as clear. Chua and Liu (2015) studied the maximization newsvendor problem using MPT. They find that optimal order quantity decreases in demand variability for zero salvage value. Begen et al. (2016) conducted the analysis to examine the impacts of supply uncertainty, demand uncertainty and uncertainty reduction efforts on optimal order quantity and total cost through MPT. Levy (1977) pointed out that the mean-preserving transformation can not explain the situation that demand mean and variability changed together. The author introduced a new transformation of a random variable, which includes location parameter

$r$ and scale parameter $\alpha$, where parameter $r$ can measure the effect of different mean, parameter $\alpha$ can measure the effect of variability. Yu and Li (2018) studied the stochastic comparison of the generalized mean-preserving transformations including Levy (1977)'s and Zhu and Wu (2014)'s transformations, and applied the obtained results to solve inventory optimization problem. They showed that the level of demand variability, the size of average demand and the sale price of products will have an impact on the profit in newsvendor model. Usually the optimal profit will decrease when demand variability increases, however, it may be increase when demand variability increases under some conditions, such as sale price of the products is low.

### 2.4    The Newsvendor Problem Considering Stockout Cost

For the newsvendor problem in the presence of the loss of goodwill stockout opportunity cost, Choi et al. (2008) and Wu et al. (2009) both provide some more analytical results. Choi et al. (2008) extend Choi et al. (2001) and consider the newsvendor problem under the MV framework with decision makers possessing different risk attitude (namely, risk-averse, risk-neutral, and risk-seeking). They found that when we include an explicit stockout penalty cost in the problem, the variance of profit becomes more complicated. If this stockout cost is huge, the optimal order quantities for risk-averse and risk-seeking decision makers will leave the corresponding efficient regions (i.e., the efficient regions when the stockout cost equals zero). In particular, Wu et al. (2009) analytically prove that when demand is distributed following a continuous power distribution, the optimal stocking quantity under an MV newsvendor model will exceed the risk-neutral case's optimal "fractile" quantity. Yu (2014a) studied the effect of demand variability on expected cost (profit) of minimization cost (maximization profit) newsvendor problems by using stochastic dominance and variability orderings. They showed that there exists a certain demand distribution, the optimal expected cost (profit) decrease (increase) when demand variability increases for those shortage penalty cost that is larger than a certain value.

## 3    Stochastic Monotonicity of the Mean-CVaRs

In this section, we characterize the stochastic monotonicity of two special mean-risk models, i.e., mean-CVaR models, by two methods: one is stochastic dominance, another is the transformation with two parameters presented by Levy (1977).

### 3.1    Stochastic Dominance and Mean-Risk Models

In this subsection, we provide several relations between stochastic dominance and mean-risk models, and show that these mean-risk models are consistent or $\xi$-consistent with the ascending stochastic dominance (abbreviated as ASD) for risk averters or descending stochastic dominance (abbreviated as DSD) for risk lovers. These characterizations will be applied to analyze inventory decision problem with risk preference decision maker in Sect. 4.2.

Stochastic dominance is defined by introducing a partial order in the space of real random variables. Consider two continuous random variables $X$ and $Y$ with distribution functions $F_X(\cdot)$ and $F_Y(\cdot)$ having a continuously differentiable densities $f_X(\cdot)$ and $f_Y(\cdot)$ on an interval $[\underline{\ell}, \bar{\ell}]$, where $\underline{\ell}$ or $\bar{\ell}$ may be finite or infinite taking $-\infty$ or $\infty$, respectively, in late case, the corresponding closed interval is written as open interval.

The first performance function $F_X^{(1)}(\cdot)$ is defined as the right-continuous cumulative distribution function itself:

$$F_X^{(1)}(t) = F_X(t) = \mathbb{P}\{X \le t\}, \ t \in [\underline{\ell}, \bar{\ell}]. \tag{1}$$

The first degree stochastic dominance (FSD) is defined as follows

$$Y \ge_{FSD} X \iff F_X^{(1)}(t) - F_Y^{(1)}(t) \ge 0 \text{ for all } t \in [\underline{\ell}, \bar{\ell}]. \tag{2}$$

The second performance function $F_X^{(2)}(\cdot)$ is given by areas below the distribution function $F_X(\cdot)$:

$$F_X^{(2)}(t) = \mathbb{E}[(t - X)_+] = \int_{\underline{\ell}}^{t} F_X(x)dx, \ x \in [\underline{\ell}, \bar{\ell}]. \tag{3}$$

and defines the ascending stochastic dominance (abbreviated as ASD) for risk averters (see Li and Wong 1999), equivalently, the second degree stochastic dominance (see Hadar and Russel 1969; Hanoch and Levy 1969):

$$Y \ge_{ASD} X \iff F_X^{(2)}(t) - F_Y^{(2)}(t) \ge 0 \text{ for all } x \in [\underline{\ell}, \bar{\ell}]. \tag{4}$$

Levy (1992) provided the necessary condition for the ASD relation

$$Y \ge_{ASD} X \implies \mu_Y \ge \mu_X. \tag{5}$$

From the definition of ASD, it is easily to see that

$$Y \ge_{ASD} X \implies Var(Y) \le Var(X) \text{ and } Cv(Y) \le Cv(X). \tag{6}$$

where $Cv(X) = Var(X)/(\mathbb{E}[X])^2$, $Cv(Y) = Var(Y)/(\mathbb{E}[Y])^2$ are coefficient of variation of $X$ and $Y$, respectively.

**Remark 1.** There are three definitions of "more riskier" associated with ASD when two random variables $X$ and $Y$ have identical means, and showed the equivalence of the following three definitions (see Rothschild and Stiglitz 1970):

(i) $Y \ge_{ASD} X$.
(ii) $\mathbb{E}[u(Y)] \ge \mathbb{E}[u(X)]$ for all increasing concave function $u(\cdot)$.
(iii) $X = Y + Z$ ($X$ equal to $Y$ plus a noise $Z$ with $\mathbb{E}[Z|Y] = 0$).

The third equivalent definitions of Rothschild and Stiglitz (1970) may help us understanding why this is a variability ordering for the ASD.

For decision making under uncertainty, the ascending stochastic dominance for risk averters is more important, it is equivalent to the comparison of two second quantile functions $F_X^{(2)}(\cdot)$ and $F_Y^{(2)}(\cdot)$. If $Y \geq_{ASD} X$, then $Y$ is preferred to $X$ within all risk-averse preference models that prefer larger outcomes. Hence, it is a matter of primary importance that an approach to the comparison of random outcomes be consistent with the ascending stochastic dominance relation. This paper focuses on the relation between ASD and the two mean-risk models which the risk measures are the conditional expected values of low or high profit.

In the following we consider the relation between ASD and usual mean-risk model. Mean-risk approaches are based on comparing two scalar characteristics, the first is $\mu$ which represents the expected outcome (reward), and the second is $\tau$ which represents some measure of risk.

**Definition 1.** We say that the mean-risk model $(\mu_X, \tau_X)$ is consistent with the ASD, if the following relation holds

$$Y \geq_{ASD} X \implies \mu_Y \geq \mu_X \text{ and } \tau_Y \leq \tau_X. \tag{7}$$

**Remark 2.** From the definition of ASD, we see that the mean-risk model with the risk measure defined as the expected shortfall below some fixed target $t$, i.e., $\tau_X(t) = \mathbb{E}[(t - X)_+]$ is consistent with the ASD, where $(z)_+ = \max\{z, 0\}$. Another risk measure is variance $Var(X) = \mathbb{E}[(X - \mathbb{E}[X])^2]$, the mean-risk model $(\mu_X, Var(X))$ is consistent with the ASD.

An important advantage of mean-risk approaches is the possibility of a pictorial tradeoff analysis. Denoted $\xi$ by a trade-off coefficient between the risk and the mean, we may compare two real values $\mu_X - \xi\tau_X$ and $\mu_Y - \xi\tau_Y$. The following relation holds

$$\mu_Y \geq \mu_X \text{ and } r_Y \leq r_X \implies \mu_Y - \xi\tau_Y \geq \mu_X - \xi\tau_X \text{ for all } \xi > 0. \tag{8}$$

In this paper we show that some mean-risk models $(\mu_X, \tau_X)$ are $\xi$-consistent with the ASD, where $\xi$ positive constant, that is, $\xi > 0$ (see Definition 1.2 of Ogryczak and Ruszczyński 2002).

**Definition 2.** We say that the mean-risk model $(\mu_X, \tau_X)$ is $\xi$-consistent with the ASD model, if there exists a positive constant $\xi > 0$ such that for all $X$ and $Y$, the following relation holds

$$Y \geq_{ASD} X \implies \mu_Y - \xi\tau_Y \geq \mu_X - \xi\tau_X. \tag{9}$$

Based on the fact that the $Y \geq_{ASD} X \implies \mu_Y \geq \mu_X$ (see Levy 1992). This Definition is consistent with that given by Ogryczak and Ruszczyński (1999).

**Remark 3.** An example of the mean-risk model is that the risk measure takes the absolute semideviation $\bar{\delta}_X$ given by

$$\bar{\delta}_X = \mathbb{E}[(\mu_X - X)_+] = \int_{-\infty}^{\mu_X} (\mu_X - x) dF_X(x), \tag{10}$$

then the mean-risk model $(\mu_X, \bar{\delta}_X)$ is 1-consistent with ASD (see Ogryczak and Ruszczyński 2002), where $(z)_+ = \max\{z, 0\}$. Another example of the mean-risk model is that the risk measure takes the conditional expected values of high profit, the mean-risk model $(\mu_X, -CVaR_\eta^{(1/\eta)}(X))$ in (39) is $(\tilde{\lambda} - 1)/(1/\eta - 1))$-consistent with ASD when $\tilde{\lambda} \geq 1$. (see Theorem 1(ii)).

From (9), for $0 < \zeta \leq \xi$, we have

$$\mu_Y \geq \mu_X \text{ and } \mu_Y - \xi\tau_Y \geq \mu_X - \xi\tau_X \Longrightarrow \mu_Y \geq \mu_X \text{ and } \mu_Y - \zeta\tau_Y \geq \mu_X - \xi\tau_X. \quad (11)$$

That is, $\xi$-consistency implies $\zeta$-consistency for all $0 < \zeta \leq \xi$. Hence, (9) may be interpreted as the consistency with ASD of the mean-risk model, provided that the trade-off coefficient is bounded from above by $\xi$.

## 3.2   Characterization of Mean-CVaR by Stochastic Dominance

In this subsection, we characterize two mean-CVaRs in (38) and (39) by using the relation between the mean-risk models and stochastic dominance in above subsection, and show that the two mean-CVaR models are $\delta$-consistent with the ascending stochastic dominance for risk averters or the descending stochastic dominance for risk lovers, where $\delta$ is a constant determined by risk preference coefficient $\tilde{\lambda}$.

Firstly, we provide the properties of quantile dominance and the Lorenz curve.

Consider the quantile model of stochastic dominance (Levy and Kroll 1978). Let $X$ be a continuous random variable with support $[\ell, \bar{\ell}]$ and let $F_X(\cdot)$ be its distribution function, i.e., $F_X(x) = \mathbb{P}\{X \leq x\}$.

The first quantile function $F_X^{-1} : (0, 1) \to [\ell, \bar{\ell}]$ corresponding to a real random variable $X$ is defined as the left continuous inverse of the cumulative distribution function (Pflug 2000)

$$F_X^{-1}(\eta) = \inf\{x : F_X(x) \geq \eta\}, \ 0 < \eta < 1. \quad (12)$$

When the inverse $F_X(\cdot)$ does not exist, one can obtain a solution $F_X^{-1}(\cdot)$ via a simple numerical calculation (Silver et al. 1998).

Given $\eta \in (0, 1)$ the number $q_X(\eta)$ is called a $\eta$-quantile of the random variable $X$ if $Pr\{X < q_X(\eta)\} \leq \eta \leq Pr\{X \leq q_X(\eta)\}$.

From the definition of FSD we have

$$Y \geq_{FSD} X \Longleftrightarrow F_Y^{-1}(\eta) \geq F_X^{-1}(\eta) \text{ for all } 0 < \eta < 1. \quad (13)$$

**Remark 4.** Value-at-Risk (VaR) defined as the maximum loss at a specified confidence level $\eta$ is a widely used quantile risk measure. For a fixed level $\eta$, $0 < \eta < 1$, define the VaR as the $\eta$-quantile (see Pflug 2000):

$$VaR_\eta(X) = F_X^{-1}(\eta). \quad (14)$$

To obtain quantile measures which are consistent with the ASD, Ogryczak and Ruszczynski (2002) introduce the second quantile function $F_X^{(-2)} : (0, 1) \rightarrow [\underline{\ell}, \overline{\ell}]$ defined as

$$F_X^{(-2)}(\eta) = \int_0^\eta F_X^{-1}(\theta)d\theta \text{ for } 0 < \eta < 1, \tag{15}$$

where $F_X^{(-2)}(0) = 0$.

**Remark 5.** The second quantile function $F_X^{(-2)}(\cdot)$ in (15) can be rewritten as

$$F_X^{(-2)}(\eta) = \int_{\underline{\ell}}^{F_X^{-1}(\eta)} (\eta - F_X(x))dx + \eta\underline{\ell}, \text{ for } 0 < \eta < 1, \tag{16}$$

which is called generalized TTT transform, denoted by $\widetilde{T}_X(\cdot)$ (see Yu 2014a).

**Remark 6.** The function $F_X^{(-2)}(\cdot)$ is well defined for any random variable $X$ satisfying the condition $\mathbb{E}[X] < \infty$. The function $F_X^{(-2)}(\cdot)$ is convex. The graph of $F_X^{(-2)}(\cdot)$ is called the absolute Lorenz curve or ALC diagram for short. The Lorenz curves are used for inequality ordering (Gastwirth 1971; Arnold 1986; Muliere and Scarsini 1989) of positive random variables, relative to their (positive expectations). Such a Lorenz curve $L_X(\gamma) = F_X^{-2}(\gamma)/\mu_X$, is convex and increasing. The main application of the ALC diagram is the analysis of risk and safety measures using quantiles of the distribution of the random outcome.

The following Proposition 1 provides basic properties for the second quantile function $F_X^{(-2)}(\cdot)$ in (15), which will be used to characterize the mean-CVaRs in (38) and (39). It can be obtained from (16), we omitted its proof.

**Proposition 1.** Consider a continuous random variable $X$. We have

(i) The second quantile function $F_X^{(-2)}(\eta)$ in (15) is increasing convex function of $\eta$ in interval $(0, 1)$ and

$$\frac{\partial F_X^{(-2)}(\eta)}{\partial \eta} = F_X^{-1}(\eta) \text{ for all } 0 < \eta < 1. \tag{17}$$

(ii) $F_X^{(-2)}(0) = 0$, $F_X^{(-2)}(1) = \mathbb{E}[X]$, and

$$0 \leq F_X^{(-2)}(\eta) \leq \mathbb{E}[X] \text{ for all } 0 < \eta < 1. \tag{18}$$

The following Lemma 1 shows that the conjugate function $F_X^{(-2)}(\cdot)$ in (15) may fully characterize the ASD relation. See Theorem 3.2 of Ogryczak and Ruszczyński (2002) or Theorem 3 in Levy (1992).

**Lemma 1.** For two random variables $X$ and $Y$ with the second quantile function $F_X^{(-2)}(\cdot)$ and $F_Y^{(-2)}(\cdot)$, respectively. Then

$$Y \geq_{ASD} X \iff F_Y^{(-2)}(\eta) \geq F_X^{(-2)}(\eta) \text{ for all } 0 < \eta < 1. \tag{19}$$

Ogryczak and Ruszczyński (2002) provided another representation of the second quantile function. Let $\eta \in (0,1)$ and suppose $F_X^{(-1)}(\cdot)$ exists such that $F_X(x) = \eta$. Then

$$F_X^{(-2)}(\eta) = \eta \mathbb{E}[X|X \leq F_X^{(-1)}(\eta)]. \tag{20}$$

**Remark 7.** From Corollary 3.3(ii) of Ogryczak and Ruszczyński (2002), The second quantile function $F_X^{(-2)}(\cdot)$ is equivalent to the maximization problem

$$\max_{z \in \mathbb{R}}(\eta z - \mathbb{E}[(z - X)_+]), \tag{21}$$

that is,

$$F_X^{(-2)}(\eta) = \max_{z \in \mathbb{R}}(\eta z - \mathbb{E}[(z - X)_+]), \tag{22}$$

which is convex conjugate of convex function $\varphi(z) = \mathbb{E}[(z - X)_+]$.

For a fixed level $\eta$, $0 < \eta < 1$, the relation (19) in Lemma 1 can be rewritten in the form

$$Y \geq_{ASD} X \iff F_Y^{(-2)}(\eta)/\eta \geq F_X^{(-2)}(\eta)/\eta \text{ for all } 0 < \eta < 1. \tag{23}$$

which is the safety measure

$$CVaR_\eta^{(1/\eta)}(X) = \frac{F_X^{(-2)}(\eta)}{\eta}, \ 0 < \eta < 1. \tag{24}$$

This measure has a equivalent definition as follows

$$CVaR_\eta^{(1/\eta)}(X) = \max_{z \in \mathbb{R}}\{z - \frac{1}{\eta}\mathbb{E}[(z - X)_+]\}, \tag{25}$$

(see e.g., Rockafellar and Uryasev 2000, 2002).

We call $CVaR_\eta^{(1/\eta)}(\cdot)$ in (25) the maximization conditional value at risk where $\eta$ reflects the degree of risk aversion, i.e., a lower value implies a higher degree of risk aversion and a value of $\eta = 1$ implies risk neutrality. Combining (24) and (20), the maximization conditional value at risk in (25) can be treated as the conditional expectation of the revenue $X$ when the revenue is not more than $\eta$-VaR, i.e

$$CVaR_\eta^{(1/\eta)}(X) = \mathbb{E}[X|X \leq F_X^{-1}(\eta)]. \tag{26}$$

The following Proposition 2 provides basic properties for the maximization conditional value at risk in (24), which will be used to characterize the mean-CVaRs in (38) and (39). Its proof is similar as Proposition 1, we omitted it.

**Proposition 2.** Consider a continuous random variable $X$ with support $[\underline{\ell}, \overline{\ell}]$. We have

(i) The maximization conditional value at risk in (24) is increasing function of $\eta$ in interval $(0, 1]$ and

$$\frac{\partial CVaR_\eta^{(1/\eta)}(X)}{\partial \eta} = \frac{1}{\eta^2} \int_{\underline{\ell}}^{F_X^{-1}(\eta)} F_X(x)dx > 0 \text{ for all } 0 < \eta < 1. \quad (27)$$

(ii) $\lim_{\eta \to 1} CVaR_\eta^{(1/\eta)}(X) = \mathbb{E}[X]$, and

$$\lim_{\eta \to 0} CVaR_\eta^{(1/\eta)}(X) = \lim_{\eta \to 0} F_X^{-1}(\eta) = \underline{\ell}. \quad (28)$$

(iii) $CVaR_\eta^{(1/\eta)}(X)$ satisfies

$$\underline{\ell} < CVaR_\eta^{(1/\eta)}(X) \le \mathbb{E}[X] \text{ for all } 0 < \eta < 1. \quad (29)$$

Another conditional value at risk called the minimization conditional value at risk. For a fixed level $\eta$, $0 < \eta < 1$, the minimization conditional value at risk $CVaR_\eta^{(0)}(\cdot)$ is defined as the solution of an optimization problem

$$CVaR_\eta^{(0)}(X) = \min_{z \in \mathbb{R}}\{z + \frac{1}{1 - \eta}\mathbb{E}[(X - z)_+]\}, \quad (30)$$

(see e.g., Uryasev and Rockafellar (1999); Pflug (2000)).

Uryasev and Rockafellar (1999) have shown that for smooth distribution $F_X(\cdot)$ the minimization conditional value at risk $CVaR_\eta^{(0)}(X)$ equals the conditional expectation of $X$, given that $X \ge VaR_\eta(X)$, i.e

$$CVaR_\eta^{(0)}(X) = \mathbb{E}[X|X \ge F_X^{-1}(\eta)\}. \quad (31)$$

It is easily to show that $CVaR_\eta^{(0)}(\cdot)$ given in (30), can be represented by the function $F_X^{(-2)}(\cdot)$, that is

$$CVaR_\eta^{(0)}(X) = \frac{\mathbb{E}[X] - F_X^{(-2)}(\eta)}{1 - \eta}, \quad 0 < \eta < 1. \quad (32)$$

The following Proposition 3 provides basic properties for the minimization CVaR in (30), which will be used to characterize the mean-CVaRs in (38) and (39). Its proof is similar as Proposition 2, we omitted it.

**Proposition 3.** Consider a continuous random variable $X$ with support $[\underline{\ell}, \bar{\ell}]$. We have

(i) The minimization conditional value at risk $CVaR_\eta^{(0)}(\cdot)$ in (30) is increasing function of $\eta$ in interval $(0, 1]$ and

$$\frac{\partial CVaR_\eta^{(0)}(X)}{\partial \eta} = \frac{1}{(1 - \eta)^2} \int_{F_X^{-1}(\eta)}^{\bar{\ell}} F_X(x)dx > 0 \text{ for all } 0 < \eta < 1. \quad (33)$$

(ii) $\lim_{\eta \to 0} CVaR_\eta^{(0)}(X) = \mathbb{E}[X]$, and

$$\lim_{\eta \to 1} CVaR_\eta^{(0)}(X) = \lim_{\eta \to 1} F_X^{-1}(\eta) = \bar{\ell}. \tag{34}$$

(iii) $CVaR_\eta^{(0)}(X)$ satisfies

$$\mathbb{E}[X] \leq CVaR_\eta^{(0)}(X) \leq \bar{\ell} \text{ for all } 0 < \eta < 1. \tag{35}$$

Combining (31) and (25), it is easily to show that the minimization conditional value at risk $CVaR_\eta^{(0)}(\cdot)$ in (30) and the maximization conditional value at risk $CVaR_\eta^{(1/\eta)}(\cdot)$ in (25) satisfy the following relation:

$$\eta CVaR_\eta^{(1/\eta)}(X) + (1 - \eta)CVaR_\eta^{(0)}(X) = \mathbb{E}[X], \ 0 < \eta < 1. \tag{36}$$

Now, we consider the mixture conditional value at risk, denoted as $CVaR_\eta^{(\lambda/\eta)}(\cdot)$, it is a convex combination of conditional expected values of low in profit (25) and high profit (30), respectively, and is given by (see e.g., Jammernegg and Kischka 2007)

$$CVaR_\eta^{(\lambda/\eta)}(X) = \lambda CVaR_\eta^{(1/\eta)}(X) + (1 - \lambda)CVaR_\eta^{(0)}(X), \tag{37}$$

where $0 \leq \lambda \leq 1$ and $0 < \eta < 1$.

**Remark 8.** According to Jammernegg and Kischka (2007), the two risk parameters $\lambda$ and $\eta$ all reflect the degree of risk aversion, a larger value of $\lambda$ implies a higher degree of risk aversion and a lower value of $\eta$ implies a higher degree of risk aversion. Based on this property, Yu (2014a) introduced a risk preference coefficient $\tilde{\lambda} := \lambda/\eta$, $0 \leq \tilde{\lambda} \leq 1/\eta$, $\tilde{\lambda}$ reflects the degree of risk preference of decision maker, (i) when $\tilde{\lambda} = 1$, it represents risk neutral, corresponding reflects to the expected value $\mathbb{E}[X]$; (ii) when $1 < \tilde{\lambda} \leq 1/\eta$, it represents risk aversion, including the conditional expected values of low profit $CVaR_\eta^{(1/\eta)}(X)$; (iii) risk taking when $0 \leq \tilde{\lambda} < 1$, including the conditional expected values of high profit $CVaR_\eta^{(0)}(X)$. This property of the risk preference coefficient $\tilde{\lambda}$ is consistent with the submodular property of the risk preference coefficient $\tilde{\lambda}$ with respect to $\tilde{\lambda}$ in Theorem 3(iii).

The following Proposition 4 show that the mixture conditional value at risk in (37) can be rewritten as two new mean-risk models, called the first and second type mean conditional value at risks. Part (i) and (ii) of Proposition 4 can be proved by combining (37) and (36), Part (iii) of Proposition 4 can be proved by (39) and (24), we omitted it proof.

**Proposition 4.** Consider a continuous random variable $X$ with support $[\underline{\ell}, \bar{\ell}]$, let $F_X(\cdot)$ be its distribution function. Then for $0 \leq \tilde{\lambda} \leq 1/\eta$, the mixture conditional value at risk $CVaR_\eta^{(\lambda/\eta)}(X)$ corresponding to $X$ is given in (37) can be rewritten as the following equivalent forms:

(i)

$$IMCVaR_\eta^{(\tilde\lambda)}(X) = \tilde\lambda\Big(\mathbb{E}[X] - \big(1 - \frac{1}{\tilde\lambda}\big)CVaR_\eta^{(0)}(X)\Big), \qquad (38)$$

which is called the first type mean conditional value at risk.

(ii)

$$IIMCVaR_\eta^{(\tilde\lambda)}(X) = \frac{\frac{1}{\eta} - \tilde\lambda}{\frac{1}{\eta} - 1}\Big(\mathbb{E}[X] - \frac{\tilde\lambda - 1}{\frac{1}{\eta} - \tilde\lambda}\big(-CVaR_\eta^{(1/\eta)}(X)\big)\Big), \quad (39)$$

which is called the second type mean conditional value at risk.

**Remark 9.** When $\tilde\lambda = 1$, two type mean conditional value at risk in Proposition 4 become the expected value $\mathbb{E}[X]$, which corresponds to risk-neutral case; when $1 < \tilde\lambda \le 1/\eta$, three type mean conditional value at risk in Proposition 4 is a function of the conditional expected values of low profit or the second quantile function, which corresponds to risk-aversion case; when $0 \le \tilde\lambda < 1$, three type mean conditional value at risk in Proposition 4 is a function of the conditional expected values of high profit, which corresponds to risk-taking case.

**Remark 10.** The first type mean conditional value at risk in (38) can be rewritten into a function of the second quantile function as follows:

$$IMCVaR_\eta^{(\tilde\lambda)}(X) = \frac{1}{1 - \eta}\Big[(1 - \eta\tilde\lambda)\mathbb{E}[X] + (\tilde\lambda - 1)F_X^{(-2)}(\eta)\Big], \ 0 < \eta < 1, \ (40)$$

where $F_X^{(-2)}(\cdot)$ is given in (15).

The following Theorem 1 provides relations between the ascending stochastic dominance for risk averters and the mean-risk models in (39). Theorem 2 can be obtained by (40) and Lemma 1, we omitted its proof.

**Theorem 1 (Characterization of the Mean-CVaR by ASD).** For two continuous random variables $X$ and $Y$ with support $[\ell, \bar\ell]$. For $1 \le \tilde\lambda \le 1/\eta$, we have

(i) the mean-risk model $\big(\mu_X, -CVaR_\eta^{(1/\eta)}(X)\big)$ in (39) is consistent with ascending stochastic dominance for risk averters.

(ii) the mean-risk model $\big(\mu_X, -CVaR_\eta^{(1/\eta)}(X)\big)$ in (39) is $(\tilde\lambda - 1)/(1/\eta - 1)$-consistent with ascending stochastic dominance for risk averters.

**Remark 11.** Theorem 1(i) shows when $1 \le \tilde\lambda \le 1/\eta$, the mean-risk model in (39) taking $-CVaR_\eta^{(1/\eta)}(\cdot)$ as a risk measure is consistent with ASD, this result was analyzed by Ogryczak and Ruszczyński (2002), but the authors did not consider the $\xi$-consistent property in Theorem 1(ii), where $\xi = (\tilde\lambda - 1)/(1/\eta - 1)$.

In order to analyze whether the mean-CVaR in (38) is consistent with some stochastic ordering relation, we provide another stochastic dominance, that is, descending stochastic dominance (abbreviated as DSD) for risk lovers, it is defined as follows:

**Definition 3 (Ross and Schechner 1984; Song 1994).** For two continuous random variables $X$ and $Y$ with support $[\ell, \bar{\ell}]$. $Y$ is said to be smaller than $X$ in the sense of descending stochastic dominance for risk lovers, denoted by $Y \leq_{DSD} X$, if and only if $\mathbb{E}[u(Y)] \leq \mathbb{E}[u(X)]$ for all increasing convex functions $u(\cdot)$.

Stoyan (1983) and Ross and Schechner (1984) showed that the descending stochastic dominance for risk lovers is equivalent to a simple condition:

**Lemma 2 (Stoyan 1983; Ross and Schechner 1984).**

$$Y \leq_{DSD} X \iff \mathbb{E}[(Y - t)_+] \leq \mathbb{E}[(X - t)_+] \text{ for all } x \in [\ell, \bar{\ell}]. \qquad (41)$$

For two nonnegative random variables having the same mean, Ross (1983) showed that (9) is equivalent to a stronger condition:

**Lemma 3 (Ross 1983).** If $X$ and $Y$ are nonnegative random variables such that $\mathbb{E}[Y] = \mathbb{E}[X]$, then $Y \leq_{DSD} X$ if and only if $\mathbb{E}[u(Y)] \leq \mathbb{E}[u(X)]$ for all convex functions $u(\cdot)$, that is, $Y$ is less than $X$ in the sense of convex ordering, denoted by $Y \leq_{CX} X$.

**Remark 12.** The descending stochastic dominance for risk lovers is weaker than (implied by) the convex ordering. The following two properties may aid us in understanding why the descending stochastic dominance for risk lovers is a variability ordering.

**Lemma 4 (Brumelle and Vickson 1975).** For two continuous random variables $X$ and $Y$ with support $[\ell, \bar{\ell}]$. We have

(i) $Y \leq_{DSD} X$ and $\mathbb{E}[Y] = \mathbb{E}[X]$ implies $Var(Y) \leq Var(X)$;
(ii) (The coupling property) $Y \leq_{DSD} X$ if and only if there exists a random variable $Z$, with $E(Z|Y) \geq 0$ almost surely, such that $X =_d Y + Z$. That is, $X$ is noisier than $Y$.

Combining Lemma 2 and the definition of the minimization CVaR in (31), it is easily to prove the following imply relation about the increasing convex ordering:

**Proposition 5.** For two continuous random variables $X$ and $Y$ with support $[\ell, \bar{\ell}]$. We have

(i) For all $0 < \eta < 1$,

$$Y \leq_{DSD} X \implies CVaR_\eta^{(0)}(Y) \leq CVaR_\eta^{(0)}(X) \text{ and } \mathbb{E}[Y] \leq \mathbb{E}[X]. \quad (42)$$

(ii) If $\mathbb{E}[Y] = \mathbb{E}[X]$, then $Y \leq_{DSD} X$ if and only if $\mathbb{E}[u(Y)] \leq \mathbb{E}[u(X)]$ for all convex functions $u(\cdot)$, that is, $Y$ is larger than $X$ in the sense of convex ordering, denoted by $Y \leq_{CX} X$.
(iii) If $\mathbb{E}[Y] = \mathbb{E}[X]$, then

$$Y \leq_{DSD} X \implies CVaR_\eta^{(0)}(Y) \leq CVaR_\eta^{(0)}(X) \text{ and } Var(Y) \leq Var(X). \quad (43)$$

**Remark 13.** Li and Wong (1999) introduced two stochastic dominance concepts: ascending stochastic dominance (ASD) for risk averters and descending stochastic dominance (DSD) for risk lovers. The two stochastic orderings were called to be increasing concave ordering (equivalently, DSD) and increasing convex ordering defined by Stoyan (1983). Based on the relations between the two class of stochastic orderings and two class of conditional value-at-risks in (23) and Proposition 6(i), we know that the conditional expected values of low profit in (25) corresponds to risk-aversion case, and the conditional expected values of high profit in (30) corresponds to risk-taking case from the perspective of decision making.

In the following we will analyze whether the two mean conditional value at risk in (38) and (39) hold under ASD or increasing convex ordering.

The following Theorem 2 provides relations between the descending stochastic dominance for risk lovers and the mean-risk models in (38) and Theorem 1 can be proved by (38) and Proposition 6(i), we omitted its proof.

**Theorem 2 (Characterization of the Mean-CVaR by DSD).** For two continuous random variables $X$ and $Y$ with support $[\underline{\ell}, \overline{\ell}]$. For $0 \leq \tilde{\lambda} < 1$, we have

(i) the mean-risk model $\left(\mu_X, -CVaR_\eta^{(0)}(\cdot)\right)$ in (38) is consistent with the descending stochastic dominance for risk lovers.

(ii) the mean-risk model $\left(\mu_X, -CVaR_\eta^{(0)}(\cdot)\right)$ in (38) is $(1/\tilde{\lambda} - 1)$-consistent with the descending stochastic dominance for risk lovers.

In the following we analyze several properties of the two type of mean-CVaRs in (38) and (39), including the monotonicity with respect to the risk preference coefficient; the upper/lower bounds; the subadditivity and convexity; the loss-aversion. Based on the equivalence of the two type of mean-CVaRs in (38) and (39), we only analyze the first type mean-CVaR in (38).

The following Theorem 3 provides the monotonicity of the mean-CVaR in (38) with respect to the risk preference coefficient $\tilde{\lambda}$ and upper/lower bounds.

**Theorem 3 (Monotonicity and Upper/Lower Bounds of the Mean-CVaR).** Let $X$ be a continuous random variable with support $[\underline{\ell}, \overline{\ell}]$ and let $F_X(\cdot)$ be its distribution function. The first type mean conditional value at risk $IMCVaR_\eta^{(\tilde{\lambda})}(\cdot)$ is given by (38).

(i) $IMCVaR_\eta^{(\tilde{\lambda})}(\cdot)$ is decreasing function of $\tilde{\lambda}$ in interval $[0, 1/\eta]$ and $IMCVaR_\eta^{(\tilde{\lambda})}(X)$ satisfies

$$CVaR_\eta^{(1/\eta)}(X) \leq IMCVaR_\eta^{(\tilde{\lambda})}(X) \leq CVaR_\eta^{(0)}(X) \text{ for all } 0 \leq \tilde{\lambda} \leq 1/\eta. \tag{44}$$

(ii) If $0 \leq \tilde{\lambda} < 1$, then $IMCVaR_\eta^{(\tilde{\lambda})}(\cdot)$ is increasing function of $\eta$ in interval $(0, 1)$; if $1 \leq \tilde{\lambda} \leq 1/\eta$, then $IMCVaR_\eta^{(\tilde{\lambda})}(\cdot)$ is decreasing function of $\eta$ in interval $(0, 1)$; and

$$\frac{\partial IMCVaR_\eta^{(\tilde{\lambda})}(X)}{\partial \eta} = \frac{1 - \tilde{\lambda}}{(1 - \eta)^2} \int_{F_X^{-1}(\eta)}^{\bar{\ell}} (1 - F_X(x)) dx \text{ for } 0 < \eta < 1, \text{ (45)}$$

(iii) $IMCVaR_\eta^{(\tilde{\lambda})}(\cdot)$ is submodular function of $(\tilde{\lambda}, \eta)$ for all $0 \le \tilde{\lambda} \le 1/\eta$ and $0 < \eta < 1$.

(iv) If $1 \le \tilde{\lambda} \le 1/\eta$, then $IMCVaR_\eta^{(\tilde{\lambda})}(X) \le \mathbb{E}[X]$; if $0 \le \tilde{\lambda} < 1$, then $IMCVaR_\eta^{(\tilde{\lambda})}(X) > \mathbb{E}[X]$.

*Proof.* Part (i): From (40), we get

$$\frac{\partial IMCVaR_\eta^{(\tilde{\lambda})}(X)}{\partial \tilde{\lambda}} = \frac{\eta}{1 - \eta} \left( \frac{F_Y^{(-2)}(\eta)}{\eta} - \mathbb{E}[X] \right)$$

$$\le 0 \text{ for } 0 < \eta < 1, \tag{46}$$

where the inequality holds due to the fact in Proposition 2(iii), which leads to the monotonicity and lower and upper bounds of $IMCVaR_\eta^{(\tilde{\lambda})}(X)$.

Part (ii): Calculating first derivatives of $\eta$ on both sides of the Eq. (40), we get Eq. (46), which leads to the monotonicity result.

Part (iii): From (46), which leads to the desired result.

Part (iv) follows by Part (i). $\qquad\square$

In order to obtain the subadditivity and convexity of the mean conditional value at risk in (38), the following Lemma 5 provides the subadditivity and convexity of the minimization CVaR in (30). Its proof see Pflug (2000).

**Lemma 5 (Pflug 2000) (Subadditivity and Convexity of the Minimization CVaR).** Let $X$ be a continuous random variable with support $[\underline{\ell}, \bar{\ell}]$ and let $F_X(\cdot)$ be its distribution function. $CVaR_\eta^{(0)}(\cdot)$ is given in (30).

(i) $CVaR_\eta^{(0)}(X)$ is translation-equivariant, i.e. $CVaR_\eta^{(0)}(X + a) = CVaR_\eta^{(0)}(X) + a$.

(ii) $CVaR_\eta^{(0)}(X)$ is positively homogeneous, i.e. $CVaR_\eta^{(0)}(kX) = k \cdot CVaR_\eta^{(0)}(X)$, if $k > 0$.

(iii) $CVaR_\eta^{(0)}(X)$ is closed under linear transform, i.e. $CVaR_\eta^{(0)}(kX + a) = k \cdot CVaR_\eta^{(0)}(X) + a$, if $k > 0$.

(iv) $CVaR_\eta^{(0)}(X)$ satisfies the subadditivity, i.e., for two random variables $X_1$ and $X_2$, we have

$$CVaR_\eta^{(0)}(X_1 + X_2) \le CVaR_\eta^{(0)}(X_1) + CVaR_\eta^{(0)}(X_2). \tag{47}$$

(v) For two random variables $X_1$ and $X_2$, $CVaR_\eta^{(0)}(\cdot)$ satisfies the convexity, i.e., for any $0 \le \lambda \le 1$, we have

$$CVaR_\eta^{(0)}(\lambda X_1 + (1 - \lambda)X_2) \le \lambda CVaR_\eta^{(0)}(X_1) + (1 - \lambda)CVaR_\eta^{(0)}(X_2). \tag{48}$$

The following Theorem 4 provides the subadditivity and convexity and upper/lower bounds of the mean-CVaR in (38). Theorem 4 can be shown by using Lemma 5 and (38), we omitted its proof.

**Theorem 4 (Subadditivity and Convexity of the Mean-CVaR).** Let $X$ be a continuous random variable with support $[\underline{\ell}, \overline{\ell}]$.

(i) $IMCVaR_\eta^{(\tilde{\lambda})}(X)$ is translation-equivariant, i.e. $IMCVaR_\eta^{(\tilde{\lambda})}(X + a) = IMCVaR_\eta^{(\tilde{\lambda})}(X) + a$.

(ii) $IMCVaR_\eta^{(\tilde{\lambda})}(X)$ is positively homogeneous, i.e. $IMCVaR_\eta^{(\tilde{\lambda})}(kX) = k \cdot IMCVaR_\eta^{(\tilde{\lambda})}(X)$, if $k > 0$.

(iii) $IMCVaR_\eta^{(\tilde{\lambda})}(X)$ is closed under linear transform, i.e. $IMCVaR_\eta^{(\tilde{\lambda})}(kX + a) = k \cdot IMCVaR_\eta^{(\tilde{\lambda})}(X) + a$, if $k > 0$.

(iv) (a) If $0 \leq \tilde{\lambda} \leq 1$, then $IMCVaR_\eta^{(\tilde{\lambda})}(X)$ satisfies the subadditivity, i.e., for two random variables $X_1$ and $X_2$, we have

$$IMCVaR_\eta^{(\tilde{\lambda})}(X_1 + X_2) \leq IMCVaR_\eta^{(\tilde{\lambda})}(X_1) + IMCVaR_\eta^{(\tilde{\lambda})}(X_2). \quad (49)$$

(b) If $1 \leq \tilde{\lambda} \leq 1/\eta$, then $IMCVaR_\eta^{(\tilde{\lambda})}(W)$ satisfies the inverse subadditivity, i.e., for two random variables $X_1$ and $X_2$, we have

$$IMCVaR_\eta^{(\tilde{\lambda})}(X_1 + X_2) \geq IMCVaR_\eta^{(\tilde{\lambda})}(X_1) + IMCVaR_\eta^{(\tilde{\lambda})}(X_2). \quad (50)$$

(v) For two random variables $X_1$ and $X_2$, we have

(a) If $0 \leq \tilde{\lambda} \leq 1$, then $IMCVaR_\eta^{(\tilde{\lambda})}(\tau X_1 + (1-\tau)X_2)$ satisfies the convexity, i.e., for any $0 \leq \lambda \leq 1$, we have

$$IMCVaR_\eta^{(\tilde{\lambda})}(\lambda X_1 + (1 - \lambda)X_2) \leq \lambda IMCVaR_\eta^{(\tilde{\lambda})}(X_1)$$
$$+ (1 - \lambda)IMCVaR_\eta^{(\tilde{\lambda})}(X_2). \quad (51)$$

(b) If $1 \leq \tilde{\lambda} \leq 1/\eta$, then $IMCVaR_\eta^{(\tilde{\lambda})}(X)$ satisfies the concavity, i.e.

$$IMCVaR_\eta^{(\tilde{\lambda})}(\tau X_1 + (1 - \tau)X_2) \geq \tau IMCVaR_\eta^{(\tilde{\lambda})}(X_1)$$
$$+ (1 - \tau)IMCVaR_\eta^{(\tilde{\lambda})}(X_2). \quad (52)$$

The following Theorem 5 shows that the mean conditional value at risk $IMCVaR_\eta^{(\tilde{\lambda})}(\cdot)$ in (38) has the loss-aversion characteristic.

**Theorem 5 (Loss-Aversion Characteristic of the Mean-CVaR).** Let $X$ be a continuous random variable with support $[\underline{\ell}, \overline{\ell}]$. $IMCVaR_\eta^{(\tilde{\lambda})}(\cdot)$ is given in (38). Let

$$IMCVaR_\eta^{(\tilde{\lambda})}(X) = \mathbb{E}[u(X)], \quad (53)$$

then

$$u(X) = \begin{cases} (\tilde{\lambda} + \frac{1-\tilde{\lambda}}{1-\eta})(X - z) + z, & X - z \geq 0, \\ \tilde{\lambda}(X - z) + z, & X - z < 0, \end{cases} \tag{54}$$

where $z = F_W^{-1}(\eta)$, $0 < \eta < 1$. When $\tilde{\lambda} > 1$, the mean conditional value at risk $IMCVaR_\eta^{(\tilde{\lambda})}(\cdot)$ in (38) has the loss-aversion property, in this case, $\tilde{\lambda} > \tilde{\lambda} + (1 - \tilde{\lambda})/(1 - \eta)$.

*Proof.* Combining (38) and (30), which leads to the desired results.    □

### 3.3  Characterization of Mean-CVaR by Levy's Transformation

In this subsection, we analyze the stochastic monotonicity of the Mean-CVaR in (38) based on the transformation with two parameters, this generalized transformation introduced by Levy (1977), which has one main different with mean-preserving transformation (abbreviated as MPT) used by Gerchak and Mossman (1992): MPT characterize the effect of demand uncertainty reduction through one single scalar, but Levy's transformation (abbreviated as LT) in (55) can study the situations which mean and variability of market demand change together.

First, we provide the definition of Levy's transformation. Let $X$ be a continuous random variable with support $[\ell, \bar{\ell}]$ and mean $\mu_X$ and variance $Var(X)$, let $F_X(\cdot)$ be its distribution function. Define a family of random variables as (Levy 1977):

$$X_{\alpha,r} =_d \alpha X + (1 - \alpha)r, \ 0 < \alpha \leq 1, \tag{55}$$

where parameters $\alpha$ and $r$ satisfy: $0 < \alpha \leq 1$ and $r > 0$, "$=_d$" means "has the same distribution as" (Rothschild and Stiglitz 1970), $r$ represents certain demand level in a certain market environment (see Levy 1977 *page 232*, he called $r$ as riskless interest rate in finance). In fact, $\mathbb{E}[X_{\alpha,r}] - \mathbb{E}[X] = (1 - \alpha)(r - \mu_X) \geq 0$ if and only if $r \geq \mu_X$, that is, when $r \geq \mu_X$, $\mathbb{E}[X_{\alpha,r}] \geq \mathbb{E}[X]$; when $r = \mu_X$, $\mathbb{E}[X_{\alpha,r}] = \mathbb{E}[X]$ and when $r < \mu_X$, $\mathbb{E}[X_{\alpha,r}] < \mathbb{E}[X]$.

The parameter $\alpha$ in (55) reflects demand variability. In fact, $Var(X_{\alpha,r}) = \alpha^2 Var(X) \leq Var(X)$, the coefficient of variation $Cv(X_{\alpha,r}) = Var(X_{\alpha,r})/(\mathbb{E}[X_{\alpha,r}])^2 = Var(X)/[\mu_X + (\frac{1}{\alpha} - 1)r]^2$ is increasing in $\alpha$ in interval $(0, 1]$, and it satisfies: $0 < Cv(X_{\alpha,r}) \leq Cv(X)$ for all $0 < \alpha \leq 1$ and $r > 0$.

**Remark 14.** In particular, (i) if $r = \mu_X$, the LT in (55) is changed to MPT which was used by Gerchak and Mossman (1992) to solve newsvendor problem. Qin et al. (2011) analyze the case in which marketing effort effects demand uncertainty with MPT. They found that demand mean increase due to marketing effort may leads to the increase about the optimal order quantity. But the effect about demand variability reduction is not as clear. In reality, we use $\mu_X$ denoting the demand mean of a certain product throughout the country, (e.g. telephone, bicycle et al.); (ii) if $r > \mu_X$, then $r$ denotes the mean demand of this product in developed regions (e.g. first-tier cities); (iii) if $r < \mu_X$, $r$ denotes the mean demand of this product in less developed area (e.g. remote region).

Using LT, we can not only study the effect through the parameter $\alpha$, but also can analysis the situation that demand mean is changed. Therefore, LT can be seen as a new tool to study the influence of demand variability on retailer's inventory decisions and utility of inventory decision maker.

The common link of the ASD is that they are generated as successive applications of mean-preserving or mean-reducing single crossing operations between distribution functions, it provides a sufficient condition to assure the mean-preserving increase in risk (MPIR) and ASD (see Diamond and Stiglitz 1974).

In order to obtain the sufficient condition for ASD, we provide the notation of sign function. Let $\varphi(\cdot)$ be a real valued function defined in interval $[\ell, \bar{\ell}] \subset \mathbb{R}$ and let

$$S(\varphi) = S(\varphi(\cdot)) = \sup S[\varphi(x_1), \varphi(x_2), ..., \varphi(x_n)] \tag{56}$$

where the supremum is extended over all sets $x_1, x_2, ..., x_n$, $(x_i \in [\ell, \bar{\ell}])$, $n$ is arbitrary but finite, and $S(y_1, y_2, ..., y_n)$ is the number of sign changes of the indicated sequence, zero terms being discarded (Karlin 1968, pp. 20).

We say that the functions $f$ and $g$ crosse each other $k$ times if $S(f - g) = k$, $k = 0, 1, 2, ....$ They cross each other at most $k$ times if $S(f - g) \leq k$. If $f$ has an integral function $F$, then $S(F) \leq S(f) + 1$ (Karlin 1968, pp. 310–311).

Let $X$ and $Y$ be two random variables with continuous distribution function $F_X(\cdot)$ and $F_Y(\cdot)$ with support $[\ell, \bar{\ell}] \subset \mathbb{R}$, where $\ell$ or $\bar{\ell}$ can be $-\infty$ or $\infty$, respectively.

**Definition 4 (Single crossing see Chateauneuf, Cohen and Meilijson 2004).** $X$ is said to be single crosses $Y$ if there exists $x_0 \in (\ell, \bar{\ell})$, such that

$$F_X(x) - F_Y(x) \gtreqqless 0 \quad \text{resp. for } x \lesseqqgtr x_0, \tag{57}$$

$$\text{which denoted by } S(F_X(\cdot) - F_Y(\cdot)) = 1 \text{ with sign sequence } +, -. \tag{58}$$

**Remark 15.** Definition 5 provides a class of stochastic order, called cut-criterion ordering (Karlin and Novikoff 1963; Whitt 1985). If $S(F_X(\cdot) - F_Y(\cdot)) = 1$ with sign sequence $+, -$, then $Y$ is said to be smaller than $X$ according to cut-criterion ordering, denoted by $Y \leq_{cut} X$.

Whitt (1985) showed the relations between the ASD, DSD, convex ordering and the cut-criterion ordering as follows:

**Lemma 6 (Whitt 1985).** Let $X$ and $Y$ be two continuous random variables with support $[\ell, \bar{\ell}]$ and let $\mathbb{E}[X]$ and $\mathbb{E}[Y]$ be means of $X$ and $Y$, respectively.

(i) If $\mathbb{E}[Y] \geq \mathbb{E}[X]$, then $S(F_X(\cdot) - F_Y(\cdot)) = 1$ with sign sequence $+, - \Rightarrow$ $Y \leq_{ASD} X$.

(ii) If $\mathbb{E}[Y] \leq \mathbb{E}[X]$, then $S(F_X(\cdot) - F_Y(\cdot)) = 1$ with sign sequence $+, - \Rightarrow$ $Y \leq_{DSD} X$.

(iii) If $\mathbb{E}[Y] = \mathbb{E}[X]$, then $S(F_X(\cdot) - F_Y(\cdot)) = 1$ with sign sequence $+, - \Rightarrow$ $Y \leq_{CX} X$, that is $\mathbb{E}[u(Y)] \leq \mathbb{E}[u(X)]$ for all convex function $u(\cdot)$.

*Proof.* See Whitt (1985). □

One of the important characteristic associated with LT is that it implies three stochastic orderings including convex ordering, increasing concave ordering, increasing convex ordering, and can analyze complex optimization problem based on the two parameters of LT.

The following Proposition 6 characterizes the variability of the LT $X_{\alpha,r}$ in (55) by three stochastic orderings in Lemma 5. In particular, taking $\alpha_1 = \alpha$ and $\alpha_2 = 1$ in Proposition 6, we get Corollary 1, which can used to analyze the effect of demand variability reduction parameter $\alpha$ about the transformation LT. The results in Corollary 1 can be obtained by in Proposition 6, we omitted its proof.

**Proposition 6.** Consider a family of random variables $X_{\alpha_i,r} = \alpha_i X + (1-\alpha_i)r$, $i = 1,2$, where $X$ is a continuous random variable with support $[\ell, \bar{\ell}]$. Let $F_X(\cdot)$ be distribution function of $X$, $F_{X_{\alpha_i,r}}(\cdot)$ be distribution function of $X_{\alpha_i,r}$, $i = 1,2$. Denoting $Cv(X)$ the coefficient of variation for the random variable $X$, that is, $Cv(X) = Var(X)/(\mathbb{E}[X])^2$. Then

(i) $S(F_{X_{\alpha_2,r}}(\cdot) - F_{X_{\alpha_1,r}}(\cdot)) = 1$ with sign sequence $+,-$ (i.e., $X_{\alpha_1,r} \leq_{cut} X_{\alpha_2,r}$) for all $0 < \alpha_1 < \alpha_2$ and all $r > 0$.

(ii) $\mathbb{E}[X_{\alpha_1,r}] \geq \mathbb{E}[X_{\alpha_2,r}]$ if and only if $r \geq \mu_X$, $Cv(X_{\alpha_1,r}) \leq Cv(X_{\alpha_2,r})$ for all $0 < \alpha_1 < \alpha_2$ and $r > 0$.

(iii) If $r \geq \mu_X$, then $X_{\alpha_1,r} \geq_{ASD} X_{\alpha_2,r}$ and $Cv(X_{\alpha_1,r}) \leq Cv(X_{\alpha_2,r})$.

(iv) If $r \leq \mu_X$, then $X_{\alpha_1,r} \leq_{DSD} X_{\alpha_2,r}$ and $Cv(X_{\alpha_1,r}) \leq Cv(X_{\alpha_2,r})$.

(v) If $r = \mu_X$, then $X_{\alpha_1,r} \leq_{CX} X_{\alpha_2,r}$ and $Var(X_{\alpha_1,r}) \leq Var(X_{\alpha_2,r})$.

*Proof.* Part (i): Since $F_{X_{\alpha_i,r}}(x) = \mathbb{P}\{X_{\alpha_i,r} \leq x\} = F_X(\frac{x-r}{\alpha_i} + r)$, where $i = 1,2$. We have $F_{X_{\alpha_2,r}}(x) - F_{X_{\alpha_1,r}}(x) = F_X(\frac{x-r}{\alpha_2} + r) - F_X(\frac{x-r}{\alpha_1} + r) > 0$ if and only if $\frac{x-r}{\alpha_2} - \frac{x-r}{\alpha_1} = \frac{\alpha_2-\alpha_1}{\alpha_1\alpha_2}(r - x) > 0$, that is, $0 < t < t_0 = r$. By Definition 5, which leads to the desired results.

Part (ii): By calculation we have $\mathbb{E}[X_{\alpha_1,r}] - \mathbb{E}[X_{\alpha_2,r}] = (\alpha_2 - \alpha_1)(r - \mu_X) \geq 0$ if and only if $r \geq \mu_X$, and $Var(X_{\alpha_i,r}) = (\alpha_i)^2 Var(X)$, $Cv(X_{\alpha_i,r}) = Var(X_{\alpha_i,r})/(\mathbb{E}[X_{\alpha_i,r}])^2 = Var(X)/(\mu_X + (1/\alpha_i - 1)r)^2$ is increasing function of $\alpha_i$, which get the inequality corresponding to the coefficient of variation for $X$.

Part (iii)–(v) can be obtained by combining Part (i) and (ii) and Lemma 5. □

**Corollary 1.** Consider LT $X_{\alpha,r}$ in (55), where $X$ be a continuous random variable with support $[\ell, \bar{\ell}]$. Denoting $Cv(X)$ the coefficient of variation for the random variable $X$. For all $0 < \alpha \leq 1$,

(i) If $r \geq \mu_X$, then $X_{\alpha,r} \geq_{ASD} X$ and $Cv(X_{\alpha,r}) < Cv(X)$.

(ii) If $r < \mu_X$, then $X_{\alpha,r} \leq_{DSD} X$ and $Cv(X_{\alpha,r}) < Cv(X)$.

(iii) If $r = \mu_X$, then $X_{\alpha,r} \leq_{CX} X$ and $Var(X_{\alpha,r}) < Var(X)$.

**Remark 16.** Li and Wong (1999) obtained similar results as in Corollary 1, but the authors did not consider the variability of random variable, e.g., variance or coefficient of variation, at same time, the authors did not analyze the stochastic monotonicity of the LT in (55) in the sense of ASD and DSD as in Proposition 6.

The following Theorem 6 provides a sufficient condition to compare the second quantile function $F_X^{(-2)}(\cdot)$ in (15) corresponding to two family of random variables associated with LT in (55).

**Theorem 6 (Characterization of the first and second quantile functions by LT).** Consider a family of random variables $X_{\alpha_i,r} = \alpha_i X + (1 - \alpha_i)r$, $i = 1, 2$, where $X$ is a continuous random variable with support $[\ell, \bar{\ell}]$. The second quantile function $F_X^{(-2)}(\cdot)$ corresponding to $X$ is given in (15). Let $F_{X_{\alpha_i,r}}^{(-2)}(\cdot)$ denote the second quantile function corresponding to $X_{\alpha_i,r}$, $i = 1, 2$.

(i) $F_{X_{\alpha_i,r}}^{-1}(\eta) = \alpha_i F_X^{-1}(\eta) + (1 - \alpha_i)r$, and it is increasing function of $\alpha_i$ in interval $(0, 1)$ if and only if $r \leq F_X^{-1)}(\eta)$, we have

$$F_{X_{\alpha_1,r}}^{-1}(\eta) - F_{X_{\alpha_2,r}}^{-1}(\eta) \lesseqqgtr 0 \text{ resp. for } r \lesseqqgtr F_X^{-1)}(\eta). \tag{59}$$

(ii) (a) $F_{X_{\alpha_1,r}}^{(-2)}(\eta) = \alpha_i F_X^{(-2)}(\eta) + (1 - \alpha_i)r$.

(b) If $r \geq \mu_X$, then $F_{X_{\alpha_1,r}}^{(-2)}(\eta)$ is decreasing function of $\alpha_i$ in interval $(0, 1)$, we have $F_{X_{\alpha_1,r}}^{(-2)}(\eta) \geq F_{X_{\alpha_2,r}}^{(-2)}(\eta)$ for all $0 < \eta < 1$.

(c) If $\ell \leq r < \mu_X$, then there exists $\eta_0 \in (0, 1)$, such that

$$F_{X_{\alpha_1,r}}^{(-2)}(\eta) - F_{X_{\alpha_2,r}}^{(-2)}(\eta) \gtreqqless 0 \text{ resp. for } \eta \lesseqqgtr \eta_0, \tag{60}$$

(d) If $0 < r < \ell$, then $F_{X_{\alpha_1,r}}^{(-2)}(\eta)$ is increasing function of $\alpha_i$ in interval $(0, 1)$, we have $F_{X_{\alpha_1,r}}^{(-2)}(\eta) \leq F_{X_{\alpha_2,r}}^{(-2)}(\eta)$ for all $0 < \eta < 1$.

(iii) If $0 < r \leq \mu_X$, then $\mathbb{E}[X_{\alpha_i,r}] - F_{X_{\alpha_i,r}}^{(-2)}(\eta)$ is increasing function of $\alpha_i$ in interval $(0, 1]$, we have

$$\mathbb{E}[X_{\alpha_1,r}] - F_{X_{\alpha_1,r}}^{(-2)}(\eta) - \left(\mathbb{E}[X_{\alpha_2,r}] - F_{X_{\alpha_2,r}}^{(-2)}(\eta)\right) \leq 0, \text{ for all } 0 < \eta < 1. \tag{61}$$

*Proof.* Part (i): The expression of $F_{X_{\alpha_i,r}}^{-1}(\eta)$ can be obtained by replacing $X$ with $X_{\alpha_i,r}$. The monotonicity of $F_{X_{\alpha_i,r}}^{-1}(\eta)$ based on the fact $\partial F_{X_{\alpha_i,r}}^{-1}(\eta)/\partial \alpha_i = F_X^{-1}(\eta) - r \geq 0$ if and only if $r \leq F_X^{-1)}(\eta)$.

Part (ii): The expression of $F_{X_{\alpha_i,r}}^{(-2)}(\eta)$ can be proved by similar method as in part (i). To show the monotonicity of $F_{X_{\alpha_i,r}}^{(-2)}(\eta)$, we note that $\partial F_{X_{\alpha_i,r}}^{(-2)}(\eta)/\partial \alpha_i = \eta(F_X^{(-2)}(\eta)/\eta - r) = \eta(CVaR_\eta^{(1/\eta)}(X) - r)$, based on the upper and lower bounds of $CVaR_\eta^{(1/\eta)}(\cdot)$ in Proposition 2(iii), which leads to the desired results.

Part (iii): Since $\mathbb{E}[X_{\alpha_i,r}] - F_{X_{\alpha_i,r}}^{(-2)}(\eta) = (1 - \eta)(\alpha_i CVaR_\eta^{(0)}(X) + (1 - \alpha_i)r)$, so $\partial \mathbb{E}[X_{\alpha_i,r}] - F_{X_{\alpha_i,r}}^{(-2)}(\eta)/\partial \alpha_i = (1 - \eta)(CVaR_\eta^{(0)}(X) - r) \leq 0$ if $0 < r \leq \mu_X$ by Proposition 3(iii). □

The following Corollarys 2 and 3 provide a sufficient condition to compare two maximization CVaR in (24) (minimization CVaR in (24) corresponding to two family of random variables associated with LT in (55). Corollarys 2 and 3 can be proved from Theorem 6, we omitted their proof.

**Corollary 2.** Consider a family of random variables $X_{\alpha_i,r} = \alpha_i X + (1 - \alpha_i)r$, $i = 1, 2$, where $X$ is a continuous random variable with support $[\ell, \bar{\ell}]$. The maximization conditional value at risk $CVaR_\eta^{(1/\eta)}(\cdot)$ corresponding to $X$ is given in (24).

(i) If $r \geq \mu_X$, then $CVaR_\eta^{(1/\eta)}(X_{\alpha_1,r}) \geq CVaR_\eta^{(1/\eta)}(X_{\alpha_2,r})$ for all $0 < \eta < 1$.
(ii) If $\ell \leq r < \mu_X$, then there exists $\eta_0 \in (0, 1)$, such that

$$CVaR_\eta^{(1/\eta)}(X_{\alpha_1,r}) - CVaR_\eta^{(1/\eta)}(X_{\alpha_2,r}) \gtreqless 0 \text{ resp. for } \eta \lesseqgtr \eta_0, \qquad (62)$$

**Corollary 3.** Consider a family of random variables $X_{\alpha_i,r} = \alpha_i X + (1 - \alpha_i)r$, $i = 1, 2$, where $X$ is a continuous random variable with support $[\ell, \bar{\ell}]$. The maximization conditional value at risk $CVaR_\eta^{(0)}(\cdot)$ corresponding to $X$ is given in (30). If $0 < r \leq \mu_X$, then

$$CVaR_\eta^{(0)}(X_{\alpha_1,r}) - CVaR_\eta^{(0)}(X_{\alpha_2,r}) \leq 0 \text{ for } 0 < \eta < 1. \qquad (63)$$

The following Corollary 4 provides a sufficient condition to compare the first type mean conditional value at risk in (38) corresponding to two family of random variables associated with **LT** in (55). Corollary 4 can be proved by combining (38) and Theorem 6, we omitted its proof.

**Corollary 4.** Consider a family of random variables $X_{\alpha_i,r} = \alpha_i X + (1 - \alpha_i)r$, $i = 1, 2$, where $X$ is a continuous random variable with support $[\ell, \bar{\ell}]$. The maximization conditional value at risk $CVaR_\eta^{(0)}(\cdot)$ corresponding to $X$ is given in (30).

(i) If $1 \leq \tilde{\lambda} \leq 1/\eta$ and $r \geq \mu_X$ and

$$IMCVaR_\eta^{(\tilde{\lambda})}(X_{\alpha_1,r}) \geq IMCVaR_\eta^{(\tilde{\lambda})}(X_{\alpha_2,r}) \text{ for all } 0 < \eta < 1, \qquad (64)$$

(ii) If $0 \leq \tilde{\lambda} \leq 1$ and $\ell \leq r \leq \mu_X$, then

$$IMCVaR_\eta^{(\tilde{\lambda})}(X_{\alpha_1,r}) \leq IMCVaR_\eta^{(\tilde{\lambda})}(X_{\alpha_2,r}) \text{ for all } 0 < \eta < 1, \qquad (65)$$

(iii) If $0 < r < \ell$ and $0 \leq \tilde{\lambda} \leq 1/\eta$, then

$$IMCVaR_\eta^{(\tilde{\lambda})}(X_{\alpha_1,r}) \leq IMCVaR_\eta^{(\tilde{\lambda})}(X_{\alpha_2,r}) \text{ for all } 0 < \eta < 1, \qquad (66)$$

**Remark 17.** Based on the relation between the LT and the stochastic dominance in Corollary 1, the results in Corollary 4(i) and (ii) based on LT are consistent with that in Theorem 2 based on stochastic dominance, and we find that the approach in Corollary 4(i) and (ii) are more simple than in Theorem 2, since the former only depends on two parameters and the conditions are more easily checked.

# 4   Applications to Stochastic Inventory Systems

In this section we model a risk-neutral newsvendor problem with stockout cost and a newsvendor problem with mean-CVaR constrain and stockout cost, and analyze their performance measures and stochastic monotonicity and upper/lower bounds.

## 4.1   The Risk-Neutral Newsvendor Problem

In this subsection, we model a risk-neutral newsvendor problem considering stockout cost, analyze the properties of optimal solution and profit, and the effect of demand uncertainty on order quantity and profit.

We consider a risk-neutral newsvendor selling short-life-cycle products with uncertain demand. At the beginning of the selling season, the newsvendor initially orders $y$ products at a unit cost $c$ from a supplier and sells at a retail price $p > c$ during the selling season. Demand $X_{\alpha,r}$ is stochastic and depends on two parameters $r$ and $\alpha$, which has the transformation form $X_{\alpha,r} =_d \alpha X + (1 - \alpha)r$, $0 < \alpha \le 1$ in (55) such that $\alpha\ell + (1 - \alpha)r \ge 0$, where $X$ is a random variable with support being continuous interval $I = [\underline{\ell}, \overline{\ell}]$. Here $r$ represents certain demand level in a certain market environment (see Levy 1977 page 232, he called $r$ as riskless interest rate in finance).

If realized demand $x$ is higher than $y$, then unit shortage cost penalty $s > 0$ is incurred on $x - y$ units. If realized demand $x$ is lower than $y$, then the newsvendor salvages $y - x$ unsold products at a unit value $v < c$. As with most of the newsvendor models, we assume $F_X(\cdot)$ is continuous, differentiable, invertible, and strictly increasing over $I$. we assume that $p > c > v$.

The parameter $\alpha$ in (55) reflects demand variability. According to Proposition 7(iii)–(v), demand $X_{\alpha,r}$ is more variable (in the sense of ASD, increasing convex ordering or convex ordering) when $\alpha$ increases for all $r > 0$. For example, $X_{\alpha,r}$ is a uniform random variable with distribution function $F_{X_{\alpha,r}}(x) = (x - (1 - \alpha)r)/\alpha$, its support is $[(1 - \alpha)r, (1 - \alpha)r + \alpha]$, mean of demand $X_{\alpha,r}$ is $(1 - \alpha)r + \alpha/2$, variance of demand $X_{\alpha,r}$ is $\alpha^2/12$. This uniform demand $X_{\alpha,r}$ can be represented as $X_{\alpha,r} =_d \alpha U + (1 - \alpha)r$, where $U$ is a uniform random variable with support $[0, 1]$. Another example is, $X_{\alpha,r}$ is a exponential random variable with with distribution function $F_{X_{\alpha,r}}(x) = 1 - \exp\{-(x - (1 - \alpha)r)/\alpha\}$, its support $[(1 - \alpha)r, \infty)$, mean of demand $X_{\alpha,r}$ is $(1 - \alpha)r + \alpha$, variance of demand $X_{\alpha,r}$ is $\alpha^2$. This exponential demand $X_{\alpha,r}$ can be represented as $X_{\alpha,r} =_d \alpha U + (1 - \alpha)r$, where $U$ is a uniform random variable with support $[0, 1]$.

Let $y$ denote the order quantity and $\Pi(y, X_{\alpha,r})$ the profit. $\Pi(y, X_{\alpha,r})$ depends on $y$ and the stochastic demand $X_{\alpha,r}$ and is given by:

$$\Pi(y, X_{\alpha,r}) = p \ \min(y, X_{\alpha,r}) + v(y - X_{\alpha,r})_+ - s(X_{\alpha,r} - y)_+ - cy$$
$$= (p + s - c)y - (p + s - v)(y - X_{\alpha,r})_+ - sX_{\alpha,r}, \qquad (67)$$

where $(x)_+ = \max\{x, 0\}$, the third term in the right-hand side of the first equation represents an artificial penalty for opportunity cost, and $s$ is often set to be 0. In particular, taking $\alpha = 1$ in (67), it is identical to the expression in Khouja (1999).

The objective of risk neutral newsvendor is to decide the order quantity $y$ and maximize his expected profit at the end of selling period, which is given by

$$\max_{y \geq 0} \pi_{X_{\alpha,r}}(y) = \mathbb{E}[\Pi(y, X_{\alpha,r})], \tag{68}$$

where $\Pi(y, X_{\alpha,r})$ is given in (67).

Let $y^c_{X_{\alpha,r}} = \arg\max_{y \geq 0} \pi_{X_{\alpha,r}}(y)$ denote the optimal order quantity of the maximization newsvendor problem (68). The following Proposition 7 provides the unique optimal order quantity and the optimal expected profit for problem (68).

**Proposition 7.** Let $X$ be a continuous random variable with support $[\ell, \overline{\ell}]$, its cumulative distribution function is denoted by $F_X(\cdot)$, assume that its inverse of the distribution function $F_X^{-1}(\cdot)$ exists. The demand $X_{\alpha,r}$ is given in (55). For the maximization newsvendor problem (68).

(i) The optimal order quantity $y^c_{X_{\alpha,r}}$ is given by

$$y^c_{X_{\alpha,r}} = \alpha F_X^{-1}(\rho) + (1 - \alpha)r, \tag{69}$$

where

$$\rho = (p + s - c)/(p + s - v), \tag{70}$$

is called the critical ratio in newsvendor problem with stockout cost.

(ii) The optimal expected profit $\pi_{X_{\alpha,r}}(y^c_{X_{\alpha,r}})$ is given by

$$\pi_{X_{\alpha,r}}(y^c_{X_{\alpha,r}}) = (p + s - v)[\alpha F_X^{(-2)}(\rho) - \frac{\rho - \rho^0}{1 - \rho^0}\alpha\mu_X + \frac{\rho^0(1 - \rho)}{1 - \rho^0}(1 - \alpha)r], \tag{71}$$

where $F_X^{(-2)}(\cdot)$ is given in (15), $\rho^0$ is a special case in (70) when $s = 0$, that is,

$$\rho^0 = (p - c)/(p - v), \tag{72}$$

which is called the critical ratio in newsvendor problem without stockout cost.

(iii) In particular, taking $\alpha = 1$, we have

$$\pi_{X_{1,r}}(y^c_{X_{1,r}}) = (p + s - v)\left[F_X^{(-2)}(\rho) - \frac{\rho - \rho^0}{1 - \rho^0}\mu_X\right] \tag{73}$$

is the optimal profit when demand is $X_{1,r} = X$, which the demand variability is up to the largest. On the other hand, taking $\alpha = 0$, we have

$$\pi_{X_{0,r}}(y^c_{X_{0,r}}) = (p - c)r \tag{74}$$

is the optimal profit when demand is deterministic $X_{0,r} = r$, in this case, the order quantity is $r$.

*Proof.* Part (i): $\pi_{X_{\alpha,r}}(y)$ in (68) can be written as

$$\pi_{X_{\alpha,r}}(y) = (p+s-c)y - \alpha(p+s-v)\int_{\underline{\ell}}^{\frac{y-r}{\alpha}+r} F_X(x)dx - s[\alpha\mu_X + (1-\alpha)r]. \quad (75)$$

Finding the first derivative of the equation expected profit $\pi_{X_{\alpha,r}}(y)$ in (68) on both sides with respect to $y$, we have

$$\frac{\partial \pi_{X_{\alpha,r}}(y)}{\partial y} = p+s-c - (p+s-v)F_X\left(\frac{y-r}{\alpha}+r\right). \quad (76)$$

From Eq. (99), we see that $\pi_{X_{\alpha,r}}(y)$ in (68) is strictly concave function of $y$ and the optimal solution of Problem (68) can be obtained by solving $\partial \pi_{X_{\alpha,r}}(y)/\partial y = 0$, which leads to the desired results

Part (ii): Combining (69) and (68), The optimal expected profit in (71).

Part (iii) can be obtained by taking $\alpha = 1$ and $\alpha = 0$ in (71).     □

Next, we characterize the monotonicity of optimal profit about newsvendor problem and stochastic comparison of newsvendor problem based on the LT. The following Theorem 7 provides the effect of demand variability about optimal order quantity in risk-neutral newsvendor problem.

**Theorem 7.** Let $X$ be a continuous random variable with support $[\underline{\ell}, \overline{\ell}]$, its cumulative distribution function is denoted by $F_X(\cdot)$, assume that its inverse of the distribution function $F_X^{-1}(\cdot)$ exists. The demand $X_{\alpha,r}$ is given in (55). For the maximization newsvendor problem (68).

(i) The optimal order quantity $y^c_{X_{\alpha,r}}$ is increasing function of the stockout cost $s$, and $y^c_{X_{\alpha,r}} \geq y^c_{X_{\alpha,r}}|_{s=0}$.

(ii) The optimal order quantity $y^c_{X_{\alpha,r}}$ is increasing function of $\alpha$ if and only if $\rho \geq \rho_0 = F_X(r)$.

(iii) For all $r > 0$, the optimal expected profit $\pi_{X_{\alpha,r}}(y^c_{X_{\alpha,r}})$ is decreasing function of the unit stockout cost $s$ in interval $[0,\infty)$ for all $\rho \in (0,1)$.

(iv) (a) When $r \geq \mu_X$, the optimal expected profit $\pi_{X_{\alpha,r}}(y^c_{X_{\alpha,r}})$ is decreasing function of $\alpha$ for all $\rho \in (0,1)$.

(b) When $0 < r < \mu_X$, there exist $\underline{\rho} \in (0,1)$ (equivalently, $\underline{s} \in (0,\infty)$) such that the optimal expected profit $\pi_{X_{\alpha,r}}(y^c_{X_{\alpha,r}})$ is increasing function of $\alpha$ for $\rho \in (0,\underline{\rho})$ (equivalently, $s \in (0,\underline{s})$); is decreasing function of $\alpha$ for $\rho \in (\underline{\rho},1)$ (equivalently, $s \in (\underline{s},\infty)$), where $\underline{\rho}$ satisfies:

$$F_X^{(-2)}(\underline{\rho}) - \frac{\underline{\rho}-\rho^0}{1-\rho^0}\mu_X - \frac{\rho^0(1-\underline{\rho})}{1-\rho^0}r = 0, \quad (77)$$

equivalently, $\underline{s}$ satisfies:

$$F_X^{(-2)}(\underline{\rho}) - \frac{\underline{\rho}-\rho^0}{1-\rho^0}\mu_X - \frac{\rho^0(1-\underline{\rho})}{1-\rho^0}r = 0, \quad (78)$$

where $\underline{\rho} = (p+\underline{s}-c)/(p+\underline{s}-v)$.

*Proof.* Part (i): From (70) in Proposition 7, it is easily to see that $\rho$ is increasing function of the stockout cost $s$, and the optimal order quantity $y^c_{X_{\alpha,r}}$ is a increasing function of $\rho$, which leads to the desired result.

Part (ii): Finding the first derivative of the optimal order quantity $y^c_{X_{\alpha,r}}$ on both sides with respect to $\alpha$, we have

$$\frac{\partial y^c_{X_{\alpha,r}}(y)}{\partial \alpha} = F_X^{-1}(\rho) - r, \tag{79}$$

which leads to the desired result.

Part (iii): Finding the first derivative of the optimal expected profit $\pi_{X_{\alpha,r}}(y^c_{X_{\alpha,r}})$ in (71) on both sides with respect to $s$, we have

$$\frac{\partial \pi_{X_{\alpha,r}}(y^c_{X_{\alpha,r}})}{\partial s} = -\alpha \int_{F_X^{-1}(\rho)}^{\bar{\ell}} (1 - F_X(x)) dx$$
$$< 0, \text{ for all } r > 0, \ 0 < \alpha \leq 1, \ \rho \in (0,1), \tag{80}$$

which leads to the desired result.

Part (iv): Finding the first derivative of the optimal expected profit $\pi_{X_{\alpha,r}}(y^c_{X_{\alpha,r}})$ in (71) on both sides with respect to $\alpha$, we have

$$\frac{\partial \pi_{X_{\alpha,r}}(y^c_{X_{\alpha,r}})}{\partial \alpha} = (p+s-v)\left(F_X^{(-2)}(\rho) - \frac{\rho - \rho^0}{1 - \rho^0}\mu_X - \frac{\rho^0(1-\rho)}{1-\rho^0}r\right)$$
$$= (p+s-v)\psi(\rho,r), \tag{81}$$

where

$$\psi(\rho,r) = F_X^{(-2)}(\rho) - \frac{\rho - \rho^0}{1 - \rho^0}\mu_X - \frac{\rho^0(1-\rho)}{1-\rho^0}r. \tag{82}$$

By calculation, we have

$$\psi(0,r) = \frac{\rho^0}{1-\rho^0}(\mu_X - r), \ \psi(1,r) = 0, \tag{83}$$

$$\frac{\partial \psi(\rho,r)}{\partial \rho} = F_X^{-1}(\rho) - \frac{\mu_X}{1-\rho^0} + \frac{\rho^0}{1-\rho^0}r \text{ is increasing function of } \rho, \text{ for } \rho \in (0,1), \tag{84}$$

$$\lim_{\rho \to 1} \frac{\partial \psi(\rho,r)}{\partial \rho} = \bar{\ell} - \frac{\mu_X}{1-\rho^0} + \frac{\rho^0}{1-\rho^0}r > 0 \text{ for enough larger } \bar{\ell}. \tag{85}$$

Based on those properties of function $\psi(\rho,r)$ in (83)–(85) which leads to the desired results. □

The following Corollary 5 provides the upper and lower bounds for the optimal order quantity $y^c_{X_{\alpha,r}}$ in (69) and the optimal expected profit $\pi_{X_{\alpha,r}}(y^c_{X_{\alpha,r}})$ in (71), those results can be obtained from the monotonicities of corresponding functions in Theorem 7.

**Corollary 5.** Let $X$ be a random variable satisfying conditions in Theorem 7. Then

(i) For all $r > 0$ and $0 < \alpha \le 1$, the optimal order quantity $y^c_{X_{\alpha,r}}$ satisfies

$$y^c_{X_{\alpha,r}} \lessgtr y^c_{X_{1,r}} \text{ resp. for } \rho \gtrless \rho_0. \tag{86}$$

(ii) For all $r > 0$, the optimal expected profit $\pi_{X_{\alpha,r}}(y^c_{X_{\alpha,r}})$ satisfies

$$\pi_{X_{\alpha,r}}(y^c_{X_{\alpha,r}}) \le \pi_{X_{\alpha,r}}(y^c_{X_{\alpha,r}})|_{s=0}, \text{ for all } 0 < \alpha \le 1 \text{ and } 0 < \rho < 1, \tag{87}$$

where $\pi_{X_{\alpha,r}}(y^c_{X_{\alpha,r}})|_{s=0}$ is the optimal expected profit corresponding to the newsvendor problem without stockout cost (i.e., $s = 0$).

(iii) (a) When $r \ge \mu_X$, the optimal expected profit $\pi_{X_{\alpha,r}}(y^c_{X_{\alpha,r}})$ satisfies

$$\pi_{X_{1,r}}(y^c_{X_{1,r}}) \le \pi_{X_{\alpha,r}}(y^c_{X_{\alpha,r}}) \le \pi_{X_{0,r}}(y^c_{X_{0,r}})$$
$$= (p-c)r, \text{ for } 0 < \alpha \le 1 \text{ and } 0 < \rho < 1, \tag{88}$$

that is,

$$\frac{\pi_{X_{1,r}}(y^c_{X_{1,r}})}{p-c} \le \frac{\pi_{X_{\alpha,r}}(y^c_{X_{\alpha,r}})}{p-c} \le r, \text{ for } 0 < \alpha \le 1 \text{ and } 0 < \rho < 1. \tag{89}$$

(b) When $0 < r < \mu_X$, the optimal expected profit $\pi_{X_{\alpha,r}}(y^c_{X_{\alpha,r}})$ satisfies

$$\pi_{X_{1,r}}(y^c_{X_{1,r}}) \lessgtr \pi_{X_{\alpha,r}}(y^c_{X_{\alpha,r}}) \lessgtr \pi_{X_{0,r}}(y^c_{X_{0,r}})$$
$$= (p-c)r \text{ resp. for } \rho \gtrless \underline{\rho}, \tag{90}$$

that is,

$$\frac{\pi_{X_{1,r}}(y^c_{X_{1,r}})}{p-c} \lessgtr \frac{\pi_{X_{\alpha,r}}(y^c_{X_{\alpha,r}})}{p-c} \lessgtr r \text{ resp. for } \rho \gtrless \underline{\rho}(equivalently, s \gtrless \underline{s}), \tag{91}$$

for $0 < \alpha \le 1$, where $\underline{\rho}$ is determined by Eq. (77) (equivalently, $\underline{s}$ is determined by Eq. (78)).

**Remark 18.** If $r = \mu_X$, the results in Corollary 4(i) are similar to the results in Gerchak and Mossman (1992). Gerchak and Mossman get the results by studying the minimization newsvendor problem with mean-preserving transformation. Corresponding to the results in Yu (2014a), minimization newsvendor has the consistency with maximization newsvendor, which leads to the desired results. We get the general results under different parameter $r$ about maximization newsvendor problem.

**Remark 19.** For the results in Corollary 4(i), we have for some insights for entrepreneurs: for the enterprises which in the first-tier cities where per capita consumption is higher than general consumption level, they would better increase their optimal order quantity if they are trying to reduce the variability of demand. But for the enterprises which in the remote region where per capita consumption is lower than general consumption level, they would better decrease their optimal order quantity if they are trying to reduce the variability of demand.

In the following we analyze the effect of reducing demand variability on the optimal profit for the risk-neutral newsvendor. We define the effect level of reducing demand variability as a function $\Delta\pi^c(\alpha, r, \rho)$ which is a difference between the optimal profit corresponding to demand $X_{\alpha,r}$ in (55) and $X_{1,r}$. That is,

$$\Delta\pi^c(\alpha, r, \rho) = \pi_{X_{\alpha,r}}(y^c_{X_{\alpha,r}}) - \pi_{X_{1,r}}(y^c_{X_{1,r}}), \tag{92}$$

where the optimal profit $\pi^c(\alpha, r, \rho)$ is given in (71).

By calculation, $\Delta\pi^c(\alpha, r, \rho)$ in (92) can be rewritten as

$$\Delta\pi^c(\alpha, r, \rho) = (1 - \alpha)(p + s - v)\psi(\rho, \alpha)$$
$$= (1 - \alpha)(p - c)[r - B_X(\rho)], \quad \text{for all } 0 < \alpha \le 1 \text{ and } r > 0, \tag{93}$$

where $\psi(\rho, \alpha)$ is given by (82) and

$$B_X(\rho) = \frac{1 - \rho^0}{\rho^0} \cdot \frac{F_X^{(-2)}(\rho) - \frac{\rho - \rho^0}{1 - \rho^0}\mu_x}{1 - \rho}$$
$$= \frac{\pi_{X_{1,r}}(y^c_{X_{1,r}})}{p - c}. \tag{94}$$

The following Theorem 8 provides the effect of demand variability about optimal order quantity in risk-neutral newsvendor problem. Part (ii) and (iii) can be shown using similar method as in Theorem 7(iii), it can be use another method to prove, the related results are identical.

**Theorem 8.** Let $X$ be a continuous random variable with support $[\underline{\ell}, \overline{\ell}]$, its cumulative distribution function is denoted by $F_X(\cdot)$, assume that its inverse of the distribution function $F_X^{-1}(\cdot)$ exists. The demand $X_{\alpha,r}$ is given in (55). For the maximization newsvendor problem (68).

(i) $\Delta\pi^c(\alpha, r, \rho)$ is increasing function of the unit stockout cost $s$ in interval $[0, \infty)$ and it satisfies:

$$\Delta\pi^c(\alpha, r, \rho^0) \le \Delta\pi^c(\alpha, r, \rho) \text{ for } \rho \in (0, 1) \text{ (equivalently } s \ge 0). \tag{95}$$

(ii) If $r \ge \mu_x$, then $\Delta\pi^c(\alpha, r, \rho)$ is decreasing function of $\alpha$ in interval $(0, 1)$ for all $\rho \in (0, 1)$ (equivalently $s \ge 0$).

(iii) If $0 < r < \mu_X$, there exist $\underline{\rho} \in (0,1)$ (equivalently $\underline{s} \in (0,\infty)$) such that $\Delta\pi^c(\alpha, r, \rho)$ is decreasing function of $\alpha$ in interval $(0,1)$ for all $\rho \in (\underline{\rho}, 1)$ (equivalently $s \geq \underline{s}$); $\Delta\pi^c(\alpha, r, \rho)$ is increasing function of $\alpha$ in interval $(0,1)$ for all $\rho \in (0, \underline{\rho})$ (equivalently $0 < s < \underline{s}$), where $\underline{\rho}$ is given in (77), $\underline{s}$ is given in (78).

*Proof.* Part (i): Finding the first derivative of the optimal order quantity $y^c_{X_{\alpha,r}}$ on both sides with respect to $\alpha$, we have

$$\frac{\partial y^c_{X_{\alpha,r}}(y)}{\partial \alpha} = F_X^{-1}(\rho) - r, \tag{96}$$

which leads to the desired result.

Part (ii): Finding the first derivative of the optimal expected profit $\pi_{X_{\alpha,r}}(y^c_{X_{\alpha,r}})$ in (71) on both sides with respect to $s$, we have

$$\frac{\partial \pi_{X_{\alpha,r}}(y^c_{X_{\alpha,r}})}{\partial s} = -\alpha \int_{F_X^{-1}(\rho)}^{\bar{\ell}} (1 - F_X(x))dx$$
$$< 0, \text{ for all } r > 0,\ 0 < \alpha \leq 1,\ \rho \in (0,1), \tag{97}$$

which leads to the desired result.

Part (iii): Finding the first derivative of the optimal expected profit $\pi_{X_{\alpha,r}}(y^c_{X_{\alpha,r}})$ in (71) on both sides with respect to $\alpha$, we have

$$\frac{\partial \pi_{X_{\alpha,r}}(y^c_{X_{\alpha,r}})}{\partial \alpha} = (p+s-v)\left(F_X^{(-2)}(\rho) - \frac{\rho - \rho^0}{1 - \rho^0}\mu_X - \frac{\rho^0(1-\rho)}{1-\rho^0}r\right)$$
$$= (p+s-v)\psi(\rho, r), \tag{98}$$

where

$$\psi(\rho, r) = F_X^{(-2)}(\rho) - \frac{\rho - \rho^0}{1 - \rho^0}\mu_X - \frac{\rho^0(1-\rho)}{1-\rho^0}r. \tag{99}$$

By calculation, we have

$$\psi(0, r) = \frac{\rho^0}{1 - \rho^0}(\mu_X - r),\ \psi(1, r) = 0, \tag{100}$$

$$\frac{\partial \psi(\rho, r)}{\partial \rho} = F_X^{-1}(\rho) - \frac{\mu_X}{1 - \rho^0} + \frac{\rho^0}{1 - \rho^0}r \text{ is increasing function of } \rho,$$
$$\text{for } \rho \in (0,1), \tag{101}$$

$$\lim_{\rho \to 1} \frac{\partial \psi(\rho, r)}{\partial \rho} = \bar{\ell} - \frac{\mu_X}{1 - \rho^0} + \frac{\rho^0}{1 - \rho^0}r > 0 \text{ for enough larger } \bar{\ell}. \tag{102}$$

Based on those properties of function $\psi(\rho, r)$ in (100)–(102) which leads to the desired results. $\square$

## 4.2 Newsvendor Problem with Mean-CVaR Constrain

The classical newsvendor maximizes expected profit. Within the expected utility theory (and other normative decision theories), this is equivalent to the assumption of risk-neutral behaviour: expected profit derived from the optimal order quantity $\mathbb{E}[\Pi(y_X^*, X)]$ is considered indifferent to the random profit $\Pi(y_X^*, X)$. In decision theory risk-aversion decision maker is characterized by the fact that the expected value $\mathbb{E}[\Pi(y_X^*, X)]$ is preferred to the random variable $\Pi(y_X^*, X)$ whereas for risk-taking behavior $\Pi(y_X^*, X)$ is preferred to $\mathbb{E}[\Pi(y_X^*, X)]$.

Consider the newsvendor model studies in Sect. 4.1, the assumptions in this section are identical as Sect. 4.1 except the newsvendor is risk-aversion or risk-taking by the mean-CVaR given by in (38). The newsvendor who faces a random demand $X_{\alpha,r} =_d \alpha X + (1-\alpha)r$, $0 < \alpha \leq 1$, which is given in (55), where $X$ is a continuous random variable with support $[\ell, \bar{\ell}]$. The newsvendor's objective is to decide the order quantity $y$ to maximize his utility, which can be expressed as

$$max_{y \geq 0} \pi_{X_{\alpha,r}}(y) = IMCVaR_\eta^{(\tilde{\lambda})}(\Pi(y, X_{\alpha,r})), \tag{103}$$

where the newsvendor's random profit $\Pi(y, X_{\alpha,r})$ is given in (67), the mean conditional value-at-risk $IMCVaR_\eta^{(\tilde{\lambda})}(\cdot))$ is given in (38).

The next Theorem 9 provides the optimal order quantity and optimal expected profit in newsvendor model with stockout cost and mix-CVaR criterion.

**Theorem 9.** Let $X$ be a continuous random variable with support $[\ell, \bar{\ell}]$, its cumulative distribution function is denoted by $F_X(\cdot)$, assume that its inverse of the distribution function $F_X^{-1}(\cdot)$ exists. For $0 < \alpha \leq 1$, $r > 0$ and $0 \leq \tilde{\lambda} \leq \frac{1}{\eta}$,

(i) The optimal order quantity $y_{X_{\alpha,r}}^*$ satisfies

$$\rho - F_X(A(\alpha, y_{X_{\alpha,r}}^*)) - \frac{\tilde{\lambda} - 1}{\frac{1}{\eta} - \tilde{\lambda}} \left[\frac{1}{\eta} F_X(A(\alpha, \frac{z_{X_{\alpha,r}}^* + (c-v)y_{X_{\alpha,r}}^*}{p - v})) - \rho\right]$$

$$= 0, \tag{104}$$

where $z_{X_{\alpha,r}}^*$ satisfies

$$F_X(A(\alpha, \frac{z_{X_{\alpha,r}}^* + (c-v)y_{X_{\alpha,r}}^*}{p - v})) - \eta$$

$$= F_X(A(\alpha, \frac{(p+s-c)y_{X_{\alpha,r}}^* - z_{X_{\alpha,r}}^*}{s})) - 1, \tag{105}$$

here $\rho$ is given in (70), and

$$A(\alpha, y) = \frac{y - (1-\alpha)r}{\alpha}. \tag{106}$$

(ii) The optimal utility profit $\pi_{X_{\alpha,r}}(y^*_{X_{\alpha,r}})$ is given by

$$
\pi_{X_{\alpha,r}}(y^*_{X_{\alpha,r}}) = \tilde{\lambda}\Big[(p+s-c)y^*_{X_{\alpha,r}} - \alpha(p+s-v)\int_\ell^{A(\alpha,y^*_{X_{\alpha,r}})} F_X(x)dx
$$

$$
- s(\alpha\mu_X + (1-\alpha)r)\Big] \tag{107}
$$

$$
- (\tilde{\lambda}-1)\Big[z^*_{X_{\alpha,r}} - \frac{\alpha(p+s-v)}{1-\eta}\int_\ell^{A(\alpha,y^*_{X_{\alpha,r}})} F_X(x)dx
$$

$$
+ \frac{\alpha(p-v)}{1-\eta}\int_\ell^{A(\alpha,\frac{z^*_{X_{\alpha,r}}+(c-v)y^*_{X_{\alpha,r}}}{p-v})} F_X(x)dx
$$

$$
+ \frac{\alpha s}{1-\eta}\int_\ell^{A(\alpha,\frac{(p+s-c)y^*_{X_{\alpha,r}}-z^*_{X_{\alpha,r}}}{s})} F_X(x)dx\Big]\Big],
$$

where $y^*_{X_{\alpha,r}}$ satisfies (104) and $z^*_{X_{\alpha,r}}$ satisfies (105).

*Proof.* Part (i): Combining problem (103) and the definition of minimization CVaR in (30), problem (103) can be rewritten as

$$
\max_{y\geq 0}\min_{z\in\mathbb{R}}\Big\{\tilde{\lambda}\mathbb{E}[\Pi(y,X_{\alpha,r})] - (\tilde{\lambda}-1)\varphi^{(0)}_{X_{\alpha,r}}(y,z)\Big\}, \tag{108}
$$

where

$$
\varphi^{(0)}_{X_{\alpha,r}}(y,z) = z + \frac{1}{1-\eta}\mathbb{E}\Big[\big((p+s-c)y - (p+s-v)(y-X_{\alpha,r})_+ - sX_{\alpha,r} - z\big)_+\Big]
$$

$$
= z + \frac{\alpha(p-v)}{1-\eta}\int_\ell^{A(\alpha,y)}\Big(x - A(\alpha,\frac{z+(c-v)y}{p-v})\Big)_+ dF_X(x) \tag{109}
$$

$$
+ \frac{\alpha s}{1-\eta}\int_{A(\alpha,y)}^{\bar{\ell}}\Big(A(\alpha,\frac{(p+s-c)y-z}{s}) - x\Big)_+ dF_X(x).
$$

(a) When $z \geq (p-c)y$, we have

$$
\varphi^{(0)}_{X_{\alpha,r}}(y,z) = z, \tag{110}
$$

it is clear that $\varphi^{(0)}_{X_{\alpha,r}}(y,z)$ is strictly increasing function of $z$ in interval $((p-c)y,\infty)$.

(b) When $z < (p-c)y$, by calculation we get

$$
\varphi^{(0)}_{X_{\alpha,r}}(y,z) = z - \frac{\alpha(p-v)}{1-\eta}\int_{A(\alpha,\frac{z+(c-v)y}{p-v})}^{A(\alpha,y)} F_X(x)dx
$$

$$
+ \frac{\alpha s}{1-\eta}\int_{A(\alpha,y)}^{A(\alpha,\frac{(p+s-c)y-z}{s})} F_X(x)dx. \tag{111}
$$

Taking the first and second derivatives two sides of in (111) with respect to $z$, we find that $\varphi_{X_{\alpha,r}}^{(0)}(y,z)$ is strictly convex function of $z$ in interval $(-\infty,(p-c)y)$, and note that $\frac{\partial \varphi_{X_{\alpha,r}}^{(0)}(y,z)}{\partial z}\big|_{z=(p-c)y-}=1>0$. We know that the optimal $z_{X_{\alpha,r}}^*$ exists, which satisfies (105).

Combining (105) and (108), the expected utility of newsvendor can be rewritten as

$$
\pi_{X_{\alpha,r}}(y) = \tilde{\lambda}\Big[(p+s-c)y - \alpha(p+s-v)\int_{\ell}^{A(\alpha,y)} F_X(x)dx\Big] - s\big(\alpha\mu_X + (1-\alpha)r\big)
$$

$$
- (\tilde{\lambda}-1)\Big[z_{X_{\alpha,r}}^* - \frac{\alpha(p-v)}{1-\eta}\int_{A(\alpha,\frac{z_{X_{\alpha,r}}^*+(c-v)y}{p-v})}^{A(\alpha,y)} F_X(x)dx \qquad (112)
$$

$$
+ \frac{\alpha s}{1-\eta}\int_{A(\alpha,y)}^{A(\alpha,\frac{(p+s-c)y-z_{X_{\alpha,r}}^*}{s})} F_X(x)dx\Big].
$$

Taking the first derivative in two sides of (112) with respect to $y$, we have

$$
\frac{\partial \pi_{X_{\alpha,r}}(y,\tilde{\lambda})}{\partial y} = (p+s-v)\frac{\frac{1}{\eta}-\tilde{\lambda}}{\frac{1}{\eta}-1}\Big[\rho - F_X(A(\alpha,y))
$$

$$
+ \frac{\tilde{\lambda}-1}{\frac{1}{\eta}-\tilde{\lambda}}\big[\rho - \frac{1}{\eta}F_X\big(A(\alpha,\frac{z_{X_{\alpha,r}}^*+(c-v)y}{p-v})\big)\big]\Big]. \qquad (113)
$$

From (113), it is easy to see $\pi_{X_{\alpha,r}}(y)$ is unimodal function of $y$ in interval $(0,\infty)$. So there exists an unique solution $y_{X_{\alpha,r}}^*$, which satisfies (104) for all $0\le\tilde{\lambda}\le1$. By using the similar method, we can get the optimal order quantity $y_{X_{\alpha,r}}^*$ using maximization CVaR for $1\le\tilde{\lambda}\le\frac{1}{\eta}$ based on the second type mean conditional value at risk $IIMCVaR_\eta^{(\tilde{\lambda})}(\cdot)$ in (39).

Part (ii): Combining $y_{X_{\alpha,r}}^*$ in (112) and (108), which leads to the desired result for $0\le\tilde{\lambda}\le1$. Similar method can be used for the case $1\le\tilde{\lambda}\le1/\eta$. □

**Remark 20.** From the proof of Theorem 8, we can easy to check, when $s=0$, we have $z_{X_{\alpha,r}}^*$ in (105) satisfies

$$
F_X\Big(A(\alpha,\frac{z_{X_{\alpha,r}}^*+(c-v)y_{X_{\alpha,r}}^*}{p-v})\Big) = \eta, \qquad (114)
$$

and $y_{X_{\alpha,r}}^*$ in (104) satisfies

$$
\rho - \frac{\tilde{\lambda}-1}{\frac{1}{\eta}-\tilde{\lambda}}(1-\rho^0) - F_X(A(\alpha,y_{X_{\alpha,r}}^*)) = 0. \qquad (115)
$$

Hence, from (105) and (114), we have

$$
F_X\Big(A(\alpha,\frac{z_{X_{\alpha,r}}^*+(c-v)y_{X_{\alpha,r}}^*}{p-v})\Big) \lesseqgtr \eta \quad\text{resp. for}\quad s\gtreqless0. \qquad (116)
$$

The following Corollary 6 provides several relations and inequalities between three functions $A(\alpha, y^*_{X_{\alpha,r}})$ and $A(\alpha, (z^*_{X_{\alpha,r}} + (c-v)y^*_{X_{\alpha,r}})/(p-v))$ in (104), $A(\alpha, ((p+s-c)y^*_{X_{\alpha,r}} - z^*_{X_{\alpha,r}})/s)$ in (105), which will be used to proof the upper and lower bounds for the optimal order quantity.

**Corollary 6.** Let $X$ be a continuous random variable with support $[\underline{\ell}, \overline{\ell}]$, its cumulative distribution function is denoted by $F_X(\cdot)$. For $0 < \alpha \leq 1$, $r > 0$ and $0 \leq \tilde{\lambda} \leq \frac{1}{\eta}$,

(i) For $s > 0$, $A(\alpha, y^*_{X_{\alpha,r}})$ and $A(\alpha, \frac{z^*_{X_{\alpha,r}} + (c-v)y^*_{X_{\alpha,r}}}{p-v})$ in (104), $A(\alpha, \frac{(p+s-c)y^*_{X_{\alpha,r}} - z^*_{X_{\alpha,r}}}{s})$ in (105) satisfies

$$(p-v)A(\alpha, \frac{z^*_{X_{\alpha,r}} + (c-v)y^*_{X_{\alpha,r}}}{p-v}) + sA(\alpha, \frac{(p+s-c)y^*_{X_{\alpha,r}} - z^*_{X_{\alpha,r}}}{s})$$
$$= (p+s-v)A(\alpha, y^*_{X_{\alpha,r}}). \tag{117}$$

(ii) For $s > 0$,

$$A(\alpha, \frac{(p+s-c)y^*_{X_{\alpha,r}} - z^*_{X_{\alpha,r}}}{s}) = A(\alpha, y^*_{X_{\alpha,r}}) + \Big(A(\alpha, y^*_{X_{\alpha,r}})$$
$$- A(\alpha, \frac{z^*_{X_{\alpha,r}} + (c-v)y^*_{X_{\alpha,r}}}{p-v})\Big) > A(\alpha, y^*_{X_{\alpha,r}}). \tag{118}$$

(iii)

$$F_X\Big(A(\alpha, \frac{z^*_{X_{\alpha,r}} + (c-v)y^*_{X_{\alpha,r}}}{p-v})\Big) > \eta - 1 + F_X(A(\alpha, y^*_{X_{\alpha,r}})). \tag{119}$$

*Proof.* Part (i) follows by checking the three functions in (104) and (105).

Part (ii): From (117), and note that $y^*_{X_{\alpha,r}} - (z^*_{X_{\alpha,r}} + (c-v)y^*_{X_{\alpha,r}})/(p-v) = [(p-c)y^*_{X_{\alpha,r}} - z^*_{X_{\alpha,r}}]/(p-v) > 0$, which leads to the desired result.

Part (iii): Combining (118) and (105), which leads to the desired result. $\square$

The following Corollary 7 provides the optimal order quantity and expected utility in problem (103) for demand $X_{1,r} = X$, where demand $X_{\alpha,r}$ is given in (55). Corollary 1 can be obtained by taking $\alpha = 1$ in Theorem 9.

**Corollary 7.** Consider problem (103) with demand if $X_{1,r}$ is given in (55) for $\alpha = 1$, where $X$ is a continuous random variable with support $[\underline{\ell}, \overline{\ell}]$, its cumulative distribution function is denoted by $F_X(\cdot)$. For $0 \leq \tilde{\lambda} \leq \frac{1}{\eta}$.

(i) The optimal order quantity $y^*_X$ satisfies

$$\rho - F_X(y^*_X) - \frac{\tilde{\lambda} - 1}{\frac{1}{\eta} - \tilde{\lambda}} \frac{1}{\eta} [\frac{1}{\eta} F_X(\frac{z^*_X + (c-v)y^*_X}{p-v}) - \rho] = 0, \tag{120}$$

where $z_X^*$ satisfies

$$F_X(\frac{(p+s-c)y_X^* - z_X^*}{s}) - F_X(\frac{z_X^* + (c-v)y_X^*}{p-v}) = 1 - \eta, \tag{121}$$

here $\rho$ is given in (70) and $\rho^0$ is given in (72).

(ii) The optimal utility profit $\pi_X(y_X^*)$ satisfies

$$\begin{aligned}
\pi_X(y_X^*) = &\tilde{\lambda}\Big[(p+s-c)y_X^* - \alpha(p+s-v)\int_\ell^{y_X^*} F_X(x)dx - s\mu_x\Big] \\
&- (\tilde{\lambda} - 1)\Big[z_X^* - \frac{(p+s-v)}{1-\eta}\int_\ell^{y_X^*} F_X(x)dx \\
&+ \frac{(p-v)}{1-\eta}\int_\ell^{\frac{z_X^*+(c-v)y_X^*}{p-v}} F_X(x)dx + \frac{s}{1-\eta}\int_\ell^{\frac{(p+s-c)y_X^*-z_X^*}{s}} F_X(x)dx\Big],
\end{aligned} \tag{122}$$

where $y_X^*$ is given in (120) and $z_X^*$ is given in (121).

The following Corollary 8 provides upper and lower bounds of the optimal order quantity in (104) according to the newsvendor model with or without stockout cost.

**Corollary 8.** Let $X$ be a continuous random variable with support $[\ell, \bar{\ell}]$, its cumulative distribution function is denoted by $F_X(\cdot)$. the optimal order quantity $y_{X_{\alpha,r}}^*$ given in (104). For $0 < \alpha \leq 1$, $r > 0$ and $0 \leq \tilde{\lambda} \leq \frac{1}{\eta}$,

(i) When $1 < \tilde{\lambda} \leq 1/\eta$, the optimal order quantity $y_{X_{\alpha,r}}^*$ satisfies: $y_{X_{\alpha,r}}^* > y_{X_{\alpha,r}}^*|_{s=0}$.

(ii) When $0 \leq \tilde{\lambda} < 1$, the optimal order quantity $y_{X_{\alpha,r}}^*$ satisfies $y_{X_{\alpha,r}}^* \leq y_{X_{\alpha,r}}^*|_{s=0}$.

*Proof.* Part (i): Combining (104) and (114), we know the function $F_X(A(\alpha, (z_{X_{\alpha,r}}^* + (c-v)y_{X_{\alpha,r}}^*)/(p-v)))$ in (104) satisfies: $F_X(A(\alpha, (z_{X_{\alpha,r}}^* + (c-v)y_{X_{\alpha,r}}^*)/(p-v))) = \eta$ when $s = 0$ and $F_X(A(\alpha, (z_{X_{\alpha,r}}^* + (c-v)y_{X_{\alpha,r}}^*)/(p-v))) < \eta$ when $s > 0$. Hence, the part $\big[F_X(A(\alpha, (z_{X_{\alpha,r}}^* + (c-v)y_{X_{\alpha,r}}^*)/(p-v)))/\eta - \rho\big]$ in middle brackets of (104) is strictly smaller than $[1 - \rho]$. When $1 < \tilde{\lambda} \leq 1/\eta$, the third part on the left of the Eq. (104) become $(\tilde{\lambda} - 1)[1 - \rho]/(1/\eta - \tilde{\lambda})$, which is the third part on the left of the equation corresponding the optimal order quantity in (115), which leads to the desired result.

Part (ii) follows by similar method as that in part (i). □

**Remark 21.** For the newsvendor problem in the presence of the loss of goodwill stockout opportunity cost, Choi et al. (2008) considered the newsvendor problem under the MV framework with decision makers possessing different risk attitude (namely, risk-averse, risk-neutral, and risk-seeking). They found that when we include an explicit stockout penalty cost in the problem, the variance of profit becomes more complicated. Wu et al. (2009) analytically proved that when demand is distributed following a continuous power distribution, the optimal stocking quantity under an MV newsvendor model will exceed the risk-neutral case's optimal "fractile" quantity. Corollary 8 provides the upper and lower bounds of the optimal order quantity in (104) for any demand distributions, and show that the optimal order quantity of the newsvendor model without stockout cost is lower bound when $\tilde{\lambda} > 1$; and upper bound when $0 \leq \tilde{\lambda} < 1$. This upper bound or lower bound properties of the optimal order quantity for the risk-aversion case ($\tilde{\lambda} > 1$) is consistent with that for the risk-neutral case ($\tilde{\lambda} = 1$). Since the optimal order quantity for the risk-neutral case is increasing function of the stockout cost $s$, that is $y^c_{X_{\alpha,r}} \geq y^c_{X_{\alpha,r}}|_{s=0}$. However, in the risk-taking case ($0 < \tilde{\lambda} < 1$), from Corollary 8, we have $y^*_{X_{\alpha,r}} \leq y^*_{X_{\alpha,r}}|_{s=0}$, which contrary to the result in the risk-neutral case.

The following Corollary 9 provides another upper and lower bounds for the optimal order quantity $y^*_{X_{\alpha,r}}$ in (104), which are the optimal order quantity corresponding to the risk-neutral case (i.e., $\tilde{\lambda} = 1$).

**Corollary 9.** Let $X$ be a continuous random variable with support $[\underline{\ell}, \bar{\ell}]$, its cumulative distribution function is denoted by $F_X(\cdot)$. For $0 < \alpha \leq 1$ and $r > 0$,

(i) When $1 < \tilde{\lambda} \leq 1/\eta$,
   (a) The optimal order quantity $y^*_{X_{\alpha,r}}$ has upper bound, that is, $y^*_{X_{\alpha,r}} \leq \bar{y}_{X_{\alpha,r}}$, where $\bar{y}_{X_{\alpha,r}}$ is given by

$$\bar{y}_{X_{\alpha,r}} = \alpha F_X^{-1}\left(\rho - \frac{\tilde{\lambda}-1}{\tilde{\lambda}}\left(\rho - \frac{\tilde{\lambda}(1-\eta)}{1-\eta\tilde{\lambda}}\right)\right) + (1-\alpha)r. \qquad (123)$$

   (b) The upper bound $\bar{y}_{X_{\alpha,r}}$ is decreasing function of $\tilde{\lambda}$, hence $\bar{y}_{X_{\alpha,r}} \leq \bar{y}_{X_{\alpha,r}}|_{\tilde{\lambda}=1} = \alpha F_X^{-1}(\rho) + (1-\alpha)r$.
   (c) The optimal order quantity $y^*_{X_{\alpha,r}}$ has a common upper bound, that is, $y^*_{X_{\alpha,r}} \leq y^*_{X_{\alpha,r}}|_{\tilde{\lambda}=1} = \alpha F_X^{-1}(\rho) + (1-\alpha)r$.

(ii) When $0 \leq \tilde{\lambda} < 1$,
   (a) The optimal order quantity $y^*_{X_{\alpha,r}}$ has lower bound, that is, $y^*_{X_{\alpha,r}} \geq \underline{y}_{X_{\alpha,r}}$, where $\underline{y}_{X_{\alpha,r}}$ is given by

$$\underline{y}_{X_{\alpha,r}} = \alpha F_X^{-1}\left(\rho + \frac{1-\tilde{\lambda}}{\tilde{\lambda}}\left(\rho - \frac{\tilde{\lambda}(1-\eta)}{1-\eta\tilde{\lambda}}\right)\right) + (1-\alpha)r. \qquad (124)$$

   (b) The lower bound $\underline{y}_{X_{\alpha,r}}$ is decreasing function of $\tilde{\lambda}$, hence $\underline{y}_{X_{\alpha,r}} \geq \underline{y}_{X_{\alpha,r}}|_{\tilde{\lambda}=1} = \alpha F_X^{-1}(\rho) + (1-\alpha)r$.
   (c) The optimal order quantity $y^*_{X_{\alpha,r}}$ has a common lower bound, that is, $y^*_{X_{\alpha,r}} \geq y^*_{X_{\alpha,r}}|_{\tilde{\lambda}=1} = \alpha F_X^{-1}(\rho) + (1-\alpha)r$.

*Proof.* Part (i): From (119) in Corollary 6, we know that $F_X\big(A(\alpha, (z^*_{X_{\alpha,r}} + (c - v)y^*_{X_{\alpha,r}})/(p - v))\big) > \eta - 1 + F_X(A(\alpha, y^*_{X_{\alpha,r}}))$, combining this inequality and Eq. (104), we get the upper bound in (123) when $1 < \tilde{\lambda} \le 1/\eta$. We note that the function in (123) is decreasing function of $\tilde{\lambda}$, this means that The optimal order quantity $y^*_{X_{\alpha,r}}$ has a common upper bound, it is The optimal order quantity in risk-neutral case (corresponding to $\tilde{\lambda} = 1$).

Part (ii) holds by similar method as that in part (i). □

## 5  Numerical Examples

To illustrate our results, we assume demand $X_{\alpha,r}$ in (55) is uniform random variable or exponential random variable. The following three examples are used to explain the obtained results in Theorem 3, Corollarys 4 and 9.

The following Example 1 explains the monotonicity and upper/lower bounds of the Mean-CVaR in (38) in Theorem 3.

**Example 1.** Consider a uniform random variable $X$ with distribution function $F_{X_{\alpha,r}}(x) = (x - a)/(b - a)$, its support is $[a, b]$, mean $\mu_X = (a + b)/2$, variance of demand $Var(X) = (b - a)^2/12$. Its inverse distribution function $F_X^{-1}(\eta) = a + (b - a)\eta$, the second quantile function of $X$ is $F_X^{(-2)}(\eta) = a\eta + (b - a)\eta^2/2$. The mean-CVaR in (38) is $IMCVaR_\eta^{(\tilde{\lambda})}(X) = (a + b)/2 + (b - a)\eta/2 - (b - a)\eta\tilde{\lambda}/2$. It is clear that $IMCVaR_\eta^{(\tilde{\lambda})}(X)$ is decreasing function of $\tilde{\lambda}$ in interval $[0, 1/\eta]$. On the other hand, $\partial IMCVaR_\eta^{(\tilde{\lambda})}(X)/\partial\eta = (b - a)(1 - \tilde{\lambda})/2 > 0$ if and only if $\tilde{\lambda} < 1$, which leads to the desired result in part (ii). Part (iii) and (iv) in Theorem 3 can be easily checked.

The following Example 2 explains the Monotonicity of the Mean-CVaR in (38) with respect to the demand variability in Corollary 4.

**Example 2.** Consider a exponential random variable $X_{\alpha,r}$ with distribution function $F_{X_{\alpha,r}}(x) = 1 - \exp\{-(x - (1 - \alpha)r)/\alpha\}$, its support $[(1 - \alpha)r, \infty)$, mean of demand $X_{\alpha,r}$ is $(1 - \alpha)r + \alpha$, variance of demand $X_{\alpha,r}$ is $\alpha^2$. This exponential random variable $X_{\alpha,r}$ can be represented as $X_{\alpha,r} =_d \alpha X + (1 - \alpha)r$, where $X$ is a exponential random variable with support $[0, \infty)$ with mean $\mu_x = 1$. By calculation, we get the inverse distribution function $F_{X_{\alpha,r}}^{-1}(\eta) = (1 - \alpha)r + \alpha(-\ln(1 - \eta))$, the second quantile function of $X_{\alpha,r}$ is $F_{X_{\alpha,r}}^{(-2)}(\eta) = (1 - \alpha)\eta + \alpha\eta + \alpha(1 - \eta)(\ln(1 - \eta))$. The mean-CVaR in (38) is $IMCVaR_\eta^{(\tilde{\lambda})}(X_{\alpha,r}) = (1 - \alpha)r + (\tilde{\lambda} - 1)\alpha\ln(1 - \eta)$. $\partial IMCVaR_\eta^{(\tilde{\lambda})}(X_{\alpha,r})/\partial\alpha = 1 - r - (\tilde{\lambda} - 1)(-\ln(1 - \eta)) < 0$ if and only if $(\tilde{\lambda} - 1)(-\ln(1 - \eta)) > 1 - r$, this condition holds if $1 \le \tilde{\lambda} \le 1/\eta$ and $r \ge \mu_x = 1$, which is the result in part (i). On the other hand, $\partial IMCVaR_\eta^{(\tilde{\lambda})}(X_{\alpha,r})/\partial\alpha = 1 - r - (\tilde{\lambda} - 1)(-\ln(1 - \eta)) > 0$ if and only if $(1 - \tilde{\lambda})(-\ln(1 - \eta)) > r - 1$, this condition holds if $0 \le \tilde{\lambda} < 1$ and $r < \mu_x = 1$, which is the result in part (ii). The result in part (iii) follows based on the fact,

$\partial IMCVaR_\eta^{(\tilde{\lambda})}(X_{\alpha,r})/\partial\alpha > 0$ for all $r < 1 + (1 - \tilde{\lambda})(-\ln(1 - \eta))$, that is, the value of $r$ is small enough.

The following Example 3 provides solution and expected utility for the problem (103) when demand $X_{\alpha,r}$ in (55) is a uniform random variable.

**Example 3.** Suppose demand $X_{\alpha,r}$ in problem (103) is a uniform random variable with support $[(1 - \alpha)r, (1 - \alpha)r + \alpha]$, where $X$ is a uniform random variable with support $[0, 1]$ and distribution function $F_X(x) = x, x \in [0, 1]$. The distribution function of $X_{\alpha,r}$ is $F_{X_{\alpha,r}}(x) = (x - (1 - \alpha)r)/\alpha$, its mean $(1 - \alpha)r + \alpha/2$, and variance $\alpha^2/12$. By calculation, the optimal order quantity is given by

$$y^*_{X_{\alpha,r}} = \alpha\left(\rho - (1 - \frac{1}{\tilde{\lambda}})\frac{p - c}{p + s - v}\right) + (1 - \alpha)r \qquad (125)$$

The optimal expected utility is given by

$$pi_{X_{\alpha,r}}(y^*_{X_{\alpha,r}}) = \tilde{\lambda}\Big[\alpha[(p + s - c)A(\alpha, y^*_{X_{\alpha,r}})$$
$$- \frac{1}{2}(p + s - v)(A(\alpha, y^*_{X_{\alpha,r}}))^2 - s\mu_x] + (p - c)(1 - \alpha)r\Big]$$
$$(1 - \tilde{\lambda})\Big[\alpha A(\alpha, y^*_{X_{\alpha,r}}) - \frac{1}{2}\frac{\alpha(p - v)s(1 - \eta)}{p + s - v} + (1 - \alpha)r\Big], \quad (126)$$

where

$$A(\alpha, y^*_{X_{\alpha,r}}) = \rho - (1 - \frac{1}{\tilde{\lambda}})\frac{p - c}{p + s - v}. \qquad (127)$$

It is easily checked that the optimal order quantity in (125) is increasing function of $s$ when $1 < \tilde{\lambda} \leq 1/\eta$, is decreasing function of $s$ when $0 \leq \tilde{\lambda} < 1$, which is the results of Corollary 8. It is clear that the optimal order quantity in (125) is decreasing function of $\tilde{\lambda}$ in interval $[0, 1/\tilde{\lambda}]$, which is the results of Corollary 9.

# 6  Conclusion Remarks

In this paper, we provide a unified framework for analyzing the effects of demand variability, decision maker's risk preference and stockout cost on replenishment policy and expected utility in stochastic inventory systems, and show that this framework of stochastic inventory systems is interesting, difficult and challenging. We describe and analyze two mean-risk models with CVaR as risk measure corresponding to a large-class measures in risk management, and specifically, we show that the two mean-CVaR measures are consistent with the ascending stochastic dominance for risk averters and descending stochastic dominance for risk lovers respectively. We show that similar results can be obtained by using a transformation of a random variable with a location and a scalar parameters.

Through some analysis, we see that the solutions of the newsvendor problem with mean-CVaR consideration is complex when stockout cost is positive $s > 0$.

We obtain the upper and lower of the optimal order quantity, and show that the risk-takers order more than risk-neutral, the risk-neutral order more than risk-averter. We obtain another upper and lower of the optimal order quantity with respect to the stockout cost, more specifically, we show that the risk-takers leads to less order quantity than the classic solution when a shortage penalty parameter is set to be zero; and the risk-averters leads to more order quantity than the classic solution when a shortage penalty parameter is set to be zero.

We hope that the methodology and results of this paper can be applicable in the study of more general stochastic models and stochastic optimization problems by means of stochastic dominance and transformation of a random variable. Along these lines, there are a number of interesting areas for potential future research, for example:

- Developing effective algorithms for computing the optimal order quantity and expected utility for these multi-objective stochastic optimization problems;
- analyzing supply chain system with single risk-neutral manufacturer and single risk-preference retailer with mean-CVaR consideration.
- considering leader-follower supply chain game with mean-CVaR consideration, the Stackelberg equilibrium need be analyzed by using the multi-objective optimization approach and stochastic dominance or transformation of random variable; and
- discussing supply chain network with multiple risk-preference manufacturers or/and multiple risk-preference retailers using mean-CVaR criterion.

**Acknowledgements.** This research was supported by the National Natural Science Foundation of China under grant Nos. 71932002, 71571010, 71672004 and the Great Wall Scholar Training Program of Beijing Municipality under grant No. CIT & TCD20180305.

# References

Agrawal, V., Seshadri, S.: Impact of uncertainty and risk aversion on price and order quantity in the newsvendor problem. Manuf. Serv. Oper. Manag. **2**, 410–423 (2000a)

Agrawal, V., Seshadri, S.: Risk intermediation in supply chains. IIE Trans. **32**, 819–831 (2000b)

Arnold, B.C.: A class of hyperbolic lorenz curves. Sankhya B **48**, 427–436 (1986)

Arrow, K.J.: Essays in the Theory of Risk-Bearing. Markham Publishing Company, Chicago (1971)

Artzner, P., Delbaen, F., Eber, J.M., et al.: Coherent measures of risk. Math. Finan. **9**(3), 203–228 (1999)

Banting, P.M., Blenkhorn, D., Dion, P.A.: Perceived consequences of business product stockouts in canada and the U.S.A. In: Sirgy, M.J., Bahn, K.D., Erem, T. (eds.) Proceedings of the 1993 World Marketing Congress. DMSPAMS, pp. 29–33. Springer, Cham (2015). https://doi.org/10.1007/978-3-319-17323-8_9

Bawa, V.S.: Research bibliography-stochastic dominance: a research bibliography. Manag. Sci. **28**(6), 698–712 (1982)

Begen, M.A., Pun, H., Yan, X.: Supply and demand uncertainty reduction efforts and cost comparison. Int. J. Prod. Econ. **180**, 125–134 (2016)

Bhattacharjee, M.C.: Some generalized variability orderings among life distributions with reliability applications. J. Appl. Probab. **28**(2), 374–383 (1991)

Bolton, G.E., Katok, E.: Learning by doing in the newsvendor problem: a laboratory investigation of the role of experience and feedback. Manuf. Serv. Oper. Manag. **10**(3), 519–538 (2008)

Brumelle, S.L., Vickson, R.G.: A unified approach to stochastic dominance. In: Stochastic Optimization Models in Finance, pp. 101–113. Academic Press, Boston (1975)

Camerer, C.F.: Prospect theory in the wild: evidence from the field. In: Kahneman, D., Tversky, A. (eds.) Choices, Values, and Frames, pp. 288–300. Cambridge University Press, Cambridge (2001)

Caine, G.J., Plaut, R.H.: Optimal inventory policy when stockouts alter demand. Naval Res. Logist. Q. **23**(1), 1–13 (1976)

Chateauneuf, A., Cohen, M., Meilijson, I.: Four notions of mean-preserving increase in risk, risk attitudes and applications to the rank-dependent expected utility model. J. Math. Econ. **40**(5), 547–571 (2004)

Chen, F., Federgruen, A.: Mean-variance analysis of basic inventory models. Technical manuscript, Columbia University (2000)

Chen, X., Sim, M., Simichi-levi, D., Sun, P.: Risk Aversion in Inventory Management. Working paper in MIT, Cambridge (2003)

Choi, T.M., Li, D., Yan, H.: Mean-variance analysis for the newsvendor problem. IEEE Trans. Syst. Man Cybern. Part A Syst. Hum. **38**, 1169–1180 (2008)

Choi, T.M., Li, D., Yan, H.: Newsvendor problem with mean-variance objectives. In: Proceedings of the 5th International Conference on Optimization: Techniques and Applications, vol. 4, pp. 1860–1867 (2001)

Chua, G.A., Liu, Y.: On the effect of demand randomness on inventory, pricing and profit. Oper. Res. Lett. **43**(5), 514–518 (2015)

Denuit, M., Dhaene, J., Goovaerts, M., et al.: Actuarial Theory for Dependent Risks: Measures. Orders and Models. Wiley, Chichester (2005)

Diamond, P.A., Stiglitz, J.E.: Increases in risk and in risk aversion. J. Econ. Theory **8**, 337–360 (1974)

Eeckhoudt, L., Gollier, C., Schlesinger, H.: The risk-averse (and prudent) newsboy. Manag. Sci. **41**(5), 786–794 (1995)

Fishburn, P.C.: Decision and Value Theory. Wiley, New York (1964)

Fishburn, P.C.: Lexicographic orders, utilities and decision rules: a survey. Manag. Sci. **20**(11), 1442–1471 (1974)

Fisher, M.A., Raman, A.: Reducing the cost of demand uncertainty through accurate response to early sales. Oper. Res. **44**, 87–99 (1996)

Gastwirth, J.L.: A general definition of the Lorenz curve. Econom. J. Econom. Soc. **39**(6), 1037–1039 (1971)

Gerchak, Y., Mossman, D.: On the effect of demand randomness on inventories and costs. Oper. Res. **40**(4), 804–807 (1992)

Gotoh, J., Takano, Y.: Newsvendor solutions via conditional value-at-risk minimization. Eur. J. Oper. Res. **179**, 80–96 (2007)

Guo, X., Wong, W.K.: Multivariate stochastic dominance for risk averters and risk seekers. RAIRO Oper. Res. **50**(3), 575–586 (2016)

Hardy, G.H., Littlewood, J.E., Polya, G.: Inequalities. Cambridge University Press, Cambridge (1934)

Hadar, J., Russel, W.R.: Rules for ordering choices involving risk. Am. Econ. Rev. **59**, 25–34 (1969)

Hadar, J., Russell, W.R.: Stochastic dominance and diversification. J. Econ. Theory **3**(3), 288–305 (1971)

Hammond, J.S.: Simplifying the choice between uncertain prospects where preference is nonlinear. Manag. Sci. **20**(7), 1047–1072 (1974)

Hanoch, G., Levy, H.: The efficiency analysis of choices involving risk. Rev. Econ. Stud. **36**, 335–346 (1969)

Hardy, G.H., Littlewood, J.E., Polya, G.: Inequalities. Cambridge University Press, Cambridge (1959)

Kahn, J.A.: Why is production more volatile than sales? Theory and evidence on the stockout-avoidance motive for inventory holding. Quart. J. Econ. **107**, 481–510 (1992)

Katariya, P.A., Cetinkaya, S., Tekin, E.: On the comparison of risk-neutral and risk-averse newsvendor problems. J. Oper. Res. Soc. **65**, 1090–1107 (2014)

Jammernegg, W., Kischka, P.: Risk-averse and risk-taking newsvendors: a conditional expected value approach. RMS **1**(1), 93–110 (2007)

Karamata, J.: Sur une inegalité relative aux fonctions convexes. Publ. Math. l'Univ. Belgrade **1**, 145–158 (1932)

Karlin, S., Novikoff, A.: Generalized convex inequalities. Pac. J. Math. **13**(4), 1251–1279 (1963)

Karlin, S.: Total Positivity. Stanford University Press, Stanford (1968)

Keren, B., Pliskin, J.S.: A benchmark solution for the risk-averse newsvendor problem. Eur. J. Oper. Res. **174**, 1643–1650 (2006)

Khouja, M.: The single-period (news-vendor) problem: literature review and suggestions for future research. Omega **27**, 537–553 (1999)

Lebreton, M.: Stochastic dominance: a bibliographical rectification and restatement of Whitmore's throrem. Math. Soc. Sci. **13**(1), 73–79 (1987)

Lehmann, E.L.: Ordered families of distributions. Ann. Math. Stat. **26**, 399–419 (1955)

Levy, H., Kroll, Y.: Ordering uncertain options with borrowing and lending. J. Finan. **33**(2), 553–574 (1978)

Levy, H.: Stochastic dominance and expected utility: survey and analysis. Manag. Sci. **38**, 555–593 (1992)

Levy, H.: The problem of causal contingency: the meaning of intensionality in physical theory. Br. J. Clin. Pract. **31**(10), 17–31 (1977)

Li, C.K., Wong, W.K.: Extension of stochastic dominance theory to random variables. RAIRO-Oper. Res. **33**(4), 509–524 (1999)

Li, Q., Atkins, D.: On the effect of demand randomness on a price/quantity setting firm. IIE Trans. **37**(12), 1143–1153 (2005)

Ma, L., Liu, F., Li, S., Yan, H.: Channel bargaining with risk-averse retailer. Int. J. Prod. Econ. **139**, 155–167 (2012)

Markowitz, H.M.: Portfolio selection. J. Finan. **7**, 77–91 (1952)

Markowitz, H.: Portfolio Selection: Efficient Diversification of Investments. John Wiley, New York (1970)

Marshall, A.W., Olkin, I.: Inequalities: Theory of Majorization and Its Applications. Academic Press, New York (1979)

Muliere, P., Scarsini, M.: A note on stochastic dominance and inequality measures. J. Econ. Theory **9**(2), 314–323 (1989)

Ogryczak, W.L., Ruszczyński, A.: Dual stochastic dominance and related mean-risk models. SIAM J. Optim. **13**(1), 60–78 (2002)

Ogryczak, W., Ruszczyński, A.: From stochastic dominance to mean-risk models: semideviations as risk measures. Eur. J. Oper. Res. **116**(1), 33–50 (1999)

Patsuris, P.: Christmas sales: the worst growth in 33 years. Forbes, 30 October 2001

Pflug, G.C.: Some remarks on the value-at-risk and the conditional value-at-risk. In: Uryasev, S.P. (ed.) Probabilistic Constrained Optimization, pp. 272–281. Springer, Boston (2000). https://doi.org/10.1007/978-1-4757-3150-7_15

Qin, Y., Wang, R., Vakharia, A.J., et al.: The newsvendor problem: review and directions for future research. Eur. J. Oper. Res. **213**(2), 361–374 (2011)

Qiu, R., Shang, J., Huang, X.: Robust inventory decision under distribution uncertainty: a CVaR-based optimization approach. Int. J. Prod. Econ. **152**, 13–23 (2014)

Quirk, J.P., Saposnik, R.: Admissibility and measurable utility functions. Rev. Econ. Stud. **29**(2), 140–146 (1962)

Rabin, M.: Psychology and economics. J. Econ. Lit. **36**, 11–46 (1998)

Rabin, M.: Risk aversion and expected-utility theory: a calibration theorem. Econometrica **68**, 1281–1292 (2000)

Rockafellar, R.T., Uryasev, S.: Conditional value-at-risk for general loss distributions. J. Bank. Finan. **26**(7), 1443–1471 (2002)

Rockafellar, R.T., Uryasev, S.: Optimization of conditional value-at-risk. J. Risk **2**, 21–42 (2000)

Ross, S.M., Schechner, Z.: Some reliability applications of the variability ordering. Oper. Res. **32**(3), 679–687 (1984)

Ross, S.M.: Stochastic Processes. Wiley, New York (1983)

Rothschild, M., Stiglitz, J.: Increasing risk: a definition and its economic consequences. Cowles Foundation for Research in Economics, Yale University (1969)

Rothschild, M., Stiglitz, J.E.: Increasing risk i: a definition. J. Econ. Theory **2**, 225–243 (1970)

Schwartz, B.L.: A new approach to stockout penalties. Manag. Sci. **12**, B-538–B-544 (1966)

Schweitzer, M., Cachon, G.: Decision bias in the newsvendor problem with a known demand distribution: experimental evidence. Manag. Sci. **46**, 404–420 (2000)

Schoemaker, P.: The expected utility models: its variants, purposes, evidence and limitations. J. Econ. Lit. **20**, 529–563 (1982)

Shaked, M., Shanthikumar, J.G.: Stochastic Orders and Their Applications. Academic Press, Boston (1994)

Silver, E.A., Pyke, D.F., Peterson, R.: Inventory Management and Production Planning and Scheduling. Wiley, New York (1998)

Song, J.S.: The effect of lead time uncertainty in a simple stochastic inventory model. Manag. Sci. **40**, 603–612 (1994)

Sriboonchitta, S., Wong, W.K., Dhompongsa, D., Nguyen, H.T.: Stochastic Dominance and Applications to Finance. Risk and Economics. Chapman and Hall/CRC, Boca Raton, Florida (2009)

Stoyan, D.: Comparison Methods for Queues and Other Stochastic Processes. Wiley, New York (1983)

Sun, Q., Dong, Y., Xu, W.: Effects of higher order moments on the newsvendor problem. Int. J. Prod. Econ. **146**(1), 167–177 (2013)

Tobin, J.: Estimation of relationships for limited dependent variables. Econometrica J. Econometric Soc. **26**(1), 24–36 (1958)

Uryasev, S., Rockafellar, R.T.: Optimization of conditional value-at-risk. Department of Industrial & Systems Engineering, University of Florida (1999)

Whitmore, G.A., Findlay, M.C.: Stochastic Dominance: An Approach to Decision-Making Under Risk. D.C.Heath, Lexington (1978)

Whitt, W.: Uniform conditional variability ordering of probability distributions. J. Appl. Probab. **22**, 619–633 (1985)

Wu, J., Li, J., Wang, S., et al.: Mean-variance analysis of the newsvendor model with stockout cost. Omega **37**(3), 724–730 (2009)

Wu, M., Zhu, S.X., Teunter, R.H.: The risk-averse newsvendor problem with random capacity. Eur. J. Oper. Res. **231**, 328–336 (2013)

Xu, X., Meng, Z., Shen, R.: A tri-level programming model based on conditional value-at-risk for three-stage supply chain management. Comput. Ind. Eng. **66**, 470–475 (2013)

Yu, H.B., Wang, Y.: Effect of uncertainty in inventory systems with mixture CVaR constrain. Oper. Res. Manag. Sci. **23**(1), 20–25 (2014a). (in Chinese)

Yu, H.B.: Impact of risk preference and demand uncertainty on inventory system with sales effort and replenishment decision. J. Ind. Eng. Eng. Manag. **28**(4), 69–74 (2014b). (in Chinese)

Yu, H.B., Li, Y.: Stochastic comparison of the generalized mean-preserving transformations and its applications to inventory decisions and optimization. Oper. Res. Manag. Sci. **27**(3), 32–40 (2018). (in Chinese)

Yu, H.B.: Stochastic monotonicity and comparability of Markov chains with block-monotone transition matrices and their applications to queueing systems. Stoch. Models **33**(4), 551–571 (2017)

Yu, H.B.: Stochastic monotonicity of mixture conditional value-at-risk and its applications to inventory system. Acta Math. Appl. Sinica **37**(5), 805–823 (2014c). (in Chinese)

Yu, H.B., Yang, F., Li, Y.: Stochastic comparison of inventory systems under supply leadtime and demand uncertainty. Oper. Res. Manag. Sci. **27**(5), 76–83 (2018). (in Chinese)

Yu, H.B.: Effect of demand uncertainty on newsvendor problems with minimization cost and maximization profit. Syst. Eng. Theory Pract. **34**(7), 1756–1768 (2014a). (in Chinese)

Zhu, W.S., Wu, Z.P.: The stochastic ordering of mean-preserving transformations and its applications. Eur. J. Oper. Res. **239**, 802–809 (2014)

# A Simple Survey for Supply Chain Finance Risk Management with Applications of Blockchain

Jian Li, Yajing Wang, Yongwu Li[✉], and Quan-Lin Li

School of Economics and Management, Beijing University of Technology,
Beijing 100124, China
{lijiansem,liyw}@bjut.edu.cn, wangyajing@emails.bjut.edu.cn,
liquanlin@tsinghua.edu.cn

**Abstract.** Supply chain finance (SCF) is an important solution that optimizes cash flow in order to lower financing costs and improve business efficiency especially for small and medium-sized enterprises (SMEs). Risk management is the essential requirement of SCF. In recent years, the digital economy is developing rapidly worldwide and holds huge potential for entrepreneurs and SMEs. In the digital economy scenario, digitalization of supply chains is also becoming increasingly dynamic. Blockchain technology is regarded as a potential means of digitalization for supply chains and could play an important role in supply chain finance risk management. This paper first reviews the literature of supply chain finance risk management and provides some disadvantages of traditional supply chain finance risk management. Then we survey the new perspective for supply chain finance risk management based on blockchain technology. In particular, blockchain can increase the information transparency of the supply chain, thereby reducing the credit risk of SMEs financing and the operational risk in SCF. Categorization and analysis of the literature, it provides an important perspective for future research of supply chain finance risk management based on blockchain technology in the era of digital economy.

**Keywords:** Supply chain finance · Risk management · Digital economy · Blockchain technology · Credit risk · Operational risk

## 1 Introduction

With the development of economic globalization and network, the comparative advantages among different companies, countries and even different regions within a country are constantly explored and strengthened. As a consequence, the supply chain is no longer a single chain, but becomes an intricate network, which involves many coordination and interaction activities between enterprises. The status of these coordination and interaction activities directly affects the service, quality and effectiveness of the supply chain. In the operation of modern

© Springer Nature Singapore Pte Ltd. 2019
Q.-L. Li et al. (Eds.): Cao Festschrift 2019, CCIS 1102, pp. 116–133, 2019.
https://doi.org/10.1007/978-981-15-0864-6_5

supply chain, the following steps are needed to realize the delivery of logistics. Firstly, processing enterprises need to purchase raw materials from raw material enterprises, process them into parts, and then sell them to component suppliers. After component suppliers produce parts, they sell them to finished products enterprises. Then finished products enterprises sell their finished products to distributors and retailers, which ultimately sell goods to consumers. However, in this series of activities, there is a capital gap as the funds of firm expenditure and income occur at different times. For instance, when providing goods to a downstream retailers (usually large buyer, such as Wal-Mart), suppliers, especially small ones, often bear a long payment delay after delivery, depending on the agreed payment time and product quality. Therefore, for small and medium-sized enterprises (SMEs) with insufficient capital, some "cost depressions" have become the bottleneck restricting the development of supply chain. Moreover, the economic downturn and lack of asset guarantees caused a considerable reduction in loans, accompanied by high borrowing costs [32]. In this context, the focus of supply chain research and exploration has gradually shifted to the supply chain finance (SCF) to improve the efficiency of capital flow [60]. Sufficient working capital is more and more important for enterprises in the fierce competition, especially for SMEs with great development opportunities but restricted by cash flow.

Supply chain finance is a kind of management behavior and process integrating logistics operation, commercial operation and financial management. It closely links buyers, sellers, third party logistics and financial institution in trade, realizes the function of using supply chain logistics to activate cash flow, and at the same time uses cash flow to simulate supply chain logistics [60]. The ultimate objective is to improve the efficiency of the supply chain. With the widespread application of SCF, the number of scientific papers focusing on SCF has increased in the last decade. However, through the contrast with the definition of SCF in the existing literatures, it is found that the problem is addressed from different perspectives, that is, the finance-oriented and supply chain-oriented perspectives [24]. More specifically, the former emphasizes on the financial aspects and considers the SCF as a set of financial solutions [6]. The latter pays more attention to the collaboration among the supply chain members, especially the inventory optimization [52]. For more recent literature on SCF, we can refer to [29, 38, 73, 76] etc.

Supply chain finance risk refers to the possibility that funding providers will suffer losses due to the influence of various unpredictable uncertainties in the financing process of supply chain enterprises, which will cause the deviation between the actual and expected returns of the financial products in the supply chain, or that the assets can not be recovered. In 2000, due to a fire at its chip supplier, Ericsson lost $400 million, and its market share dropped from 12% to 9% [49]. In 2008, as the financial crisis broke out, numerous high-profit companies went bankrupt one after another and the global economy was in a downturn. The examples go on and on. However, no matter the risk caused by natural disasters or human factors, the final loss is difficult to estimate.

SCF approach is a set of solutions that optimizes cash flow by lending to funders through credit trade, accounts receivable, pledged inventory or order financing. Firms, especially the most vulnerable, once cannot repay the funding provider in time due to the credit risk caused by uncertain factors, in that way the funding provider will suffer great losses. Therefore, numerous scholars and practitioners begin to pay attention to the risk management of supply chain finance. Specifically, supply chain finance risk mainly include credit risk, market risk, legal risk and operational risk. The core of supply chain finance risk management includes four key factors: risk identification, risk measurement and evaluation, risk monitoring and early warning and risk management. This paper focuses on risk management, including risk management of supply chain finance via loan-to-value ratio, contracts coordination and financing schemes, and financial derivatives.

In recent years, the digital economy is developing rapidly worldwide and holds huge potential for entrepreneurs and SMEs. Digital economy, also sometimes called the Internet Economy, New Economy, or Web Economy, refers to an economy that is based on digital computing technologies. The term "Digital Economy" was first mentioned in Japan by a Japanese professor and research economist in the midst of Japan's recession of the 1990s. In the west the term followed and was coined in [17] that was among the first books to consider how the Internet would change the way to do business. In the digital economy scenario, digitalization of supply chains is becoming increasingly dynamic. Blockchain technology is regarded as a potential means of digitalization for supply chains [40]. In particular, blockchain can increase the information transparency of the supply chain and play an important role in supply chain finance risk management. For example, blockchain can reduce the credit risk of SMEs financing and the operational risk in SCF. It provides an important perspective for future research of supply chain finance risk management based on blockchain technology in the era of digital economy.

This paper first reviews the literature of supply chain finance risk management and provides some disadvantages of traditional supply chain finance risk management. Then we survey the new perspective for supply chain finance risk management based on blockchain technology. Based on previous studies, not only is an innovative general framework to reduce credit risk proposed, but also supplements the SCF risk management system.

The remainder of this paper is organized as follows. We review the research of the supply chain finance risk management focusing on loan-to-value ratio in Sect. 2. We review the research on supply chain finance risk management from another perspective-improving the supply chain performance in Sect. 3 and on supply chain finance risk management via financial derivatives, especially options contracts in Sect. 4. Section 5 is devoted to present a new perspective for supply chain finance risk management: blockchain technology and review some recent preliminary explorations in this field.

## 2  Supply Chain Finance Risk Management via Loan-to-Value Ratio

Due to the good liquidity and marketability of inventory, in the last few years, the financing of inventory pledge has become the focus of attention and research in academia. Unlike traditional bank loans, which are mainly secured by real estate mortgage or third-party guarantee companies, the financing of inventory pledge utilizes movable property in real trade behaviors between enterprises and upstream and downstream to obtain loans from banks and other financial institutions. Therefore, the development of inventory financing business can realize the "multi-win" of financial institutions, financing enterprises and third-party logistics. However, everything has two sides. It brings tremendous benefits, but also risks and losses. Currently, the most significant factor affecting the inventory pledge financing risk is the price risk, which is mainly caused by the fluctuation of the market value of the pledges. The credit risk of the enterprises turn into the liquidity risk of collateral in the inventory pledge financing [3,15,18]. Therefore, it is significant to evaluate and manage the collateral risk not only for reducing the bank losses but also for the development of this business.

With these pressing needs, the scholars and practitioners have made many beneficial explorations on risk management of pledges. Through literature review, this section focuses on the research of key risk control indicator, loan-to-value rate. Moreover, its reasonable determination plays an important role in the smooth development of inventory financing business and the effective reduction of risk. Loan-to-value ratio is the ratio of loans and collateral value. The indicators related to loan-to-value ratio include the discount rate or haircut rate, the loan amount and the quantity of the collateral required by the loan, among which the discount rate or haircut rate is the value of the collateral minus the amount of the loan divided by the value of the collateral, i.e. 1-loan-to-value. Furthermore, the final setting of the loan-to-value rate is closely related to the business model, characteristics of guarantee inventory, default probability of enterprises, supervision mode and loan interest rate, etc. which can reflect the risk status of supply chain financial business in a comprehensive way. Hence, the determination and optimization of loan-to-value rate has crucial theoretical significance and broad application value.

Cossin *et al.* [16] propose a general framework for determining collateral risk control in repurchase transactions. By establishing a credit risk model, the risk level of collateral is determined by taking into account the intrinsic risk parameters of collateral: mark-to-market and haircut. Jokivuolle and Peura [33] present a risk debt model, in which the value of collateral may be related to the probability of default. This model can not only estimate the expected default loss, but also determine the loan-to-value rate as a loan criterion. Li *et al.* [44] analyze the optimization of the pledge rate that the inventory price obeys various random distributions. Using the "subject + debt" strategy, it establishes a pledge rate decision-making model to avoid downsides risks, and points out that under the static pledge situation, as long as the distribution of the market price of the pledge at the end of the financing period is known, an analytical formula of the pledge rate can be obtained.

With the rapid development of modern financial risk management technology, the researchers have made considerable progress in using risk management tools to manage inventory financing. For example, the VaR (value at risk) method for measuring market risk was first proposed in 1993. Subsequently, the risk control model of risk metrics proposed by Morgan [34] for calculating the value of risk has been widely adopted by numerous financial institutions. The VaR approach is essentially a statistical measurement of the pricing fluctuations of the object. It estimates the statistical distribution or probability density function of the future price changes of the object according to the historical data. The mathematical definition of VaR is:

$$Prob(\Delta p > VaR) = 1 - c \tag{1}$$

Where, $\Delta p$ is the loss of the object in the target period $\Delta t$, and $c$ is the confidence level. Taking pledges as the object, the following describes the basic calculation process of VaR. Assuming that $\omega_0$ is the original value of the pledge at the beginning of the loan; $Y$ is the rate of change in value of the pledge during the target period; $\mu$ is the expected value of rate of change; $\sigma$ is the standard deviation of rate of change, then the value of the pledge at the end of the loan period $\omega$ is

$$\omega = \omega_0(1 + Y). \tag{2}$$

Further, assuming that at the confidence level $c$, the minimum value change rate of the pledges during the target period is $Y^*$, then the minimum repayment value $\omega^*$ of the pledges at the confidence level is

$$\omega^* = \omega_0(1 + Y^*). \tag{3}$$

Thus, the relative VaR of the pledge in the target period can be defined as:

$$VaR = E(\omega) - \omega^* = -\omega_0(Y^* - \mu). \tag{4}$$

The absolute VaR of the pledge in the target period can be defined as:

$$VaR = \omega_0 - \omega^* = -\omega_0 Y^*. \tag{5}$$

From the above, as long as the value of $\omega^*$ or $Y^*$ at the confidence level $c$ is calculated, the relative VaR and absolute VaR can be obtained.

The measurement methods of VaR can be basically divided into two categories. The first category is the local valuation method including moving average, GARCH model, analysis variance-covariance approach and so on; The second category is the full valuation method such as historical simulation, monte carole simulation and bootstrap. Establishing a model with the formula AR(1)-GARCH(1,1)-GED, He et al. [36] put forward the method of dividing the pledge period into different risk windows to set the dynamic pledge rate. He et al. [28] propose a new measurement method for extreme risk of futures price (value at risk and conditional value at risk) according to inventory financing portfolio and time interval of dynamic impawn rate, and introduce monte carlo simulation method to conclude that inventory portfolio can disperse financing risk.

In recent years, in addition to the above methods to determine the loan-to-value rate, some innovative ideas have been proposed. Zhang et al. [72] investigate the determination of the loan-to-value rate in an idea of option pricing, that is, the present value of the pledge equals the value of the put option. Simple empirical method or VaR method to determine the impawn rate may lead to valuation failure, thus increasing the risk of banks or financial institutions. This provides a new direction for the accurate solution of the pledge rate.

On the one hand, in addition to proposing methods and frameworks to determine and optimize the pledge rate, on the other hand, some literatures also involve the research on the impact of the pledge rate under the constraint of capital on logistics operation decision of enterprises. Buzacott and Zhang [3] introduce asset-based financing into production decision-making for the first time, and analyze how the setting of interest rate and loan-to-value rate affected the profitability of enterprises by establishing a newsboy model. Dada and Hu [19] analyze the decision-making of capital-constrained enterprises to borrow and order seasonal inventory at a given lending rate based on the newsboy model. Caldentey and Haugh [5] also study the decisions of retailers and manufactures with capital constraints in the stackelberg game and analyze the supply chain contract design when retailers can make financial hedging. In conclusion, there are still a lot of papers considering the impact of pledge rate on decision-making, thus they are not listed one by one.

Although the current research has provided some methods to evaluate the collateral, there are still some other risks. For instance, financial institutions such as banks cannot determine whether the inventories are repeatedly pledged, which requires financial institutions to spend more manpower, material and financial resources to verify. Scholars still need to explore new ways to address this problem.

## 3 Supply Chain Finance Risk Management via Contracts Coordination and Financing Schemes

The first section summarizes the research status of the risk control indicator loan-to-value rate on the control of supply chain financial risk. This section will conclude the exploration and achievements of researchers on supply chain financial risk management from another perspective, that is, improving the supply chain performance. It is highlighted that suppliers or retailers with capital constraints or both are subject to capital constraints. By introducing contracts or choosing good financing schemes, the performance of the supply chain can be improved, so as to reduce risk. Similar to the traditional supply chain without capital constraint, Kanda and Deshmukh [37] show that supply chain members attempt to coordinate by using contracts for better management of supplier-buyer relationships and risk management.

As we all know, to develop great supply chain finance business is inseparable from the normal operation of its supply chain. Once the supply chain is subject to fluctuation or disruption, enterprises with constraint capital are very likely to have operational risk and credit risk, causing huge losses to financial institutions. Consequently, the risk of supply chain finance must be closely related to the risk in supply chain operation and management. Before describing the outstanding contributions made by researchers to supply chain risk management, it is necessary to introduce the most basic newsvendor model. Newsvendor model plays an irreplaceable role in supply chain operation decision.

Consider a two-echelon supply chain consisting of one retailer and one supplier. Assuming that market demand $D$ is random and continuous, its distribution function is $F$ and density function is $f$. The wholesale price of goods is $w$; the retail price is $p$; the residual value is $v$, and the penalty for short supply is $s$. The retailer orders goods with quantity $Q$ from the supplier, so the retailer's expected revenue is

$$\Pi(Q) = E[p\min(Q, D) + v\max(Q - D, 0) - s\max(Q - D, 0) - wQ]. \quad (6)$$

The retailer needs to determine the optimal order quantity $Q$ to maximize the expected revenue, i.e.,

$$\max_{Q \geq 0} \Pi(Q). \quad (7)$$

By solving the optimization problem, the optimal order quantity $Q^*$ is obtained:

$$Q^* = F^{-1}(\frac{p + s - w}{p + s - v}). \quad (8)$$

Thus, the optimal expected revenue of retailer $\Pi^*(Q^*)$ can be abtained.

With the frequent occurrence of risks and heavy losses, supply chain risk management has attracted attention from both academia and practitioners of operations management. A various set of supply disruption examples have been provided by Chopra and Sodhi [7], sheffi [57] and Wu et al. [67]. There are diverse mitigation techniques that deal with material flow risk such as multiple sourcing [2], alternative sourcing and backup production options [1,70], flexibility [61,64], and supplier selection [20]. In particular, Tang [58] provides a review of supply risk for reference. In addition, a number of papers study financial flow risk of supply chain [67]. For instance, Li et al. [45] and Goh et al. [25] research the exchange rate risk which greatly affects the enterprises' supplier selection, market development and other operation decisions. Papadakis [50] studys the price and cost risk which is bound up with exchange rate. The financial strength of supply chain partners is also considered. Tang [58] points out that the vulnerability of financial strength of s supply chain partner may easily affect the whole supply chain network. Finally, except that the above material flow risk and financial flow risk, stream of research is related to information flow risk, which focuses on information accuracy [41,48], information system security and disruption [21] and information outsourcing [14,48]. The literature of supply chain risk management is more mature than that of supply chain finance risk management, so it is not detailed.

Recently, researchers have found that the operation decisions and capital decisions of enterprises are inseparable, because in reality, enterprises often suffer capital constrains, and their development is greatly influenced by bond financing, bank loans, risk fund and other external investments. Moreover, Xu and Birge [68] study the decision of a capital constrained buyer, and show that integrating financial and operational decisions can improve significantly the enterprises' value. Thus, a number of papers related to coordination with capital constrained supply chain are proposed in the supply chain finance system.

Dada and Hu [19] consider a supply chain with a capital constrained retailer, and derive a non-linear loan schedule that coordinates the supply chain with a wholesale price contract. By considering the quantity discount contracts, buy-back contracts, two-part tariff contracts, revenue-sharing contracts and mark-down allowance contracts, Lee and Rhee [42,43] show that these contracts can coordinate the supply chain with trade credit financing rather than the bank credit financing. In addition, Zhang et al. [71] explore the issue of supply chain coordination by considering trade credit and default risk. They study a modified quantity discount based on order quantity and advance payment and find that the existence of an appropriate amount of advance and discount can coordinate the supply chain. Moreover, trade credit can not only solve financing needs for enterprises with capital constrain, but also plays other functions other than financing. Long et al. [46] and Rui and Lai [54] successively propose that trade credit can be used as a guarantee for product quality. However, a single contract has its risk, such as double marginal effect. Therefore, in order to overcome the shortcomings of single contract and give full play to the unique advantages of different contracts, scholars adopt joint contract, which not only plays an important role in controlling the risk of capital provider, but also can better guarantee the revenue of enterprises in the chain, so as to coordinate the supply chain more effectively [35,55,69]. Cai et al. [4] proposes revenue sharing and supplier subsidy contract, and show that compared with single revenue sharing contract, joint contract can realize pareto improvement.

Another stream of literatures that comparing various financing schemes to obtain which one is the best have been studied. For instance, Chen and Cai [8] study the value of third party logistics (3PL) firms in budget-constrained supply chain, and find that all partners can be better off under 3PL financing rather than supplier credit financing. Tunay and Zhu [63] study BIF (large buyers intermediate between banks and suppliers with capital constraint to guarantee financing for suppliers) in a three-way decentralized game between the supplier, the buyer, and the bank with supplier defects and endogenous buyer determined interest rate and wholesale prices. Comparing the performance of two different financial schemes with asymmetric information, they show that by allocating risk away from the supplier and towards the buyer, BIF can improve supply chain performance.

## 4  Supply Chain Financing Risk Management via Financial Derivatives

It is well known that financial derivatives are important tools for risk management. A derivative is a financial contract whose value depends on one or more basic assets or indices-underlying assets. The basic types of contracts include forward, futures, swaps and options. Further, financial derivatives also include hybrid financial instruments with multiple characteristics of forward, futures, swaps and options. Originally, the main purpose of early participants in derivatives markets is to hedge against forward risks. In fact, financial derivatives, such as futures, options, forward, swaps and so on, have become an effective tool to manage and reduce the risk of market participants.

To the best of our knowledge, the first contribution to the study of supply chain risk management using derivatives is Ritchken and Tapiero [53], in which they introduced the option contract into inventory management for hedging against price and quantity uncertainty. Option contracts can provide the buyer with right to buy a certain amount of goods from suppliers at a fixed price in the future. The aim of this paper is to use options as hedging tools to manage risks arising from uncertain prices and demand. Next, let's briefly describe the model in this article to show the risk management approach that uses option contracts for hedging against price and quantity uncertainty in inventory procurement.

Consider a single period inventory model. Assumed that $v_0$ unit goods with per unit price of $S_0$ are acquired at $i = 0$, and stored to period $i = 1$. Let $u_0$ be the number of options with strike price $\pi$ at $i = 1$ are purchased at $i = 0$, the option price is $C_0$. $S_1$ and $d_1$ (random variables) represent the price of goods and order quantities in period $i = 1$. Denoted $h_s(v_0)$ and $h_c(u_0)$ by the present value of the inventory and option carrying charge. The decision make' optimization problem at $i = 0$ is

$$\min_{V_0, U_0} \mathbf{E} U(F(v_0, u_0; S_1, d_1)), \qquad (9)$$

where $F(v_0, u_0) = S_0 v_0 + h_s(v_0) + C_0 u_0 + h_c(u_0) - \delta C_1 u_0 + \delta S_1 v_1$ is the net present value of cost, $U$ is a utility function and $\delta$ is the discount rate, and $v_1 = d_1 - v_0$, $v_1$ is the number of units goods purchased in period $i = 1$, $C_1$ is the inherent value of the option contract in period $i = 1$, i.e., $C_1 = \max\{0, S_1 - \pi\}$.

Option contracts have been increasingly employed as a popular strategy by supply chain firms to hedge against the risk of unanticipated demand. Golovachkina [26] designs a supply chain contract by using option in which a competitive supplier offer an incentive-compatible contract to a manufacturer. Furthermore, under information asymmetry about the buyer's valuation premium for the suppliers' product, Pei et al. [51] built a model to study the pricing problem of contracts for a supplier of an industrial good in the presence of spot trading, in which they provided an approach to the design of procurement contracts combining buyers' private valuations, spot market trading, information asymmetry, and general pricing structures for option contracts.

Schummer and Vohra [56] study the mechanism-design problem for a supply chains, in which a single buyer with an unknown future demand need to procure purchase options for a homogeneous good. Wu and Kleindorfer [66] develop a framework for analyzing business-to-business (B2B) transactions and supply chain management based on integrating contract procurement markets with spot markets using capacity options and forwards. The key feature of this paper is the competition and interaction among multiple sellers with heterogeneous technologies. Option contracts are of increasing importance in practice and this paper showed that options contracts were having fundamental impacts on both B2B contracting as well as the operational decisions that flow from it.

Fu *et al.* [22] study a single-product periodic-review inventory system, in which the demand is random and price-dependent. A company purchased a single product from a range of supply sources, including a set of options contracts (suppliers) with different flexibility and costs, and a spot market with uncertain prices. By exercising the option reserved from the supplier and ordering from the spot market when needed, the company made three decisions in each period, namely, the selling price of the product, the quantity of option reserved and the replenishment of inventory under the objective of maximizing the total expected profit over a finite planning horizon.

Gaur and Seshadri [23] address the problem of hedging inventory risk using the newsvendor model when demand is correlated with the price of a financial asset. In both the mean-variance framework and the more general utility-maximization framework, they showed how to construct optimal static hedging strategies that minimize the variance of profit for a given inventory level and increase the expected utility for a risk-averse decision maker. They also analyzed the impact of hedging on the expected utility of the decision maker and on the optimal inventory decision. The results showed that for both risk-neutral and risk-averse decision makers, (1) hedging reduces the variance of profit and increases expected utility, the reduction in the variance of profit is directly proportional to the correlation of demand with the price of the asset; (2) it provides an incentive to a risk-averse decision maker to order a quantity that is closer to the expected value-maximizing quantity; (3) the hedging transactions do not require additional investment.

The option contract can also coordinate the supply chain. Considering that cooperative relationships are becoming more and more prevalent in supply chains under the current industry environment, Zhao *et al.* [74] take a cooperation approach to solve the coordination issues for manufacturer-retailer supply chains using option contracts. Taking into account supply chain members risk preferences and negotiating powers, they developed an option contract model by using the wholesale price mechanism as a benchmark. The results showed that, compared with the benchmark as wholesale pricing mechanism, option contracts can coordinate the supply chain and achieve Pareto-improvement.

Chen *et al.* [9] investigate a one-period two-echelon supply chain, which is composed of a risk-neutral supplier that produces short life-cycle products and a loss-averse retailer that orders from the supplier via option contracts and sells to end-users with stochastic demand in the selling season. When a single retail season begins, the retailer can obtain goods by purchasing and exercising call options. The results showed that, (1) the loss-averse retailer may order less than, equal to, or more than the risk-neutral retailer; (2) the loss-averse retailer's optimal order quantity may increase in retail price and decrease in option price and exercise price, which is different from the case of a risk-neutral retailer; (3) there always exists a Pareto contract as compared to the non-coordinating contracts.

Considering a service requirement, Chen and Shen [10] and Chen *et al.* [11] examine the effect of options contracts and bidirectional option contracts on a two echelon supply chain consisting of a supplier and a retailer, respectively. Their study showed that the service level with (bidirectional) option contracts is equivalent to that without them when the service requirement is binding, while the service level with (bidirectional) option contracts is higher than that without them when the service requirement is not binding. In the presence of (bidirectional) option contracts and a service requirement, a distribution-free coordination condition was as well proposed to achieve the Pareto improvement.

Hua *et al.* [31] develope a Stackelberg game to analyze the joint ordering and financing problems in a two-echelon supply chain based on the option contract, in which a capital-constrained retailer ordered via the option contract from a single large supplier due to uncertain market demand. The retailer can apply for either a bank loan or trade credit from the supplier whenever necessary to maintain a reasonable capital level to pay for option orders because of its limited capital. The results showed that the supplier should always finance the retailer at the risk-free interest rate in the presence of the retailer's bankruptcy risk. Meanwhile, under trade credit, the supply chain's efficiency is improved (decreased) when the production cost is high (low). Furthermore, the results showed that the supplier's relationship concern can improve the supply chain's efficiency and the retailer's revenue most of the time, but increase the retailer's bankruptcy risk when the production cost is high, implying that the supplier's attempt to help the retailer eventually harms its long-run survival.

In conclusion, there are two mainly disadvantages about present risk management of SCF. First, the current risk management methods lack risk control before financing. The funding providers or the financing party usually use the above three risk management methods to reduce the risk after determining the financing. Second, risk management of SCF is realized through loan-to-value rate, and contracts coordination and financing schemes, which is essentially the risk sharing among members of the supply chain, third-party logistics companies

## 5   The Combination of Emerging Blockchain Technology and SCF

In 2008, Nakamoto [47] proposes a scenario design for how to build a trusted trading network without relying on authoritative third-party institutions, such as banks. For the first time, the concept of blockchain technology is well known. Blockchain technology is a new distributed infrastructure and computing method, which uses blockchain data structure to verify and store data, uses distributed node consensus algorithm to generate and update data, uses cryptography to ensure data transmission and access security, and uses smart contract composed of automated script code to program and operate data. More concretely, blockchain is a growing list of records, called blocks, which are cryptographically linked. Each block contains a cryptographic hash of the previous block, a timestamp, and transaction data. Due to its design, the blockchain is resistant to modification of the data. A blockchain is an open, distributed ledger that can record transactions between two parties efficiently and in a verifiable and permanent way. With a blockchain, many people can write entries into a record of information, and a community of users can control how the record of information is amended and updated [13].

Because of the above characteristics, blockchain technology has received growing attentions from both academia and industry in the past few years. For instance, the Economist defined blockchain technology as a 'trusted machine' in November 2015. Due to the huge potential of blockchain technology, the model of 'blockchain +' is also being studied continuously. This section mainly focuses on the application of blockchain technology in supply chain management and SCF. Streams of papers with blockchain-based supply chain management are as follows. Blockchain-based applications using radio frequency identification (RFID), the Internet of things(IoT), and tracking sensors could make breakthroughs in supply chain management and logistics [59, 62]. Key supply chain management objectives such as cost, quality, sustainability, flexibility and risk management can be improved according to the use of blockchain to increase transparency of information in supply chain activities [39]. Zhu and Kouhizadeh [75] propose that blockchain technology with traceability, transparency and so on may help address information challenges to reduce the risk for rational product deletion. On the other hand, the SCF based on blockchain technology has attracted extensive attention. Chod et al. [12] identify a crucial benefit of blockchain adoption-by opening a window of transparency into operations of a firm, blockchain technology furnishes the ability to secure favorable financing terms at lower signaling costs. In addition, more SCF platforms based on blockchain technology have emerged, aiming at solving the problems of information asymmetry, low efficiency of financing and high cost of financing in SCF [65]. Hofmann et al. [30] try to discover possible opportunities from the application of the blockchain technology to SCF financing solutions, particularly in approved payables financing. The results show that blockchains and distributed ledgers technology can bring substantial benefits to all parties involved in SCF transactions by speeding up the processes and reducing the overall cost of financing schemes.

## 5.1   The Significance of Blockchain Technology to SCF

Because of the small scale of operation, lack of fixed assets, poor risk resistance and low credit rating of banks, financial institutions usually need to invest a very high human cost for reducing the risk of financing. However, cost is always the core of business practice. Therefore, SMEs are generally difficult to obtain bank loans at low interest rates like competitive enterprises or even impossible to get. To develop financing services for SMEs and reduce the risk of financing, specifically, financial institutions need to investigate the real operation of supply chains to ensure that participants, trading results and documents are the basis of real asset transactions. A large number of documents auditing, risk control index estimation and establishment, as well as high operating costs, make it impossible for Banks to fully and effectively carry out supply chain finance, which greatly hinders the formation of scale economy and scope economy of Banks.

In addition, the goal of SCF is to cover the financing of SMEs in an all-round way. Under the credit guarantee of core enterprises, most of the financing objects of SCF are 'primary suppliers' and 'primary distributors'. However, the financing needs of a large number of secondary and tertiary suppliers and distributors are still unable to meet.

These painful points have always been difficult points to overcome for scholars and practitioners. However, with the emergence of blockchain technology, researchers believe that blockchain technology, due to its characteristics, can well make up for the existing deficiencies of SCF business, delivering credit, demonstrating authenticity at low cost and better serving SMEs. More concretely, financial institutions can use transparent data on the chain and judge the authenticity of trade by means of cross-validation, so as to decide whether to finance enterprises or not. The SCF model under the blockchain structure can greatly improve the operation efficiency of the capital market, reduce fraud in financial transactions, and lower transaction costs. This will lead to a series of theoretical and applied innovations in the field of SCF.

## 5.2    New Perspective for Risk Management: Blockchain Technology

Risk management of SCF is realized through loan-to-value rate, contracts coordination and financing schemes, which is essentially the risk sharing among members of the supply chain, third-party logistics companies and financial institutions. However, with the rise of blockchain technology, researchers have found that supply chain finance risk management can be realized from a new perspective. Compared with risk sharing under information asymmetry, the new risk management measure is more to increase the information transparency among the members involved in SCF to avoid the generation of risk.

In the process of supply chain operation, all information is stored separately. For example, financing enterprises have information of corporate operation and capital, logistics companies have information of logistics and warehouse receipt pledge, and Banks have information of financial products. Information dispersion and false information make it necessary for financial institutions to spend a great many costs when evaluating the credit status of financing enterprises. This situation leads to the difficulty and high cost of financing for financing enterprises, and further hinders financial institutions from carrying out supply chain financial business.

Nowadays, practitioners can build a platform for the gathering participants involved in SCF based on blockchain technology. In addition, tax authorities, electric power authorities and legal authorities should be incorporated into the platform to make full use of the information of authoritative institutions. Members store information related to the transaction content on the platform. As long as the data uploaded to the platform is authentic, the non-tamper ability of blockchain can ensure that the data on the platform will always be true. When the financing enterprise proposes the financing demand to the financial institution, the financial institution can make statistics and analysis of all the information related to the financing enterprise on the platform by using the big data technology. According to the data obtained, the bank judges the financing risk and thus avoid huge losses.

## 6  Conclusions and Future Research

Risk management has become an essential tool in dealing with the risk issues in supply chain finance. This paper surveys the applications of traditional risk disposal to supply chain finance and reviews the existing literature including supply chain finance risk management via loan-to-value ratio, contracts coordination and financing schemes, and financial derivatives. However, because of the small scale of operation, lack of fixed assets, poor risk resistance and low credit rating of banks, the cost for reducing the risk of supply chain finance via the traditional risk disposal is very high. Specifically, financial institutions need to investigate the real operation of supply chains to ensure that participants, trading results and documents are the basis of real asset transactions. In addition, the financing demand of a large number of secondary and tertiary suppliers and distributors are still unable to meet.

The development of the digital economy, in which the key factor of production is the data in digital form, can significantly improve the effectiveness of different types of production, technology, equipment, storage, sale, delivery of goods and services. In the digital economy scenario, digitalization of supply chains is also becoming increasingly dynamic. One of the tools of the digital economy, allowing to provide all the necessary conditions and mechanisms for technology is the blokchain technology. Blockchain technology can alleviate or even resolve the business pain point of SCF, delivering credit and demonstrating authenticity at low cost. The SCF model under the block chain structure can greatly improve the operation efficiency of the capital market, reduce fraud in financial transactions, and lower transaction costs. This will lead to a series of theoretical and applied innovations in the field of SCF.

However, we still face some difficulties and challenges. For instance, how to ensure that the data uploaded to the platform is authentic? How to motivate participants to join the platform? Some researchers, such as He et al. [27], propose a real incentive mechanism for distributed P2P applications based on blockchain, which uses encrypted currencies such as bitcoin to motivate users to cooperate. Therefore, the use of this technology needs higher technical conditions. Besides, credit risk and operational risk can be reduced through blockchain technology. However, how to reduce the market risk more effectively? This is something we need to think about constantly in the future.

**Acknowledgement.** The authors particularly acknowledge the financial supports from the Research Base of Beijing Modern Manufacturing Development at Beijing University of Technology and the Research Center of Blockchain at Beijing University of Technology. Jian Li was supported by National Natural Science Foundation of China under grant Nos. 71932002, 71571010 and by Great Wall Scholar Training Program of Beijing Municipality under grant No. CIT&TCD20180305. Yongwu Li was supported by the Beijing Natural Science Foundation under Grant No. 9192001.

# References

1. Babich, V.: Vulnerable options in supply chains: effects of supplier competition. Nav Res. Logistics **53**(7), 656–673 (2006)
2. Babich, V., Burnetas, A.N., Ritchken, P.H.: Competition and diversification effects in supply chains with supplier default risk. Manuf. Serv. Oper. Manag. **9**(2), 123–146 (2007)
3. Buzacott, J.A., Zhang, R.Q.: Inventory management with asset-based financing. Manage. Sci. **50**(9), 1274–1292 (2004)
4. Cai, J., Hu, X., Tadikamalla, P.R., et al.: Flexible contract design for VMI supply chain with service-sensitive demand: Revenue-sharing and supplier subsidy. Eur. J. Oper. Res. **261**(1), 143–153 (2017)
5. Caldentey, R., Haugh, M.B.: Supply contracts with financial hedging. Oper. Res. **57**(1), 47–65 (2009)
6. Camerinelli, E.: Supply chain finance. J. Payments Strategy Syst. **3**(2), 114–128 (2009)
7. Chopra, S., Sodhi, M.S.: Supply-chain breakdown. MIT Sloan Manag. Rev. **46**(1), 53–61 (2004)
8. Chen, X., Cai, G.G.: Joint logistics and financial services by a 3PL firm. Eur. J. Oper. Res. **214**(3), 579–587 (2011)
9. Chen, X., Hao, G., Li, L.: Channel coordination with a loss-averse retailer and option contracts. Int. J. Prod. Econ. **150**, 52–57 (2014)
10. Chen, X., Shen, Z.J.: An analysis of a supply chain with options contracts and service requirements. IIE Trans. **44**(10), 805–819 (2012)
11. Chen, X., Wan, G., Wang, X.: Flexibility and coordination in a supply chain with bidirectional option contracts and service requirement. Int. J. Prod. Econ. **193**, 183–192 (2017)
12. Chod J., Trichakis N., Tsoukalas G., Aspegren H., Weber M.: Blockchain and the value of operational transparency for supply chain finance. Social Science Electronic Publishing (2017)
13. Christidis, K., Devetsikiotis, M.: Blockchains and smart contracts for the internet of things. IEEE Access **4**, 2292–2303 (2016)
14. Christopher, M., Lee, H.: Mitigating supply chain risk through improved confidence. Int. J. Phys. Distrib. Logistics Manag. **34**(5), 388–396 (2004)
15. Cossin, D., Hricko, T.: A structural analysis of credit risk with risky collateral: a methodology for haircut determination. Econ. Notes **32**(2), 243–282 (2003)
16. Cossin, D., Huang, Z., Auron-Nerin, D., et al.: A framework for collateral risk control determination. In: ECB Working Paper (2003)
17. Crawford, W.: The Digital Economy: Promise and Peril in the Age of Networked Intelligence: by Don Tapscott. McGraw-Hill, New York (1996)

18. Cruz, J.M., Nagurney, A., Wakolbinger, T.: Financial engineering of the integration of global supply chain networks and social networks with risk management. Naval Res. Logistics (NRL) **53**(7), 674–696 (2006)
19. Dada, M., Hu, Q.: Financing newsvendor inventory. Oper. Res. Lett. **36**(5), 569–573 (2008)
20. Deng, S.J., Elmaghraby, W.: Supplier selection via tournaments. Prod. Oper. Manag. **14**(2), 252–267 (2005)
21. Finch, P.: Supply chain risk management. Supply Chain Manag. Int. J. **9**(2), 183–196 (2004)
22. Fu, Q., Zhou, S.X., Chao, X., Lee, C.Y.: Combined pricing and portfolio option procurement. Prod. Oper. Manag. **21**(2), 361–377 (2012)
23. Gaur, V., Seshadri, S.: Hedging inventory risk through market instruments. Manufact. Serv. Oper. Manag. **7**(2), 103–120 (2005)
24. Gelsomino, L.M., Mangiaracina, R., Perego, A., Tumino, A.: Supply chain finance: a literature review. Int. J. Phys. Distrib. Logistics Manag. **46**(4), 348–366 (2016)
25. Goh, M., Lim, J.Y.S., Meng, F.: A stochastic model for risk management in global supply chain networks. Eur. J. Oper. Res. **182**(1), 164–173 (2007)
26. Golovachkina, N.: Supplier-manufacturer relationships under forced compliance contracts. Manufact. Serv. Oper. Manag. **5**(1), 67–69 (2003)
27. He, Y., Li, H., Cheng, X., Liu, Y., Yang, C., Sun, L.M.: A blockchain based truthful incentive mechanism for distributed P2P applications. IEEE Access **6**, 27324–27335 (2018)
28. He, J., Wang, J., Jiang, X., Chen, X., Chen, L.: The long-term extreme price risk measure of portfolio in inventory financing: an application to dynamic impawn rate interval. Complexity **20**(5), 17–34 (2015)
29. Hofmann, E., Belin, O.: Supply Chain Finance Solutions. Springer, Heidelberg (2011). https://doi.org/10.1007/978-3-642-17566-4
30. Hofmann, E., Strewe, U.M., Bosia, N.: Supply Chain Finance and Blockchain Technology. SF. Springer, Cham (2018). https://doi.org/10.1007/978-3-319-62371-9
31. Hua, S., Liu, J., Cheng, T.C.E., Zhai, X.: Financing and ordering strategies for a supply chain under the option contract. Int. J. Prod. Econ. **208**, 100–121 (2019)
32. Ivashina, V., Scharfstein, D.: Bank lending during the financial crisis of 2008. J. Financ. Econ. **97**(3), 319–338 (2010)
33. Jokivuolle, E., Peura, S.: Incorporating collateral value uncertainty in loss given default estimates and loan-to-value ratios. Eur. Fin. Manag. **9**(3), 299–314 (2003)
34. Jorion P.: Value at risk (2000)
35. Jörnsten, K., Nonås, S.L., Sandal, L., Ubøe, J.: Mixed contracts for the newsvendor problem with real options and discrete demand. Omega **41**(5), 809–819 (2013)
36. He, J., Jiang, X.L., Wang, J., Zhu, D.L., Zhen, L.: VaR methods for the dynamic impawn rate of steel in inventory financing under autocorrelative return. Eur. J. Oper. Res. **223**(1), 106–115 (2012)
37. Kanda, A., Deshmukh, S.G.: Supply chain coordination: perspectives, empirical studies and research directions. Int. J. Prod. Econ. **115**(2), 316–335 (2008)
38. Khojasteh, Y.: Supply Chain Risk Management. Springer, Singapore (2017). https://doi.org/10.1007/978-981-10-4106-8
39. Kshetri, N.: 1 Blockchain's roles in meeting key supply chain management objectives. Int. J. Inf. Manage. **39**, 80–89 (2018)
40. Korpela, K., Hallikas, J., Dahlberg, T.: Digital supply chain transformation toward blockchain integration. In: Proceedings of the 50th Hawaii International Conference on System Sciences

41. Lee, H.L.: The triple-A supply chain. Harvard Business Rev. **82**(10), 102–113 (2004)
42. Lee, C.H., Rhee, B.D.: Coordination contracts in the presence of positive inventory financing costs. Int. J. Prod. Econ. **124**(2), 331–339 (2010)
43. Lee, C.H., Rhee, B.D.: Trade credit for supply chain coordination. Eur. J. Oper. Res. **214**(1), 136–146 (2011)
44. Li, Y., Feng, G., Xu, Y.: Research on loan-to-value ratio of inventory financing under randomly-fluctuant price. Syst. Eng. Theor. Pract. **12**, 42–48 (2007)
45. Li, L., Porteus, E.L., Zhang, H.: Optimal operating policies for multiplant stochastic manufacturing systems in a changing environment. Manage. Sci. **47**(11), 1539–1551 (2001)
46. Long, M.S., Malitz, I.B., Ravid, S.A.: Trade credit, quality guarantees, and product marketability. Financ. Manag., 117–127 (1993)
47. Nakamoto, S.: Bitcoin: a peer-to-peer electronic cash system (2008)
48. Nishat Faisal, M., Banwet, D.K., Shankar, R.: Information risks management in supply chains: an assessment and mitigation framework. J. Enterp. Inf. Manag. **20**(6), 677–699 (2007)
49. Norrman, A., Jansson, U.: Ericsson's proactive supply chain risk management approach after a serious sub-supplier accident. Int. J. Phys. Distrib. Logistics Manag. **34**(5), 434–456 (2004)
50. Papadakis, I.S.: Financial performance of supply chains after disruptions: an event study. Supply Chain Manag. Int. J. **11**(1), 25–33 (2006)
51. Pei, P.P., Simchi-Levi, D., Tunca, T.: Sourcing flexibility, spot trading, and procurement contract structure. Oper. Res. **59**(3), 578–601 (2011)
52. Pfohl, H.C., Gomm, M.: Supply chain finance: optimizing financial flows in supply chains. Logistics Res. **1**(3–4), 149–161 (2009)
53. Ritchken, P.H., Tapiero, C.S.: Contingent claims contracting for purchasing decisions in inventory management. Oper. Res. **34**(6), 864–870 (1986)
54. Rui, H., Lai, G.: Sourcing with deferred payment and inspection under supplier product adulteration risk. Prod. Oper. Manag. **24**(6), 934–946 (2015)
55. Sarathi, G.P., Sarmah, S.P., Jenamani, M.: An integrated revenue sharing and quantity discounts contract for coordinating a supply chain dealing with short life-cycle products. Appl. Math. Model. **38**(15–16), 4120–4136 (2014)
56. Schummer, J., Vohra, R.: Auctions for procuring options. Oper. Res. **51**(1), 41–51 (2003)
57. Sheffi, Y.: The Resilient Enterprise: Overcoming Vulnerability for Competitive Advantage. MIT Press Books, Cambridge (2005). 1
58. Tang, C.S.: Robust strategies for mitigating supply chain disruptions. Int. J. Logistics Res. Appl. **9**(1), 33–45 (2006)
59. Tian, F.: An agri-food supply chain traceability system for China based on RFID & blockchain technology. In: 2016 13th International Conference on Service Systems and Service Management (ICSSSM) (2016)
60. Timme, S., Williams-Timme, C.: The financial-SCM connection. Supply Chain Manag. Rev. **4**(2), 33–40 (2000)
61. Tomlin, B., Wang, Y.: On the value of mix flexibility and dual sourcing in unreliable newsvendor networks. Manufact. Serv. Oper. Manag. **7**(1), 37–57 (2005)
62. Toyoda, K., Mathiopoulos, P.T., Sasase, I., Ohtsuki, T.: A novel blockchain-based product ownership management system (POMS) for anti-counterfeits in the post supply chain. IEEE Access **5**, 17465–17477 (2017)
63. Tunca, T.I., Zhu, W.: Buyer intermediation in supplier finance. Manage. Sci. **64**(12), 5631–5650 (2017)

64. Van Mieghem, J.A.: Commissioned paper: capacity management, investment, and hedging: review and recent developments. Manufact. Serv. Oper. Manag. **5**(4), 269–302 (2003)
65. Wang, J.S., Li, L., He, Q.S., Yu, X., Liu, Z.B.: Research on the application of block chain in supply chain finance. DEStech Trans. Comput. Sci. Eng. (2017)
66. Wu, D.J., Kleindorfer, P.R.: Competitive options, supply contracting, and electronic markets. Manage. Sci. **51**(3), 452–466 (2005)
67. Wu, J., Li, J., Chen, J., Zhao, Y.X., Wang, S.Y.: Risk management in supply chains. Int. J. Revenue Manag. **5**(2/3), 157–204 (2011)
68. Xu, X., Birge, J.R.: Joint production and financing decisions: modeling and analysis (2004)
69. Yan, J., Wang, X., Cheng, H., Huang, L.: Study on the coordination contract in supply chain under trade credit based on risk compensation. Chaos Solitons Fractals **89**, 533–538 (2016)
70. Yang, Z., Aydın, G., Babich, V., Beil, D.R.: Supply disruptions, asymmetric information, and a backup production option. Manage. Sci. **55**(2), 192–209 (2009)
71. Zhang, Q., Dong, M., Luo, J., Segerstedt, A.: Supply chain coordination with trade credit and quantity discount incorporating default risk. Int. J. Prod. Econ. **153**, 352–360 (2014)
72. Zhang, R., Zhang, J., Xu, S.: Determining pledged loan-to-value ratio: an option pricing perspective. Financ. Innov. **1**(1), 16 (2015)
73. Zhao, L., Huchzermeier, A.: Supply Chain Finance. Springer, Cham (2018). https://doi.org/10.1007/978-3-319-76663-8
74. Zhao, Y., Wang, S., Cheng, T.C.E., Yang, X., Huang, Z.: Coordination of supply chains by option contracts: a cooperative game theory approach. Eur. J. Oper. Res. **207**(2), 668–675 (2010)
75. Zhu, Q., Kouhizadeh, M.: Blockchain technology, supply chain information, and strategic product deletion management. IEEE Eng. Manag. Rev., 1 (2019)
76. Zsidisin, G.A., Henke, M. (eds.): Revisiting Supply Chain Risk. Springer, Cham (2019). https://doi.org/10.1007/978-3-030-03813-7

# A Survey for Stochastic Decomposition in Vacation Queues

Naishuo Tian[1], Xiuli Xu[1(✉)], Zhanyou Ma[1], Shunfu Jin[2], and Wei Sun[3]

[1] School of Science, Yanshan University, Qinhuangdao 66004, Hebei,
People's Republic of China
tiannsh@ysu.edu.cn, xxl-ysu@163.com, mzhy55@163.com
[2] School of Information Science and Engineering, Yanshan University,
Qinhuangdao 066004, Hebei, People's Republic of China
jsf@ysu.edu.cn
[3] School of Economics and Management, Qinhuangdao 066004, Hebei,
People's Republic of China
wsun@ysu.edu.cn

**Abstract.** The study of vacation queue started in the 1970s. Up to now, it has made abundant achievements, formed a theoretical framework with stochastic decomposition as the core, and has been applied to many fields. This paper gives a comprehensive overview of the research results and analysis methods of vacation queue, including its applications in the communication networks.

The essence of vacation queue is that the service may be interrupted. As early as 1982, Jinhua Cao and Kan Cheng studied a kind of repairable queueing system, which was the earliest work involving vacation queue at home. From then on, a group of domestic scholars began to engage in the study of vacation queue, and achieved a series of important achievements, which are also described in this paper.

**Keywords:** Vacation queue · Working vacation queue · Stochastic decomposition

## 1 Introduction

Since the pioneering work of Erlang [1], queueing theory has been developed over almost 100 years and become an active branch of operations research and applied probability theory. The analysis of classical queue model has formed a complete theoretical system, and widely applied to various areas, such as, communication, manufacture, traffic, service, management, military, and others. In modern times, queueing theory has successfully applied to the system design and performance analysis of high-tech field, such as, computer networks, wireless communication, flexible manufacturing, e-commerce, online finance, and others. The fundamental cause of keeping queuing theory in perpetuity is that it is connected with the latest technological development in each period. The excellent monographs on the classical queueing theory have been published, see [2–6], etc.

© Springer Nature Singapore Pte Ltd. 2019
Q.-L. Li et al. (Eds.): Cao Festschrift 2019, CCIS 1102, pp. 134–158, 2019.
https://doi.org/10.1007/978-981-15-0864-6_6

Vacation queue is the generalization of classical queue. In many practical applications, the server may become unavailable for a period of time due to a variety of reasons. This period of server absence is called "vacation time". For example, the damage and repair of service facilities; routine maintenance of service facilities in order to ensure service efficiency once the system becomes empty; the server works on some supplementary jobs in its idle time when the system load is lower; highway traffic jam; the airport was closed by fog, all cases may result in the server vacation. Vacation queue reflects the fact that the service may be interrupted. On the other hand, various vacation behaviors may provide more flexibility for the optimization design and operation control of the system. Therefore, the study of vacation queue has become the research topic of stochastic models, and has obtained rich theoretical achievements and the practical application achievements.

This paper systematic comments on the research achievements and analysis methods of the vacation queue, and its applications in the performance analysis of the communication networks.

Though priority queue, polling queue, repairable queue, all can be included in the vacation queue, but it is generally regarded that the research on vacation queue started from Levy and Yachiali [7]. From the view of effectively utilize idle time, the authors introduced the term "vacation queue" and studied a kind of M/G/1 vacation queue using embedded Markov chain, and derived the stationary queue length distribution and waiting time distribution. Another important contribution of literature [7] is to prove that the stationary queue length and waiting time in this M/G/1 model can be decomposed into the sum of two independent random variables, where one part is the stationary queue length and waiting time in the classical M/G/1 queue without vacation; the other is additional queue length and additional delay caused by vacation behaviors. This kind of "independent sum" structure of stationary indices is called "stochastic decomposition" in literatures. Subsequently, it has been proved that the stochastic decomposition structure is a universal rule in various vacation queues, and it has important role in the analysis of vacation queue. Firstly, the stationary distribution of classical queue without vacation is known, the stochastic decomposition makes the analysis of vacation queue to focus on the study of additional queue length and additional delay. Secondly, stochastic decomposition results may clearly show the effect of various vacation polices on the stationary performance indices of classical queueing system. Therefore, the stochastic decomposition brings great convenience to theoretical analysis and practical applications.

From 1975 to 1985, the early research on vacation queue focused on the M/G/1 type queue models with variety of vacation policies. Using different methods, many researchers investigated the M/G/1 queue with multiple vacations, single vacation, setup times, N-policy, etc. In 1986, a long survey written by Doshi [8] detailed commented the early research results and analysis methods of vacation queue on the first issue of the journal "Queueing System". This survey greatly prompted the study of vacation queue. Almost simultaneously, another survey written by Teghen [9] was published in the journal "European Journal of Operational Research", introducing the early research achievements from the view of service control. Tian et al. [10] introduced the matrix-geometric solution method into the study of vacation queue, analyzed a kind of GI/M/1 vacation model and obtained the stationary distribution and its stochastic decomposition. Subsequently, Chatterjee and Mukherjee [11] studied the GI/M/1 type vacation queue using different method. Doshi [12] discussed the stochastic decomposition structure in a general GI/G/1 vacation queue. In about 1990, there are many research works on multiserver vacation queue. The earlier works see Levy and Yachili [13] and Igaki [14]. Then, Tian et al. [15–18] investigated the M/M/c queue with various vacation policies by quasi-birth-and-death process method. The related works see Tian and Li [19], Xu and Zhang [20] and Tian and Zhang [21–23]. The analysis of the GI/M/c type vacation queue can refer to Tian and Zhang [24], Chao and Zhao [25], etc.

Different from single server vacation queue, the stationary index distributions in a multiserver vacation queue is complicated, and show no stochastic decomposition rule. However, Tian, Li and Cao [26] revealed the conditional stochastic decomposition structure in the multiserver vacation queue. The number of customers in queue under the condition the servers are all busy can be decomposed into the sum of two independent random variables, where one variable is the number of customers in a corresponding classical multiserver queue without vacation given that the servers are all busy, the other is additional queue length caused by vacation. When an arrival finds all the server busy, the conditional waiting time was proved to satisfy the similar conditional stochastic decomposition. In fact, for a single server vacation queue, the conditional queue length and conditional waiting time given that the server is busy also satisfied this kind of conditional stochastic decomposition rule. Dishi [27] discussed the conditional distribution in the M/G/1 vacation model. Only because the unconditional distributions of stationary queue length and waiting time exist simple stochastic decomposition structure, which conceals the study of conditional stochastic decomposition. The stochastic decomposition of conditional queue length and waiting time is the common rule for both single server vacation queue and multiserver vacation queue.

In this years, the queueing theory monograph published has included the analysis of vacation queue. The analysis of all kinds of M/G/1 type (Geom/G/1-type) vacation queues can refer to Bose [28] and Takagi [29]. The book edited by Dshalalow [30] introduced the queueing system where vacation queue was listed in the queue model with the state dependent on parameter. Tian and Zhang [31] took stochastic decomposition as the main line to construct the theoretical framework of vacation queue. This monograph dedicated on vacation models, and included the study of not only various M/G/1 type vacation queues but also GI/M/1 type and multiserver vacation queue, moreover, introduced the optimization design and operation control method, and some application examples. The main contributions of this paper is to provide a comprehensive overview for the stochastic decomposition structures of vacation queues.

The remainder of this paper is organized as follows. In Sect. 2, we describe various vacation policies. Subsequently, we respectively give the stochastic decomposition results for the M/G/1 vacation queue with non-exhaustive service, the M/G/1 vacation queue with exhaustive service, GI/M/1-type vacation queue, multi-server vacation queue and working vacation queue from Sect. 3 to Sect. 7. Finally, we present the conclusions.

## 2    Vacation Policies

A vacation queuing system is formed by introducing appropriate vacation policy into a classical queuing system. Vacation policy indicates the vacation startup rule and vacation termination rule. The vacation strategies frequently appeared in the literature are cited.

One class is called exhaustive service: once the server begins to serve the customer, the server will continuously serve customers until the system becomes empty. The vacation starts at the end of busy period. The vacation policies with exhaustive service are as follows.

(1) Multiple Vacation (MV). Once there is no customer in the system, the server begins to take a vacation of random length $V$. The server will take another i.i.d vacation if there is no customer waiting in the system at a vacation completion instant. The server terminates vacation period and begins a new busy period until there are customers in the system at a vacation completion instant.

(2) Single Vacation (SV). Once the system becomes empty, the server begins to take a vacation of random length $V$. If there are customers in the system at a vacation completion instant, the server begins to serve customers; Otherwise, the server stays in idle.

(3) Setup Time (ST). The service facility is closed once a busy period completes, the system will be restarted to serve the customer after a setup (preheat) time of duration $V$ when a customer arrives.

(4) $N$-policy Vacation (NV). Once the system becomes empty, the server begins to take a vacation, a new busy period is started until there are $N$ arrivals in the system.

The other class is called nonexhaustive service. The server can take a vacation even when the system is not empty. The classical nonexhaustive service includes the following polices:

(1) Gated Service (GS). When the server returns from a vacation and begins a service period, it serves continuously only those customers present at that time, deferring the service of all customers that arrive during the service period until after the completion of the next vacation.

(2) Limited Service (LS). The amount of work done in serving customers during a given service period is limited. For example, the number of customers serves continuously is not large than $M$ ($M$-limited) or the total work time

length of a service period is not large than $T$ ($T$-limited). Whenever the service limit is reached, the server starts a vacation regardless of the number of customers in the system.

(3) Decrementing Service (DS). Once the server resumes queue service after a vacation, it keeps serving customers until the amount of work (total customer service time) is smaller than the amount of work at the beginning of the busy period, and then it takes a vacation. For example, the server may start a vacation when the number of customers in the system becomes $M$ less than the number of customers when the busy period started.

(4) Bernoulli Schedule (BS). The server takes a vacation of duration $V$ with probability $p$ and serves another customer, if any, with probability $1 - p$ after each customer service.

(5) Repairable Queue (RQ). The operating service may be damaged, and it continues to serve the customer after repair, taking the repair time as server's vacation.

The above illustrates some common vacation policies for the sake of narrative. In fact, we may present various vacation policies according to application requirements. Many special vacation policies in literatures are not mentioned here.

## 3   M/G/1 Vacation Queue with Exhaustive Service

Consider a classical M/G/1 queue, where customers arrive according to a Poisson process with rate $\lambda$, and service times are general distributed random variables with LST $b(s)$ and mean $\mu^{-1}$. Assume that the service order is first-come-first-served (FCFS). Let $Q$ and $W$ be the stationary queue length and waiting time, respectively. Denote the probability generating function (PGF) of $Q$ and the Laplace-Stieltjes transform (LST) of $W$ by $Q(z)$ and $W(s)$, respectively. When $\rho = \lambda\mu^{-1} < 1$, it is well known that

$$Q(z) = \frac{(1 - \rho)(1 - z)b(\lambda(1 - z))}{b(\lambda(1 - z)) - z} \tag{1}$$

$$W(s) = \frac{(1 - \rho)s}{s - \lambda(1 - b(s))} \tag{2}$$

and the means

$$E(Q) = \rho + \frac{\lambda^2 b^{(2)}}{2(1 - \rho)}, \quad E(W) = \frac{\lambda b^{(2)}}{2(1 - \rho)}$$

where $b^{(2)}$ is the second order of the service time. This is the famous Pallaczak-Khinthine formulae.

Introducing some vacation policy with exhaustive service, vacation time $V$ follows a general distribution, denote its LST by $v(s)$. The embedded Markov chain method is a simple and effective tool to deal with this kind of $M/G/1$ vacation queue. Denote the number of customers in the system at the $n$th customer departure instant by $L_n$, then $\{L_n, n \geq 1\}$ is an embedded Markov chain of the queueing process, and

$$L_{n+1} = \begin{cases} L_n - 1 + A, \ L_n \geq 1, \\ Q_b - 1 + A, \ L_n = 0 \end{cases}$$

where $A$ is the number of arrivals in a service time. $Q_b$ is the number of customers present in the system at the beginning of a busy period, its distribution is dependent on vacation policy. Note that all regeneration points (embedded instants) are in busy period, therefore, comparing $\{L_n, n \geq 1\}$ in vacation queue with the embedded chain in the classical $M/G/1$ queue, only the first row of their transition probability matrix is different. Moreover, based on PASTA property, the stationary queue length distribution of $\{L_n, n \geq 1\}$ is the queue length distribution in any instant for the Poisson arrival queue.

Now, denote the PGF of the stationary number of customers in the system, $Q_v$, and the LST of the stationary waiting time, $W_v$, by $Q_v(z), W_v(s)$, respectively. Using embedded Markov chain method, for various $M/G/1$ vacation queues with exhaustive service, we obtain the stochastic decomposition structure

$$Q_v(z) = Q(z)Q_d(z), \quad W_v(s) = W(s)W_d(s)$$

where $Q(z), W(s)$ are the PGF of the stationary queue length and the LST of the stationary waiting time in the classical $M/G/1$ queue without vacation, respectively, which is derived by Eqs. (1) and (2). $Q_d(z), W_d(s)$ are the PGF of the additional queue length and the LST of the additional waiting time, respectively. For the $M/G/1$ queue with multiple vacations, single vacation or setup times, they are as follows

$$Q_d(z) = \begin{cases} \dfrac{1 - v(\lambda(1-z))}{\lambda E(V)(1-z)}, & \text{(MV)} \\[2ex] \dfrac{1 - v(\lambda(1-z)) + v(\lambda)(1-z)}{[v(\lambda) + \lambda E(V)](1-z)}, & \text{(SV)} \\[2ex] \dfrac{1 - zv(\lambda(1-z))}{[1 + \lambda E(V)](1-z)}, & \text{(ST)} \end{cases}$$

and

$$W_d(s) = \begin{cases} \dfrac{1 - v(s)}{E(V)s}, & \text{(MV)} \\[2ex] \dfrac{sv(\lambda) + \lambda(1 - v(s))}{[v(\lambda) + \lambda E(V)]s}, & \text{(SV)} \\[2ex] \dfrac{\lambda - (\lambda - s)v(s)}{[1 + \lambda E(V)]s}, & \text{(ST)} \end{cases}$$

From this kind of stochastic decomposition structure, it is easy to derive various performance indices of vacation model. For example, the mean of $Q_v$ is

$$E(Q_v) = \rho + \frac{\lambda^2 b^{(2)}}{2(1-\rho)} + \begin{cases} \dfrac{\lambda E(V^2)}{2E(V)}, & \text{(MV)} \\[2mm] \dfrac{\lambda^2 E(V^2)}{2[v(\lambda) + \lambda E(V)]}, & \text{(SV)} \\[2mm] \dfrac{2\lambda E(V) + \lambda^2 E(V^2)}{2[1 + \lambda E(V)]}, & \text{(ST)} \end{cases}$$

For the M/G/1 queue with N-policy, vacation time can be regard as the sum of $N$ interarrivals, and $Q_b = N$. The conditional waiting time $W_v'$ given the arrival occurs in a busy period can be decomposed into independent sum $W_v' = W + W_d'$, and

$$Q_d(z) = \frac{1 - z^N}{N(1 - z)},$$
$$W_d'(s) = \frac{\mu(1 - b^N(s))}{Ns}.$$

Levy and Yechiali [7] firstly studied the M/G/1 (MV) and M/G/1 (SV). Subsequently, some researchers investigated the M/G/1 vacation queue with exhaustive service using different methods. Such as, Fuhrmann [32], Doshi [33], Harris and Marchal [34], Levy and Kleinrock [35], Takagi [36], Lee [37], Lee et al. [38], Keilson and Servi [39], Keilson and Ramaswami [40], and others.

Various exhaustive service M/G/1 type queues were investigated successively, and the stochastic decomposition of stationary indices were derived. Heymann [41] discussed the M/G/1 queue with $T$-policy, once the busy period completes, the server leaves the system for a certain length of time $T$, it begins to serve the arrival or enter the idle state decided by the number of customers in the system when it returns. Balachandran and Tijms [42] examined a kind of M/G/1 queue with $D$-policy, the server leaves the system after a busy period, it returns until the work accumulation (total service time of waiting customers) reaches a threshold $D$. Artalejo [43, 44] analyzed a class of M/G/1 queue with $N$-policy and $D$-policy. Yadin and Naor [45] introduced $N$-policy of the setup times and closed times into the M/G/1 queue with the portable server. Lee and Srinivason [46], Baba [47] studied the batch arrival M/G/1 vacation queue. Tian [48] introduced a multiple adaptive vacation (MAV) policy, the number of vacations after a busy period was controlled by a integer random variable $H$ and arrival process. Specially, $H = 0$ and $H = 1$ correspond to (MV)-policy and (SV)-policy, respectively. The stationary distributions and their stochastic decomposition of the M/G/1 (MAV) were derived in [48].

Other study is not on special vacation policy but to give the stochastic decomposition directly by taking various vacation policies satisfying some requirement as a whole. The related researches see Fuhrmann and Cooper [49], Shanthikumar [50], Li and Zhu [51], and others.

Recently, Sun et al. [52, 53] not only discussed the stationary distribution but also economic behaviors of some M/G/1 vacation queues, and analyzed Nash equilibrium policies under competitive mechanism. More related researches see Su and Li [54], Wang [55].

The analysis of exhaustive service M/G/1 type vacation queue has extended corresponding discrete time Geom/G/1 queue. The discrete time queue is more suitable for modeling analysis of computer networks and communication systems, which refers to Brunnel and Kin [56], Woodward [57], Kobayashi and Konheim [58], and others. Zhang and Tian [59] investigated the Geom/G/1 (MAV). The Geom/G/1 (MV) and Geom/G/1 (SV) can be regarded as special cases of [59]. The researches of Geom/G/1 vacation queue see Fiems and Bruneel [60,61], Bruneel [62,63], Alfa [64,65], and others. The analysis of Geom/G/1 vacation queue refers to Takagi [29] (volume 3), and Tian and Zhang [31].

## 4   M/G/1 Vacation Queue with Non-exhaustive Service

The embedded Markov chain method does not apply to the M/G/1 vacation queue with non-exhaustive service. The regeneration cycle method is a powerful tool to deal with this kind of model, and refers to Heyman and Sovel [66], or Ross [67].

Under the non-exhaustive vacation policy, there are customers in the system at the service period ending instant and starting instant. Let $Q_b$ be the number of customers present in the system at the beginning of a service period, with the PGF $Q_b(z)$, which is determined by vacation policy.

Taking gated service (GS) for example, when the server finishes services of $Q_b$ customers in the system, it starts to take a vacation of duration $V$. If there is no customer at a vacation completion instant, the server takes another vacation. The server starts a new service period until there are customers in the system at some vacation completion instant, which is called gated service and multiple vacation (GS, MV) policy. Similarly, we can define (GS, SV) policy. If $\rho < 1$, using regeneration cycle method, we obtain the following stochastic decomposition. $Q_v$ and $W_v$ can be decomposed into the sum of three independent random variables

$$Q_v(z) = Q(z)Q_d(z)Q_r(z) \quad W_v(s) = W(s)W_d(s)W_r(s).$$

$Q(z), W(s)$ are given by Eqs. (1) and (2). For the M/G/1 (GS, MV), we have

$$Q_d(z) = \frac{1 - v(\lambda(1-z))}{\lambda E(V)(1-z)},$$
$$Q_r(z) = Q_b(b(\lambda(1-z))),$$
$$W_d(s) = \frac{1 - v(s)}{E(V)s},$$
$$W_r(s) = Q_b(b(s)).$$

where it is complicated for the distribution of $Q_b$, but it is easy to derive the function equation satisfied by its PGF $Q_b(z)$, and get the mean of $Q_b$

$$E(Q_b) = \frac{\lambda E(V)}{1 - \rho}.$$

Now, from the above stochastic decomposition, we obtain the means in the M/G/1 (GS, MV)

$$E(Q_v) = \rho + \frac{\lambda^2 b^{(2)}}{2(1-\rho)} + \frac{\lambda E(V^2)}{2E(V)} + \frac{\lambda \rho E(V)}{1-\rho},$$

$$E(W_v) = \frac{\lambda b^{(2)}}{2(1-\rho)} + \frac{E(V^2)}{2E(V)} + \frac{\rho E(V)}{1-\rho}.$$

Similarly, for the gated service and single vacation M/G/1 (GS, SV), we have

$$E(Q_b) = \frac{v(\lambda) + \lambda E(V)}{1-\rho}.$$

and

$$Q_d(z) = \frac{1 - v(\lambda(1-z)) + v(\lambda)(1-z)}{[v(\lambda) + \lambda E(V)](1-z)},$$

$$Q_r(z) = Q_b(b(\lambda(1-z))),$$

$$W_d(s) = \frac{v(\lambda)s + \lambda(1 - v(s))}{[v(\lambda) + \lambda E(V)]s},$$

$$W_r(s) = Q_b(b(s)).$$

The expected queue length and waiting time in the M/G/1 (GS, SV) respectively are

$$E(Q_v) = \rho + \frac{\lambda^2 b^{(2)}}{2(1-\rho)} + \frac{\lambda^2 E(V^2)}{2[v(\lambda) + \lambda E(V)]} + \frac{[v(\lambda) + \lambda E(V)]\rho}{1-\rho},$$

$$E(W_v) = \frac{\lambda b^{(2)}}{2(1-\rho)} + \frac{\lambda E(V^2)}{2[v(\lambda) + \lambda E(V)]} + \frac{v(\lambda) + \lambda E(V)}{\mu(1-\rho)}.$$

Regeneration cycle method can be used to deal with various non-exhaustive service M/G/1 vacation queue and obtained similar stochastic decomposition. Levy [68] introduced a Binomial distributed gated service policy. There are $Q_b$ customers in the system at the beginning of the service period, the number of customers served in this service period follows a Binomial distribution with parameter $(Q_b, p)$. Other related researches of GS see Levy [68], Browne et al. [69,70], Altman [71,72], Bacot et al. [73], Aktman [74], Choi et al. [75], and others. The studies of discrete time Geom/G1 queue with gated service refer to Ishizaki [76], fiems [77], and others.

Limited service (LS) M/G/1 vacation model was presented by Leung and Eisenberg [78,79]. Lee [80,81] studied the limited space (LS) M/G/1/N queue. Levy [68] investigated Bernoulli gated service model, was called Bernoulli limited service system in literature. Takagi and Leung [82] discussed the Geom/G/1 vacation queue with (LS)-policy. Ma and Tian [83,84] introduced MAV policy into the non-exhaustive service system, and studied the Geom/G/1 (LS,MAV) and (GS, MAV). Decrementing service (DS) M/G/1 vacation model referred to Takagi [85], this paper also studied the model with Bernoulli process. Other works involved Bernoulli process see Keilson and Servi [86,87],

Servi [88], Ramaswami and Servi [89], Choi and Park [90], Tedijanto [91], Wortman et al. [92], Kumar [93], Madan [94], and others.

Taking the damage and maintenance as server vacation, the repairable queue can be regarded as vacation queue with non-exhaustive service policy, where the vacation may take place in the service time randomly. Cao and Cheng [95], Cao [96] considered the repairable M/G/1 queue by introducing "the generalized service time", not only obtained the stationary indices but also reliability indices including frequency of damage, availability, etc. Wang, Cao and Li [97] analyzed the retrial M/G/1 repairable queue.

Bischof [98] investigated the M/G/1 type vacation queue with many vacation polices. Variety of M/G/1, Geom/G/1 vacation queue with exhaustive and non-exhaustive service refer to the monographs Takagi [29], Tian and Zhang [31].

## 5    GI/M/1-Type Vacation Queue

For a classical GI/M/1 queue, the interarrival time follows a general distribution, with the LST $a(s)$ and the mean $\lambda^{-1}$. The service time follows an exponential distribution with parameter $\mu$. The service order is an FCFS discipline.

Let $L_n(n \geq 1)$ be the number of customers in the system just before the $n$th arrival instant. When $\rho < \lambda\mu^{-1} < 1$, denote the stationary queue length and stationary waiting time just before the $n$th arrival instant by $L$ and $W$, with the PGF $L(z)$ and the LST $W(s)$, respectively. We have

$$L(z) = \frac{1-r}{1-rz}, \quad W(s) = \frac{(1-r)(\mu+s)}{\mu(1-r)+s} \tag{3}$$

where $r$ is the unique root of the equation $z = a(\mu(1-z))$ in the interval $(0,1)$.

Compared to the M/G/1 type vacation queue, the study of the GI/M/1 type vacation queue was very late. Tian et al. [10] detailed analyzed the GI/M/1 (MV), where the vacation time $V$ follows an exponential distribution with parameter $\theta$. Now, $\{L_n, n \geq 1\}$ can't completely indicate the state of the system, we need to distinct the arrival occurs in busy period or vacation. Let

$$J_n = \begin{cases} 0, & \text{the } n\text{th arrival occurs in busy period,} \\ 1, & \text{the } n\text{th arrival occurs in vacation.} \end{cases}$$

$\{(L_n, J_n), n \geq 1\}$ is a two-dimensional embedded Markov chain, and the transition probability matrix has a block structure. Now, every numerical element of the transition probability matrix of $\{L_n, n \geq 1\}$ in a classical GI/M/1 queue is corresponding to a $2 \times 2$ sub-block, then, it is called the GI/M/1 type structural matrix. The matrix analysis method developed by Netus is used to deal with $\{(L_n, J_n), n \geq 1\}$. This method has become a powerful tools to analyze the complex stochastic model, see Netus [99, 100], Latouch and Rammaswami [101], Li [102], Tian and Yue [103], and others.

Using the matrix-geometric solution method ([99]) of the GI/M/1 type structural matrix, the stationary distribution of $\{(L_n, J_n), n \geq 1\}$ is obtained, the stochastic decomposition of the stationary queue length $L_v$ and the stationary waiting time $W_v$ at the arrival instant are derived. Both $L_v$ and $W_v$ can be decomposed into the sum of two independent random variables, that is, $L_v(z) = L(z)L_d(z)$, $W_v(s) = W(s)W_d(s)$. $L(z)$ and $W(s)$ are the PGF and LST of corresponding variables in the classical GI/M/1 queue without vacation, given by Eq. (3). The PGF $L_d(z)$ of additional queue length and the LST $W_d(s)$ of additional delay respectively are

$$L_d(z) = \sigma \frac{1 - zr + z\beta(r - a(\theta))}{1 - za(\theta)},$$

$$W_d(s) = \frac{\theta}{\theta + s}$$

where both $\beta$ and $\sigma$ are constants, and

$$\beta = \frac{\theta}{\theta - \mu(1 - a(\theta))},$$

$$\sigma = \frac{1 - a(\theta)}{1 - r + \beta(r - a(\theta))}.$$

It was proved if $\rho < 1$ then $\beta(r - a(\theta)) > 0$ in [10]. Expanding $L_d(z)$, we obtain the distribution of $L_d$

$$\begin{cases} P\{L_d = 0\} = \sigma, \\ P\{L_d = k\} = \sigma \frac{\mu}{\theta}(1 - a(\theta))\beta(r - a(\theta))[a(\theta)]^{k-1}, \quad k \geq 1. \end{cases}$$

The probabilities of an arrival occurs in the busy period or vacation in steady state respectively are

$$P\{J = 0\} = \frac{\beta(r - a(\theta))}{1 - r + \beta(r - a(\theta))},$$

$$P\{J = 0\} = \frac{1 - r}{1 - r + \beta(r - a(\theta))}.$$

From the above stochastic decomposition, we can derive the expectation formulae in the GI/M/1 (MV)

$$E(L_v) = \frac{r}{1 - r} + \frac{\mu\beta(r - a(\theta))}{\theta[1 - r + \beta(r - a(\theta))]},$$

$$E(W_v) = \frac{r}{\mu(1 - r)} + \frac{1}{\theta}.$$

Moreover, if the interarrival time follows a general distribution, the stationary distribution of $\{L_n, n \geq 1\}$ is not the same with the stationary queue length distribution at any time. Using semi-Markov chain method, [10] gave the stationary queue length distribution at any time in the GI/M/1 (MV).

Tian [104] discussed the GI/M/1 (SV) in parallel, and obtained the stationary analysis and the stochastic decomposition.

Chatterjee and Mukherjee [11] studied the GI/M/1 (MV) using the different method. Tian and Zhang [105] investigated the corresponding discrete time GI/Geom/1 (MV) and got the stationary distributions and the stochastic decomposition. Zhang and Tian [106] detailed analyzed the GI/M/1 queue with $N$-policy. Tain [107], Tian and Zhang [108] studied the GI/M/1 queue with PH vacation or setup times. Dukhovny [109] connected vacation queue with Reimann boundary value problem, and investigated the $GI^X/M^X/1$ queue with batch arrival and batch service by a special method. Karaesmen and Gupta [110] examined the limited space GI/M/1/N vacation queue using supplementary variable method, and obtained the stationary queue length distribution and the loss probability. Laxmi and Gupta [111] considered a class of vacation queue with batch service, limited space and a general distributed arrival. Ke [112] studied the GI/M/1 vacation queue with multiple vacation and $N$-policy based on the supplementary variable. Machihara [113] discussed the GI/SM/1 queue with a Markov service and vacation dependent on service time. The literatures on the GI/M/1 type queue are relatively less than those on the M/G/1 type queue. No research work on the non-exhaustive service GI/M/1 vacation queue is seen. The analysis of GI/M/1 type vacation model referred to Tian and Zhang [31].

## 6 Multi-server Vacation Queue

In a multiserver vacation queue, vacation policies have more complicated changes. All servers may enter and complete vacation state synchronously, is called synchronous (SY) vacation, or may start and terminate vacation individually, is called asynchronous (ASY) vacation. If vacation time follows an exponential distribution or PH distribution in the M/M/c queue with various vacation polices, quasi-birth-and-death process (QBD) and matrix-geometric solution method are ideal analysis tools.

Consider an M/M/c system with arrival rate $\lambda$, service rate $\mu$, and FCFS service order. When $\rho < 1$, denote the stationary queue length and waiting time by $Q$ and $W$, respectively. In order to derive the conditional stochastic decomposition, denote

$$Q^{(c)} = (Q - c | \text{all } c \text{ servers are busy}),$$
$$W^{(c)} = (W | \text{the arrival finds all servers busy}).$$

where $Q^{(c)}$ is the number of waiting customers in steady state given that all servers are busy, $W^{(c)}$ is the conditional waiting time given that a customer arrives at a state when all the servers are busy, denote corresponding PGF and LST by $Q^{(c)}(z)$ and $W^{(c)}(s)$, respectively. For the classical M/M/1 queue, it is easy to prove

$$Q^{(c)}(z) = \frac{1-\rho}{1-\rho z}, \quad W^{(c)}(s) = \frac{c\mu(1-\rho)}{s + c\mu(1-\rho)}. \tag{4}$$

Introducing some sort of vacation policy, the corresponding denotations in the M/M/c queue are added subscript $v$.

Considering the following vacation policy: once all the servers become empty, $c$ servers begin to take a vacation of duration $V$ together. If the system has no customer at the end of a vacation, the servers take another vacation; If there are $j \geq 1$ customers waiting in the system, all the servers terminate the vacation synchronously. If $1 \leq j \leq c$, $j$ servers begin to serve the customers, other $c - j$ servers enter the idle state and serve the arrival at any time. This kind of vacation policy is called synchronous single vacation policy (SY, SV). Assume that $V$ follows an exponential distribution with parameter $\theta$.

For the M/M/c (SY,MV), let $Q_v(t)$ be the number of customers in the system at time $t$, $J(t) = 1$ means all servers take a vacation at time $t$, otherwise, $J(t) = 0$. $\{(Q_v(t), J(t)), t \geq 0\}$ is a two dimensional QBD. Using QBD and matrix-geometric solution method, we obtain not only various stationary distribution but also the following conditional stochastic decomposition

$$Q_v^{(c)}(z) = Q^{(c)}(z)Q_d(z), \quad W_v^{(c)}(s) = W^{(c)}(s)W_d(s).$$

which indicate the conditional random variables $Q_v^{(c)}$ and $W_v^{(c)}$ in the M/M/c vacation model can decomposed into the sum of two independent random variables, one is conditional random variables $Q^{(c)}$ and $W^{(c)}$ in the classical M/M/c queue, and $Q_v^{(c)}(z)$ and $W_v^{(c)}(s)$ are given by Eq. (4). $Q_d$ and $W_d$ are called the additional queue length and additional delay, their PGF and LST respectively are

$$Q_d(z) = \frac{1}{\sigma}\left[\frac{1}{(c-1)!}\left(\frac{\lambda}{\mu}\right)^{c-1}\Psi + \left(\frac{\lambda}{\lambda+\theta}\right)^{c-1}\frac{\theta+\lambda}{\theta+\lambda(1-z)}\right],$$

$$W_d(s) = 1 - \frac{1}{\sigma}\frac{\lambda}{\theta}\left(\frac{\lambda}{\lambda+\theta}\right)^{c-1} + \frac{c\mu}{\sigma}\left(\frac{\lambda}{\lambda+\theta}\right)^{c}\left(s + \frac{\lambda}{\theta+\theta}c\mu\right)^{-1}.$$

where both $\Psi$ and $\sigma$ are constants, and

$$\Psi = \sum_{i=0}^{c-2}i!\left(\frac{\mu}{\lambda+\theta}\right)^{i},$$

$$\sigma = \frac{1}{(c-1)!}\left(\frac{\lambda}{\mu}\right)^{c-1}\Psi + \left(\frac{\lambda}{\lambda+\theta}\right)^{c-1}\frac{\lambda+\theta}{\theta}.$$

It is easy to verify that additional queue length $Q_d$ follows a modified geometric distribution, and additional delay $W_d$ follows a modified exponential distribution.

From the stochastic decomposition, we derive the expectations of conditional variables in the M/M/c (SY, MV)

$$E(Q_v^{(c)}) = \frac{1}{1-\rho} + \frac{1}{\sigma}\left(\frac{\lambda}{\lambda+\theta}\right)^{c-1}\frac{\lambda(\lambda+\theta)}{\theta^2},$$

$$E(W_v^{(c)}) = \frac{1}{c\mu(1-\rho)} + \frac{1}{\sigma}\left(\frac{\lambda}{\lambda+\theta}\right)^{c-1}\frac{\lambda(\lambda+\theta)}{c\mu\theta^2}$$

For the M/M/c (SY, SV) and M/M/c (SY, ST) queue, we also give various stationary distributions and the similar conditional stochastic decomposition, the details refer to Tian [114,115].

The asynchronous vacation can be divided into multiple vacation (ASY, MV), single vacation (ASY, SV), setup time (ASY, ST), etc. The QBD method can deal with the M/M/c queue with the above polices. Now, $J(t) = 0, 1, \cdots, c$ denote the number of the busy servers at time $t$, $\{(Q(t), J(t)), t \geq 0\}$ is a $c + 1$-dimensional QBD. Using the matrix-geometric solution method of QBD process, we derive the detailed analysis, and obtain the conditional stochastic decomposition structure.

$$Q_d(z) = \frac{1}{\sigma} \{\beta + z\, \boldsymbol{\delta}(\boldsymbol{I} - z\boldsymbol{H})^{-1}\boldsymbol{\eta}\},$$
$$W_d(s) = \frac{\beta}{\sigma} + \frac{c\mu}{\sigma}\boldsymbol{\delta}\,[s\boldsymbol{I} - c\mu(\boldsymbol{H} - \boldsymbol{I})]^{-1}\,\boldsymbol{\eta}.$$

where $\beta$ and $\sigma$ are constants, $\boldsymbol{\delta}$ and $\boldsymbol{\eta}$ are $c$-dimensional row vectors and $c$-dimensional column vector, respectively. $\boldsymbol{H}$ is $c \times c$ random matrix, all these depend on explicit vacation policy, and

$$\sigma = \beta + \boldsymbol{\delta}(\boldsymbol{I} - \boldsymbol{H})^{-1}\boldsymbol{\eta}.$$

$c + 1$-order block matrix

$$\boldsymbol{R} = \begin{bmatrix} \boldsymbol{H} & \boldsymbol{\eta} \\ \boldsymbol{0} & \rho \end{bmatrix}$$

is the minimal non-negative solution of matrix quadratic equation. In all kinds of vacation polices, by solving the matrix quadratic equation to determine and calculate these vectors and matrices, the details refer to Tian et al. [16]. Tian, Li and Cao [26], Tian [115]. From the above stochastic decomposition, it is easy to get the means of conditional random variables in vacation models

$$E(Q_v^{(c)}) = \frac{\rho}{1 - \rho} + \frac{1}{\sigma}\boldsymbol{\delta}(\boldsymbol{I} - \boldsymbol{H})^{-2}\boldsymbol{\eta},$$
$$E(W_v^{(c)}) = \frac{1}{c\mu}\left[\frac{\rho}{1 - \rho} + \frac{1}{\sigma}\boldsymbol{\delta}(\boldsymbol{I} - \boldsymbol{H})^{-2}\boldsymbol{\eta}\right].$$

The earliest work on the multiserver vacation model was done by Levy and Yechiali [13], the preliminary results were derived using the classical birth-and-death process. Igaki [14] considered the M/M/c queue with multiple vacation and $N$-policy. Vinod [116] firstly proposed QBD method to study the M/M/c queue, and gave the numerical method to determined stationary indices. Tian and Li [19] investigated the M/M/c queue with synchronous vacation, where the vacation times follow a PH distribution. Tian et al. [16] obtained the stationary distribution and the stochastic decomposition of the M/M/c queue with various asynchronous vacations. Tian, Li and Cao [26] constructed the uniform theoretical framework of the conditional stochastic decomposition for the M/M/c queues with synchronous and asynchronous vacations using matrix-geometric solution method. Zhang and Tian [23] introduced partial server' vacation policy, that is, there are no more than $d(1 \leq d \leq c)$ servers taking vacation and at least $c - d$ servers on duty(serving or stay in idle) in the system at any time. The analysis of the M/M/c queue with partial server' synchronous vacation or asynchronous vacation and their conditional stochastic decomposition refer to Zhang and Tian [22], Tian and Zhang [21], Xu and Zhang [20], and others.

Bardhan [117] studied a class of GI/M/c queue with interruptible service by the diffusion approximation. Chao and Zhao [24] discussed the GI/M/c vacation queue, taking synchronous vacation and asynchronous vacation as "the space vacation" and "the server vacation", respectively. Tian et al. [15] analyzed the GI/M/c queue with various synchronous vacations and got the conditional stochastic decomposition. Tian and Zhang [25] generalized the exponentially distributed vacation time to the PH distributed vacation time, and obtained the stationary distribution and the conditional stochastic decomposition of the GI/M/c vacation queue.

The analysis of variety of M/M/c and GI/M/c vacation queue refered to the monograph published by Tian and Zhang [31].

## 7    Working Vacation Queue

Servi and Finn [118] introduced a semi-vacation policy, is called working vacation (WV). For a classical M/M/1 queue with arrival rate $\lambda$ and service rate $\mu_b$, once the system becomes empty, the server begins to take a vacation of duration $V$, and $V$ follows an exponential distribution with parameter $\theta$. In a vacation period, the serve does not stop service completely and serves the arrival at a lower service rate $\mu_v$. If there is no customer in the system at the end of a vacation, the server takes another vacation; Otherwise, the server changes service rate from $\mu_v$ to $\mu_b$ and begins a busy period. This is called the multiple working vacation (MWV) policy. Similarly, we can define the single working vacation (SWV). Obviously, the working vacation queue is the generalization of the general vacation queue.

Literature [118] discussed the M/M/1 (MWV) using classical birth-and-death process, and obtained the stationary distribution results. Based on QBD and the matrix-geometric solution method, Liu, Xu and Tian [119] restudied the M/M/1 (MWV), derived various index distributions and revealed the stochastic decomposition rule. Denote the stationary queue length and waiting time by $Q_v$ and $D_v$, and their PGF and LST by $Q_v(z)$ and $D_v(s)$, respectively, then

$$Q_v(z) = Q(z)Q_d(z), \quad D_v(s) = D(s)D_d(s)$$

where $Q(z)$ and $D(s)$ are the PGF and LST of the stationary queue length and sojourn time, respectively, and

$$Q(z) = \frac{1-\rho}{1-z\rho}, \quad D(s) = \frac{\mu_b(1-\rho)}{s+\mu_b(1-\rho)}. \tag{5}$$

The PGF of the additional queue length and the LST of the additional sojourn time respectively are

$$Q_d(z) = K\left[1 - r + r\left(1 - \frac{\mu_v}{\mu_b}\right)\frac{z(1-r)}{1-zr}\right],$$

$$D_d(s) = K\left[\frac{\mu_v}{\mu_b}(1-r) + \left(1 - \frac{\mu_v}{\mu_b}\right)\frac{\sigma}{s+\sigma}\right].$$

where $r, \sigma, K$ are constants, and

$$r = \frac{1}{2\mu_v}\left[\lambda + \theta + \mu_v - \sqrt{(\lambda + \theta + \mu_v)^2 - 4\lambda\mu_v}\right],$$

$$\sigma = \frac{\lambda}{r}(1 - r), \qquad K = \left[1 - r + r\left(1 - \frac{\mu_v}{\mu_b}\right)\right]^{-1}.$$

It is proved that the additional queue length $Q_d$ follows a modified geometric distribution and the additional sojourn time $D_d$ follows a modified exponential distribution.

Tian et al. [120] studied the M/M/1 (SWV) model and gave the similar stochastic decomposition, and

$$Q_d(z) = K^*\left[\frac{\lambda + \theta}{\lambda}(1 - r) + \left(1 - \frac{\mu_v}{\mu_b}\right)\frac{zr}{1 - zr}\right],$$

$$D_d(s) = K^*\left[\left(\frac{\theta}{\lambda} + \frac{\mu_v}{\mu_b}\right)(1 - r) + \left(1 - \frac{\mu_v}{\mu_b}\right)\frac{\sigma}{s + \sigma}\right].$$

where

$$K^* = \left[\frac{\lambda + \theta}{\lambda}(1 - r) + r\left(1 - \frac{\mu_v}{\mu_b}\right)\right]^{-1}.$$

Li and Tian [121] introduced a interruptible working vacation policy: the number of customers reaches a threshold in a working vacation period, then the server interrupts the vacation, changes service rate from $\mu_v$ to $\mu_b$ and begins a regular busy period. In the M/M/1 queue with interruptible multiple working vacations, [121] gave the stationary distributions and stochastic decomposition results of $Q_v$ and $D_v$. Xu, Zhang and Tian [122] investigated the M/M/1 models with single working vacation and setup times. Tian Ma and Liu [123] extended the analysis of M/M/1 (WV) to discrete time queue, detailed analyzed the Geom/Geom/1 (MWV) queue and obtained its stochastic decomposition.

The GI/M/1 (WV) firstly studied by Baba [124], using matrix analysis method to get the stationary queue length distribution at the arrival instant and at any instant. Li [125] further obtained the stationary waiting time distribution, and proved the stochastic decomposition structure of the stationary queue length $L_v$ and waiting time $W_v$ at arrival instant. Both can be decomposed into the sum of two independent random variables: $L_v = L + L_d, W_v = W + W_d$, where $L$ and $W$ are the corresponding stationary random variables in the classical GI/M/1 queue, their PGF and LST are given by Eq. (3). The PGF of additional queue length $L_d$ and the LST of the additional delay $W_d$ respectively are

$$L_d(z) = K\frac{1 - rz + \alpha(r - h)z}{1 - hz},$$

$$W_d(s) = \frac{\mu_b(1 - h)}{s + \mu_b(1 - h)}.$$

where $r, h$ are the unique roots of equations $z = a(\mu_b(1 - z))$ and $z = a[\theta + \mu_v(1 - z)]$ in the interval $(0,1)$, respectively, Both $\alpha$ and $K$ are constants

$$\alpha = \frac{\theta}{\theta - (\mu_b - \mu_v)(1 - h)},$$
$$K = \frac{1 - h}{1 - r + \alpha(r - h)}$$

It is proved if $\rho = \lambda\mu^{-1} < 1$ then $\alpha(r - h) > 0$.

There are many researches on the GI/M/1 type working vacation queue. Banik et al. [126] investigated the limited space GI/M/1/N (MWV), and gave stationary indices and computational method. Li et al. [127] introduced interruptible working vacation into the GI/M/1 type queue, and got the stationary distributions of $L_v$ and $W_v$ and their stochastic decompositions. Yu et al. [128] examined the $GI^X/M^b/1$ multiple working vacation queue with batch arrivals and batch services using supplementary variable method. Chae et al. [129] discussed the GI/M/1 queue with single working vacation and corresponding discrete time GI/Geom/1 queue, derived the distributions of the stationary queue length and sojourn time. Li and Tian [130] gave the stationary distributions and stochastic decompositions of $L_v$ and $W_v$ by structural matrix analysis. Laxmi et al. [131] introduced interruptible working vacation into the GI/M(n)/1/N queue with service depend on states, and obtained the stationary distributions with supplementary variable method. Li and Tian [132] analyzed the discrete time GI/Geom/1 queue with interruptible working vacation and got the stationary index distributions and their stochastic decompositions.

It is more difficult to analyze the M/G/1 type working vacation queue. Kim and Chae [133] firstly presented the M/G/1 (MWV), gave the calculation method of the queue length distribution. Wu and Takagi [134] computed the stationary indices of the M/G/1 (MWV) using embedded Markov chain and the contour integral of complex function. Yi et al. [135] discussed the discrete time Geom/G/1 model with disaster and multiple working vacations. In fact, the M/G/1 type structural matrix analysis (see Netus [100] or Li [102]) is the ideal tool to analyze the M/G/1 type working vacation queue. Li et al. [136] firstly used structural matrix analysis to deal with the M/G/1 (MWV), obtained not only the expressions of the stationary queue length, waiting time and busy period distribution, but also the stochastic decomposition structure. Furthermore, Li et al. [137] analyzed the discrete time Geom/G/1 (MWV) in parallel. Zhang and Hou [138] studied the MAP/G/1 queue with interruptible working vacation, got the stationary analysis by supplementary variable and structural matrix methods. Gao and Liu [139] investigated the M/G/1 queue with single interruptible working vacation and Benoulli schedule, obtained stationary indices by structural matrix analysis.

Tian et al. [140] assembled plenty of achievements of working vacation queue using matrix analysis method in the long survey. Various discrete time working vacation queues referred to Tian, Xu and Ma [141]. Li [125] used the structural matrix method to analyze kinds of working vacation queues in the monograph.

# 8   Related Applications

The study of vacation queue enriches the theory and method of stochastic model analysis, and it has been widely used in various high-tech fields.

Kuehn [142] studied time division multiple access link in computer networks based on the $M^X/G/1$ (MV) with batch arrival and multiple vacations, and gave the calculation method of its performance indices, and indicated that time sharing system is superior to frequency sharing system from the view of average response time. Gavish and Sumita [143] used vacation queue to model channel and disk subsystem, obtained the calculation and comparison of various indices for universal IBM-3330 and IBM-3380. Zhang [144] introduced the $M/G/1$ queue with two kinds of vacations, and applied analysis results to the optimization design of a class of flexible manufacturing system. Chafir and Silio [145] analyzed the performance analysis of the ring LAN based on the $M/G/1$ queue with Bernoulli vacation policy. Due to dynamic setup and removal of VCC in ATM networks, Hassan et al. [146] regarded VCC as the $M/G/1$ model with setup times and closed-down period, obtained calculation formulae of performance indices, such as, setup rate, vacancy rate and response time, etc. Considering the correlation structure of information arrival process and limited buffer capacity in ATM networks, Niu et al. [147,148] used the $MAP/G/1/N$ models with setup and closed-down to analyze VCC more accurately. The discrete time queue is more suitable to model VCC because IP block transmission in ATM networks is a kind of discrete time phenomenon. Jin et al. [149–151] systematic developed the analysis method of VCC based on the discrete time vacation queue. Jin and Huo [152] established theory foundation and analysis method of VCC in the monograph on VCC modeling analysis. Shen [153] applied the $M/M/c$ queue with partial server' vacation to operation analysis of e-commerce system. Liu et al. [154,155] constructed the hybrid service access control model in wireless communication networks by the discrete time vacation queue. Huo et al. [156, 157] introduced a class of multiple vacations queue with closure mechanism to model sleep mode of IEEE 8.2.16e access protocol in the mobile communication networks, and obtained energy-saving efficiency, average delay, etc, and presented a cost structure to determine the optimal sleep time. Li, Liu and Tian [137] applied the discrete time working vacation queue to performance analysis of ethernet fiber network by configuring different service rates through the router. Similar configuration problems also occurred in stochastic networks, such as, e-commerce, call center, city power supply, traffic dispersion, etc, hoping to analyze by working vacation queue.

Due to limitations of the authors' research interests and knowledge, only a few applied studies are cited here. A large number of successful cases usually published in specialized journals in various application fields, are beyond our knowledge.

## 9  Conclusions

Vacation queues have become the important component of complex stochastic models analysis, plays a crucial role in the theory analysis and applications in the operations research and management science. Taking stochastic decomposition as the main line, we give a systematic review on the research results of vacation queue, hoping to bring convenience for further research and wider application.

**Acknowledgements.** The authors would like to thank the anonymous referees for their constructive comments that help us to improve the present paper, and thank the support from Natural Science Foundation of Hebei province, China (No. A2019203313) and Key Project of Scientific Research in Higher Education of Hebei Province (Natural Sciences Class), China (No. ZD2019079).

## References

1. Erlang, A.K.: Solution of some problems in the theory of probabilities of significance in automatic telephone exchanges. Post Off. Electr. Eng. J. **10**, 189–197 (1917)
2. Saaty, T.L.: Elements of Queue Theory. McGraw Hill, New York (1961)
3. Takacs, L.: Introduction to the Theory of Queues. Oxford University Press, New York (1962)
4. Kleinrock, L.: Queueing Systems, vol. 1–2. Wiley, New York (1975)
5. Gross, D., Harris, C.M.: Fundamentals of Queueing Theory. Wiley, New York (1985)
6. Hsu, G.H.: Stochastic Service System. Science Press, Beijing (1980)
7. Levy, Y., Yechiali, U.: Utilization of idle time in an M/G/1 queueing system. Manag. Sci. **22**, 202–211 (1975)
8. Doshi, B.T.: Queueing systems with vacation-a survey. Queueing Syst. **1**, 29–66 (1986)
9. Teghem, J.: Control of the service process in queueing system. Eur. J. Oper. Res. **23**, 141–158 (1986)
10. Tian, N., Zhang, D., Cao, C.: The GI/M/1 queue with exponential vacations. Queueing Syst. **5**, 331–344 (1989)
11. Chatterjee, U., Mukherjee, S.: GI/M/1 queue with server vacation. J. Oper. Res. Soc. **41**, 83–87 (1990)
12. Doshi, B.: A note on stochastic decomposition in a GI/G/1 queue with vacation times or set-up times. J. Appl. Probab. **22**, 419–428 (1985)
13. Levy, Y., Yechiali, U.: An M/M/s queue with servers' vacations. INFOR **14**, 153–163 (1976)
14. Igaki, N.: Exponential two server queue with N-policy and multiple vacations. Queueing Syst. **10**, 279–294 (1992)
15. Tian, N., Zhang, Z.: The steady theory for GI/M/c queue with synchronous vacations. OR Trans. **5**(1), 70–80 (2001). (in Chinese)
16. Tian, N., Gao, Z., Zhang, Z.: The equilibrium theory for queueing system M/M/c queue with asynchronous vacations. Acta Mathematicae Applicatae Sinica **24**(2), 185–194 (2001). (in Chinese)
17. Tian, N.: The M/M/c queue with synchronous times of phase type set-up. Acta Mathematicae Applicatae Sinica **20**(2), 275–281 (1997). (in Chinese)

18. Tian, N., Hou, Y.: The PH structure in queueing systems M/M/c with synchronous vacations. Mathematica Applicata **1**, 5–8 (1998). (in Chinese)
19. Tian, N., Li, Q.: The M/M/c queue with phase type synchronous vacations. J. Syst. Sci. Math. Sci. **63**(1), 7–16 (2000)
20. Xu, X., Zhang, Z.G.: The analysis of multi-server queue with single vacation and an $(e, d)$ policy. Perform. Eval. **63**(8), 825–838 (2006)
21. Tian, N., Zhang, Z.G.: A two threshold vacation policy in multi-server queueing systems. Eur. J. Oper. Res. **168**, 153–163 (2006)
22. Zhang, Z.G., Tian, N.: Analysis on queueing systems with synchronous vacation of partial servers. Perform. Eval. **52**, 282–296 (2003)
23. Zhang, Z.G., Tian, N.: Analysis of queueing systems with synchronous single vacation for some servers. Queueing Syst. **45**, 161–175 (2003)
24. Chao, X., Zhao, Y.: Analysis of multi-server queue with station and server vacation. Eur. J. Oper. Res. **110**, 392–406 (1998)
25. Tian, N., Zhang, Z.G.: Stationary distributions of GI/M/c queue with PH type vacations. Queueing Syst. **49**, 341–351 (2003)
26. Tian, N., Li, Q., Cao, J.: Conditional stochastic decomposition in M/M/c queue with server vacation. Stoch. Models **14**, 367–377 (1999)
27. Doshi, B.: Conditional and unconditional distribution for M/G/1 type queues with server vacations. Queueing Syst. **7**, 229–252 (1990)
28. Bose, S.K.: An Introduction to Queueing Systems. Kluwer Academic, Dordrecht (2001)
29. Takagi, H.: Queueing Analysis, a Foundation of Performance Evaluation, vol. 1–2. North-Holland, Amsterdam (1993)
30. Dshalalow, J.: Frontiers in Queueing. CRC Press, New York (1997)
31. Tian, N., Zhang, Z.G.: Vacation Queueing Models-Theory and Applications. International Series in Operations Research & Management Science, vol. 93. Springer, New York (2006). https://doi.org/10.1007/978-0-387-33723-4
32. Fuhrmann, S.: A note on the M/G/1 queue with server vacations. Oper. Res. **32**, 1368–1373 (1984)
33. Doshi, B.: An M/G/1 queue with variable vacations. In: Proceedings International Conference on Performance Modeling, Sophia Antipolis, France (1985)
34. Harris, C., Marchal, W.: State dependence in M/G/1 server vacation models. Oper. Res. **36**, 560–565 (1989)
35. Levy, H., Kleinrock, I.: A queue with starter and a queue with vacations: delay analysis by decomposition. Oper. Res. **34**, 426–436 (1986)
36. Takagi, H.: Time-dependent process of M/G/1 with exhaustive service. J. Appl. Probab. **29**, 418–424 (1992)
37. Lee, T.: M/G/1/N queue with vacation time and exhaustive service discipline. Oper. Res. **32**, 774–784 (1984)
38. Lee, H., Lee, S., Park, J.: Analysis of $M^X$/G/1 queue with N-policy and multiple vacations. J. Appl. Probab. **31**, 476–496 (1994)
39. Keilson, J., Servi, L.: Dynamics of the M/G/1 vacation model. Oper. Res. **35**, 575–582 (1987)
40. Keilson, J., Ramaswami, R.: The backlog and deletion-time process for M/G/1 vacation models with exhaustive servece discipline. J. Appl. Probab. **25**, 404–412 (1988)
41. Heyman, D.: The T-policy for the M/G/1 queue. Manag. Sci. **23**, 775–778 (1977)
42. Balachandran, K., Tijms, H.: On the D-policy for the M/G/1 queue. Manag. Sci. **21**, 1073–1076 (1975)

43. Artalejo, J.: Some results on the M/G/1 queue with N-policy. Asia-Pac. J. Oper. Res. **15**, 147–157 (1998)
44. Artalejo, J.: On the M/G/1 queue with D-policy. Appl. Math. Model. **25**, 1055–1069 (2001)
45. Yadin, M., Naor, P.: Queueing systems with a removable server. Oper. Res. **14**, 393–405 (1963)
46. Lee, H., Srinivasan, M.: Control policies for the $M^X$/G/1 queueing system. Manag. Sci. **35**, 707–712 (1989)
47. Baba, Y.: On the $M^X$/G/1 queue with vacation time. Oper. Res. Lett. **5**, 93–98 (1986)
48. Tian, N.: The queue M/G/1 with multiple adaptive vacation. Mathematica Applicata **4**, 12–18 (1992). (in Chinese)
49. Fuhrmann, S., Cooper, R.: Stochastic decompositions in the M/G/1 queue with generalized vacations. Oper. Res. **32**, 1117–1129 (1985)
50. Shanthi Kumar, J.: On stochastic decomposition in M/G/1 type queue with generalized server vacation. Oper. Res. **36**, 516–569 (1988)
51. Li, H., Zhu, Y.: On M/G/1 queue with exhaustive service and generalized vacations. Adv. Appl. Probab. **27**, 510–531 (1995)
52. Sun, W., Guo, P., Tian, N.: Equilibrium threshold stratigies in observable queueing systems with setup/closedown times. Central Eur. J. Oper. Res. **18**(3), 241–268 (2010)
53. Sun, W., Li, S.: Equilibrium and optimal balking strategies of customers in markov queue with multiple vacations and N-policy. Appl. Math. Model. **40**, 284–301 (2016)
54. Sun, W., Li, S.: Stochastic Service System under the Vision of Economics. Publishing House of Electronics Industry, Beijing (2017). (in Chinese)
55. Wang, J.: Fundament of Queuing Game Theory. Science Press, Beijing (2016). (in Chinese)
56. Brunnel, H., Kim, B.: Discrete-Time Models for Communication System Including ATM. Kluwer Academic Publishers, Boston (1993)
57. Woodward, M.: Communication and Computer Networks: Modeling with Discrete-Time Queues. IEEE Computer Society Press, Los Alamitos (1994)
58. Kobayashi, H., Konheim, A.: Queueing models for computer communications analysis. IEEE Trans. Commun. **25**, 1–9 (1977)
59. Zhang, Z., Tian, N.: Diserete time Geom/G/1 queue with multiple adaptive vacations. Queuing Syst. **38**, 419–429 (2001)
60. Fiems, D., Bruneel, H.: Discrete time queueing systems with vacations governed by geometrically distributed times. In: Proceedings Africom, Fifth International Conference on Communication Systems (2001)
61. Fiems, D., Bruneel, H.: Analysis of a diserete time queueing system with timed vacations. Queueing Syst. **42**, 243–254 (2002)
62. Bruneel, H.: Analysis of discrete-time buffer with single server output, subject to interruption process. In Performance 1984, pp. 103–115. Elsevier, Amsterdam (1984)
63. Bruneel, H.: Analysis of an infinite buffer system with random server interruption. Comput. Oper. Res. **11**, 373–386 (1994)
64. Alfa, A.S.: A discrete MAP/PH/1 queue with vacations and exhaustive service. Oper. Res. Lett. **18**, 31–40 (1995)
65. Alfa, A.S.: Vacation models in discrete time. Queueing Syst. **44**, 5–30 (2003)
66. Heyman, D., Sobel, M.: Stochastic Models in Operations Research. McGraw-Hill Publishing, New York (1982)

67. Ross, S.M.: Stochastic Processes. Wiley, New York (1997)
68. Levy, H.: Analysis of cyclic polling systems with binomial-gated service. In: Hasegawa, T., Takagi, H., Takahashi, Y. (eds.) Performance of Distributed and Paralled System, pp. 127–139. North-Holland, Amsterdam (1989)
69. Browne, S., Coffman, E., Gilbert, E., Wright, E.: Gated exhaustive parallel service. Probab. Eng. Inform. Sci. **6**, 217–239 (1992)
70. Browne, S., Kella, O.: Parallel service with vacations. Oper. Res. **43**, 870–878 (1995)
71. Altman, E., Blabc, H., Khamisy, A., Yechiali, U.: Gated-type polling systems with walking and switch-in times. Stoch. Models **10**, 741–763 (1994)
72. Altman, E., Khamisy, A., Yechiali, U.: On elevator polling with globally gated regime. Queueing Syst. **11**, 85–90 (1992)
73. Bacot, J., Dshalalow, J.: A bulk input queueing system with batch gated service and multiple vacation policy. Math. Comput. Model. **34**, 873–886 (2001)
74. Altman, E.: Stochastic recursive equations with applications to queue with dependent vacations. Ann. Oper. Res. **112**, 43–61 (2002)
75. Choi, B., Kim, B., Choi, S.: An M/G/1 queue with multiple type of feedback gated vacations and FIFS policy. Comput. Oper. Res. **30**, 1289–1309 (2003)
76. Ishizaki, F., Takine, T., Hasegawa, T.: Analysis of a discrete-time queue with gated priority. Perform. Eval. **23**, 121–143 (1995)
77. Fiems, D., Vuyst, S., Bruneel, H.: The combined gated exhaustive vacation system in discrete time. Perform. Eval. **49**, 221–239 (2002)
78. Leung, K., Eisenberg, M.: A single queue with vacations and gated time-limited service. In: IEEE INFOCOM, vol. 89, pp. 899–906 (1989)
79. Leung, K., Eisenberg, M.: A single queue with vacations and non-gated time-limited service. In: IEEE INFOCOM, vol. 90, pp. 277–283 (1990)
80. Lee, T.: M/G/1/N queue with vacation time and limited service discipline. Perform. Eval. **9**, 181–190 (1989)
81. Lee, T.: Analysis of infinite server polling system with correlated input process and state dependent vacations. Eur. J. Oper. Res. **115**, 392–412 (1990)
82. Takagi, H., Leung, K.: Analysis of a discrete-time queueing system with time-limited service. Queueing Syst. **18**, 183–197 (1994)
83. Ma, Z., Tian, N.: Pure limited service Geom/G/1 queue with multiple adaptive vacations. J. Comput. Inform. Syst. **9**(3), 515–521 (2005)
84. Ma, Z., Tian, N.: M/G/1 gated service system with multiple adaptive vacation. J. Comput. Inform. Syst. **1**(4), 985–991 (2005)
85. Takagi, H.: Mean message waiting time in a symmetric polling system. In: Glenbe (ed.) Performance 1984, pp. 293–302. Elsevier Science Publishers (1985)
86. Kelison, J., Servi, L.: Oscillating random walk models for GI/G/1 vacation systems with Bernoulli schedules. J. Appl. Probab. **23**, 790–802 (1986)
87. Keilson, J., Servi, L.: Bloking probabilities for M/G/1 vacation system with occupancy level dependent schedules. Oper. Res. **37**, 134–140 (1989)
88. Servi, L.: Average delay approximation of M/G/1 cyclic queue with Bernoulli schedules. IEEE J. Sel. Areas Commun. **4**, 813–822 (1986). SAC-4
89. Ramaswami, R., Servi, L.: The busy period of the M/G/1 vacation model with a Bernoulli schedule. Stoch. Models **4**, 507–521 (1988)
90. Choi, B., Park, K.: The M/G/1 retrial queue with Bernoulli schedule. Queueing Syst. **7**, 219–228 (1990)
91. Tedijanto, E.: Exact results for the cyclic service queue with a Bernoulli schedule. Perform. Eval. **11**, 107–115 (1990)

92. Wortman, M., Desney, R., Kiessler, P.: The M/G/1 Bernoulli feedback queue with vacations. Queueing Syst. **9**, 353–364 (1991)
93. Kumar, B., Arivudainnambi, D.: The M/G/1 retrial queue with Bernoulli schedules and general retrial times. Comput. Math. Appl. **43**, 15–30 (2002)
94. Madan, K., Abu-Dayyeh, W., Taiyyan, F.: A two server queue with Bernoulli schedules and a single vacation policy. Appl. Math. Comput. **145**, 59–171 (2003)
95. Cao, J., Cheng, K.: Analysis of M/G/1 queueing system with repairable service station. Acta Mathematicae Applicatae Sinica **5**(2), 113–127 (1982). (in Chinese)
96. Cao, J.: Ananlysis of the service equipment repairable machine service model. J. Math. Res. Appl. **4**, 89–96 (1985). (in Chinese)
97. Wang, J., Cao, J., Li, Q.: Reliability analysis of the retrial queue with server breakdowns and repeairs. Queueing Syst. **38**, 363–380 (2001)
98. Bischof, W.: Analysis of M/G/1 queues with setup time and vacations under six different service disciplines. Queueing Syst. **39**, 265–301 (2001)
99. Neuts, M.: Matrix-Geometric Solutions in Stochastic Models. Johns Hopkins University Press, Baltimore (1981)
100. Neuts, M.: Structured Stochastic Matrices of M/G/1 Type and Their Applications. Marcel Dekker, New York (1989)
101. Latouche, G., Rammaswami, V.: Introduction to Matrix Analytic Methods in Stochastic Modeling. ASA-SCAM Series on Applied Probability, New York (1999)
102. Li, Q.: Constructive Computation in Stochastic Models With Applications. RG-Factorization. Springer, Heidelberg (2010). https://doi.org/10.1007/978-3-642-11492-2
103. Tian, N., Yue, D.: Quasi Birth and Death Process and Matrix Geometric Solution. Science Press, Beijing (2002). (in Chinese)
104. Tian, N.: The GI/M/1 queueing system with single exponential vacation. J. Syst. Sci. Complexity **13**(1), 1–9 (1993). (in Chinese)
105. Tian, N., Zhang, Z.G.: The discrete time GI/Geom/1 queue with multiple vacations. Queueing Syst. **40**, 283–294 (2002)
106. Zhang, Z.G., Tian, N.: The N-threshold for the GI/M/1 queue. Oper. Res. Lett. **32**, 77–84 (2004)
107. Tian, N.: The GI/M/1 queue with phase type vacations. Acta Mathematicae Applicatae Sinica **16**(4), 452–461 (1993). (in Chinese)
108. Tian, N., Zhang, Z.G.: A note on GI/M/1 queue with PH type setup times or server vacations. INFOR **41**, 341–351 (2004)
109. Dukhovny, A.: Vacations in $GI^X/M^X/1$ systems an Riemann boundary value problems. Queueing Syst. **217**, 351–366 (1997)
110. Karaesmen, F., Gupta, S.: The finite capacity GI/M/1/N queue with server vacations. J. Oper. Res. Soc. **47**, 817–828 (1996)
111. Laxmi, P., Gupta, U.: On the finite-buffer bulk service queue with general independent arrival. Oper. Res. Lett. **25**, 957–967 (1999)
112. Ke, J.: The analysis of a general input queue with N-policy and exponential vacations. Queueing Syst. **45**, 135–160 (2003)
113. Machihara, F.: A GI/SM/1 queue with vacations depending on service times. Stoch. Models **11**, 671–690 (1995)
114. Tian, N.: Progress of the multiserver vacation queueing system. J. Yanshn Univ. **22**(1), 32–35 (1998). (in Chinese)
115. Tian, N.: Stochastic Service System with Vacation. Beking University Press, Beijing (2001). (in Chinese)
116. Vinod, B.: Exponential queue with server vacations. J. Oper. Res. Soc. **37**, 1007–1014 (1986)

117. Bardhan, I.: Diffusion approximations for GI/M/s queue with service interruptions. Oper. Res. Lett. **13**, 175–182 (1993)
118. Servi, L., Finn, S.: M/M/1 queue with working vacations (M/M/1/MV). Perform. Eval. **50**, 41–52 (2001)
119. Liu, W., Xu, X., Tian, N.: Stochastic decompositions in the M/M/1 queue with working vacations. Oper. Res. Lett. **35**, 595–600 (2007)
120. Tian, N., Zhao, X., Wang, K.: The M/M/1 queue with single working vacation. Int. J. Inf. Manage. Sci. **19**(4), 621–634 (2008)
121. Li, J., Tian, N.: The M/M/1 queue with working vacations and vacation interruption. JSSSE **16**(1), 121–127 (2007)
122. Xu, X., Zhang, Z., Tian, N.: The M/M/1 queue with single working vacation and setup times. Int. J. Oper. Res. **6**(3), 420–434 (2009)
123. Tian, N., Ma, Z., Liu, M.: The discrete time Geom/Geom/1 queue with multiple working vacations. Appl. Math. Model. **32**, 2941–2953 (2008)
124. Baba, Y.: Analysis of a GI/M/1 queue with multiple working vacations. Oper. Res. Lett. **33**, 201–209 (2005)
125. Li, J.: Structural Matrix Method and Working Vacation Queue. Science Press, Beijing (2016). (in Chinese)
126. Banik, A., Gupta, U., Pathak, S.: On the GI/M/1/N queue with multiple working vacations-analytic analysis and computation. Appl. Math. Model. **31**, 1701–1710 (2007)
127. Li, J., Tian, N., Ma, Z.: Performance analysis of GI/M/1 queue with working vacations and vacation interruption. Appl. Math. Model. **32**, 2715–2730 (2008)
128. Yu, M., Tang, Y., Fu, Y.: Steady state analysis of the GI/M/1/L queue with multiple working vacations and partial batch rejection. Comput. Ind. Eng. **56**, 1243–1253 (2009)
129. Chae, K., Lim, D., Yang, W.: The GI/M/1/L queue and the GI/Geom/1 queue both with single working vacation. Perform. Eval. **66**, 356–367 (2009)
130. Li, J., Tian, N.: Performance analysis of a GI/M/1 queue with single working vacation. Appl. Math. Comput. **267**, 4960–4971 (2011)
131. Laxmi, P., Goswami, V., Suchitra, V.: Analysis of GI/M(n)/1/N queue with single working vacation and vacation interruption. Int. J. Math. Comput. **7**(4), 475–481 (2013)
132. Li, J., Tian, N.: The discrete time GI/Geom/1 queue with working vacation and vacation interruption. Appl. Math. Comput. **185**, 1–10 (2007)
133. Kim, J., Chae, K.: Analysis of queue-length distribution of the M/G/1 queue with working vacations. In: Hawaii International Conference on Statistics and Related Fields (2003)
134. Wu, D., Takagi, H.: M/G/1 queue with working vacations. Perform. Eval. **63**, 654–681 (2006)
135. Yi, X., Kim, J., Chae, K.: The Geom/G/1 queue with disaster and multiple working vacations. Stoch. Models **23**, 537–549 (2007)
136. Li, J., Tian, N., Zhang, Z., Luh, H.: Analysis of the M/G/1 queue with exponentially working vacations-a matrix analytic approach. Queueing Syst. **61**, 139–166 (2009)
137. Li, J., Liu, W., Tian, N.: Steady state analysis of a discrete time batch arrival queue with working vacations. Perform. Eval. **67**, 897–912 (2010)
138. Zhang, M., Hou, Z.: Performance analysis of MAP/G/1 queue with working vacations and vacation interruption. Appl. Math. Model. **35**, 1551–1560 (2011)
139. Gao, S., Liu, Z.: An M/G/1 queue with single working vacation and vacation interruption under Bernoulli schedule. Appl. Math. Model. **37**, 1564–1579 (2013)

140. Tian, N., Li, J., Zhang, Z.: Matrix analytic method and working vacation queues-a survey. Int. J. Inf. Manage. Sci. **20**, 603–633 (2009)

141. Tian, N., Xu, X., Ma, Z.: Discrete Time Queueing Theory. Science Press, Beijing (2008). (in Chinese)

142. Kuehn, P.: Multiqueue systems with non-exhaustive cyclic service. Bell Syst. Tech. J. **58**, 671–698 (1979)

143. Gavish, B., Sumita, U.: Analysis of channed and disk subsystems in computer systems. Queueing Syst. **3**(1), 1–23 (1988)

144. Zhang, Z.G., Vickson, R., Eenige, M.: Optimal two threshold policies in an M/G/1 queue with two vacation type. Perform. Eval. **29**, 63–80 (1997)

145. Ghafir, H., Silio, C.: Performance analysis of a multiple access ring network. IEEE Trans. Commun. **41**, 1494–1506 (1993)

146. Hassan, M., Atiquzzaman, M.: A delayed vacation model of an M/G/1 queue with setup time and its application to SVCC-based ATM networks. IEICE Trans. Commun. **E80–B**, 317–323 (1997)

147. Niu, Z., Takahashi, Y., Endo, N.: Performance evaluation of SVC-based IP-over-ATM networks. IEICE Trans. Commun. **E81–B**, 948–957 (1998)

148. Niu, Z., Takahashi, Y.: A finite capacity queue with exhaustive vacation/close-down/setup time and Markov arrival processes. Queueing Syst. **31**, 1–23 (1999)

149. Jin, S., Tian, N.: Performance evaluation of virtual channel switching system based on discrete time queue. J. China Inst. Commun. **25**(6), 58–68 (2004). (in Chinese)

150. Jin, S., Tian, N.: Performance evaluation of connection oriented user initiated session based on discrete time queueing. Syst. Eng. Electron. **27**, 931–935 (2005). (in Chinese)

151. Jin, S., Yue, W., Tian, N.: Performance analysis of ARQ schems in selfsimilar traffic. Technical report of IEICE, pp. 41–46 (2007)

152. Jin, S., Huo, Z.: Performance Analysis of Switched Virtual Channels. Publishing House of Electronics Industry, Beijing (2007). (in Chinese)

153. Shen, L.: Performance Analysis and Optimization of E-Commerce System Based on Partial Server' Vacation. Doctoral Dissertation of Yanshan University (2005). (in Chinese)

154. Liu, M., Ma, Z., Tian, N.: Performance of admission control for multi-traffic in wireless communication network base on discrete time queue. J. China Inst. Commun. **27**, 230–234 (2006). (in Chinese)

155. Liu, M., Tian, N.: Modeling of base station in wireless networks with finite user population discrete time delay and loss system. J. Beijing Univ. Posts Telecommun. **30**, 27–31 (2007). (in Chinese)

156. Huo, Z., Yue, W., Tian, N., Jin, S.: Modelling and performance evaluation for the sleep mode in IEEE 802.16e wireless networks. In: Proceeding 11th IEEE International Conference on Communication Systems, Guangzhou, pp. 1140–1144 (2008)

157. Huo, Z., Jin, S., Tian, N., Wang, Y.: Modelling and performance evaluation of the sleep mode in IEEE 802.16e wireless networks with selfsimilar traffic. J. China Univ. Posts Telecommun. **66**(4), 34–41 (2009)

# Reliability Models

# Performance Degradation Analysis of Doppler Velocity Sensor Based on Inverse Gaussian Process and Poisson Shock

Yixuan Geng, Shaoping Wang[✉], Jian Shi, and Weijie Wang

School of Automation Science and Electrical Engineering, Beihang University,
XueYuan Road No. 37, 100083 Beijing, China
gengyixuanbuaa@163.com, shaopingwang@vip.sina.com

**Abstract.** The degraded failure of on-board Doppler Velocity Sensor (DVS), which achieves non-contact velocity measurement based on Doppler principle, can be mainly attributed to the aging of microwave modules and deviation of the radar emission angle. For the microwave modules, active devices such as Gunn diodes are prior in degradation with respect to other passive devices, with the phase noise expanding monotonically. On the other hand, the emission angle of antenna deviates due to the metro vibration. In view of the actual working condition of metro, the DVS may also suffer external shocks during the natural degradation process, which is mixed with the natural degradation by model of compound Poisson process in this paper. In view of the non-reversibility of degradation, the inverse Gaussian process is chosen to describe the gradual degradation of DVS. In addition, given the inherent and postnatal differences among individual products, such as the dislocation of active devices induced during the thermos-compression bonding and individual installation error of antenna, the drift coefficients in the model are randomized. On this basis, the impact of external shock is introduced into the reliability analysis competing with the natural degradation of components. Finally, through parameters estimation of virtual degradation testing data by simulation, the methodology is demonstrated.

**Keywords:** Degradation · Process · DVS · Inverse Gaussian process · Poisson shock · Randomized coefficients

## 1 Introduction

Degradation is the accumulation of product damage over time, which is usually inevitable and irreversible in the field of engineering. From the perspective of reliability, when the accumulated damage hits a certain threshold, whether constant or random, the degradation level is thought to be unacceptable and thus, the system can be regarded as failed. Different from the reliability assessment based on the single point failure-time, performance degradation analysis focus on monitoring the dynamic deterioration process, characterizing the underlying failure of system and predicting the remaining useful life (RUL) of products. In view of this, according to some literatures, for instance [1, 2], the degradation level can product more reliability information than

© Springer Nature Singapore Pte Ltd. 2019
Q.-L. Li et al. (Eds.): Cao Festschrift 2019, CCIS 1102, pp. 161–175, 2019.
https://doi.org/10.1007/978-981-15-0864-6_7

the time-to-failure data. In recent years, the degradation analysis has been accepted and booming in various fields, including the reliability tests [3–5], reliability analysis [6–9] and fault prognostics [10–13]. In practice, the performance of on-board Doppler Velocity Sensor (DVS), which achieves non-contact velocity measurement based on Doppler principle, degrades gradually due to the aging of microwave module [14] and deviation of the radar emission angle [15, 16], as well as suffering from random shocks.

In the field of reliability research, the accumulated deterioration of product performances over time is usually represented by the degradation curve. In some practical cases, the degradation curve of product can be drawn based on the mass historical data collected, such as the curve of bearings. However, due to the limited sample size and significant individual variation, it is hard to measure the degradation degree of DVS under actual working conditions accurately with an intuitive indicator. In this context, we tried to build a probability model to depict its degradation process in the reliability analysis of DVS.

Different from the degradation analysis on account of separate parts, the degradation of DVS is a more complex procedure combined with random shocks. According to reliability engineers in favor of stochastic modeling, the degradation of product can be regarded to be induced by a sequence of exogenous shocks [10, 17]. Generally speaking, the degradation of DVS can be modeled by a stochastic process on account of its inherent randomness and additivity of degradation. Common stochastic process models include Wiener process, Gamma process and IG process [17]. Notwithstanding the wide application of well-studied Wiener and Gamma process [18–20], the IG process has the advantages of both monotonic increments and flexibility of incorporating random and explanatory variables [21–23]. Although it used to be doubted for the parameters lacking of physical meaning, according to [22], the inverse Gaussian process can be interpreted physically as a limiting compound Poisson process with non-negative increments, in this context, it is also fit for modeling the degradation of components subjecting to complex failure mechanisms. With these nice properties, it fits the requirements of modeling the monotone degradation process of DVS greatly.

In this study, the stochastic degradation model of DVS with coupled factors is proposed, taking both the aging of components and random shocks into account. Considering the individual heterogeneity of products and actual working condition, the unobservable factors such as internal defects are included into the model by randomizing the drift parameter. Besides, the random external shock is modeled with a compound Poisson shock process and mixed with the gradual degradation to estimate the overall reliability of DVS. To deal with the individual difference, the Bayesian method is used to estimate the parameters adaptively.

The remainder of the article is organized as follows. In Sect. 2, the gradual degradation due to the aging of microwave modules and the deviation of antenna angle, the reason to choose IG process, as well as the necessity to induce random shocks are briefly discussed. Based on the discussion, Sect. 3 introduces the basic model of IG process, the incorporation of random effects, the compound Poisson process model of external shocks as well as relevant statistical inferences. Section 4 focuses on the parameter estimation based on Bayesian framework and uses simulated data to illustrate the reliability assessment of proposed method. Finally, the article is concluded in Sect. 5, then the prospect for further research is discussed.

# 2 Degradation Analysis of DVS

## 2.1 Fault Mechanism Analysis of Metro DVS

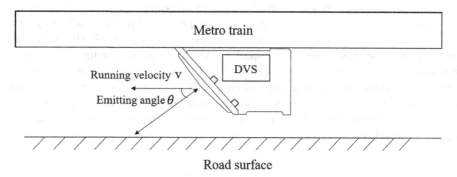

**Fig. 1.** Schematic of metro DVS

DVS is one of the core equipment of train control system and its performance degradation is severely unfavorable to the safe operation. As given in Fig. 1, in the metro system, the DVS is installed at the bottom of metro train and emits the microwave with a certain frequency towards the road surface with a pre-specified emitting angle $\theta$. The running velocity of metro train is measured based on the frequency difference between the emitted and the received microwaves, which is dependent on the reflecting condition and vibration to some extent. As a complex electronic instrument composed by multi electron devices, in the successive operation of metro, the working temperature of DVS rises with time inevitably and may reach over 100 °C, giving rise to the accelerated aging of microwave module and drift of antenna. In addition, the DVS suffers from vibration and shocks in three directions due to the sudden acceleration and deceleration, road bump, as well as electromagnetic inferences from external environment and other equipment on the train. In this context, besides the gradual degradation attributed to the aging of microwave modules and deviation of the radar emission angle, the DVS also suffers from the sudden increase of degradation level caused by some random shocks, which manifests as the decrease of velocity measurement and even blackout status of sensor. The mechanism of gradual degradation and the model of random shocks are introduced briefly in the following section.

## 2.2 Gradual Degradation Analysis of DVS and the Influence on Velocity Measurements

The DVS achieves non-contact velocity measurement based on Doppler principle. When the length of electromagnetic wave emitted by the microwave module is $\lambda$, the emitting angle between the horizontal and antennal direction is $\theta$ and the Doppler frequency extracted by corresponding algorithm is $f_d$, the velocity of metro can be calculated by (1) as below.

$$v = \frac{\lambda f_d}{2 \cos \theta} \qquad (1)$$

According to [14–16], the gradual degradation of DVS can be mainly attributed to the aging of microwave modules and deviation of emission angle. For the microwave modules, active devices such as Gunn diodes are prior in degradation with respect to other passive devices, with the phase noise expanding monotonically. When the increase of noise magnitude with respect to that under normal status hits a certain threshold, the microwave module can be considered as failed. In addition, in the practical operation, the emitting angle deviates due to the external vibration, bringing non-irreversible error into the measure results. Assuming that the deviation of emitting angle is $\delta_\theta$ while the real velocity remains $v$, then the Doppler frequency extracted $f_d'$ can be obtained by

$$f_d' = \frac{2v \cos(\theta + \delta_\theta)}{\lambda} \qquad (2)$$

Meanwhile, the velocity measured $v'$ can be obtained by

$$v' = \frac{\lambda f_d'}{2 \cos \theta} \qquad (3)$$

Based on the inference above, corresponding deviation in velocity measurement $\Delta v$ can be calculated by

$$\Delta v = \frac{v - v'}{v} = 1 - \frac{\frac{\lambda f_d'}{2 \cos \theta}}{\frac{\lambda f_d}{2 \cos \theta}} = 1 - \frac{f_d'}{f_d} = 1 - \frac{\cos(\theta + \delta_\theta)}{\cos \theta} \qquad (4)$$

It shows that with the monotonic deviation of antenna angel, the velocity measure error also increases intuitively.

### 2.3    Model Selection of Random Shocks

In the operation of metro, besides the normal operating condition, the DVS also suffers from random external shocks, including the high-temperature, vibration, electromagnetic disturbances, and so on. Besides the gradual degradation, electromagnetic disturbance and vibration shocks may also introduce sudden increase of degradation, which has been evidenced by the historical record information. In the engineer field, the random shocks can be divided into five types generally, among which the extreme shock has been paid special attention [24–26]. Considering that the extreme shock causes product failure only when its size is beyond a certain threshold, it may be the most suitable to depict the electromagnetic and vibratory shocks the DVS suffers. However, to describe the temporal and spatial randomness of shocks, in the next section, the shocks are modeled with a compound Poisson process and incorporated into the stochastic process model to assess the overall reliability of DVS. In this paper, the final failure of DVS is attributed to the competition between gradual degradation and random shocks.

## 3 Degradation Model Based on Inverse Gaussian Process and Shock Model Based on Compound Poisson Process

Based on analysis above, putting aside the random electromagnetic and vibratory shocks, firstly the gradual degradation of DVS is modeled with an IG process in the following section. The basic model and corresponding assumptions are given as below.

### 3.1 Introduction of IG Process

As a branch of stochastic process, the IG process $\{Y(t), t \geq 0\}$ satisfying the following characteristics:

- $Y(t)$ has independent increments, that is, $Y(t_2) - Y(t_1)$ and $Y(s_2) - Y(s_1)$ are independent for $\forall t_2 > t_1 \geq s_2 > s_1 > 0$;
- $Y(t) - Y(s)$ follows an IG distribution $IG\left(\mu(\Lambda(t) - \Lambda(s)), \eta[\Lambda(t) - \Lambda(s)]^2\right)$, for $\forall t > s \geq 0$, where $\Lambda(t)$ is the monotonically increasing shape function.

When the mean and shape parameter of distribution are given by $a, b > 0$, the probability density function (PDF) of $IG(a, b)$ can be denoted as

$$f_{IG}(y; a, b) = \sqrt{\frac{b}{2\pi y^3}} \cdot \exp\left[-\frac{b(y-a)^2}{2a^2 y}\right], y > 0 \tag{5}$$

while the cumulative distribution function (CDF) can be obtained by integration as

$$F_{IG}(y; a, b) = \Phi\left[\sqrt{\frac{b}{y}}\left(\frac{y}{a} - 1\right)\right] + e^{\frac{2b}{a}}\Phi\left[-\sqrt{\frac{b}{y}}\left(\frac{y}{a} + 1\right)\right], y > 0 \tag{6}$$

where $\Phi(\cdot)$ is the standard normal CDF; under this assumption, the mean and variance of $y$ can be calculated by $a$ and $a^3/b$ respectively. To simplify the form of formulas while not losing generality, it is assumed that the initial values of $\Lambda(\cdot)$ and $Y(\cdot)$ at zero point time are both zero, then $Y(t)$ subjects to a distribution $IG\left(\mu\Lambda(t), \eta\Lambda(t)^2\right)$, whose mean is $\Lambda(t)$ and the variance is $\mu^3\Lambda(t)/\eta$. In view of this assumption and (5), the PDF of degradation quantity $Y(t)$ can be denoted as

$$f_{IG}\left(y; \mu\Lambda(t), \eta\Lambda^2(t)\right) = \sqrt{\frac{\eta\Lambda^2(t)}{2\pi y^3}} \cdot \exp\left[-\frac{\eta(y - \mu\Lambda(t))^2}{2\mu^2 y}\right], y > 0 \tag{7}$$

and its CDF can be obtained referring to (6) as

$$F_{IG}\left(y; \mu\Lambda(t), \eta\Lambda^2(t)\right) = \Phi\left[\sqrt{\frac{\eta}{y}}\left(\frac{y}{\mu} - \Lambda(t)\right)\right] + e^{\frac{2\eta\Lambda(t)}{\mu}}\Phi\left[-\sqrt{\frac{\eta}{y}}\left(\frac{y}{\mu} + \Lambda(t)\right)\right] \quad (8)$$

Corresponding to the concept of failure time in traditional reliability analysis but slightly differing, for the degradation model, the failure time $T_D$ is usually defined as the time when the accumulated degradation exceeds a certain threshold $D$ for the first time. Given this, the reliability function, namely the possibility that $Y(t) < D$ is denoted as

$$R\left(t; \mu\Lambda(t), \eta\Lambda^2(t)\right) = P\left(Y(t) < D; \mu\Lambda(t), \eta\Lambda^2(t)\right)$$
$$= \Phi\left[\sqrt{\frac{\eta}{D}}\left(\frac{D}{\mu} - \Lambda(t)\right)\right] + e^{\frac{2\eta\Lambda(t)}{\mu}}\Phi\left[-\sqrt{\frac{\eta}{D}}\left(\frac{D}{\mu} + \Lambda(t)\right)\right] \quad (9)$$

In the formulae above, the parameter $\mu$, which is usually called the drift parameter in literature, denotes the degradation rate while the parameter $\eta$ can't be interpreted with a certain physical meaning. However, according to [22], the IG process is a limit of compound Poisson processes with a certain kind of shock size distribution. Besides, because of the inverse relation between it and the Wiener process, it is apparent that the IG process enjoys greater flexibility comparing to other ones when the random effects need to be incorporated in the model, which is explained more specifically in the following part.

## 3.2    Incorporation of Random Effects

Due to the diverse working condition of metros, such as the traffic flow and ventilation conditions, as well as the inherent heterogeneity of electronic components and installed error, the dispersion of degradation rates within the DVS population can't be ignored simply, which requires the incorporation of random effects into the IG degradation model. In this paper, the random effects mainly manifests as the variation of degradation rate and they are embodied into the IG process in accordance with the inverse relation between these two stochastic processes. By letting $\mu$ subject to a truncated normal distribution $TN(\omega, \kappa^{-2})$, Ye and Chen [22] introduced a random drift IG process taking the dispersion of degradation rate into consideration. However, considering that under this assumption the variance $\mu^3\Lambda(t)/\eta$ is also dependent of $\mu$, the randomized IG process model is not only randomized in drift in a strict sense. In this paper, considering the engineering-driven case, the negativity of $\mu$ is not of critical concern, thus according to [22], to denote the differences among individual products, the parameter $\mu$ is assumed to follow a normal distribution $N(\omega, \kappa^2)$, whose PDF is

$$g_\mu\left(\mu|\omega, \kappa^2\right) = \frac{1}{\sqrt{2\pi}\kappa}\exp\left(-\frac{(\mu - \omega)^2}{2\kappa^2}\right) \quad (10)$$

Then, the PDF of degradation can be obtained by integrating the marginal density in the entire possible range with respect to $\mu$ as

$$f_{RD}(y|\omega,\kappa,\Lambda(t),\eta) \approx \int_{\mu>0} f_{IG}(y|\mu\Lambda(t),\eta\Lambda^2(t))g_\mu(\mu|\omega,\kappa^2)d\mu$$

$$= \int_{\mu>0} \sqrt{\frac{\eta\Lambda^2(t)}{2\pi y^3}} \cdot \exp\left[-\frac{\eta(y-\mu\Lambda(t))^2}{2\mu^2 y}\right] \frac{1}{\sqrt{2\pi\kappa}}\exp\left(-\frac{(\mu-\omega)^2}{2\kappa^2}\right)d\mu \tag{11}$$

When the threshold of failure is set to be $D$, the reliability function of randomized IG process is

$$R_{RD}(t|\omega,\kappa,\Lambda(t),\eta) \approx \int_{\mu>0} R(t|\mu\Lambda(t),\eta\Lambda^2(t))g_\mu(\mu|\omega,\kappa^2)d\mu$$

$$= \int_{\mu>0}\left[\Phi\left[\sqrt{\frac{\eta}{D}}\left(\frac{D}{\mu}-\Lambda(t)\right)\right] + e^{\frac{2\eta\Lambda(t)}{\mu}}\Phi\left[-\sqrt{\frac{\eta}{D}}\left(\frac{D}{\mu}+\Lambda(t)\right)\right]\right] \tag{12}$$

$$\cdot \frac{1}{\sqrt{2\pi\kappa}}\exp\left(-\frac{(\mu-\omega)^2}{2\kappa^2}\right)d\mu$$

### 3.3   The Compound Poisson Model of Random Shocks

For the on-board DVS, according to the historical failure record and empirical information, the random shocks, which are mainly the electromagnetic disturbances and strong vibration, are random both temporally and spatially. As a consequence, they can be incorporated into the reliability analysis with a compound Poisson process, which can be seen as a series of jumps with random arrival time and magnitude [27]. To depict the temporal randomness, the arrival of shock is modeled as a homogeneous Poisson process with adjacent intervals subjecting to exponential distribution, while the intensity parameter of Poisson process is denoted by $\lambda$ and the total number of shocks arrived before any non-negative time point $t$ is represented by $S(t)$. The probability of $S(t) = k$ is

$$P\{S(t) = k\} = \frac{(\lambda t)^k}{k!}e^{-\lambda t} \tag{13}$$

On the other hand, considering practical engineering cases, the magnitudes of each shock $Y_i$ are set as independent identically distributed and subjects to a normal distribution $Y_i \sim N(\rho,\sigma^2)$. In this paper, given the modeling object DVS, it is supposed that the shock comes at time $t$ will result in the failure of DVS when its magnitude exceeds a threshold $H$, otherwise the system is not impacted at all. Under this assumption, the survival probability of DVS at each shock is

$$P(Y_i < H) = \Phi\left(\frac{H-\rho}{\sigma}\right) \tag{14}$$

Considering the impact of external shocks during the operation of DVS, in this paper the possible correlation between the shock process and gradual degradation are not taken into account, therefore these two processes can be regarded as totally independent. In view of this, the final reliability of the DVS at time point $t$ can be given by

$$
\begin{aligned}
R(t) &= P\{Y(t) < D, S(t) = 0\} \\
&\quad + \sum_{i=1}^{S} P\{Y(t) < D, Y_1 < H, \cdots, Y_{S(t)} < H, S(t) = i\} \\
&= P\{Y(t) < D|S(t) = 0\} \cdot P\{S(t) = 0\} \\
&\quad + \sum_{i=1}^{S} P^i(Y_i < H) \cdot P\{Y(t) < D|S(t) = i\} \cdot P\{S(t) = i\} \\
&= R(t|\omega, \kappa, \eta)e^{-\lambda t} \\
&\quad + \sum_{i=1}^{S} \Phi^i\left(\frac{H-\rho}{\sigma}\right) \cdot R(t|\omega, \kappa, \eta) \cdot \frac{(\lambda t)^i}{i!} e^{-\lambda t}
\end{aligned}
\tag{15}
$$

Based on analysis above, the overall reliability model of DVS is built, in which the random shocks and gradual degradation process determines the reliability level of DVS jointly.

## 4   Parameter Estimation Based on Bayesian Framework

### 4.1   Likelihood Function of IG Process

To estimate the parameter of IG degradation models, it is assumed that the degradation observations for $n$ samples at $m$ discrete time points are obtained as $Y(t)$, with the observations noted by $t_{ij}(i = 1, 2, \cdots, n; j = 1, 2, \cdots, m)$. After specifying the observation time points, which separate from each other with the same interval in this paper, the degradation increments can be represented by $\Delta y_{ij} = Y(t_{ij}) - Y(t_{i,j-1})$, with $\Delta y_{ij} \sim IG\left(\mu_i \Delta \Lambda_{ij}, \eta \Delta \Lambda_{ij}^2\right)$, $\Delta \Lambda_{ij} = \Lambda(t_{ij}) - \Lambda(t_{i,j-1})$ and $\Lambda(t) = t$. The randomized drift $\mu$ subjects to $N(\omega, \kappa^2)$. On this basis, the likelihood of sample $i(i = 1, 2, \ldots, n)$ can be obtained by

$$
L^i = \int g_\mu\left(\mu_i|\omega, \kappa^2\right) \prod_{j=1}^{m} f\left(\Delta y_{ij}|\mu_i \Delta \Lambda_{ij}, \eta \Delta \Lambda_{ij}^2\right) d\mu
\tag{16}
$$

where $\omega$ and $\kappa$ are the hyper parameters that determine the distribution of drift parameter $\mu$. Then the overall likelihood function can be calculated by

$$
L = \prod_{i=1}^{n} L^i
\tag{17}
$$

Considering the complex form of likelihood function, it is almost impossible to maximize it in an analytical way directly. Given this, although the formula of likelihood function is given here, in actual calculation, the Markov Chain Monte Carlo (MCMC) method is used to estimate the parameters of the model based on random sampling. In terms of the Bayesian framework presented in [28], in this case the model can be regarded to have fixed parameters $\omega, \kappa, \eta$ and a randomized parameter $\mu$. To perform Bayesian analysis, the informative prior distribution for all the parameters are given by

$$\begin{cases} \omega \sim N(a_\omega, b_\omega^2) \\ \kappa \sim N(a_\kappa, b_\kappa^2) \end{cases} \{\eta \sim Uniform(a_\eta, b_\eta) \tag{18}$$

On this basis, the joint posterior distribution can be obtained according to (19)

$$p(\omega, \kappa, \eta, \mu | Y) \propto \pi(\omega)\pi(\kappa)\pi(\eta)L \tag{19}$$

where $\pi(\omega)$, $\pi(\kappa)$ and $\pi(\eta)$ denote the probability density function of $\omega$, $\kappa$ and $\eta$ respectively, that can be calculated in terms of the prior distributions given in (18).

## 4.2   Degradation Monitor of DVS

Although the degradation of DVS can be attributed to the aging of microwave and deviation of antenna angle, in practice, it is almost implausible to monitor the noise magnitude and angle deviation within the sensor directly. Besides, based on the velocity measurement, it can be seen that the microwave frequency and emitting angle of antenna are coupled and they result in the final deviation of measurements. However, the real velocity of metro can't be obtained directly, thus the velocity error can't be chosen as the degradation reference. Fortunately, along the metro railway, there are located a series of transponders separating from each other with certain distances. Considering that the trip distance is the integral of velocity, the velocity measured by DVS can be converted to trip distance, then in terms of the reference distance given by transponders, the degradation of DVS performance can be monitored indirectly.

Assuming that the distance between two transponders are $L$ and the velocity measurements during this distance with sampling interval $T$ are $\tilde{v}_k(k = 1, 2, \cdots, K)$, then without degradation, the relation between the metro velocity and trip distance is $L = \sum_{k=1}^{K} \tilde{v}_k + L'$. With the performance degradation of DVS, the difference value $L'$ increases monotonically, as a result, it can be used to measure the degradation to some extent. Given that the distances between transponders are not unique, in this paper, the ratio $L'/L$ is chosen to characterize the degradation of DVS.

## 4.3   Simulation Study

To demonstrate the reliability analysis on the basis of compound model presented, the degradation path of 20 DVS components are simulated as given in Fig. 2, in which the error percentage of distance measurement $L'/L$ is chosen to characterize the degradation level. Considering the non-homogeneity of degradation rates among components, the drift parameters of paths deviate from each other to some extent.

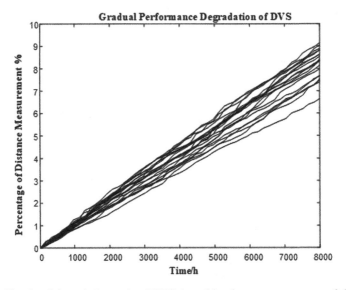

**Fig. 2.** Simulated degradation path of DVS (noted by the error percentage of distance)

Based on the discrete sample points extracted from the simulated path, the Open BUGs software is used to achieve parameter estimation via MCMC. To perform parameter estimation, 100000 samples were generated and according to empirical information the first 5000 samples were deleted. The parameter estimation result for the randomized IG model is shown in Table 1, while the distribution and boxplot of randomized drift parameters of 20 samples are shown in Figs. 3 and 4.

**Table 1.** Estimated parameter of random drift IG model.

| Parameter | Mean | True value | MC error | Median |
|---|---|---|---|---|
| $\omega$ | 9.674e-4 | 10e-4 | 2.385e-4 | 11.11e-4 |
| $\kappa$ | 4.86e-5 | 5e-5 | 2.54e-5 | 6.02e-5 |
| $\eta$ | 5.51e-5 | 5.45e-5 | 3.137e-6 | 5.114e-5 |

Based on this, the reliability function of DVS components with different $\mu$ is shown in Fig. 5, where the drift parameters are set with the estimated mean.

On the other hand, in real application the parameters of compound Poisson process should be estimated based on historical records, in this paper, the parameters are set empirically. The reliability function of compound Poisson process is given in Fig. 6 and the final reliability with different thresholds obtained by (15) is given in Fig. 7.

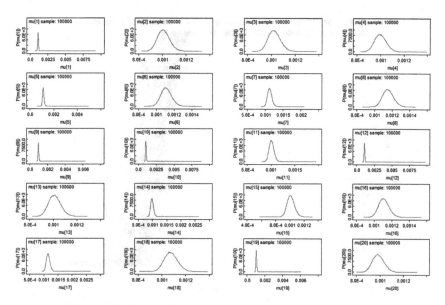

**Fig. 3.** Distribution of randomized drift parameters of 20 samples

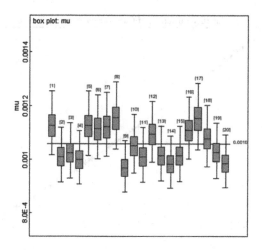

**Fig. 4.** Boxplot of drift parameter among 20 samples

From above the reliability curve, this method incorporates the gradual degradation model based on IG process and the random shocks model based on Compound Poisson Process, fusing the status monitoring data with historical empirical information. It not only gives a guide to evaluate the state of DVS, but also help in the optimization of maintain arrangement.

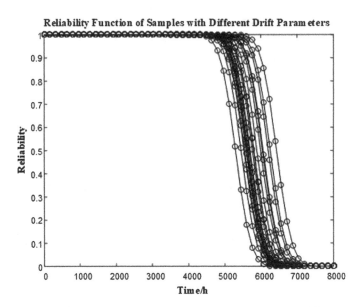

**Fig. 5.** Reliability function of samples with different drift

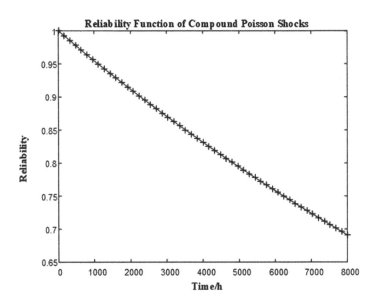

**Fig. 6.** Reliability function of compound Poisson shock

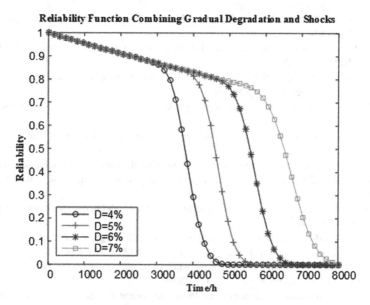

**Fig. 7.** Final reliability integrating gradual degradation and shocks

## 5 Conclusions

Besides the aging of microwave module and deviation of the radar emission angle, the degraded failure of on-board DVS is also attributed to random external shocks. In this paper, the random shocks are mixed with the stochastic process degradation by model of compound Poisson shock. In view of the non-reversibility of degradation, the inverse Gaussian process is used to model the gradual degradation of on-board DVS. Considering the inherent and postnatal differences among individual products, the drift coefficients in the model are randomized with a normal distribution. On this basis, the impact of external shock, which is modeled by a compound Poisson distribution due to its temporal and spatial randomness, is introduced into the reliability analysis competing with the natural degradation of components. Finally, through reasonable coefficients estimation of virtual degradation testing data by simulation, the reliability curve of DVS is estimated.

**Acknowledgement.** This paper was co-supported by the Natural Science Foundation of Beijing Municipality (L171003), National Natural Science Foundation of China (51620105010, 51575019), and Program 111 of China.

## References

1. Meeker, W.Q., Escobar, L.A.: Statistical Methods for Reliability Data. Wiley, New York (1998)

2. Nelson, W.: Accelerated Testing: Statistical Models, Test Plans, and Data Analysis. Wiley, New York (1990)
3. Shi, Y.: Bayesian methods for accelerated destructive degradation test planning. IEEE Trans. Reliab. **61**, 245–253 (2012)
4. Mohammadian, S.: Quantitative accelerated degradation testing: practical approaches. Reliab. Eng. Syst. **95**, 149–159 (2010)
5. Ye, Z.: Degradation-based burn-in planning under competing risks. Technometrics **54**(2), 159–168 (2012)
6. Chen, Z.: Lifetime distribution based degradation analysis. IEEE Trans. Reliab. **54**, 3–10 (2005)
7. Guida, M.: The inverse gamma process: a family of continuous stochastic models for describing state-dependent deterioration phenomena. Reliab. Eng. Syst. Saf. **120**, 72–79 (2013)
8. Liao, H.: Reliability inference for field conditions from accelerated degradation testing. Nav. Res. Logist. **53**, 576–587 (2006)
9. Wang, L.: A Bayesian reliability evaluation method with integrated accelerated degradation testing and field information. Reliab. Eng. Syst. Saf. **112**, 38–47 (2013)
10. Peng, Y.: Current status of machine prognostics in condition-based maintenance: a review. Int. J. Adv. Manuf. Technol. **50**, 297–313 (2010)
11. Gebraeel, N.: Residual-life distributions from component degradation signals: a Bayesian approach. IIE Trans. **37**(6), 543–557 (2005)
12. Wang, W.: A simulation-based multivariate Bayesian control chart for real time condition-based maintenance of complex systems. Eur. J. Oper. Res. **218**(3), 726–734 (2012)
13. Xu, Z.: Real-time reliability prediction for a dynamic system based on the hidden degradation process identification. IEEE Trans. Reliab. **57**, 230–242 (2008)
14. Sun, Z.: The error analysis and improving method of traffic radar speed gun. Chin. J. Sci. Instrument. **24**(4), 418–420 (2003)
15. Sun, D.: Beam direction correction of Doppler speed radar on locomotive. Fire Control Radar Technol. **38**(1), 48–51 (2009)
16. Lu, L.: Performance Degradation Monitoring and Speed Reading Compensation Method for On-Board Radar Speed Sensors of Trains **32**(4), 93–97 (2011)
17. Ye, Z.: Stochastic modelling and analysis of degradation for highly reliable products. Appl. Stoch. Models Bus. Ind. **31**(Special Issue), 16–32 (2015)
18. Ye, Z.: Degradation data analysis using Wiener processes with measurement errors. IEEE Trans. Reliab. **62**(4), 772–780 (2013)
19. Wang, X.: Wiener processes with random effects for degradation data. J. Multivar. Anal. **101**, 340–351 (2010)
20. Zhang, Z.: A prognostic approach for systems subject to Wiener degradation process with cumulative-type random shocks. In: 6th Data Driven Control and Learning Systems Conference, pp. 694–698. IEEE, Chongqing, China (2017)
21. Peng, C.: Inverse Gaussian processes with random effects and explanatory variables for degradation data. Technometrics **57**(1), 100–111 (2015)
22. Ye, Z.: The inverse Gaussian process as a degradation model. Technometrics **56**(3), 302–311 (2014)
23. Wang, X.: An inverse Gaussian process model for degradation data. Technometrics **52**(2), 188–197 (2010)
24. Ye, Z.: A distribution-based systems reliability model under extreme shocks and natural degradation. IEEE Trans. Reliab. **60**(1), 246–256 (2011)
25. Wang, Q.: Failure modeling and maintenance decision for GIS equipment subject to degradation and shocks. IEEE Trans. Power Delivery **32**(2), 1079–1088 (2017)

26. Si, X.: A prognostic model for degrading systems with randomly arriving shocks. In: Prognostics and System Health Management Conference, pp. 1–4. IEEE, Chengdu, China (2016)
27. Mauricio, J.: Optimal maintenance policy for a compound Poisson shock model. IEEE Trans. Reliab. **62**(1), 66–72 (2013)
28. Peng, W.: Inverse Gaussian process models for degradation analysis: a Bayesian perspective. Reliab. Eng. Syst. Saf. **130**, 175–189 (2014)

# An Elman Artificial Neural Network for Remaining Useful Life Prediction

Chong Liu[1] and Chi Zhang[2(✉)]

[1] Department of Industrial Engineering, Tsinghua University,
Beijing 100084, China
[2] School of Economics and Management, Beijing University of Technology,
Beijing 100124, China
chizhang@live.com

**Abstract.** In the last decade, the high-speed rail (HSR) has undergone rapid development and is playing a more and more important role in the transportation system of China. However, the currently adopted maintenance policy of HSR is still mainly usage-based preventive maintenance, which is quite conservative and incurs tremendous annual maintenance costs. Thus, it is necessary to conduct predictive maintenance so as to save maintenance cost as well as ensure the reliability of HSR, which requires for predicting the remaining useful life (RUL) as an essential step. As sensor technology and the 5th generation wireless technology advance, condition monitoring has been convenient and cost efficient. Based on the collected condition information data, the RUL prediction becomes possible.

In this research, we develop an Elman artificial neural network for the purpose of predicting the RUL of HSR bearings, based on the condition monitoring data. To fulfill this purpose, we firstly propose the concepts of current and cumulative state characteristics for analyzing the state monitoring data to extract and filter features that can reflect the current state of the bearings. Then, we build the Elman artificial neural network, evaluate the role cumulative state characteristics play in the model and obtain the weights and thresholds with optimal prediction performance. This way, the network structure and the neuron number of hidden layers are optimized. Experimentation based on the data set of the 2012 IEEE PHM Data Challenge demonstrates the goodness of the proposed approach.

**Keywords:** Condition monitoring · Elman neural network · High-speed rail · Remaining useful life prediction

## 1 Introduction

### 1.1 Review of RUL Estimation for Bearings

With its fast development in the last decade, the high-speed rail (HSR) has become increasingly significant for the transportation of China. Thus, it has been paramount to ensure the safe and reliable long-term operation of HSR in a cost-effective manner. However, the currently adopted maintenance policy of HSR is still mainly usage-based

© Springer Nature Singapore Pte Ltd. 2019
Q.-L. Li et al. (Eds.): Cao Festschrift 2019, CCIS 1102, pp. 176–198, 2019.
https://doi.org/10.1007/978-981-15-0864-6_8

preventive maintenance with a quite conservative policy, which cannot fully utilize the service life of critical components like bearings and has incurred tremendous annual maintenance costs. Therefore, predictive maintenance is in urgent need to lower down the maintenance cost. As an essential step of predictive maintenance, it is necessary to predict the remaining useful life (RUL), which is the length of time from a specific time point to the time of failure. Statistics show that 30% of the failures of rotating machines, 40% of the failures of induction machines and 20% of the failures of gearboxes are caused by bearings [1, 2]. Failure of a bearing can cause a chain reaction of the system, and its normal operation is critical to the safety and reliability of HSR. Thus, this research focuses on the RUL prediction for bearings.

The availability of data on component conditions used to be the bottleneck of predicting RUL precisely. Recently, the advancement of sensor technology and the 5th generation wireless technology has made it much more convenient and less costly to collect condition monitoring data, which can be used to predict the RUL of HSR components, and help managers decide when to maintain a component and at what level of maintenance. This way, the maintenance costs can be significantly reduced by avoiding improper maintenance activities.

RUL can be predicted via analyzing historical event data and condition monitoring data of a component. The prediction of RUL of bearings has been extensively studied over the past several decades. We roughly divide the existing prediction approaches into three categories: the approach based on statistical theory, the one based on physical model and the one based on machine learning. The Lunberg-Palmgren (L-P) model, the Ioannides-Harris (I-H) model, and the three-parameter Weibull distribution model are all based on statistical theory. Based on Weibull's [3] early work, Lundberg and Palmgren [4] developed L-P theory to predict the RUL of bearings. By using the parameters of bearings, the operation load, L-P model can predict the RUL of bearings with high precision in most cases. However, it cannot cover the case of extra-long bearing lifetime in endurance experiments.

As an extension of L-P theory, Ioannides and Harris [5] presented I-H theory which states that bearings wouldn't generate fatigue failure if the load is lower than the endurance fatigue limit. They introduced reliability coefficient and correction factors into the model and described the multiple effects which influence bearing life successfully. The three-parameter Weibull distribution model [6] simplifies a bearing into a mechanical system made up of outer ring, inner ring and rolling elements in series. The contact fatigue lifetime distribution is considered as a three-parameter Weibull distribution. By analyzing large amounts of bearing operating data, the parameters can be determined. But the calculation process is quite complex, which has limited its practical application to the analysis of real-world bearings.

Physical models divide the life of a bearing into three stages according to the growing process of defects: the incubation period, the stable expansion period and the unstable expansion period [7]. In the incubation period and stable expansion period, the degradation of the bearing is regular and predictable. Therefore, its life can be predicted according to the theory of fracture mechanics [7]. Physical models can be further divided into three types: the model based on crack generation, the model based on crack development, and the model combining the two periods together [8]. Cheng et al. [9] assumed that the crack of a bearing is incurred by the incorrect accumulation of particles on the moving belt and derived the formula describing the life of the crack generation. Paris and Eedogan [10] used material coefficient and stress intensity to describe the rate of crack development. Liu *et al.* [11] applied the fatigue life prediction theory based on crack development to the prediction of the crack generation stage in the latent period. They used equivalent crack size method to unify the bearing life prediction during crack generation and development stages. The advantage of bearing life prediction based on physical models is that less experimental sample data is required to establish a model describing the degradation failure, compared with the other models. Nevertheless, the limitation of physical models is that the physical structures of bearings are heterogeneous, and there doesn't exist a unified physical model to be universally applicable. Moreover, some basic assumptions in the physical models cannot be directly applied to the practical situation. For example, the assumption that the monitoring of the crack state of a bearing can be performed without affecting its normal operation usually does not hold in reality.

In recent years, the advancement of sensor and telecommunication technologies have made it more convenient and less costly for monitoring the conditions of a component. The data generated by events such as installation, failure, and replacement of components are called event data. The temperature, humidity, vibration frequency, amplitude, and X-ray image recorded by the sensors during the operation of a device can reflect the condition of the device and are called condition monitoring data. The event data and condition monitoring data can be applied to predict the degradation process of bearings as well as predict the RUL. The degradation of bearings is highly nonlinear, and its mechanism is not yet clear to us. Meanwhile, machine learning theories can be used for the modeling of highly nonlinear processes with good performance. Therefore, it has been widely used in the RUL prediction of bearings, and approaches, such as artificial neural networks, random forests, decision trees, SVM, and logistic regression, have been developed. Emanuele *et al.* [12] developed a new algorithm to predict train rolling bearing's RUL by using online support vector regression (OL-SVR). The innovation lies in the use of heuristic algorithms to optimize the accuracy of the OL-SVR model and the development of a model selection strategy that can balance the model performance and the computational resources required. However, this approach still needs to consume huge computational resources when the amount of data is large. Sutrisno *et al.* [13] proposed a method for testing bearing RUL using anomaly detection, degradation feature inference and survival time ratio estimation. They did band-pass filtering to the vibration signal of rolling bearings and generate the spectrum of the vibration signals by the Fast Fourier Transform (FFT) algorithm. Then, they predicted the RUL using the duration of the defected frequency. However, this method only achieved good prediction results for certain bearings. Ren *et al.* [14] proposed a sparse representation model to extract the intrinsic correlation of the training dataset and measure the similarity between the test dataset and the training dataset, and a stratified Hough voting process to evaluate the RUL of the test sample using continuous information of the monitoring data. In summary, the bearing life prediction method based on machine learning predicts the RUL of the bearing by extracting the characteristic parameters during the bearing operation. The common shortcomings of these models are that their parameters have no clear physical meanings, and thus, the model's interpretability is not good.

## 1.2　Main Research Contents

This research develops an Elman artificial neural network for the purpose of predicting the RUL of HSR bearings. Compared with traditional data processing methods, ANN (artificial neural network) technology has obvious advantages in dealing with fuzzy data, random data, and nonlinear data. It is especially suitable for systems with large scale, complex structure and unclear information [15]. Therefore, we study the adoption of ANN into the RUL prediction of bearings. And the Elman neural network realizes the one-step delay operator by adding the receiving layer in the hidden layer of the feedforward network and memorizes the historical output of the network. Thus, Elman neural network can dynamically update the parameters by self-feedback, which is superior for time series prediction [16]. Therefore, we choose the Elman neural network to develop the model for RUL prediction. For this purpose, we also propose the concepts of current and cumulative state characteristics and then compare and evaluate the role they play in the prediction of RUL. The condition monitoring data for model training is provided by the 2012 IEEE PHM Data Challenge. By experimentation, we obtain the weights and thresholds of Elman neural network with good predictive precision.

The following notations are used in this research:

Notations

| | |
|---|---|
| $X_t$ | RUL at time $t$ |
| $Y_t$ | The history data and condition monitoring data up to time $t$ |
| $f(x_t|y_t)$ | The probability distribution function of $X_t under Y_t$ |
| $E(x_t|y_t)$ | The expectation value of $X_t under Y_t$ |
| $U(k)$ | The external signal of Elman neural network |
| $X^c(k)$ | The hidden layer feedback signal |
| $X(k+1)$ | The output of hidden layer |
| $Y(k+1)$ | The neural network output |
| $l$ | The number of neurons in the hidden layer |
| $m$ | The number of input dimension |
| $n$ | The number of output dimension |
| $a$ | A positive integer between 0 and 10 |

Compared with existing studies, we acquire the model with better RUL prediction effects and more robustness by adding cumulative time domain indexes. The introduction of cumulative time-domain indexes can accelerate the convergence of Elman neural network and help in identifying the bearing failure modes. Thus, we achieve better RUL prediction effects and make the model more robust.

The paper is organized as follows. Section 2 builds Elman artificial neural network and optimizes the network structure from training functions and the neuron number of hidden layers. Section 3 analyzes the state monitoring data of bearings, extracts and filters features that can reflect the current state of the bearings. Section 4 compares the performance of neural networks based on the current state characteristics, the cumulative state characteristics and both types of features and evaluate the effect of cumulative time domain indexes. Section 5 presents conclusions and discussion on future improvements.

## 2   The Design of Artificial Neural Network

Since the 1980s, the research on artificial neural network (ANN) has already become a hot topic. The theory has achieved great development and has been widely applied in various fields. ANN is divided into several categories according to its structure and principle [17]: According to the flow of information while running, it can be divided into feedforward and feedback neural networks. The feedforward neural network introduces hidden layers and nonlinear transfer functions for nonlinear mapping, and its output only depends on the network input and network weight matrix, with no relation to the historical output of the network, while the input of feedback neural network also includes the feedback of model output. The state of the network neuron changes with the learning process until it reaches a steady state, which indicates the completion of model training. The feedback neural network mainly includes Hopfield neural network, Elman neural network and Boltzmann neural network.

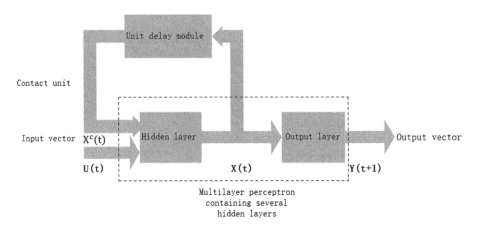

**Fig. 1.**  Elman neural network model

The Elman neural network model was proposed by Elman in 1990 [18]. To describe the proposed approach, we use $X_t$ to denote RUL at time $t$, $Y_t$ to denote the historical event data and condition monitoring data up to time $t$. The objective function of the prediction is $f(x_t|y_t)$ or $E(x_t|y_t)$, with $f(x_t|y_t)$ representing the probability distribution function of $X_t$ given $Y_t$ and $E(x_t|y_t)$ representing the expectation of $X_t$ given $Y_t$. Elman neural network is used as a recursive network model with an external signal, denoted as $U(k)$, and a hidden layer feedback signal, denoted as $X^c(k)$, as input. The feedback is realized through a context unit. The model structure [19] is shown in Fig. 1, where, $X^c(k)$ refers to the output of the context unit time $k$; $X(k+1)$ represents the output of hidden layer; $Y(k+1)$ represents the network output.

When the Elman network adopts a linear transfer function, the hidden unit, the context unit and the output unit are shown in Eqs. 1–3, respectively.

$$X(k) = \sigma_x[W_x U(k) + U_x X^c(k-1) + b_h] \tag{1}$$

$$X^c(k) = X(k-1) \tag{2}$$

$$Y(k+1) = \sigma_y[W_y X(k) + b_y] \tag{3}$$

where, $W$, $U$ and $b$ are parameter matrices and vector, $\sigma_x$ $and$ $\sigma_y$ are activation functions.

We set the mean square error (MSE) as the performance function of Elman neural network to study the proposed problem. When there are too many hidden layers in the middle of the neural network, the model may be easily over fitted, and the training speed will be slow [20]. For the prediction problem of one-dimensional output, generally only one hidden layer is needed to achieve the required prediction accuracy [19]. Thus, we set the number of hidden layers to 1. The number of neurons in the hidden layer is determined according to the empirical formula, as described in Eq. 4.

$$l = \sqrt{m+n} + a \tag{4}$$

where, $l$ refers to the number of neurons in the hidden layer, $m$ refers to the number of input dimensions, $n$ refers to the number of output dimensions, and $a$ refers to a positive integer between 0 and 10.

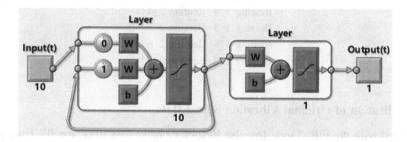

**Fig. 2.** The structure of Elman neural network model

In this study, the number of neurons in the input layer of the Elman neural network equals the number of time domain indexes we input, and the number of neurons in the output layer equals the number of output characteristics of the model, which equals 1 in the RUL prediction problem. Therefore, the final Elman neural network we built is shown in Fig. 2.

# 3  The Processing of Original Condition Monitoring Data

In this section we first design IIR (Infinite Impulse Response) elliptical band-pass filter to preprocess the original vibration signals. By moving the filtering bandpass we implemented 24 experiments and chose the appropriate interval. With the filtered data we extracted the characteristics reflecting the running state of the bearings and smoothed them to reduce the data noise.

This study uses data provided by the FEMTO-ST Association of France, and the bearing aging test was carried out on the laboratory test platform (PRONOSTIA) [21], which accelerates bearing aging under constant or varying operating conditions while on-line collecting health monitoring data (rotation speed, load force, temperature, vibration) of the experiments. In the accelerated aging test of the bearing, the failure point of the experimental bearing is defined as the timing of the bearing vibration acceleration (horizontal or vertical) to reach 20 g for the first time. The experimental data set is shown in Table 1. In all bearing accelerated aging experiments, the vibration acceleration signal is used to sample the vibration signal acceleration of the rolling bearing in the vertical and horizontal direction at intervals.

**Table 1.**  The datasets

| Datasets | Operating conditions | | |
|---|---|---|---|
| | Condition 1 | Condition 2 | Condition 3 |
| Learning set | Bearing 1_1 | Bearing 2_1 | Bearing 3_1 |
| | Bearing 1_2 | Bearing 2_2 | Bearing 3_2 |
| Test set | Bearing 1_3 | Bearing 2_3 | Bearing 3_3 |
| | Bearing 1_4 | Bearing 2_4 | |
| | Bearing 1_5 | Bearing 2_5 | |
| | Bearing 1_6 | Bearing 2_6 | |
| | Bearing 1_7 | Bearing 2_7 | |

## 3.1  Filtration of Original Vibration Signal Data

Compared with the FIR (Finite Impulse Response) band-pass filter, the IIR band-pass filter has the advantages of low signal time delay, low order and high processing speed while being able to achieve the same filtering performance. Therefore, we use IIR elliptical band-pass filtering to process the vibration acceleration of the bearings. Using the SPtool toolbox provided by MATLAB R2017b, we design the elliptical bandpass filter as follows: The input signal frequency is set as 25.6 kHz, the filter strength is set as Astop1 = 60 dB, Apass = 1, Astop1 = 80 dB, and the elliptical bandpass filter prototype is generated by the minimum order. The waveform of the original signal was obtained by the SPtool toolbox, as shown in Fig. 3. The FFT (Fast Fourier Transform) algorithm is employed to conduct the Fourier transform, and transfer time domain signals into frequency domain signals, as shown in Fig. 4.

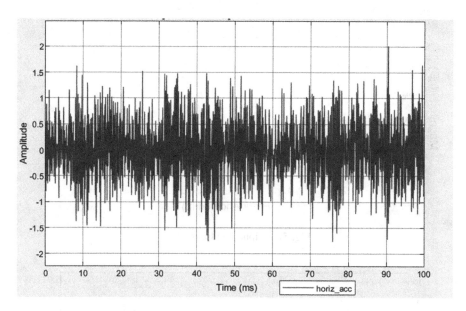

**Fig. 3.** Original vibration signal waveform

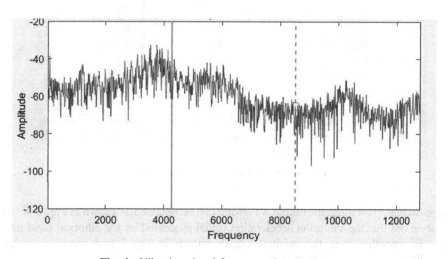

**Fig. 4.** Vibration signal frequency domain diagram

We use 500 Hz as the unit frequency band and start moving the passband from 0–500 Hz to 11.5–12 kHz to cover the full frequency of the signal, evaluated by the smoothness of the signal kurtosis curve [22]. Finally, we determined the bandpass of the elliptical bandpass filter as: Fstop1 = 5000 Hz, Fpass1 = 5500 Hz, Fpass2 = 6000 Hz, Fstop2 = 6500 Hz, as shown in Fig. 5. After the original signal is filtered by the

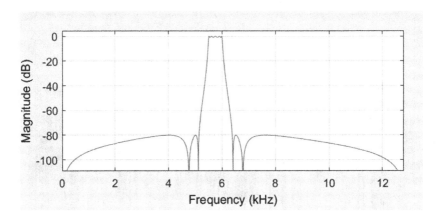

**Fig. 5.** Elliptic bandpass filter

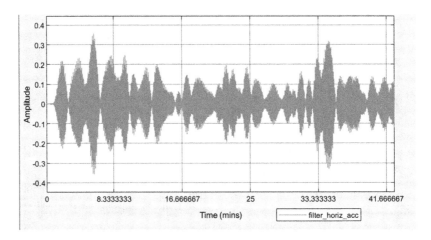

**Fig. 6.** The vibration signal after filtering

elliptical bandpass filter, the output waveform can be obtained as shown in Fig. 6. It can be observed that the vibration acceleration signal processed by the elliptical band pass filter is smoother than the original signals shown in Fig. 3 and shows a certain periodicity.

## 3.2 Extraction of Vibration Signal Characteristics

After filtering and smoothing the original vibration signals, we extract time domain indexes that can reflect the state of bearings. The advantage of prediction method by time domain indexes is that various characteristic indexes are accessible from a single signal.

The physical characteristics of the rolling bearing during operation, such as vibration speed, vibration acceleration, vibration phase, vibration frequency and internal temperature, can reflect the current state of the bearing and be used to analyze and monitor the state of the bearing. Therefore, we extract common time domain indexes of horizontal and vertical vibration acceleration signals, including range, root-mean-square (RMS), kurtosis, arcsine standard deviation, and skewness.

Range refers to the amplitude of bearing vibration signal per unit time. It can be calculated with Eq. 5.

$$X_p = max[x(t_i)] - min[x(t_i)] \tag{5}$$

where, $x(t_i)$ refers to the signal value at time $t_i$.

RMS is employed to reflect the average energy of the signal. It is often used to evaluate the wear level of bearings. The more the bearing wears, the greater the RMS value of its vibration signal. RMS can be calculated via Eq. 6.

$$\Psi_x = \sqrt{\frac{1}{N}\sum_{i=1}^{N} x^2(t_i)} \tag{6}$$

where, $x(t_i)$ reflects the state of the machine, and $N$ refers to the number of points of the discrete signal sample.

Kurtosis refers to the fourth-order central moment of the vibration signal. It is sensitive to vibration shock and is superior for the diagnosis of early surface damage in bearings. Kurtosis can be calculated via Eq. 7.

$$K = \frac{1}{N}\sum_{i=1}^{N} \left(\frac{x(t_i)-\mu}{\sigma}\right)^4 \tag{7}$$

where, $\mu$ refers to the mean of the acceleration signal, and $\sigma$ refers to the standard deviation of the acceleration signal. The normal kurtosis (zero kurtosis) of the defined signal is $K = 3$, which indicates that the bearing is operating in a good condition.

Arcsine standard deviation is derived from the trigonometric function of the vibration signal. It performs better in the manner of signal trend and scale than the other indexes. The formula of arcsine standard deviation is shown as Eq. 8:

$$STD(asin) = \sigma\left(\log\left[x_i + \sqrt{x_i^2 + 1}\right]\right) \tag{8}$$

Skewness refers to the third-order central moment of the vibration signal. It is used to measure the skew level of the signal waveform and can be obtained via Eq. 9.

$$Skew(x) = E\left[\left(\frac{x-\mu}{\sigma}\right)^3\right] \tag{9}$$

In addition, we also include the mean and standard deviation of the time domain indexes we inspect. All the time domain indexes of the bearing vibration signal are shown in Table 2. In order to obtain the time domain indexes which can reflect the state of the bearing, we evaluate the time domain indexes by the correlation coefficient, denoted as $r$, between the indexes and the bearing running time and the monotonic trend index $M$, which is defined as Eq. 10.

**Table 2.** Time domain indexes of vibration signal

| Signal | Time domain indexes | | | | | | |
|---|---|---|---|---|---|---|---|
| Horizontal | range | rms | kurtosis | stdasin | skewness | std | mean |
| Vertical | range | rms | kurtosis | stdasin | skewness | std | mean |

$$M = \left| \frac{number\ of\ \frac{d}{dx} > 0}{n-1} - \frac{number\ of\ \frac{d}{dx} < 0}{n-1} \right|, \qquad (10)$$

$$M \in [0, 1]$$

The time domain index M approaches 1 for indexes with monotonic trend, and approaches 0, otherwise. We obtained the correlation coefficient $r$ and the monotonicity index M of the time domain indexes, as shown in Table 3 and Fig. 7.

**Table 3.** The Results of time domain indexes

| \ | Traditional time domain indexes | | | | | | |
|---|---|---|---|---|---|---|---|
|  | range | rms | kurtosis | stdasin | skewness | std | mean |
| r1 | −0.45 | −0.40 | −0.54 | −0.43 | 0.64 | −0.40 | −0.01 |
| M1 | 0.06 | 0.06 | 0.20 | 0.05 | 0.10 | 0.06 | 0.03 |
| r2 | −0.42 | −0.42 | −0.31 | −0.43 | −0.06 | −0.41 | −0.06 |
| M2 | 0.07 | 0.05 | 0.16 | 0.03 | 0.01 | 0.05 | 0.02 |

It can be seen from Fig. 7 that the correlation coefficient $r$ and the monotonicity index M of the mean of the horizontal and vertical vibration signal tend to be zero, indicating that the mean of the vibration signal has no significant correlation with the bearing's RUL, and there is no significant monotonicity. Therefore, we remove the mean from the time domain indexes. As for the horizontal and vertical skewness, the monotonic index M is close to 0, indicating that there is no obvious monotonic trend in the skewness index. And the horizontal skewness index is positively correlated with the bearing running time, while the vertical skewness index is inversely correlated to the bearing running time. This is not consistent with the vertical and horizontal consistency principle compared with other time domain indexes. Therefore, we removed the skewness index from the time domain indexes. Finally, by analyzing the correlation coefficient $r$ and the monotonicity index M of the indexes, we determine to keep the range, the root mean square, the kurtosis, the arcsine standard deviation, and the standard deviation as the input of the model.

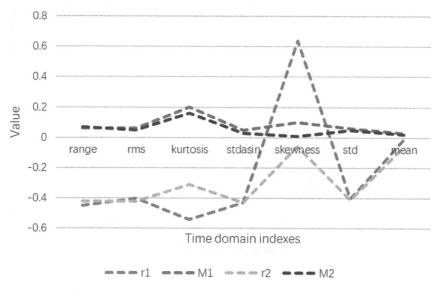

**Fig. 7.** The correlation coefficient and the monotonicity index

However, from Fig. 7, it can also be seen that the traditional time-domain indexes such as kurtosis and standard deviation show relatively weak trend and monotony, and the correlation with bearing running time is not strong enough, either. The reason of this phenomenon lies in that the operation of the bearings is usually affected by quite complex environmental factors resulting in a large amount of noise data and fluctuation of the vibration signals. Moreover, traditional time domain indexes can only reflect the state information of the rolling bearings at the current time point, neglecting historical data, which may reflect the failure modes of the bearings and play an important role in bearing RUL prediction. Therefore, in this research, we focus on the cumulative state characteristics which refer to the indexes that reflect historical operating data of the bearings and can be transformed with Eq. 11 [23].

$$CF_v = \frac{\sum_{i=1}^{N} F_v(i)}{\left| \sum_{i=1}^{N} F_v(i) \right|^{\frac{1}{2}}} \tag{11}$$

where, $\sum_{i=1}^{N} F_v(i)$ refers to the sum of current state characteristics from time 0 to $v$.

After the cumulative transformation, the trend of the time domain indexes can be greatly improved, and the stability has also been enhanced. For example, the kurtosis index of the vibration signal before and after the cumulative transformation is shown in Figs. 8 and 9, respectively. By comparing these two figures, it can be seen that after the cumulative transformation, the kurtosis index increases monotonically with a good linear trend, and the curve transition is relatively smooth.

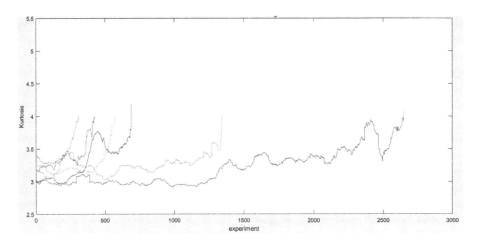

**Fig. 8.** Trend of kurtosis

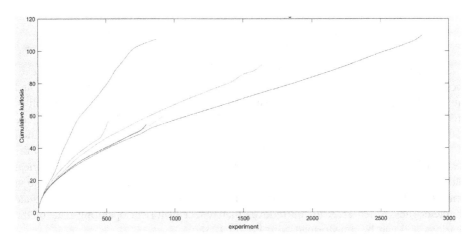

**Fig. 9.** Trend of cumulative kurtosis

**Table 4.** The cumulative state characteristics

|  | Cumulative time domain indexes | | | | |
|---|---|---|---|---|---|
| \ | range | rms | kurtosis | stdasin | std |
| r1 | −0.99 | −0.98 | −0.99 | −0.99 | −0.99 |
| M1 | 1 | 1 | 1 | 1 | 1 |
| r2 | −0.99 | −0.98 | −0.98 | −0.98 | −0.98 |
| M2 | 1 | 1 | 0.998 | 1 | 1 |

Thus, it can be concluded that the traditional time-domain indexes after cumulative transformation show better trend and stability. We also obtained the correlation coefficient $r$ and the monotonicity index M of the cumulative transformation indexes for the bearings, as shown in Table 4, from which it can be seen that the correlation coefficient between the cumulative index and the bearing RUL approaches 1 and exhibits a good linear correlation, and the monotonic index M of the cumulative index also approaches 1 and presents a quite good trend. Thus, we determine to use the cumulative indicator as an auxiliary input to help predict the bearing's RUL.

## 4    Experimentation

We first consider the traditional time domain indexes selected in Sect. 3 as the input of the Elman network, as shown in Table 5. Thus, the input dimension of the Elman neural network is 10, and the output dimension is 1. According to the empirical formula, the number of neurons in the hidden layer is determined to be integers within interval $\left[\sqrt{11}+1, \sqrt{11}+10\right]$. The Elman neural network training functions we examined are shown in Table 6. Next, we adjust the parameters of the network by training the functions and the number of hidden layer neurons.

We first fixed the number of hidden layer neurons to the default value of 10 and selected the triangular basis transfer function TANSIG as the transfer function. The maximum number of training steps was set to equal 10,000 steps with the maximum number of failed steps allowed equal to 6. By changing the training function, we obtained the determined neural network, as shown in Table 7. Figure 10 shows the convergence speed of the Elman neural network with different training functions. It can be seen that training functions TRAINBFG, TRAINCGB, TRAINLM, TRAINRP, TRAINSCG have relatively short training steps and durations. Figure 11 shows the Elman neural network obtained by different training functions. It can be observed that the network performance and training error of training functions TRAINBR and TRAINLM are significantly lower than the other training functions. Therefore, we choose training function TRAINLM as the Elman neural network training function.

**Table 5.** Traditional time domain features of vibration signals

| Signal | Traditional time domain indexes | | | | |
|---|---|---|---|---|---|
| Horizontal | range | rms | kurtosis | stdasin | std |
| Vertical | range | rms | kurtosis | stdasin | std |

Next, we fixed the training function to TRAINLM with other parameters remaining unchanged, except for the number of neurons in the hidden layer. The obtained neural networks are shown in Table 8. Figure 12 shows the convergence rate of Elman networks for different hidden layer neurons. It can be observed that as the number of neurons in the hidden layer increases, the training duration and training step length of the network also increase and they increase sharply, when the hidden layer neurons are more than 9. Figure 13 shows the Elman network training effect of different hidden layer neurons. It can be observed that the training MSE (Mean Square Error) and training error of the Elman neural network decrease with the increase of the number of neurons in the hidden layer. However, too many neurons in the hidden layer will lead to over-fitting of the model. When the number of neurons is 8, the training performance decreases to 0.00697 and the training error decreases to 469.9, both reaching the minimum point. So, we choose the number of neurons in the hidden layer to be 8, which can ensure good network training performance as well as fast convergence speed.

**Table 6.** Elman network functions to be examined

| Training functions | Function declaration |
|---|---|
| TRAINBFG | Quasi-Newton BP algorithm |
| TRAINBR | Bayesian standardized training function |
| TRAINCGB | Power-Beale conjugate gradient BP algorithm |
| TRAINGD | BP algorithm with gradient descent |
| TRAINGDM | BP algorithm with gradient descent momentum |
| TRAINLM | Levenberg-Marquardt |
| TRAINRP | Resettable BP algorithm (Rprop) |
| TRAINSCG | BP algorithm for quantifying continuous gradient |

**Table 7.** Elman network performance with different train functions

| Training function | Epoch | Time/s | Performance | Gradient | Error |
|---|---|---|---|---|---|
| TRAINBFG | 128 | 2 | 0.0214 | 0.0141 | 880.1 |
| TRAINBR | 622 | 67 | 0.00707 | 9.71E-07 | 460.8 |
| TRAINCGB | 49 | 1 | 0.032 | 0.0304 | 1088.7 |
| TRAINGD | 10000 | 34 | 0.0462 | 0.00516 | 1365.8 |
| TRAINGDM | 10000 | 33 | 0.0423 | 0.00306 | 1294.7 |
| TRAINLM | 187 | 18 | 0.0071 | 4.34E-03 | 463.5 |
| TRAINRP | 768 | 2 | 0.0277 | 0.0108 | 1023.3 |
| TRAINSCG | 53 | 1 | 0.0402 | 0.00755 | 1275.3 |

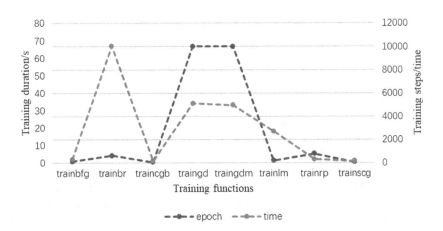

**Fig. 10.** The convergence rate of Elman neural network with different training functions

After the parameter adjustment of the model, we finally obtained the Elman neural network based on the traditional time domain indexes. The network uses TRAINLM as the training function. The number of hidden layer neurons is 8. The training performance is 0.009995 and the training error is 536.6.

The network convergence process is shown in Fig. 14. The prediction error is shown in Fig. 15. It can be observed that the Elman neural network based on the traditional time domain does not converge well. The value of error rate shows large fluctuations. Employing this network, the RUL of the test bearings is obtained, as shown in Table 9. It can be observed that the Elman neural network based on the traditional time domain indexes may result in large errors in predicting the RUL of some bearings, for example, the error rates of 7 out of 11 bearings are higher than 100%. The reason for this may be the lack of classification and identification of fault modes of the bearing.

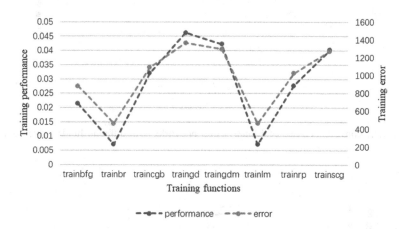

**Fig. 11.** The Elman neural network obtained by different training functions

**Table 8.** Elman network under different hidden layer neurons

| Number of neurons | Epoch | Time/s | Performance | Error |
|---|---|---|---|---|
| 4 | 30 | 1 | 0.0152 | 725.3 |
| 5 | 75 | 2 | 0.0116 | 626.6 |
| 6 | 63 | 2 | 0.0114 | 616.2 |
| 7 | 102 | 5 | 0.011 | 590.7 |
| 8 | 143 | 9 | 0.00697 | 470.0 |
| 9 | 118 | 9 | 0.00904 | 525.6 |
| 10 | 175 | 17 | 0.00887 | 524.0 |

To increase the prediction precision, we further trained the Elman neural network based on the cumulative time domain indexes, following the same process as done in the method of modeling based on the traditional time domain indexes. The cumulative time domain indexes that have been selected and normalized are shown in Table 10. The obtained Elman neural network uses TRAINLM as the training function, with the number of hidden layer neurons equal to 8. The training performance equals 0.000184 and the training error equals 25.9.

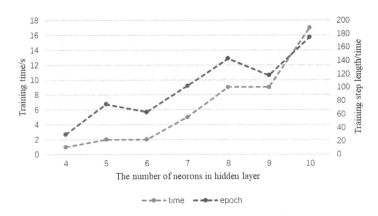

**Fig. 12.** The convergence rate of Elman neural networks with different hidden layer neurons

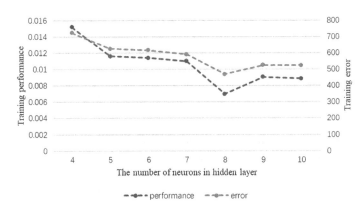

**Fig. 13.** The training effect of Elman neural networks with different hidden layer neurons

The network convergence process is shown in Fig. 16. The prediction error is shown in Fig. 17. It can be observed that the convergence effect of the Elman neural network based on the cumulative time domain indexes is better than that based on the traditional time domain indexes. Most of the absolute value of the prediction error rate

**Table 9.** The prediction accuracy of Elman network based on traditional time domain features

| Bearing | 1–3 | 1–4 | 1–5 | 1–6 | 1–7 | 2–3 | 2–4 | 2–5 | 2–6 | 2–7 | 3–3 |
|---|---|---|---|---|---|---|---|---|---|---|---|
| Lifetime | 5730 | 2890 | 1610 | 1460 | 7570 | 7530 | 1390 | 3090 | 1290 | 580 | 820 |
| Prediction | 5310 | 0 | 9190 | 5264 | 5970 | 11600 | 5437 | 6429 | 233 | 7719 | 5718 |
| Error rate | 0.07 | 1 | −4.71 | −2.61 | 0.21 | −0.54 | −2.91 | −1.08 | 0.82 | −12.3 | −5.97 |

is smaller than 5%. Using this network, the RUL of the test bearings can be predicted as shown in Table 11, from which, it can be observed that the model based on the cumulative time domain index can hardly predict the RUL of the bearing, with only 1 out of 11 bearings' prediction error rate below 100%. Thus, it is not feasible to predict the bearing RUL only by the cumulative indexes.

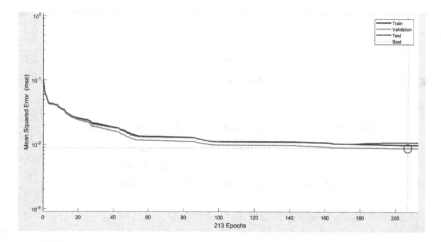

**Fig. 14.** The convergence process of Elman network based on traditional time domain indexes

In order to improve the quality of predicting the bearing's RUL, we then combine the traditional and cumulative time domain indexes together as the input of the Elman network. Following the same process to adjust the parameters as done before, we can obtain the Elman neural network based on both the traditional and cumulative time domain indexes. The network uses TRAINLM as the training function. The number of hidden layer neurons is 8. The training performance is 0.000588 and the training error is 110.1.

The network convergence process is shown in Fig. 18. The prediction error is shown in Fig. 19. It can be observed that the convergence accuracy of the Elman neural network based on both the traditional and cumulative time domain indexes is much better than that based only on traditional time domain index, and slightly worse than that based only on the cumulative time domain indexes, with the mean square error converging between 0.0001 and 0.001. Using this network, the RUL of test bearings was determined as shown in Table 12. It can be observed that the Elman neural network based on both the traditional and cumulative time domain indexes have a better effect, with 7 out of 11 bearings' RUL prediction error rate below 50%.

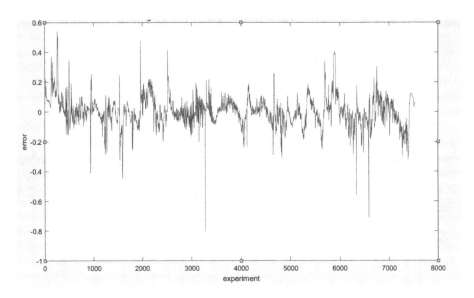

**Fig. 15.** The training error of Elman network based on traditional time domain features

**Table 10.** Cumulative time domain features of vibration signals

| Signal | Cumulative time domain indexes | | | | |
|---|---|---|---|---|---|
| Horizontal | range | rms | kurtosis | stdasin | std |
| Vertical | range | rms | kurtosis | stdasin | std |

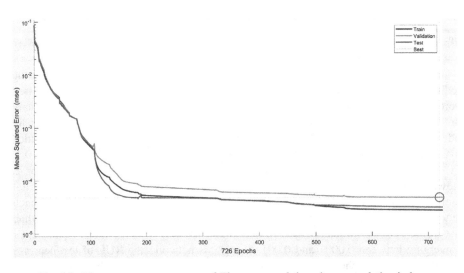

**Fig. 16.** The convergence process of Elman network based on cumulative indexes

**Table 11.** The performance of the Elman network based on cumulative indexes

| Bearing | 1–3 | 1–4 | 1–5 | 1–6 | 1–7 | 2–3 | 2–4 | 2–5 | 2–6 | 2–7 | 3–3 |
|---|---|---|---|---|---|---|---|---|---|---|---|
| Lifetime | 5730 | 2890 | 1610 | 1460 | 7570 | 7530 | 1390 | 3090 | 1290 | 580 | 820 |
| Prediction | 133 | 0.02 | 0 | 0 | 157 | 0.76 | 4021 | 3241 | 2843 | 3684 | 1828 |
| Error rate | 0.98 | 1 | 1 | 1 | 0.98 | 1 | −1.89 | −0.05 | −1.20 | −5.35 | −1.23 |

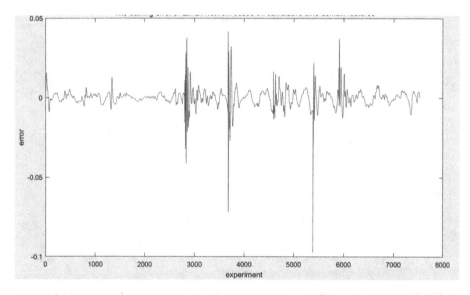

**Fig. 17.** The training error of Elman network based on cumulative time domain features

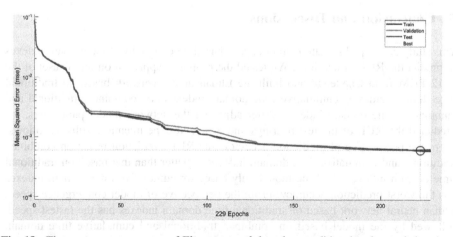

**Fig. 18.** The convergence process of Elman network based on traditional and cumulative time domain features

**Table 12.**  The prediction accuracy of Elman network based on traditional and cumulative time domain indexes

| Bearing | 1–3 | 1–4 | 1–5 | 1–6 | 1–7 | 2–3 | 2–4 | 2–5 | 2–6 | 2–7 | 3–3 |
|---|---|---|---|---|---|---|---|---|---|---|---|
| Lifetime | 5730 | 2890 | 1610 | 1460 | 7570 | 7530 | 1390 | 3090 | 1290 | 580 | 820 |
| Prediction | 5100 | 3 | 1960 | 830 | 7410 | 6740 | 2890 | 3012 | 2240 | 27003 | 486 |
| Error rate | 0.11 | 1 | −0.22 | 0.43 | 0.02 | 0.11 | −1.08 | 0.03 | −0.74 | −45.6 | 0.41 |

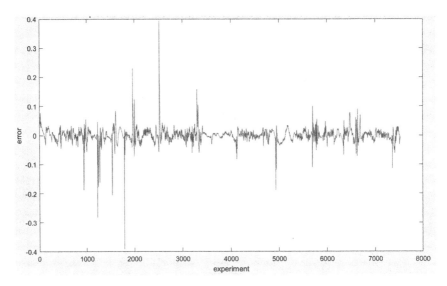

**Fig. 19.**  The prediction effort of Elman network based on traditional and cumulative time domain indexes

## 5  Conclusion and Discussions

This study developed Elman neural network based on cumulative time domain indexes to predict the RUL of bearings. We tested the proposed approach on the dataset of the 2012 PHM Data Challenge and built the Elman neural network based on traditional time domain indexes, cumulative time domain indexes and combined traditional and cumulative time domain indexes. After adjusting the network model parameters, we predicted the RUL of the test bearings and analyzed experimental results as follows: From the perspective of prediction accuracy, the RUL prediction based on combined traditional and cumulative time domain indexes is better than that based on traditional time domain indexes. And the model only based on cumulative time domain indexes has the worst prediction accuracy. From the perspective of model convergence speed, Elman neural network based on traditional time domain indexes has the fastest speed, followed by the model based on combined traditional and cumulative time domain indexes, and the one based on cumulative time domain indexes has the slowest convergence speed. From the perspective of neural network training accuracy, the model based on cumulative time domain indexes has the smallest error and the best performance, followed by the model based on both traditional and cumulative time domain indicators, while the model based on traditional time domain index Neural network has the worst training effect.

In summary, the ANN based on only traditional time domain indexes will lead to difficult convergence as well as bad training accuracy. Only some bearings have favorable RUL prediction results under this model. The model based on only cumulative time domain indexes will cause the neural network to converge slowly. Although the training effect turns better, it is prone to over-fitting and the prediction results of the model is not good. Compared with models only considering traditional time domain indexes, we can acquire the model with better RUL prediction effects and more robustness by adding cumulative time domain indexes into the model. The introduction of cumulative time-domain indexes can accelerate the convergence speed of the Elman neural network, obtain better training effect and help in identifying the bearing failure modes. Thus, we can achieve better RUL prediction effects and make the model more robust.

This study only tested the public data set of the 2012 PHM Data Challenge. However, in real industrial systems, bearings are diverse in materials and sizes. More data samples should be tested to verify whether the proposed prediction methods can be generalized. The parameters of the prediction model based on neural network do not have actual physical meanings and, thus, the model interpretability is not good. In future, time series prediction models, such as ARIMA, will be adopted and compare with the developed Elman neural network to evaluate the effect of cumulative time domain indexes.

# References

1. Schoen, R.R., Habetler, T.G., Kamran, F., Bartfield, R.G.: Motor bearing damage detection using stator current monitoring. IEEE Trans. Ind. Appl. **31**(6), 1274–1279 (1995)
2. Guangwei, M.: The committee on fatigue and fracture reliability of the committee on structural safety and reliability of the structural division. J. Struct. **108**, 3–23 (1982)
3. Weibull, W.: A Statistical Theory of the Strength of Materials IVA. Ingenioersvetenskap-sakad Handl., vol. 151 (1939)
4. Lundberg, G., Palmgren, A.: Dynamic capacity of rolling bearings. J. Appl. Mech. **16**, 165–172 (2019)
5. Ioannides, E., Harris, T.A.: A new fatigue life model for rolling bearings. J. Tribol. **107**(3), 367–377 (1985)
6. Tallian, T.: Weibull distribution of rolling contact fatigue life and deviations therefrom. ASLE Trans. **5**(1), 183–196 (1962)
7. Sadeghi, F., Jalalahmadi, B., Slack, T.S., Raje, N., Arakere, N.K.: A review of rolling contact fatigue. J. Tribol. **131**(4), 041403 (2009)
8. Dong, X.: Research on Residual Fatigue Life Analysis and Prediction of Ball Bearings. National University of Defense Technology, Changsha (2011). (Chinese in English abstract)
9. Cheng, W.H.S.T., Cheng, H.S., Mura, T., Keer, L.M.: Micromechanics modeling of crack initiation under contact fatigue. J. Tribol. **116**(1), 2–8 (1994)
10. Paris, P., Erdogan, F.: A critical analysis of crack propagation laws. J. Basic Eng. **85**(4), 528–533 (1963)

11. Liu, Y., Mahadevan, S.: Probabilistic fatigue life prediction using an equivalent initial flaw size distribution. Int. J. Fatigue **31**(3), 476–487 (2009)
12. Fumeo, E., Oneto, L., Anguita, D.: Condition based maintenance in railway transportation systems based on big data streaming analysis. Procedia Comput. Sci. **53**, 437–446 (2015)
13. Sutrisno, E., Oh, H., Vasan, A.S.S., Pecht, M.: Estimation of remaining useful life of ball bearings using data driven methodologies. In: 2012 IEEE Conference on Prognostics and Health Management, pp. 1–7. IEEE (2012)
14. Ren, L., Lv, W., Jiang, S.: Machine prognostics based on sparse representation model. J. Intell. Manuf. **29**(2), 277–285 (2018)
15. Floreano, D., Mattiussi, C.: Bio-Inspired Artificial Intelligence: Theories, Methods, and Technologies. MIT Press, Cambridge (2008)
16. Sak, H., Senior, A., Beaufays, F.: Long short-term memory recurrent neural network architectures for large scale acoustic modeling. In: Fifteenth Annual Conference of the International Speech Communication Association (2014)
17. Freeman, J.A., Skapura, D.M.: Neural Networks: Algorithms, Applications, and Programming Techniques. Addison Wesley Longman Publishing Co., Inc., Redwood City (1991)
18. Elman, J.L.: Finding structure in time. Cogn. Sci. **14**(2), 179–211 (1990)
19. Heaton, J.: Programming Neural Networks with Encog2 in C. Heaton Research, Inc. (2010)
20. Elman, J.L.: Learning and development in neural networks: the importance of starting small. Cognition **48**(1), 71–99 (1993)
21. Nectoux, P., et al.: PRONOSTIA: an experimental platform for bearings accelerated degradation tests. In: IEEE International Conference on Prognostics and Health Management, PHM 2012, pp. 1–8. IEEE Catalog Number: CPF12PHM-CDR (2012)
22. Tandon, N., Choudhury, A.: A review of vibration and acoustic measurement methods for the detection of defects in rolling element bearings. Tribol. Int. **32**(8), 469–480 (1999)
23. Javed, K.: A robust & reliable data-driven prognostics approach based on extreme learning machine and fuzzy clustering. Doctoral dissertation (2014)

# Bayesian Networks in Reliability Modeling and Assessment of Multi-state Systems

Tao Jiang[1], Yi-Xuan Zheng[1], and Yu Liu[1,2(✉)]

[1] School of Mechanical and Electrical Engineering,
University of Electronic Science and Technology of China, Chengdu 611731,
Sichuan, People's Republic of China
t_jiang@hotmail.com, beiqi.zheng@hotmail.com,
yuliu@uestc.edu.cn
[2] Center for System Reliability and Safety, University of Electronic Science and
Technology of China, Chengdu 611731, Sichuan, People's Republic of China

**Abstract.** Multi-state is a characteristic of advanced engineering systems and products. The reliability of multi-state systems (MSSs) has been received considerable attentions since the middle of 1970s. Over the last decade, Bayesian networks (BNs), as an effective and efficient reasoning tool under uncertainty, have been intensively concerned in MSS reliability modeling and assessment. This chapter presented a holistic framework for MSS reliability modeling and assessment by BNs. Firstly, the basic characteristics of MSSs and BNs are reviewed. Secondly, the detailed procedures of constructing the BN models of diverse MSSs are provided. The corresponding dynamic Bayesian network (DBN) models are also constructed to characterize the degradation profiles of MSSs over time, as well as various dependencies among components. Thirdly, a reliability assessment method by fusing multi-level observation data is developed. The results show that the reliability modeling and assessment for MSSs by BNs are effective considerably.

**Keywords:** Multi-state systems (MSSs) · Bayesian networks (BNs) ·
Reliability modeling · Reliability assessment · Parameter estimation ·
Dependency

## 1 Introduction

Multi-state is a typical characteristic of advance engineering systems and products [1–4]. Many technical systems that perform their intended tasks/missions with multiple (more than two) distinguishable states between perfectly functioning and completely failed can be regarded as multi-state systems (MSSs) [1]. The MSS reliability models, first introduced in the mid-1970s, have received considerable concerns in the past few decades, because the models can characterize complicated deterioration processes of engineering systems more precisely than that of the traditional binary-state system (BSS) reliability models [1, 5]. For example, based on the length of flank wear, the health status of a cutting tool can be classified approximately into five discrete states from the normal state (perfectly functioning) to nominally sharp (<0.1 mm), part worn (0.1–0.15 mm), severely worn (>0.15 mm), and fractured/chipped (completely failed)

© Springer Nature Singapore Pte Ltd. 2019
Q.-L. Li et al. (Eds.): Cao Festschrift 2019, CCIS 1102, pp. 199–228, 2019.
https://doi.org/10.1007/978-981-15-0864-6_9

states [6]. Another instance is that a power generating system can function at multiple levels of generating capacity [1]. Similar treatment can also be found in diverse engineering situations, e.g., manufacturing systems, networked systems, grid systems, spacecraft, and municipal infrastructure.

As both components, subsystems, and the entire system can manifest multiple states, the MSS reliability models are, therefore, much complicated. The approaches to MSS reliability modeling and assessment can be roughly classified into four categories [1].

- An extension of the Boolean models to the multi-value case. The methods based on the extension of the Boolean models is natural expansions of Boolean methods that were well implemented in BSSs, such as, for example, multi-state fault tree [7], multi-state minimal cuts/paths [8, 9], and multi-value decision diagram [10].
- Stochastic models. The stochastic models, such as homogeneous/non-homogeneous Markov [11] and semi-Markov [12, 13], are more universal to characterize the degradation processes of MSSs. However, due to the dimension damnation, the stochastic models only suit to relatively small scale MSSs because the number of system states increases dramatically with the increase in the number of components and component states. Another severe restriction to implement the stochastic models is the computational complexity, because it is inevitably to solve a system of differential equations (for homogeneous/non-homogeneous Markov) or a system of integral equations (for semi-Markov).
- Universal generating functions (UGFs). The UGF technique is effective enough that utilizes a rapid algebraic procedure to identify the state probability distribution of the entire system based on the state probability distributions of all the components [14]. However, this technique is a sort of "static" approaches that cannot characterize the dynamic degradation profiles of MSSs.
- Simulation-based methods. The degradation behaviors of most MSSs in real-world situations can be simulated by the Monte Carlo method [15]. Nevertheless, the time consumption involved in the development and execution of the simulation models are oftentimes unaffordable to achieve a high accurate result.

The recursive algorithms were also developed to evaluate the reliability of generalized multi-state $k$-out-of-$n$ systems and multi-state weighted $k$-out-of-$n$ systems [16, 17]. It was proved that the recursive algorithms can outperform the UGF approaches with or without collecting like terms for the reliability assessment of multi-state weighted $k$-out-of-$n$ systems. In addition, the degradation process of each multi-state component in an MSS can be characterized by the stochastic models, and thus the state probability distribution of the component at any particular time can be obtained. By combining the stochastic models and UGF approaches, the state probability distribution of the entire system at any particular time can be readily obtained, even for relatively large-scale systems.

Apart from the aforementioned methods and tools, Bayesian networks (BNs) [18], as a probabilistic graphical model, are capable of handling with various uncertainty problems effectively based on probabilistic information representation and inference. BNs have gained considerable popularity in MSS reliability modeling and assessment over the last decade. There is still a booming interest for using BNs in the reliability community, especially for MSS reliability modeling and assessment. This chapter will present a holistic framework for MSS reliability modeling and assessment with BNs. The contributions of this chapter are trifold: (1) the proposed framework can effectively characterize the dynamic behaviors of various MSSs; (2) the proposed framework can effectively characterize various dependencies in MSSs; (3) the proposed framework can effectively aggregate multi-level observation data to dynamically assess reliability of MSSs.

The reminder of this chapter is organized as follows. In Sect. 2, the basic characteristics of MSS and BNs are reviewed. The detailed procedures of constructing the BN models for a diversity of MSSs are provided in Sect. 3. A reliability assessment method by fusing multi-level observation data is developed in Sect. 4. A brief closure is given in Sect. 5.

## 2 Preliminaries

### 2.1 Multi-state Systems

An MSS herein is composed of $M^c$ homogenous or heterogeneous multi-state components. The states of each component are distinguished by its performance capacities or degradation levels. Suppose that component $l$ can possess $N_l^c$ mutually ordered states, then the sets of the performance capacity and state component $l$ can be denoted as $\mathbf{g}_l^c = \{g_{l,1}, g_{l,2}, \ldots g_{l,N_l^c}\}$ and $\mathbf{s}_l^c = \{1, 2, \ldots, N_l^c\}$, respectively. States 1 and $N_l^c$ are the best state and worst state of component $l$, respectively. The performance capacity and state of component $l$ at time $t$ are denoted as $G_l^c(t)$ ($G_l^c(t) \in \mathbf{g}_l^c$) and $C_l(t)$ ($C_l(t) \in \mathbf{s}_l^c$), respectively. If component $l$ sojourns in state $i$ at time $t$, i.e., $C_l(t) = i$, the performance capacity $G_l^c(t) = g_{l,i}$. In this chapter, states $\{1, 2, \ldots, N_l^c - 1\}$ are acceptable states; therefore, component $l$ is viewed as being failed if the component sojourns in state $N_l^c$.

In this chapter, the degradation profile of each component is assumed to follow a homogeneous discrete-time Markov process. Other stochastic models, such as non-homogenous Markov process and semi-Markov process, can also be adopted. As each component degrades from the best state to the worst state, the Markov model is irreducible, transient, and aperiodic. The one-step state transition matrix of the Markov model for component $l$ is represented as:

$$\mathbf{P}_l = \begin{bmatrix} p_{l,(1,1)} & p_{l,(1,2)} & \cdots & p_{l,(1,N_l^c)} \\ 0 & p_{l,(2,2)} & \cdots & p_{l,(2,N_l^c)} \\ \vdots & \vdots & \ddots & \vdots \\ 0 & 0 & \cdots & p_{l,(N_l^c,N_l^c)} \end{bmatrix},$$

where $p_{l,(i,j)} = \Pr\{C_l(t + \Delta t) = j | C_l(t) = i\}$ $(1 \leq i \leq j \leq N_l^c)$ is the state probability of component $l$ from state $i$ to state $j$ within a basic time interval $\Delta t$. The state probability distribution of component $l$ at time $t$ is denoted by a probability vector $\mathbf{p}_l(t) = [p_{l,1}(t), p_{l,2}(t), \ldots, p_{l,N_l^c}(t)]$, where $p_{l,i}(t) = \Pr\{C_l(t) = i\}$. With the known state probability distribution of component $l$ at time $t$, i.e., $\mathbf{p}_l(t)$, the state probability distribution of the component at time $t + k\Delta t$ can be computed as follows:

$$\mathbf{p}_l(t + k\Delta t) = \mathbf{p}_l(t) \cdot (\mathbf{P}_l)^k. \tag{1}$$

Based on the physical configuration and/or functional relations between components, the components in an MSS can be divided into $M^{\text{sub}}$ subgroups that are considered as $M^{\text{sub}}$ subsystems. The number of the states that subsystem $m$ and the entire system can have are $N_m^{\text{sub}}$ and $N^{\text{sys}}$, respectively. Likewise, the performance capacity and state of subsystem $m$ at time $t$ are denoted as $G_m^s(t)$ and $S_m(t)$, respectively; the performance capacity and state of the entire system at time $t$ are denoted as $G(t)$ and $S(t)$, respectively; states 1 is the best state each subsystem and the entire system; states $N_m^{\text{sub}}$ and $N^{\text{sys}}$ are the worst states of subsystem $m$ and the entire system, respectively.

The states of each subsystem and the entire system are completely determined by the state combinations of their corresponding constituents. The structure function $\phi_m(\cdot)$ that identifies the relation between subsystem $m$ and its constituents are deterministic and known; the structure function $\phi(\cdot)$ that identifies the relation between the entire system and its constituents are also deterministic and known. It is common that more than one state combination of components may result in particular subsystem and/or system state [19]. An MSS is considered reliable if the system sojourns in the acceptable states during the operation period. Therefore, the reliability of an MSS is defined as the sum of the probabilities of the system sojourning in the acceptable states.

## 2.2    Bayesian Networks

BNs [18], also known as belief networks, Bayesian belief networks, and casual networks, are inherently compact representations of multivariate statistical distribution functions. A BN contains a qualitative part, i.e., the direct acyclic graph (DAG), and a quantitative part, i.e., a set of conditional probability tables (CPTs). The DAG of a BN consists of a set of nodes denoting random variables $\{X_1, X_2, \ldots, X_n\}$ and a set of links characterizing the probabilistic dependencies among nodes. The terms *node* and *random variable* are used interchangeably hereinafter. Based on the types of all nodes, a BN can be classified into one of the three categories [20], i.e., discrete BNs, continuous BNs, and hybrid BNs. This chapter limits the treatment to the discrete BNs in which all nodes are discrete.

Each node in a BN can manifest finite mutually exclusive states. A link, as a directed edge from $X_j$ to $X_i$, represents that $X_j$ has a directed casual effect on $X_i$. Therefore, $X_j$ is considered a *parent* of $X_i$, which can be denoted as $X_j \in \mathbf{pa}(X_i)$; whereas $X_i$ is regarded as a *child* of $X_j$. Particularly, a node without any parent nodes and child nodes are called a *root* node and a *leaf* node, respectively. The DAG of a BN reflects the casual relations between all nodes, whereas the CPTs of the BN characterize the strength of these casual relations quantitatively. For a node $X_i$ with a parent set $\mathbf{pa}(X_i)$, the CPT of $X_i$, denoted as $\Pr\{X_i|\mathbf{pa}(X_i)\}$ represents the conditional probability mass function of $X_i$ under the condition of $\mathbf{pa}(X_i)$. Particularly, a set of marginal probability tables (MPTs) need to be assigned to all root nodes. An illustrative BN with six nodes is shown in Fig. 1. $X_1$ and $X_3$ are root nodes, whereas $X_5$ and $X_6$ are leaf nodes. The parent nodes of $X_2$ and $X_4$ are denoted as $\mathbf{pa}(X_2) = X_1$ and $\mathbf{pa}(X_4) = \{X_1, X_2, X_3\}$, respectively. The parent node of both $X_5$ and $X_6$ is $X_4$.

Based on the chain rule, the joint probability distribution of all the random variables in a BN can be decomposed into the product of a set of conditional probability distributions, and it is given by:

$$\Pr\{X_1, X_2, \ldots, X_n\} = \prod_{i=1}^{n} \Pr\{X_i | \mathbf{pa}(X_i)\}. \tag{2}$$

As an example, the joint probability distribution of the BN in Fig. 1 is represented as:

$$\begin{aligned} &\Pr\{X_1, X_2, \ldots, X_6\} \\ &= \Pr\{X_1\} \Pr\{X_3\} \Pr\{X_2 | X_1\} \Pr\{X_4 | X_1, X_2, X_3\} \Pr\{X_5 | X_4\} \Pr\{X_6 | X_4\} \end{aligned}. \tag{3}$$

When one or more nodes are observed/instantiated, or say, evidences are inputted

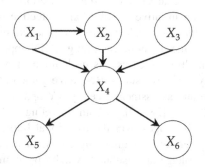

**Fig. 1.** An illustrative BN

into these nodes, BNs are capable of updating the probability distributions of other nodes without observation/instantiation/evidence via effective inference. Various efficient algorithms for exact or approximate probabilistic inference can be utilized to update the entire BN, such as variable elimination algorithm, junction tree algorithm, and Markov chain Monte Carlo (MCMC) methods. The details of the BN inference algorithms can be referred to the books by Jensen and Nielsen [18], and Koller and Friedman [21].

The BN in Fig. 1 is essentially a static model that can only represent the casual relations among nodes at a particular time instant. To characterize to the evolving behaviors of random variables over time, local models are necessary to be constructed for each unit of time. A local BN model at a particular time is called a *time slice*. Temporal links that are also directed edges are introduced to constitute a full model by connecting all the time slices in a chronological order. The full model is called a *dynamic Bayesian network (DBN)*. The detailed procedures of constructing DBN models will be elaborated in Sect. 3.2.

Numerous software can be utilized to model practical problems by BN or DBN from different aspects. An overview of available software in the literature are provided herein without pretending to be exhaustive.

- Various software with integrated and intuitive graphical interfaces are powerful and user-friendly, e.g., BayesiaLab, GeNIe, Hugin, Netica, and AgenaRisk. These software make BNs accessible to engineers without programming skills.
- A diversity of packages in different programming environments are also available, e.g., various R packages on CRAN, BNT in MATLAB, and BayesPy in Python. These packages that can make BNs manipulable are efficient, flexible, and extendable enough for engineers with proficient programming skills.
- BUGS (Bayesian inference Using Gibbs Sampling) is concerned with several flexible software that implement the approximate Bayesian inference using MCMC methods. The well-known WinBUGS, OpenBUGS, and JAGS are all a sort of BUGS software packages.

BNs can represent and characterize various uncertainties and dependencies in reliability engineering in an intuitive, flexible, and effective manner; therefore, BNs have become a very popular tool to address diverse practical reliability problems [22–27]. The reported works in the literature regarding to BN applications in BSSs can be essentially extended to multi-state cases. As each node of a BN can have multiple (more than two) mutually exclusive states, BNs have gained considerable concerns in MSS reliability modeling and assessment recently. Compared with classical reliability formalisms, such as fault trees [28–30], in both modeling and analysis features, BNs have showed significant advantages over the traditional frameworks. Therefore, BNs have been applied to a diversity of engineering cases, such as the search and rescue operations [31], medium voltage air insulated switch operation [32], axle and vehicle [33], power generating systems [19], cutter feeding control system [34, 35], water distribution system [36, 37], bridge condition modeling [38], and subsea blowout preventer [39–41].

Due to the powerful capabilities in modeling and reasoning, BNs were utilized to characterize both random and epistemic uncertainties as well as various dependencies in the context of MSSs. For example, to deal with epistemic uncertainty in reliability evaluation, BNs and DBNs were extended to evidential networks and dynamic evidential networks based on Dempster-Shafer evidence theory [34, 35, 42], respectively. Various failure dependencies between components, such as common cause failures (CCFs) [34, 43, 44] and cascading failures [45], were also modeled by BNs. As the system reliability can be updated based on BN inference algorithms if a component node is instantiated, BNs were adopted in the importance measure analysis [35, 46]. In addition to these aspects, BNs have also been extensively applied to system mainte-nance management in which BNs were used to infer the condition of a system or its components if some components and/or the entire system can be observed before maintenance decision-making [23, 47–51].

The temporal BN model of a system can be constructed to characterize the degradation/failure profile (temporal dependency) of the system. In general, temporal models can be divided into two broad categories based on the time representation, i.e., event-based approaches and time-sliced approaches. Based on the event-based approaches, Boudali and Dugan [52, 53] constructed a discrete-time BN and a continuous-time BN reliability modeling and analysis frameworks. Based on the time-sliced approaches, various DBN models were constructed to evaluate reliability of a system over time [19, 43, 54–58]. For example, Cai et al. [43] constructed a multi-phase DBN model to determine the safety integrity level of a safety instrumented system with CCFs. Jiang and Liu [19], and Xu et al. [57] developed a data-driven reliability assessment method based on DBNs by aggregating multi-level observation data. Khakzad [58] developed a DBN model to characterize the dynamic behaviors of the wildfire spread in wildland-industrial interfaces. Additionally, by decomposing an entire system model into several smaller modules, the object-oriented BNs were built up for large-scale, complex, and hierarchical systems [59–63].

To improve the modeling and inference efficiencies, various improved algorithms were proposed for BNs and DBNs, such as the topology optimization algorithm [64], dynamic discretization method [65], discretization of continuous random variables [66], compression inference algorithm [67], and improved compression inference algorithm [68].

# 3 Reliability Modeling by BNs

This section provides general procedures of constructing BN and DBN models for various typical MSSs, e.g., series systems, parallel systems, series-parallel systems, bridge systems, and phased-mission systems. Two kinds of failure dependencies among components are also considered in the BN and DBN models.

## 3.1 BN Models of Typical MSSs

The states of the components, subsystems, and the entire system of an MSS at a particular time are all inherently random variables. To characterize an MSS in the framework of BNs, the components, subsystems, and the entire system of an MSS are represented by nodes. For an MSS consisting of $M^c$ components that can be divided into $M^{sub}$ subsystems, a corresponding BN model of the system can be constructed using a set of nodes, denoted as $\Omega = \{C_1, C_2, \ldots, C_{M^c}; S_1, S_2, \ldots, S_{M^{sub}}; S\}$. In the BN model, nodes $C_l$ ($l \in \{1, 2, \ldots, M^c\}$), $S_m$ ($m \in \{1, 2, \ldots, M^{sub}\}$), and $S$ correspond to component $l$, subsystem $m$, and the entire system, respectively. Directed edges that link different nodes are added based on the relations between the states of components, subsystems, and the entire system. For an MSS with all the components being $s$-independent, node $C_l$ ($l \in \{1, 2, \ldots, M^c\}$) is a root node, whereas node $S$ is a leaf node. If subsystem $m$ (or the entire system) is composed of components $\{l_1, l_2, \ldots, l_k\}$ and subsystems $\{m_1, m_2, \ldots, m_n\}$, directed edges for nodes $\{C_{l_1}, C_{l_2}, \ldots, C_{l_k}; S_{m_1}, S_{m_2}, \ldots, S_{m_n}\}$ to node $S_m$ (or node $S$) are added into the BN model.

The CPTs and MPTs of the nodes quantifies the directed edges in a BN model. For an MSS in which all the components are $s$-independent, the MPT of each root node $C_l$ is the state probability distribution of component $l$ at a particular time. The CPTs of each node $S_m$ and leaf node $S$ can be essentially obtained by the structure function of subsystem $m$ and the entire system, respectively.

To provide the detailed procedures of constructing various BN models, several typical MSSs, i.e., series systems, parallel systems, series-parallel systems, bridge systems, and phased-mission systems, shown in Fig. 2, are used for illustration hereinafter. For a better comparison of the BN models for different system types, the components that constitute these systems are set to be the same. Each system is composed of five $s$-independent components, and the performance capacities of each component corresponding to its states are tabulated in Table 1.

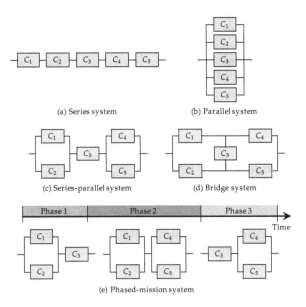

(a) Series system          (b) Parallel system

(c) Series-parallel system          (d) Bridge system

(e) Phased-mission system

**Fig. 2.** Configurations of several typical MSSs

**Table 1.** Performance capacities of each component

| Component no. | State 1 | State 2 | State 3 | State 4 |
|---|---|---|---|---|
| 1 | 7 | 3 | 0 | – |
| 2 | 8 | 4 | 0 | – |
| 3 | 11 | 8 | 3 | 0 |
| 4 | 7 | 3 | 0 | – |
| 5 | 8 | 4 | 0 | – |

### 3.1.1    BN Models of the Illustrative Series System

As all the five components are connected in series, two candidate BN models of the illustrative series system, shown in Fig. 3, can be constructed. Although both two candidate BN models are correct, candidate BN model 2 is superior to model 1 because it has a simpler CPT for each child node than that of model 1. For candidate BN model 1, all the component nodes, i.e., nodes $\{C_1, C_2, \ldots, C_5\}$, are linked to the system node, i.e., node $S$, directly; therefore, the CPT of node $S$ is a Cartesian product of $N_1^c \times N_2^c \times \ldots \times N_5^c$. For candidate BN model 2, three additional child nodes of the component nodes, i.e., nodes $S_1$, $S_2$, and $S_3$, are added, and they can avoid an oversize CPT of the system node. The dimensions of the CPTs for nodes $S_1$, $S_2$, $S_3$, and $S$ in model 2 are $N_1^c \times N_2^c$, $N_3^c \times N_1^s$, $N_4^c \times N_2^s$, and $N_5^c \times N_3^s$, respectively. In this regard, candidate BN model 2 of the illustrative series system is preferable and will be used for further analysis hereinafter. Interested readers can also find more details in [64] where a topology optimization algorithm was proposed to address the inefficiency of a con-verging BN structure.

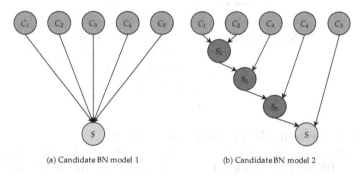

(a) Candidate BN model 1                    (b) Candidate BN model 2

**Fig. 3.** Two candidate BN models of the series and parallel systems

For any multi-state series system consisting of $n$ $s$-independent components $\{C_{l_1}, C_{l_2}, \ldots, C_{l_n}\}$, the system performance capacity at any time is determined by the performance capacities of all the components and is equal to $G(t) = \min\{G_{l_1}^c(t), G_{l_2}^c(t), \ldots, G_{l_n}^c(t)\}$. Therefore, the system state can be obtained based on the state combinations of all the components. As an example, for the series system in Fig. 2, the performance capacity of subsystem 1 at any time is completely determined by com-ponents 1 and 2. The relations between the performance capacities (states) of sub-system 1 and the corresponding state combinations of components 1 and 2 are given in Table 2. As a result, the CPT of node $S_1$ in the BN model of the series system, shown in Table 3, can be obtained. Each element in Table 3 is a conditional probability of node $S_1$ conditional on a particular state combination of nodes $C_1$ and $C_2$. In a similar manner, the CPTs of nodes $S_2$, $S_3$, and $S$ can be obtained readily. The MPT of node $C_l$ ($l \in \{1, 2, \ldots, 5\}$) are essentially the state probability distributions of component $l$ at a particular time, and it can be obtained by Eq. (1).

**Table 2.** Performance capacities and states of subsystem 1 of the series system

| Performance capacity | State of subsystem 1 | State of component 1 | State of component 2 |
|---|---|---|---|
| 7 | 1 | 1 | 1 |
| 4 | 2 | 1 | 2 |
| 3 | 3 | 2 | 1 |
| | | 2 | 2 |
| 0 | 4 | 1, 2 | 3 |
| | | 3 | 1, 2, 3 |

**Table 3.** CPT of node $S_1$ of the series system

| State of node $C_1$ | | 1 | | | 2 | | | 3 | | |
|---|---|---|---|---|---|---|---|---|---|---|
| State of node $C_2$ | | 1 | 2 | 3 | 1 | 2 | 3 | 1 | 2 | 3 |
| State of node $S_1$ | 1 | 1 | 0 | 0 | 0 | 0 | 0 | 0 | 0 | 0 |
| | 2 | 0 | 1 | 0 | 0 | 0 | 0 | 0 | 0 | 0 |
| | 3 | 0 | 0 | 0 | 1 | 1 | 0 | 0 | 0 | 0 |
| | 4 | 0 | 0 | 1 | 0 | 0 | 1 | 1 | 1 | 1 |

### 3.1.2 BN Models of the Illustrative Parallel System

For the illustrative parallel system that is composed of five $s$-independent components, two candidate BN models, shown in Fig. 3, can also be constructed. It is worth noting that the two DAGs of both the two candidate BN models for the illustrative series and parallel system are exactly the same. Likewise, candidate BN model 2 of the illustrative parallel system is preferable and will be used for further analysis hereinafter.

For any multi-state parallel system consisting of $n$ $s$-independent components $\{C_{l_1}, C_{l_2}, \ldots, C_{l_n}\}$, the system performance capacity at any time is determined by the performance capacities of all the components and is equal to $G(t) = \sum_{k=1}^{n} G_{l_i}^c(t)$. Therefore, the system state can be obtained based on the state combinations of all the components. As an example, for the parallel system in Fig. 2, the performance capacity of subsystem 1 at any time is completely determined by components 1 and 2. The relations between the performance capacities (states) of subsystem 1 and the corresponding state combinations of components 1 and 2 are given in Table 4. As a result, the CPT of node $S_1$ in the BN model of the parallel system, shown in Table 5, can be obtained. It can be seen that the CPT of node $S_1$ of the parallel system is different from the CPT of node $S_1$ of the series system. The CPTs of nodes $S_2$, $S_3$, and $S$ can be obtained readily in the same fashion.

**Table 4.** Performance capacities and states of subsystem 1 of the parallel system

| Performance capacity | State of subsystem 1 | State of component 1 | State of component 2 |
|---|---|---|---|
| 15 | 1 | 1 | 1 |
| 11 | 2 | 1 | 2 |
| | | 2 | 1 |

(continued)

**Table 4.** (*continued*)

| Performance capacity | State of subsystem 1 | State of component 1 | State of component 2 |
|---|---|---|---|
| 8 | 3 | 3 | 1 |
| 7 | 4 | 1 | 3 |
|   |   | 2 | 2 |
| 4 | 5 | 3 | 2 |
| 3 | 6 | 2 | 3 |
| 0 | 7 | 3 | 3 |

**Table 5.** CPT of node $S_1$ of the parallel system

| State of node $C_1$ | | 1 | | | 2 | | | 3 | | |
|---|---|---|---|---|---|---|---|---|---|---|
| State of node $C_2$ | | 1 | 2 | 3 | 1 | 2 | 3 | 1 | 2 | 3 |
| State of node $S_1$ | 1 | 1 | 0 | 0 | 0 | 0 | 0 | 0 | 0 | 0 |
| | 2 | 0 | 1 | 0 | 1 | 0 | 0 | 0 | 0 | 0 |
| | 3 | 0 | 0 | 0 | 0 | 0 | 0 | 1 | 0 | 0 |
| | 4 | 0 | 0 | 1 | 0 | 1 | 0 | 0 | 0 | 0 |
| | 5 | 0 | 0 | 0 | 0 | 0 | 0 | 0 | 1 | 0 |
| | 6 | 0 | 0 | 0 | 0 | 0 | 1 | 0 | 0 | 0 |
| | 7 | 0 | 0 | 0 | 0 | 0 | 0 | 0 | 0 | 1 |

### 3.1.3  BN Model of the Illustrative Series-Parallel System

Based on the system structure of the illustrative series-parallel system in Fig. 2, two candidate BN models, shown in Fig. 4, can also be constructed. Although candidate BN model 1 is intuitive, the CPTs of the child nodes in candidate BN model 2 are simplified by adding a subsystem node, i.e., node $S_2$ of candidate BN model 2. Therefore, candidate BN model 2 of the illustrative series-parallel system is preferable and will be used for further analysis hereinafter.

Corresponding to candidate BN model 2, subsystem 1 is composed of components 1 and 2 in parallel; subsystem 2 is composed of component 3 and subsystem 1 in series; subsystem 3 is composed of components 4 and 5 in parallel; the entire system consists of subsystems 2 and 3 in series. Consequently, the performance capacities of subsystems 1, 2, and 3 at time $t$ are computed as $G_1^s(t) = G_1^c(t) + G_2^c(t)$, $G_2^s(t) = \min\{G_3^c(t), G_1^s(t)\}$, and $G_3^s(t) = G_4^c(t) + G_5^c(t)$, respectively; the performance capacity of the entire system at time $t$ is computed as $G(t) = \min\{G_2^s(t), G_3^s(t)\}$. The similar analyses implemented in Tables 3 and 5 can also be done herein to obtain the CPTs of nodes $S_1$, $S_2$, $S_3$, and $S$.

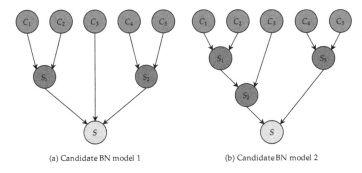

(a) Candidate BN model 1          (b) Candidate BN model 2

**Fig. 4.** Candidate BN models of the series-parallel system

### 3.1.4    BN Model of the Illustrative Bridge System

The BN model of the illustrative bridge system can be constructed based on the minimal success paths [42]. For the illustrative bridge system in Fig. 2, there exist four minimal success paths, i.e., $\{C_1, C_4\}$, $\{C_1, C_3, C_5\}$, $\{C_2, C_5\}$, and $\{C_2, C_3, C_4\}$. Therefore, the bridge system can be decomposed into two simplified sub-models, and the BN model of the bridge system can be constructed as shown in Fig. 5. In the BN model, nodes $S_3$ and $S_6$ represent sub-models 1 and 2, respectively; node $S$ represents the entire bridge system. The performance capacities of subsystems 1, 2, and 3 at time $t$ can be computed as $G_1^s(t) = \min\{G_3^c(t),\ G_5^c(t)\}$, $G_2^s(t) = G_4^c(t) + G_1^s(t)$, and $G_3^s(t) = \min\{G_1^c(t),\ G_2^s(t)\}$, respectively; the performance capacities of subsystems 4, 5, and 6 at time $t$ can be computed as $G_4^s(t) = \min\{G_3^c(t),\ G_4^c(t)\}$, $G_5^s(t) = G_5^c(t) + G_4^s(t)$, and $G_6^s(t) = \min\{G_2^c(t),\ G_5^s(t)\}$, respectively; the performance capacity of the entire system at time $t$ can be computed as $G(t) = G_3^s(t) + G_6^s(t)$. Consequently, the CPTs of all the child nodes in the BN model can be obtained readily.

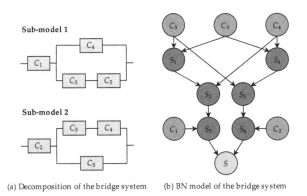

(a) Decomposition of the bridge system          (b) BN model of the bridge system

**Fig. 5.** Decomposition and BN model of the bridge system

### 3.1.5  BN Model of the Illustrative Phased-Mission Systems

The multi-state phased-mission system herein is intended to perform a mission with $H$ phases. The system may reconfigure in different phases to meet varying mission demands, resulting in a distinct system structure in each phase. If a component is suspended in a phase, the component is assumed to not deteriorate during the phase and its state remains unchanged. The number of subsystems in phase $h$ is denoted by $M_h^{\text{sub}}$ ($h \in \{1, 2, \ldots, H\}$); the numbers of the states of subsystem $m$ and the entire system in phase $h$ are denoted by $N_{m,h}^{\text{sub}}$ and $N_h^{\text{sys}}$, respectively. The performance capacities of component $l$, subsystem $m$, and the entire system at time $t$ in phase $h$ are denoted by $G_{l,h}^{c}(t)$, $G_{m,h}^{s}(t)$, and $G_h(t)$, respectively. The states of component $l$, subsystem $m$, and the entire system at time $t$ in phase $h$ are denoted by $C_{l,h}(t)$, $S_{m,h}(t)$, and $S_h(t)$, respectively. It should be noted that $S_m(t)$ hereinbefore denotes the state of subsystem $m$ at time $t$ for a general MSS, whereas $S_h(t)$ herein represents the state of the entire system at time $t$ in phase $h$ for a multi-state phased-mission system. The duration of phase $h$ is $T_h$ ($h \in \{1, 2, \ldots, H\}$) times of the basic time interval.

The system survival at the end of a phase is not only determined by the system state at the end of the phase, but also depends on whether the system can survive at the end of the last phase. Therefore, a binary-state node, denoted as $D_h$ ($h \in \{1, 2, \ldots, H\}$), is introduced herein to indicate whether the system can survive at the end of phase $h$. By linking node $D_h$ of adjacent phases by the directed edges, the probability of the system surviving in each phase can be characterized. Let $D_h = 1$ and $D_h = 2$ denote the system being in the functioning state and failure state at the end of phase $h$, respectively. Therefore, the conditional probabilities of node $D_h$ can be represented as follows:

$$\Pr\{D_h = 1 | D_{h-1}, S_h\} = \begin{cases} 1 & D_{h-1} = 1 \text{ and } S_h \text{ is acceptable} \\ 0 & \text{otherwise} \end{cases}, \tag{4}$$

$$\Pr\{D_h = 2 | D_{h-1}, S_h\} = \begin{cases} 1 & D_{h-1} = 2 \text{ or } S_h \text{ is unacceptable} \\ 0 & \text{otherwise} \end{cases}. \tag{5}$$

A set of nodes, which is denoted as $\Omega = \{\Omega_1, \Omega_2, \ldots, \Omega_H\}$, are used to construct the BN model of a phased-mission system. $\Omega_h = \{C_{1,h}, C_{2,h}, \ldots, C_{M^c,h}; S_{1,h}, S_{2,h}, \ldots, S_{M^{\text{sub}},h}; S_h; D_h\}$ ($h \in \{1, 2, \ldots, H\}$) is the set of nodes in phase $h$, where nodes $C_{l,h}$ ($l \in \{1, 2, \ldots, M^c\}$), $S_{m,h}(m \in \{1, 2, \ldots, M_h^{\text{sub}}\})$, and $S_h$ correspond to component $l$, subsystem $m$, and the entire system in phase $h$, respectively. In each phase, a local BN model can be constructed first based on the corresponding system structure. Directed edges are then added between component nodes across different phases to characterize the relations between different phases. The BN model of the illustrative phased-mission system in Fig. 2 can be constructed as shown in Fig. 6. The local BN model of phase $h$ characterizes the phased-mission system at the end of phase $h$. In phase 1, components 1, 2, and 3 are in operation, whereas components 4 and 5 are suspended. The performance capacities of subsystem 1 and the entire system at time $t$ in phase 1 are denoted as $G_{1,1}^{s}(t) = G_{1,1}^{c}(t) + G_{2,1}^{c}(t)$ and $G_1(t) = \min\{G_{3,1}^{c}(t), G_{1,1}^{s}(t)\}$, respectively. Likewise, the performance capacities of the subsystem(s) and the entire system in phases 2 and 3 can be obtained. Consequently, the CPTs of all the subsystem nodes and the system nodes in the BN model can be obtained readily.

If a component node in a phase is a root node, the MPT of the component node in the phase is essentially the state probability distribution of the corresponding component at a particular time. Nevertheless, if a component node in a phase is a child node, the CPT of the component node in the phase is essentially the state transition matrix of the corresponding component. As an example, for the BN model of the illustrative phased-mission system in Fig. 6, the CPTs of nodes $C_{1,2}$ and $C_{2,2}$ are the $T_2$-step state transition matrix of components 1 and 2, respectively; the CPTs of nodes $C_{4,3}$ and $C_{5,3}$ are the $T_3$-step state transition matrix of components 4 and 5, respectively; the CPT of node $C_{3,3}$ is the $T_3$-step state transition matrix of component 3.

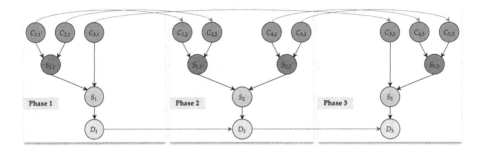

**Fig. 6.** BN model of the phased-mission system

### 3.2 DBN Models of Typical MSSs

The BN models in Sect. 3.2 are all static models that can only represent an MSS at a particular time. The DBN model of an MSS can characterize the degradation process of the MSS during the operation period. By using a time slice to represent an MSS at a particular time, the DBN model of the MSS is inherently a discrete-time model. In a DBN model, all the time slices are the repetitive BN models of an MSS at a particular time. The time interval between two adjacent time slices is a basic time interval, i.e., $\Delta t$. Suppose that the operation period is $T \cdot \Delta t$, the number of time slices is thus equal to $T + 1$. Time slice $t$ ($t \in \{0, 1, \ldots, T\}$) represents the local BN model at time $t$. A set of nodes, denoted as $\Omega = \{\Omega(0), \Omega(1), \ldots, \Omega(T)\}$, are used to construct the DBN model of an MSS. $\Omega(t) = \{C_1(t), C_2(t), \ldots, C_{M^c}(t); S_1(t), S_2(t), \ldots, S_{M^{sub}}(t); S(t)\}$ is the set of nodes in time slice $t$, where $C_l(t)$ ($l \in \{1, 2, \ldots, M^c\}$), $S_m(t)$ ($m \in \{1, 2, \ldots, M^{sub}\}$), and $S(t)$ correspond to component $l$, subsystem $m$, and the entire system at time $t$, respectively. The MPT of node $C_l(0)$ is the state probability distribution of component $l$ at the beginning of use.

A temporal link from node $C_l(t)$ ($l \in \{1, 2, \ldots, M^c\}$, $t \in \{0, 1, \ldots, T-1\}$) to node $C_l(t+1)$ is added to connect the two component nodes between two adjacent time slices, and it characterizes the degradation profiles of component $l$ within a basic time interval. The strength of the temporal link from node $C_l(t)$ to node $C_l(t+1)$ is quantified by the CPT of node $C_l(t+1)$ which is equivalent to the state transition matrix of component $l$. The illustrative systems in Fig. 2 are used herein to provide detailed procedures of constructing the DBN models of various systems.

The DBN models of the illustrative series system in an extended form and an abstract form are shown in Fig. 7. In the extended form of the DBN model, all the time slices from time slice 0 to time slice $T$ are displayed. In the abstract form of the DBN model, only a particular time slice, i.e., time slice $t$, is displayed. The number attached to each temporal link, i.e., "1", represents the number of time slices used for the temporal dependency. The number in the square box, i.e., $T+1$, represents the total number of time slices in the DBN model. The DBN models of the illustrative parallel system in an extended form and an abstract form are also shown in Fig. 7. As discussed in Sect. 3.1, although the DAGs of the DBN models for the series and parallel systems are identical, the CPTs of the subsystem nodes and system nodes, i.e., nodes $S_m(t)$ ($m \in \{1, 2, \ldots, M^{\text{sub}}\}$, $t \in \{0, 1, \ldots, T\}$) and $S(t)$ ($t \in \{0, 1, \ldots, T\}$) are distinct. In a similar manner, the abstract forms of the DBN models of the illustrative series-parallel and bridge systems are shown in Fig. 8.

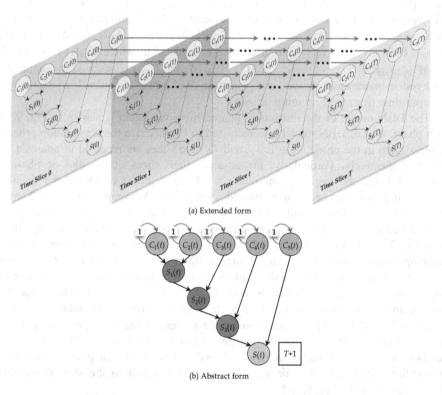

(a) Extended form

(b) Abstract form

**Fig. 7.** DBN models of the series and parallel systems

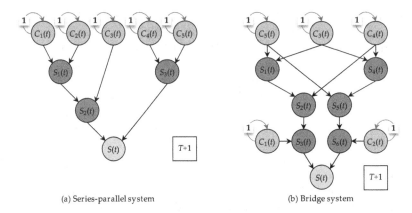

(a) Series-parallel system                    (b) Bridge system

**Fig. 8.** DBN models of the series-parallel and bridge systems

For a phased-mission system, a set of nodes, denoted as $\Omega = \{\Omega_1, \Omega_2, ..., \Omega_H\}$, are used to construct the DBN model, where $\Omega_1 = \{\Omega_1(0), \Omega_1(1), ..., \Omega_1(T_1)\}$ and $\Omega_h = \{\Omega_1(\sum_{k=1}^{h-1} T_k + 1), \Omega_h(\sum_{k=1}^{h-1} T_k + 2), ..., \Omega_h(\sum_{k=1}^{h} T_k)\}$ $(h \in \{2, 3, ..., H\})$ represent the local DBN models of the system in phase 1 and phase $h$, respectively. $\Omega_h(t) = \{C_{1,h}(t), C_{2,h}(t), ..., C_{M^c,h}(t); S_{1,h}(t), S_{2,h}(t), ..., S_{M^{sub},h}(t); S_h(t); D_h(t)\}$ $(h \in \{1, 2, ..., H\})$ represents time slice $t$ in phase $h$, where $C_{l,h}(t)$ $(l \in \{1, 2, ..., M^c\})$, $S_{m,h}(t)$ $(m \in \{1, 2, ..., M_h^{sub}\})$, $S_h(t)$, and $D_h(t)$ correspond to component $l$, subsystem $m$, the entire system, and system survival at time $t$ in phase $h$, respectively. $t$ is elapse time from the beginning of use.

The DBN model of the illustrative phased-mission system is shown in Fig. 9. In each phase, a local DBN model is constructed to characterize the degradation process of the system in the phase. $T_1 + 1$, $T_2$, and $T_3$ time slices are repeated in phases 1, 2, and 3, respectively. Particularly, the adjacent time slices at the end of phase $h$ $(h \in \{1,2\})$ and the beginning of phase $h + 1$ are depicted to shown the detailed temporal dependencies between the two adjacent phases. As components 1 and 2 are in operation in both phases 1 and 2, in time slices $T_1 + 1$ and $T_1 + 2$, only two temporal links are added to the corresponding component nodes, i.e., the directed edge from node $C_{l,1}(T_1 + 1)$ to node $C_{l,2}(T_1 + 2)$ $(l \in \{1,2\})$. Likewise, as components 4 and 5 are in operation in both phases 2 and 3, in time slices $T_1 + T_2 + 1$ and $T_1 + T_2 + 2$, only two temporal links are added to the corresponding component nodes, i.e., the directed edge from node $C_{l,2}(T_1 + T_2 + 1)$ to node $C_{l,3}(T_1 + T_2 + 2)$ $(l \in \{4,5\})$. As components 4 and 5 are suspended in phase 1, the state probability distributions of nodes $C_{4,2}(T_1 + 2)$ and $C_{5,2}(T_1 + 2)$ are actually the corresponding state probability distributions of components 4 and 5 at the beginning of use, respectively. Nevertheless, as component 3 is in operation in phases 1 and 3 and in idle in phase 2, the state probability distribution of node $C_{3,3}(T_1 + T_2 + 2)$ is equal to the state probability distribution of node $C_{3,1}(T_1 + 1)$.

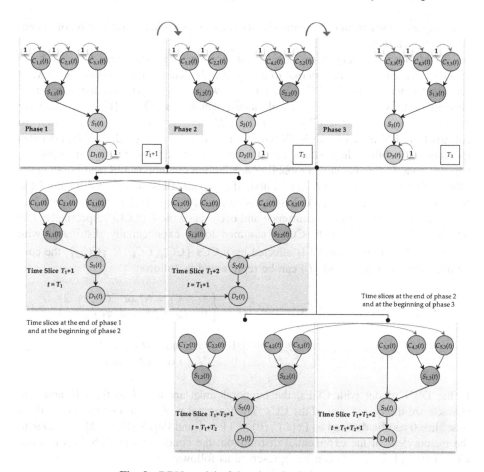

**Fig. 9.** DBN model of the phased-mission system

## 3.3    Failure Dependencies in BN and DBN Models

The components in the BN and DBN models presented in Sects. 3.1 and 3.2 are assumed to be $s$-independent. Nevertheless, in real-world situations, the failure processes of the components may be inevitably $s$-dependent. BNs are a powerful tool to cope with various dependencies and can be utilized to model failure dependencies between components during their degradation processes. Two typical failure dependencies,.i.e., CCFs [44, 69–72] and immediate failure dependence (IFD) [73, 74], are considered in the illustrative series-parallel system herein.

CCFs are the failures of multiple dependent components within a system because of a share root cause or a common cause (CC) [69, 71], such as extreme environmental conditions or human errors. The presence of CCFs tends to increase the joint failure probability of a system, contributing significantly to the overall unreliability of systems subject to CCFs. Therefore, it is crucial to incorporate CC effects into the reliability modeling and assessment of systems subject to CCFs to avoid overestimation of system reliability measures.

An MSS can be subject to CCFs because of various elementary CCs. CCs are exclusive mutually and are external to the system. In general, CCs existing in an MSS can be denoted as $\{CC_1, CC_2, \ldots, CC_{n_{CC}}\}$, where $n_{CC}$ represents the number of elementary CCs. Therefore, a set of nodes, denoted as $\Omega = \{C_1, C_2, \ldots, C_{M^c};$ $S_1, S_2, \ldots, S_{M^{sub}}; S; CC_1, CC_2, \ldots, CC_{n_{CC}}; U_1, U_2, \ldots, U_{M^c}\}$, can be used to construct the BN model of an MSS with CCFs. Node $C_l$ ($l \in \{1, 2, \ldots, M^c\}$) denotes the state probability distribution of component $l$ caused by its own degradation, whereas node $U_l$ denotes state probability distribution of component $l$ incorporating the effects of CCs. Node $U_l$ can be a null node if component $l$ is not affected by any CCs. Node $CC_k$ ($k \in \{1, 2, \ldots, n_{CC}\}$) has two states, i.e., $CC_k \in \{1, 2\}$. States 1 and 2 of node $CC_k$ represent the non-occurrence and occurrence the $k$ th CC, respectively. The inter-arrival time of the $k$ th CC is assumed to be exponentially distributed with parameter $\lambda_k^{CC}$. If component $l$ is affected by $n$ CCs $\{CC_{k_1}, CC_{k_2}, \ldots, CC_{k_n}\}$, the conditional probabilities of node $U_l$ can be represented as follows:

$$\Pr\{U_l = N_l^c | C_l; CC_{k_1}, CC_{k_2}, \ldots, CC_{k_n}\} = \begin{cases} 1 & C_l = N_l^c \text{ or } \exists CC_{k_j} = 2 \\ 0 & C_l \neq N_l^c \text{ and } \forall CC_{k_j} = 1 \end{cases}, \quad (6)$$

$$\Pr\{U_l = i | C_l; CC_{k_1}, CC_{k_2}, \ldots, CC_{k_n}\} = \begin{cases} 1 & C_l = i \text{ and } \forall CC_{k_j} = 1 \\ 0 & C_l \neq i \text{ or } \exists CC_{k_j} = 2 \end{cases}, i \neq N_l^c. \quad (7)$$

In the DBN model with CCFs, the temporal links are added to the CC nodes to characterize the occurrence of the CCs. The marginal probabilities of node $CC_k(0)$ in time slice 0 can be denoted as $\Pr\{CC_k(0) = 1\} = 1$ and $\Pr\{CC_k(0) = 2\} = 0$. Due to the memoryless of the exponential distribution, the conditional probabilities of node $CC_k(t)$ ($t \in \{1, 2, \ldots, T\}$) can be represented as follows:

$$\Pr\{CC_k(t) = 1 | CC_k(t-1)\} = 1 - \exp(-\lambda_k^{CC} \cdot \Delta t), \quad (8)$$

$$\Pr\{CC_k(t) = 2 | CC_k(t-1)\} = \exp(-\lambda_k^{CC} \cdot \Delta t). \quad (9)$$

In addition, if component $l$ is not affected by CCs, a temporal link will be added to node $C_l(t)$ from time slice $t - 1$ ($t \in \{1, 2, \ldots, T\}$) to time slice $t$. On the contrary, if component $l$ is affected by CCs, a temporal link will be added from node $U_l(t - 1)$ ($t \in \{1, 2, \ldots, T\}$) to node $C_l(t)$ as $U_l(t - 1)$ represents the actual condition of component $l$.

The illustrative series-parallel system in Fig. 2 is used herein for further analysis. Suppose that two CCs exist in the system; $CC_1$ affects components 1 and 2; $CC_2$ affects components 4 and 5. Consequently, the BN and DBN models of the illustrative series-parallel system with CCFs are shown in Fig. 10. In the BN model, node $U_3$ is omitted since component 3 is not affected by any CCs. As an example, based on Eqs. (6) and (7), the CPT of $U_1$ is tabulated in Table 6. Two time slices of the DBN model are shown in Fig. 11 to present the details of the temporal links between the CC nodes and component nodes. In the DBN model, the CPT of node $C_l(t)$ $(t \in \{1, 2, \ldots, T\})$ is always the one-step transition matrix of component $l$ regardless of the parent node of node $C_l(t)$. Furthermore, the BN and DBN models with CCFs can be extended to more generalized cases, such as probabilistic CCFs [71, 72] and the case in which a CC can manifest multiple states [43].

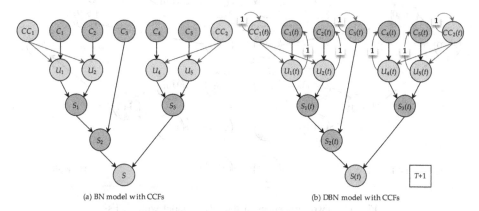

(a) BN model with CCFs          (b) DBN model with CCFs

**Fig. 10.** BN and DBN models of the series-parallel system with CCFs

**Table 6.** CPT of node $U_1$ in the BN model with CCFs

| State of node $CC_1$ | | 1 | | | 2 | | |
|---|---|---|---|---|---|---|---|
| State of node $C_1$ | | 1 | 2 | 3 | 1 | 2 | 3 |
| State of node $U_1$ | 1 | 1 | 1 | 0 | 0 | 0 | 0 | 0 |
| | 2 | 0 | 1 | 0 | 0 | 0 | 0 |
| | 3 | 0 | 0 | 1 | 1 | 1 | 1 |

IFD is common in real-world situations, and it refers to that the failure a component (influencing component) may cause immediate failures of some other components (affected components) [73, 74]. For instance, the failure of an electrical component creates a voltage spike that immediately triggers the failures of the neighboring components.

A set of nodes, denoted as $\Omega = \{C_1, C_2, \ldots, C_{M^c}; S_1, S_2, \ldots, S_{M^{sub}}; S; U_1, U_2, \ldots, U_{M^c}\}$, can be used to construct the BN model of an MSS with IFD. Node $C_l$ $(l \in \{1, 2, \ldots, M^c\})$ denotes the state probability distribution of component $l$ caused by its own degradation, whereas node $U_l$ denotes state probability distribution of component $l$ incorporating the effects of IFD. Likewise, node $U_l$ can be a null node if component $l$ is not affected by other components. Suppose that an immediate failure of affected component $l$ occurs with probability $p_l^{\mathrm{IF}}$ if component $l$ is not failed and one of its influencing components fails. If the failure of any of $n$ components $\{C_{k_1}, C_{k_2}, \ldots, C_{k_n}\}$ can cause the failure of component $l$, the conditional probability of node $U_l$ can be represented as follows:

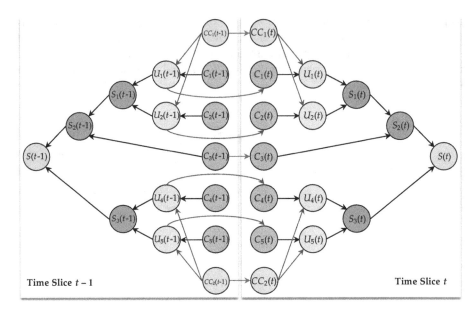

**Fig. 11.** Two time slices of the DBN model with CCFs

$$\Pr\{U_l = N_l^c | C_l; C_{k_1}, C_{k_2}, \ldots, C_{k_n}\} = \begin{cases} 1 & C_l = N_l^c \\ p_l^{\mathrm{IF}} & C_l \neq N_l^c \text{ and } \exists C_{k_j} = N_l^c, \\ 0 & C_l \neq N_l^c \text{ and } \forall C_{k_j} \neq N_l^c \end{cases} \quad (10)$$

$$\Pr\{U_l = i | C_l; C_{k_1}, C_{k_2}, \ldots, C_{k_n}\} = \begin{cases} 1 & C_l = i \text{ and } \forall C_{k_j} \neq N_l^c \\ 1 - p_l^{\mathrm{IF}} & C_l = i \text{ and } \exists C_{k_j} = N_l^c, i \neq N_l^c. \\ 0 & C_l \neq i \end{cases} \quad (11)$$

In the DBN model with IFD, if component $l$ is not affected by other components, a temporal link will be added to node $C_l(t)$ from time slice $t-1$ ($t \in \{1, 2, \ldots, T\}$) to time slice $t$. On the contrary, if the failure of any of $n$ components $\{C_{k_1}, C_{k_2}, \ldots, C_{k_n}\}$ can cause the failure of component $l$, a temporal link will be added from node $U_l(t-1)$ ($t \in \{1, 2, \ldots, T\}$) to node $C_l(t)$ since $U_l(t-1)$ represents the actual condition of component $l$.

The illustrative series-parallel system in Fig. 2 is used herein for further analysis. Suppose that the failure of component 1 can cause an immediate failure of component 2; the failure of component 4 can also lead to a failure of component 5 immediately. Consequently, the BN and DBN models of the illustrative series-parallel system with IFD are shown in Fig. 12. In the BN model, nodes $U_1$, $U_3$, and $U_4$ are omitted. As an example, based on Eqs. (10) and (11), the CPT of node $U_2$ is tabulated in Table 7.

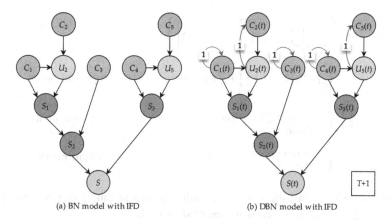

(a) BN model with IFD          (b) DBN model with IFD

**Fig. 12.** BN and DBN models of the series-parallel system with IFD

**Table 7.** CPT of node $U_2$ in the BN model with IFD

| State of node $C_1$ | 1 | | | 2 | | | 3 | | |
|---|---|---|---|---|---|---|---|---|---|
| State of node $C_2$ | 1 | 2 | 3 | 1 | 2 | 3 | 1 | 2 | 3 |
| State of node $U_2$ 1 | 1 | 1 | 0 | 0 | 1 | 0 | 0 | $1-p_2^{\text{IF}}$ | 0 | 0 |
| 2 | 0 | 1 | 0 | 0 | 1 | 0 | 0 | 0 | $1-p_2^{\text{IF}}$ | 0 |
| 3 | 0 | 0 | 1 | 0 | 0 | 1 | $p_2^{\text{IF}}$ | $p_2^{\text{IF}}$ | 1 |

# 4   Reliability Assessment by DBNs

In this section, system reliability of the preceding MSSs can be assessed based on DBN models. If no evidence is inserted, the state probability distribution of the entire system at any time can be obtained by marginalizing the system node in the corresponding time slice. If some nodes are instantiated, the state probability distribution of the system node in any time slice can be updated by BN inference algorithms. Subsequently, by defining the acceptable states of an MSS, the reliability of the entire system can be estimated for any time instant.

## 4.1   BN Inference

Let the node set $\Omega = \{\Omega(0), \Omega(1), \ldots, \Omega(T)\}$ denote the DBN model of an MSS, where $\Omega(t) = \{C_1(t), C_2(t), \ldots, C_{M^c}(t); S_1(t), S_2(t), \ldots, S_{M^{\text{sub}}}(t); S(t)\}$. The joint probability of the DBN model can be expressed as:

$$\Pr(\mathbf{\Omega}) = \Pr\left\{\bigcup_{t=0}^{T}\left[\bigcup_{l=1}^{M^c}C_l(t); \bigcup_{m=1}^{M^{sub}}S_m(t); S(t)\right]\right\}$$

$$= \prod \begin{cases} \displaystyle\prod_{l=1}^{M^c}\Pr\{C_l(0)\}\prod_{t=1}^{T}\Pr\{C_l(t)|C_l(t-1)\} & \text{component nodes} \\[2ex] \displaystyle\prod_{m=1}^{M^{sub}}\prod_{t=0}^{T}\Pr\{S_m(t)|\mathbf{pa}(S_m(t))\} & \text{subsystem nodes} \\[2ex] \displaystyle\prod_{t=0}^{T}\Pr\{S(t)|\mathbf{pa}(S(t))\} & \text{system nodes} \end{cases} \qquad (12)$$

The state probability distribution of the entire system at time $t$ can be obtained by marginalizing node $S(t)$, which is represented as [18]:

$$\Pr\{S(t)\} = \sum_{\mathbf{\Omega}\setminus S(t)}\Pr\{\mathbf{\Omega}\}. \qquad (13)$$

During the operation period, the states of some components, subsystems, and the entire system can be observed by conducting condition monitoring periodically or non-periodically. If a component, subsystem, or the entire system is observed in a particular state at a particular time, the corresponding node in the DBN model of the system is instantiated with the observed state. Suppose that $ne$ nodes in a DBN model are instantiated, the evidence of a DBN model is denoted as $\mathbf{e} = \{e_{X_1}, e_{X_2}, \ldots, e_{X_{ne}}\}$, where $e_{X_i}$ ($i \in \{1, 2, \ldots, ne\}$) denotes the evidence of node $X_i$, i.e., the observed state of a component, subsystem, or the entire system at a particular time. Consequently, when evidence $\mathbf{e}$ is inputted into a DBN model, on the basis of the Bayes formula, the posterior probability distribution of the system state at time $t$ can be obtained by marginalizing node $S(t)$, and it is represented as follows [18]:

$$\Pr\{S(t)|\mathbf{e}\} = \frac{\sum_{\mathbf{\Omega}\setminus S(t)}\Pr\{\mathbf{\Omega}, \mathbf{e}\}}{\Pr\{\mathbf{e}\}}, \qquad (14)$$

where $\Pr\{\mathbf{e}\}$ is the prior probability of evidence $\mathbf{e}$. $\Pr\{\mathbf{e}\}$ can be calculated by marginalizing the instantiated nodes, i.e., nodes $\{X_1, X_2, \ldots, X_{ne}\}$, which is represented as follows [18]:

$$\Pr\{\mathbf{e}\} = \sum_{\mathbf{\Omega}\setminus\{X_1, X_2, \ldots, X_{ne}\}}\Pr\{\mathbf{\Omega}, \mathbf{e}\}. \qquad (15)$$

Equations (13)–(15) can be calculated by various BN inference algorithms, such as variable elimination algorithm and junction tree algorithm. The details involved in the BN inference algorithms can be found in the books by Jensen and Nielsen [18], and Koller and Friedman [21]. Consequently, the state probability distribution of the entire system at time $t$ can be evaluated by Eq. (13).

## 4.2    Reliability Assessment by Aggregating Multi-level Observation Data

The degradation processes of the components, subsystems, and the entire system of an MSS can be inspected by collecting condition monitoring data from sensors that are mounted at various physical levels of the system (component level, subsystem level, and system level). Observation data can be collected from multiple levels of an MSS simultaneously or asynchronously during the operation stage [7, 19, 75–77]. If an inspection is conducted at a particular time, the state probability distribution and reliability of an MSS can be updated by aggregating multi-level observation data. Moreover, if inspections are conducted chronologically during the operation period, the state probability distribution and reliability of an MSS can be updated dynamically. An evidence in the DBN model of an MSS is essentially the collected multi-level observation data. Therefore, the state probability distribution and reliability of an MSS can be updated using Eq. (14) once an evidence is inserted into the DBN model of the MSS. More details involved in updating system reliability dynamically by observation data during the operation period can be referred to [77–79]. The illustrative systems in Fig. 2 are used herein for further analysis.

For each of the five systems in Fig. 2, i.e., the series system, parallel system, series-parallel system, bridge system, and phased-mission system, the system is considered as reliable if the performance capacity of the entire system is not less than a user demand. The user demand of the five systems are set to be 3. The one-step state transition matrixes of all the components are given as follows:

$$\mathbf{P}_1 = \begin{bmatrix} 0.9185 & 0.0564 & 0.0251 \\ 0 & 0.9608 & 0.0392 \\ 0 & 0 & 1 \end{bmatrix}, \mathbf{P}_2 = \begin{bmatrix} 0.9231 & 0.0474 & 0.0295 \\ 0 & 0.9734 & 0.0266 \\ 0 & 0 & 1 \end{bmatrix},$$

$$\mathbf{P}_4 = \begin{bmatrix} 0.9579 & 0.0290 & 0.0131 \\ 0 & 0.9724 & 0.0276 \\ 0 & 0 & 1 \end{bmatrix}, \mathbf{P}_5 = \begin{bmatrix} 0.9550 & 0.0308 & 0.0142 \\ 0 & 0.9704 & 0.0296 \\ 0 & 0 & 1 \end{bmatrix},$$

$$\mathbf{P}_3 = \begin{bmatrix} 0.9465 & 0.0286 & 0.0149 & 0.0100 \\ 0 & 0.9589 & 0.0281 & 0.0130 \\ 0 & 0 & 0.9802 & 0.0198 \\ 0 & 0 & 0 & 1 \end{bmatrix}.$$

The duration of the operation period is set at $T = 50$ units of time. For the phased-mission system, the durations of the three phases are set at $T_1 = 12$ units of time, $T_2 = 18$ units of time, and $T_3 = 20$ units of time. Consequently, for the series, parallel, series-parallel, and bridge systems, system reliabilities at time $t$, denoted as $R(t)$, are shown in Fig. 13; for the phased-mission system, system reliability at time $t$ is shown in Fig. 14.

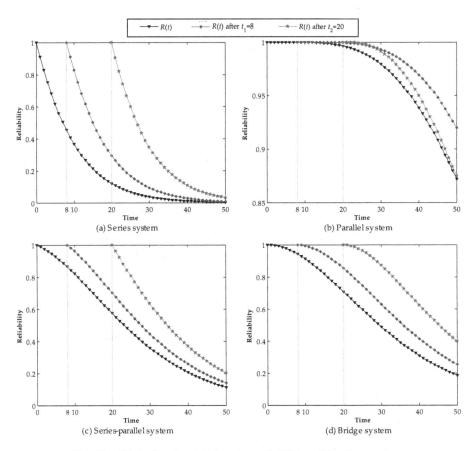

**Fig. 13.** Original and updated system reliabilities of the four systems

When one or more inspections are conducted chronologically, the system reliability of an MSS will be updated dynamically. The system-level or multi-level observation data of the five systems collected at two different time instants, i.e., $t_1 = 8$ units of time and $t_2 = 20$ units of time, are listed in Table 8. As a result, for each of the five systems, system reliability can be updated dynamically at the two inspection time instants. The updated system reliabilities of the series, parallel, series-parallel, and bridge systems are shown in Fig. 13. The updated system reliabilities of the phased-mission system are shown in Fig. 14.

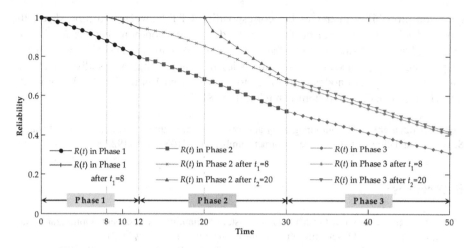

**Fig. 14.** Original and updated system reliabilities of the phased-mission system

**Table 8.** Multi-level observation data

| Systems | Observations | |
|---|---|---|
| | $t_1 = 8$ | $t_2 = 20$ |
| Series | $S(8) = 1$ | $S(20) = 2,\ C_1(20) = 1$ |
| Parallel | $S(8) = 3$ | $S(20) = 16,\ S_2(20) = 6$ |
| Series-parallel | $S(8) = 2,\ S_1(8) = 2$ | $S(20) = 4,\ C_4(20) = 2$ |
| Bridge | $S(8) = 1,\ C_3(8) = 2$ | $S(20) = 5,\ C_3(20) = 3,\ C_4(20) = 2$ |
| Phased-mission | $S_1(8) = 1,\ S_{1,1}(8) = 2$ | $S_2(20) = 3,\ S_{1,2}(20) = 3,\ C_{4,2}(20) = 2$ |

## 5 Conclusions and Discussions

In this chapter, a holistic framework for MSS reliability modeling and assessment based on BNs and DBNs is presented. The basic characteristics of MSSs and BNs are presented. The detailed procedures of constructing the BN and DBN models of various MSSs are provided. The results show that BNs and DBNs can effectively represent and characterize dependency among components in MSSs. A reliability assessment approach by aggregating multi-level observation data is developed, which can update the system reliability dynamically once an additional inspection is conducted. The reliability modeling and assessment results of five typical MSSs show that BNs and DBNs are effective considerably in terms of modeling and assessing reliability of MSSs.

A crucial premise in this chapter is that the degradation process of each component in an MSS is characterized by a homogenous Markov process. Nevertheless, in real-world situations, the degradation process of a component may follow a non-homogenous Markov process or semi-Markov process. Under such a circumstance, one can calculate the transition probability matrix of a non-homogenous Markov process or semi-Markov process between any two time instants [12, 13, 80]. By setting the transition probability matrix as the corresponding CPT between two time slices, the proposed DBN models can be further extended to the non-homogenous Markov or semi-Markov case. Additionally, the CPTs of subsystem and system nodes in this chapter are all assumed to be deterministic. It is noted that probabilistic CPTs of subsystem and system nodes correspond to a generalized BN model which can reflect imperfect knowledge of system behaviors [28, 81].

**Acknowledgements.** The authors greatly acknowledge grant support from the National Natural Science Foundation of China under contract numbers 71771039 and 71922006.

# References

1. Lisnianski, A., Levitin, G.: Multi-State System Reliability: Assessment, Optimization and Applications. World Scientific Publishing Company, Singapore (2003)
2. Kuo, W., Zuo, M.J.: Optimal Reliability Modeling: Principles and Applications. Wiley, Hoboken (2003)
3. Lisnianski, A., Frenkel, I., Ding, Y.: Multi-State System Reliability Analysis and Optimization for Engineers and Industrial Managers. Springer, London (2010). https://doi.org/10.1007/978-1-84996-320-6
4. Jiang, T., Liu, Y., Zheng, Y.X.: Optimal loading strategy for multi-state systems: Cumulative performance perspective. Appl. Math. Model. **74**, 199–216 (2019)
5. Lisnianski, A., Frenkel, I., Karagrigoriou, A.: Recent Advances in Multi-State Systems Reliability: Theory and Applications. Springer, Cham (2017). https://doi.org/10.1007/978-3-319-63423-4
6. Dimla Sr., D.E., Lister, P.M.: On-line metal cutting tool condition monitoring. II: tool-state classification using multi-layer perceptron neural networks. Int. J. Mach. Tools Manuf. **40**(5), 769–781 (2000)
7. Graves, T.L., Hamada, M.S., Klamann, R., Koehler, A., Martz, H.F.: A fully Bayesian approach for combining multi-level information in multi-state fault tree quantification. Reliab. Eng. Syst. Saf. **92**(10), 1476–1483 (2007)
8. Yeh, W.: A fast algorithm for searching all multi-state minimal cuts. IEEE Trans. Reliab. **57**(4), 581–588 (2008)
9. Lin, Y.: Network reliability of a time-based multistate network under spare routing with p minimal paths. IEEE Trans. Reliab. **60**(1), 61–69 (2011)
10. Shrestha, A., Xing, L., Dai, Y.: Decision diagram based methods and complexity analysis for multi-state systems. IEEE Trans. Reliab. **59**(1), 145–161 (2010)
11. Trivedi, K.S.: Probability and Statistics with Reliability, Queuing, and Computer Science Applications. Wiley, Hoboken (2016)
12. Limnios, N., Oprişan, G.: Semi-Markov Processes and Reliability. Springer, Heidelberg (2001). https://doi.org/10.1007/978-1-4612-0161-8
13. Limnios, N., Barbu, V.S.: Semi-Markov Chains and Hidden Semi-Markov Models toward Applications: Their use in Reliability and DNA Analysis. Springer, New York (2008). https://doi.org/10.1007/978-0-387-73173-5
14. Levitin, G.: The Universal Generating Function in Reliability Analysis and Optimization. Springer, London (2005). https://doi.org/10.1007/1-84628-245-4
15. Zio, E.: The Monte Carlo Simulation Method for System Reliability and Risk Analysis. Springer, London (2013). https://doi.org/10.1007/978-1-4471-4588-2

16. Zuo, M.J., Zhigang, T.: Performance evaluation of generalized multi-state k-out-of-n systems. IEEE Trans. Reliab. **55**(2), 319–327 (2006)

17. Li, W., Zuo, M.J.: Reliability evaluation of multi-state weighted k-out-of-n systems. Reliab. Eng. Syst. Saf. **93**(1), 160–167 (2008)

18. Jensen, F.V., Nielsen, T.D.: Bayesian Networks and Decision Graphs. Springer, New York (2007). https://doi.org/10.1007/978-0-387-68282-2

19. Jiang, T., Liu, Y.: Parameter inference for non-repairable multi-state system reliability models by multi-level observation sequences. Reliab. Eng. Syst. Saf. **166**, 3–15 (2017)

20. Scutari, M., Denis, J.-B.: Bayesian Networks with Examples in R. CRC Press, Boca Raton (2014)

21. Koller, D., Friedman, N.: Probabilistic Graphical Models: Principles and Techniques. The MIT Press, Massachusetts (2009)

22. Langseth, H., Portinale, L.: Bayesian networks in reliability. Reliab. Eng. Syst. Saf. **92**(1), 92–108 (2007)

23. Weber, P., Medina-Oliva, G., Simon, C., Iung, B.: Overview on Bayesian networks applications for dependability, risk analysis and maintenance areas. Eng. Appl. Artif. Intell. **25**(4), 671–682 (2012)

24. Mkrtchyan, L., Podofillini, L., Dang, V.N.: Bayesian belief networks for human reliability analysis: A review of applications and gaps. Reliab. Eng. Syst. Saf. **139**, 1–16 (2015)

25. Kelly, D.L., Smith, C.L.: Bayesian inference in probabilistic risk assessment—the current state of the art. Reliab. Eng. Syst. Saf. **94**(2), 628–643 (2009)

26. Cai, B., Huang, L., Xie, M.: Bayesian networks in fault diagnosis. IEEE Trans. Ind. Inform. **13**(5), 2227–2240 (2017)

27. Cai, B., et al.: Application of Bayesian networks in reliability evaluation. IEEE Trans. Ind. Inform. **15**(4), 2146–2157 (2019)

28. Bobbio, A., Portinale, L., Minichino, M., Ciancamerla, E.: Improving the analysis of dependable systems by mapping fault trees into Bayesian networks. Reliab. Eng. Syst. Saf. **71**(3), 249–260 (2001)

29. Khakzad, N., Khan, F., Amyotte, P.: Safety analysis in process facilities: comparison of fault tree and Bayesian network approaches. Reliab. Eng. Syst. Saf. **96**(8), 925–932 (2011)

30. Montani, S., Portinale, L., Bobbio, A., Codetta-Raiteri, D.: Radyban: a tool for reliability analysis of dynamic fault trees through conversion into dynamic Bayesian networks. Reliab. Eng. Syst. Saf. **93**(7), 922–932 (2008)

31. Norrington, L., Quigley, J., Russell, A., Van der Meer, R.: Modelling the reliability of search and rescue operations with Bayesian belief networks. Reliab. Eng. Syst. Saf. **93**(7), 940–949 (2008)

32. Nordgård, D.E., Sand, K.: Application of Bayesian networks for risk analysis of MV air insulated switch operation. Reliab. Eng. Syst. Saf. **95**(12), 1358–1366 (2010)

33. Morales-Nápoles, O., Steenbergen, R.D.J.M.: Analysis of axle and vehicle load properties through Bayesian networks based on Weigh-in-Motion data. Reliab. Eng. Syst. Saf. **125**, 153–164 (2014)

34. Mi, J., Li, Y.-F., Peng, W., Huang, H.-Z.: Reliability analysis of complex multi-state system with common cause failure based on evidential networks. Reliab. Eng. Syst. Saf. **174**, 71–81 (2018)

35. Xiahou, T.F., Liu, Y., Jiang, T.: Extended composite importance measures for multi-state systems with epistemic uncertainty of state assignment. Mech. Syst. Sig. Process. **109**, 305–329 (2018)

36. Francis, R.A., Guikema, S.D., Henneman, L.: Bayesian belief networks for predicting drinking water distribution system pipe breaks. Reliab. Eng. Syst. Saf. **130**, 1–11 (2014)

37. Tang, K., Parsons, D.J., Jude, S.: Comparison of automatic and guided learning for Bayesian networks to analyse pipe failures in the water distribution system. Reliab. Eng. Syst. Saf. **186**, 24–36 (2019)

38. Rafiq, M.I., Chryssanthopoulos, M.K., Sathananthan, S.: Bridge condition modelling and prediction using dynamic Bayesian belief networks. Struct. Infrastruct. Eng. **11**(1), 38–50 (2015)

39. Cai, B., Liu, Y., Liu, Z., Tian, X., Dong, X., Yu, S.: Using Bayesian networks in reliability evaluation for subsea blowout preventer control system. Reliab. Eng. Syst. Saf. **108**, 32–41 (2012)

40. Cai, B., Liu, Y., Zhang, Y., Fan, Q., Yu, S.: Dynamic Bayesian networks based performance evaluation of subsea blowout preventers in presence of imperfect repair. Expert Syst. Appl. **40**(18), 7544–7554 (2013)

41. Liu, Z., Liu, Y.: A Bayesian network based method for reliability analysis of subsea blowout preventer control system. J. Loss Prev. Process Ind. **59**, 44–53 (2019)

42. Simon, C., Weber, P.: Evidential networks for reliability analysis and performance evaluation of systems with imprecise knowledge. IEEE Trans. Reliab. **58**(1), 69–87 (2009)

43. Cai, B., Liu, Y., Fan, Q.: A multiphase dynamic Bayesian networks methodology for the determination of safety integrity levels. Reliab. Eng. Syst. Saf. **150**, 105–115 (2016)

44. Zuo, L., Xiahou, T., Liu, Y.: Reliability assessment of systems subject to interval-valued probabilistic common cause failure by evidential networks. J. Intell. Fuzzy Syst. **36**, 3711–3723 (2019)

45. Li, M., Liu, J., Li, J., Uk Kim, B.: Bayesian modeling of multi-state hierarchical systems with multi-level information aggregation. Reliab. Eng. Syst. Saf. **124**, 158–164 (2014)

46. Si, S., Cai, Z., Sun, S., Zhang, S.: Integrated importance measures of multi-state systems under uncertainty. Comput. Ind. Eng. **59**(4), 921–928 (2010)

47. Jones, B., Jenkinson, I., Yang, Z., Wang, J.: The use of Bayesian network modelling for maintenance planning in a manufacturing industry. Reliab. Eng. Syst. Saf. **95**(3), 267–277 (2010)

48. Liu, X., Zheng, J., Fu, J., Nie, Z., Chen, G.: Optimal inspection planning of corroded pipelines using BN and GA. J. Pet. Sci. Eng. **163**, 546–555 (2018)

49. Wang, X., Zhang, Y., Wang, L., Wang, J., Lu, J.: Maintenance grouping optimization with system multi-level information based on BN lifetime prediction model. J. Manuf. Syst. **50**, 201–211 (2019)

50. BahooToroody, A., Abaei, M.M., Arzaghi, E., BahooToroody, F., De Carlo, F., Abbassi, R.: Multi-level optimization of maintenance plan for natural gas system exposed to deterioration process. J. Hazard. Mater. **362**, 412–423 (2019)

51. Luque, J., Straub, D.: Risk-based optimal inspection strategies for structural systems using dynamic Bayesian networks. Struct. Saf. **76**, 68–80 (2019)

52. Boudali, H., Dugan, J.B.: A discrete-time Bayesian network reliability modeling and analysis framework. Reliab. Eng. Syst. Saf. **87**(3), 337–349 (2005)

53. Boudali, H., Dugan, J.B.: A continuous-time Bayesian network reliability modeling, and analysis framework. IEEE Trans. Reliab. **55**(1), 86–97 (2006)

54. Khakzad, N., Landucci, G., Reniers, G.: Application of dynamic Bayesian network to performance assessment of fire protection systems during domino effects. Reliab. Eng. Syst. Saf. **167**, 232–247 (2017)

55. Rebello, S., Yu, H., Ma, L.: An integrated approach for system functional reliability assessment using dynamic Bayesian network and Hidden Markov model. Reliab. Eng. Syst. Saf. **180**, 124–135 (2018)

56. Amin, M.T., Khan, F., Imtiaz, S.: Dynamic availability assessment of safety critical systems using a dynamic Bayesian network. Reliab. Eng. Syst. Saf. **178**, 108–117 (2018)

57. Xu, Z., Mo, Y., Liu, Y., Jiang, T.: Reliability assessment of multi-state phased-mission systems by fusing observation data from multiple phases of operation. Mech. Syst. Sig. Process. **118**, 603–622 (2019)

58. Khakzad, N.: Modeling wildfire spread in wildland-industrial interfaces using dynamic Bayesian network. Reliab. Eng. Syst. Saf. **189**, 165–176 (2019)

59. Weber, P., Jouffe, L.: Complex system reliability modelling with Dynamic Object Oriented Bayesian Networks (DOOBN). Reliab. Eng. Syst. Saf. **91**(2), 149–162 (2006)

60. Liu, Q., Pérès, F., Tchangani, A.: Object oriented Bayesian network for complex system risk assessment. IFAC-PapersOnLine **49**(28), 31–36 (2016)

61. Cai, B., Liu, H., Xie, M.: A real-time fault diagnosis methodology of complex systems using object-oriented Bayesian networks. Mech. Syst. Sig. Process. **80**, 31–44 (2016)

62. Sarwar, A., Khan, F., James, L., Abimbola, M.: Integrated offshore power operation resilience assessment using object oriented Bayesian network. Ocean Eng. **167**, 257–266 (2018)

63. Abimbola, M., Khan, F.: Resilience modeling of engineering systems using dynamic object-oriented Bayesian network approach. Comput. Ind. Eng. **130**, 108–118 (2019)

64. Bensi, M., Kiureghian, A.D., Straub, D.: Efficient Bayesian network modeling of systems. Reliab. Eng. Syst. Saf. **112**, 200–213 (2013)

65. Zhu, J., Collette, M.: A dynamic discretization method for reliability inference in dynamic Bayesian networks. Reliab. Eng. Syst. Saf. **138**, 242–252 (2015)

66. Zwirglmaier, K., Straub, D.: A discretization procedure for rare events in Bayesian networks. Reliab. Eng. Syst. Saf. **153**, 96–109 (2016)

67. Tien, I., Der Kiureghian, A.: Algorithms for Bayesian network modeling and reliability assessment of infrastructure systems. Reliab. Eng. Syst. Saf. **156**, 134–147 (2016)

68. Zheng, X., Yao, W., Xu, Y., Chen, X.: Improved compression inference algorithm for reliability analysis of complex multistate satellite system based on multilevel Bayesian network. Reliab. Eng. Syst. Saf. **189**, 123–142 (2019)

69. Xing, L.: Reliability evaluation of phased-mission systems with imperfect fault coverage and common-cause failures. IEEE Trans. Reliab. **56**(1), 58–68 (2007)

70. Xing, L., Levitin, G.: BDD-based reliability evaluation of phased-mission systems with internal/external common-cause failures. Reliab. Eng. Syst. Saf. **112**, 145–153 (2013)

71. Wang, C., Xing, L., Levitin, G.: Explicit and implicit methods for probabilistic common-cause failure analysis. Reliab. Eng. Syst. Saf. **131**, 175–184 (2014)

72. Wang, C., Xing, L., Levitin, G.: Probabilistic common cause failures in phased-mission systems. Reliab. Eng. Syst. Saf. **144**, 53–60 (2015)

73. Sun, Y., Ma, L., Mathew, J., Zhang, S.: An analytical model for interactive failures. Reliab. Eng. Syst. Saf. **91**(5), 495–504 (2006)

74. Dao, C.D., Zuo, M.J.: Selective maintenance for multistate series systems with $s$-dependent components. IEEE Trans. Reliab. **65**(2), 525–539 (2016)

75. Jackson, C., Mosleh, A.: Bayesian inference with overlapping data for systems with continuous life metrics. Reliab. Eng. Syst. Saf. **106**, 217–231 (2012)

76. Li, M., Hu, Q., Liu, J.: Proportional hazard modeling for hierarchical systems with multi-level information aggregation. IIE Trans. **46**(2), 149–163 (2014)

77. Liu, Y., Chen, C.: Dynamic reliability assessment for nonrepairable multistate systems by aggregating multilevel imperfect inspection data. IEEE Trans. Reliab. **66**(2), 281–297 (2017)

78. Ghasemi, A., Yacout, S., Ouali, M.S.: Evaluating the reliability function and the mean residual life for equipment with unobservable states. IEEE Trans. Reliab. **59**(1), 45–54 (2010)

79. Liu, Y., Zuo, M.J., Li, Y., Huang, H.: Dynamic reliability assessment for multi-state systems utilizing system-level inspection data. IEEE Trans. Reliab. **64**(4), 1287–1299 (2015)
80. Liu, Y.W., Kapur, K.C.: Reliability measures for dynamic multistate nonrepairable systems and their applications to system performance evaluation. IIE Trans. **38**(6), 511–520 (2006)
81. Yontay, P., Pan, R.: A computational Bayesian approach to dependency assessment in system reliability. Reliab. Eng. Syst. Saf. **152**, 104–114 (2016)

# Assessment of Reliability in Accelerated Degradation Testing with Initial Status Incorporated

Chengjie Wang, Qingpei Hu[(✉)], and Dan Yu

Academy of Mathematics and Systems Science, Chinese Academy of Sciences,
Beijing, China
wangchengjie16@mails.ucas.ac.cn,{qingpeihu,dyu}@amss.ac.cn

**Abstract.** Accelerated Degradation Test (ADT) provides effective information for reliability assessment of performance characteristic of long-life and high-reliability products. Existing typical models and analysis usually assume that the products under test are of high consistency level during the manufacturing process, which implies that the individual differences of the initial performance of the products can be ignored. However, this may not be the case, and the initial performance of the test units may have great impact on the subsequent degradation rate. Both positively related and negatively related are possible. This phenomenon can be observed in many different examples, such as the performance of inkjet printer heads. It means that reliability-related information can be obtained before accelerated degradation test. The study considers the impact of initial performance on the reliability assessment. Based on the existing typical accelerated degradation test model and analysis process, this paper introduces the initial information of the products to carry out reliability assessment and test plan. The asymptotic variance of a lifetime quantile at normal use conditions is considered to obtain the optimum test plan. Results show that the initial performance of the test units can be made use of to improve the accuracy of estimators. The impact of fisher information has been taken into account.

**Keywords:** Accelerated degradation test · Reliability assessment · Random initial degradation · Fisher information · Asymptotic variance · Test plan

## 1 Introduction

With the development of science and technology, the reliability of industrial products is getting higher and higher, and the requirement for the high quality of components and systems is also increasing. The reliability of products with long lifetime need to be evaluated in advance. There are usually very few failure data in a limited amount of time for test. This is a challenge to the traditional statistical inference method based on life test. On the other hand,

© Springer Nature Singapore Pte Ltd. 2019
Q.-L. Li et al. (Eds.): Cao Festschrift 2019, CCIS 1102, pp. 229–239, 2019.
https://doi.org/10.1007/978-981-15-0864-6_10

the development of technology also enables us to monitor the failure modes or key performance parameters of products. Such degradation information provides abundant information for the reliability evaluation of highly reliable products. By applying additional stress to the products, the degradation process can usually be accelerated, so that more degradation and failure data can be observed within the same length of test duration, and the performance of the product can be more fully understood. Degradation modelling has been applied for many years in various fields such as electronics, materials and so on. Meeker and Escobar [8] introduces the general methods of modelling, analysis and test plan of degradation data. Wu and Shao [13] use the mixed effect model to fit the degradation data. The least square method is used to estimate the parameters. The reliability of test units has also been evaluated. Weaver et al. [11] considers a linear random effect model in which the degradation rate of products follows the normal distribution. In addition to the general path model, the stochastic process model can be used to fit the degraded data, such as the Gauss process model [12] as well as the inverse Gauss process model [10].

By carefully planning the stress level, sample allocation and measuring time point of accelerated degradation test, the obtained data can be more efficient in statistical inference of product reliability, which can also reduce testing costs. Sheng-Tsaing and Hong-Fwu [9] propose a criterion for determining the termination time of the test. Yu and Tseng [15] introduce cost function and get the test plan by minimizing the cost function. Marseguerra et al. [7] present a multi-objective genetic algorithm for solving the optimal degradation test plan. In Weaver et al. [11], experimental design is also concerned with minimizing the asymptotic variance of the statistics of interest. In many practical applications, the degradation of test units does not start from zero. There are differences among individuals, such as the examples of disk error units in Meeker and Escobar [8]. The initial difference may further affect the subsequent degradation rate. Lu et al. [6] assume that the initial state and degradation rate follow the multivariate normal distribution, which is also the model to be used in this paper. This correlation may have some effects on the statistical inference process. But he did not focus on the impact of random initial degradation status. Li [4] first use the method of functional analysis to consider the related problems. The existence of optimal test plan in some given probability space has been proof. The corresponding iterated numerical optimization algorithm is also proposed, despite the convergence of the algorithm. Ye et al. [14] make a theoretical study of the initial degradation and carried out numerical simulation from the statistical perspectives.

In this paper, the random initial status has been taken into account in the modelling and analysis of accelerated degradation test, and the effects on reliability evaluation accuracy and test scheme design are studied. The degraded model is briefly introduced and the inference process is given. By analysing and comparing a group of real data, this study demonstrates the effect of initial status on degradation and its effect on product evaluation, and verifies the results of optimal design of test scheme considering initial goodness by simulation analysis.

# 2    Modelling and Inference

The acceleration model considered in this paper contains only one stress variable. The degradation data are positive and have a positive degradation rate. The degradation rate is faster at high stress level and the initial state is negatively correlated with the degradation rate. In the traditional test plan of accelerated degradation test, samples are randomly allocated to different stress levels, which we call random allocation. Here we consider a new allocation scheme, because we have the initial degradation information of the test units, that is, when $t = 0$, the degradation data of the test units is denoted as $X$. We assign the samples to the stress levels from high to low based on the order of initial states. We call this scheme as strategic allocation. The basic idea is as follows. Assuming that there are $n$ samples, $X_i$ is the $i$-th order statistic of the initial degradation, there are two stress levels in the test. The first $n_L$ samples with better initial state, i.e. the first $n_L$ samples, are allocated to the lower stress level $s_L$, and the remaining samples are allocated to the higher stress level $s_H$. Because of the negative correlation between the initial state and the degradation rate, samples with better initial state may degrade much faster, so even at low stress levels, they will provide more information of failure and performance, which is more conducive to the inference of reliability than that in random allocation. This idea is quite intuitive. Li [4] has proved that the intuitive allocation plan will actually improve the accuracy of estimation with functional theory. Though it is not an optimal plan as a whole, the plan can be easily calculated and implemented.

## 2.1    Model of ADT

For sample $i$, we record the stress level as $s_i$, the measurement time point $t_i = (t_{i1}, \cdots, t_{im_i})'$, where $m_i$ is the number of measurements besides the initial state of sample $i$, and the corresponding measurement data as $Y_i = (Y_{i1}, \cdots, Y_{im_i})'$. Thus, all the data collected are $D = \{(X_i, s_i, t_i, Y_i) \mid i = 1, 2, \cdots, n\}$. A random effect model is used in this paper to analyse the degradation data. This model was firstly introduced by Lu et al. [6]. For test units with stress level $s$, the degradation level at time $t$ is measured

$$Y(t, s) = D(t, s) + \varepsilon = b_0 + b_1 \exp(\gamma s)t + \varepsilon \tag{1}$$

where $b_0$ is the initial degradation performance, $b_1$ is the slope, indicating the degradation rate in normal use condition, $\gamma$ is the model parameter which is used to fit the effect of stress level on the degradation rate, and $\varepsilon \sim N(0, \sigma_\varepsilon^2)$ is the measurement error. In order to consider the initial quality difference among test units and its effect on degradation rate, it is assumed that $(b_0, b_1)$ follows normal distribution

$$\begin{pmatrix} b_0 \\ b_1 \end{pmatrix} \sim N\left( \begin{pmatrix} \alpha_0 \\ \alpha_1 \end{pmatrix}, \begin{pmatrix} \sigma_0^2 & \sigma_{01} \\ \sigma_{01} & \sigma_0^2 \end{pmatrix} \right)$$

the correlation coefficient $\rho = \sigma_{01}/(\sigma_0 \sigma_1)$ may be positive or negative.

In order to simplify the calculation process, we can always conduct a monotone transformation on the stress level. Assume $s_0$ to be the normal use level of the test unit and $s_H$ be the maximum stress level applicable to model (1). The following transformation is used to normalize the stress level

$$s = \frac{\psi(\tilde{s}) - \psi(\tilde{s}_0)}{\psi(\tilde{s}_H) - \psi(\tilde{s}_0)} \tag{2}$$

where $\psi(\cdot)$ is a monotone function and $\tilde{s}$ means the stress level before normalization whereas $s$ means the normalized stress level. Particularly, $\psi(\tilde{s}) = 1/\tilde{s}$ in the Arrhenius model and $\psi(\tilde{s}) = \ln(\tilde{s})$ in the power law model. After the normalization, all the stress levels involved in the tests are compressed into the interval $[0, 1]$, and the normal use level becomes 0.

## 2.2   Parameter Estimation

Model (1) is a random effect model. We use the maximum likelihood estimation method to estimate the parameters of the model. In the process, we derive the Fisher information matrix for subsequent estimation of variance. The degradation level of the initial measurement is $X = b_0 + \varepsilon \sim N(\alpha_0, \sigma_0^2 + \sigma_\varepsilon^2)$. All the measurements $(X_i, Y_i)'$ of sample $i$ also follows the multivariate normal distribution

$$\begin{pmatrix} X_i \\ Y_i \end{pmatrix} \sim N(\mu_i, \Sigma_i)$$

where $\mu_i = Z_i z_i \begin{pmatrix} \alpha_0 \\ \alpha_1 \end{pmatrix}$ $\Sigma_i = Z_i z_i \begin{pmatrix} \sigma_0^2 & \sigma_{01} \\ \sigma_{01} & \sigma_0^2 \end{pmatrix} z_i Z_i'$, and $I_{m_i}$ represents the identity matrix with order $m_i$

$$Z_i = \begin{pmatrix} 1 & 1 & \cdots 1 \\ 0 & t_{i1} & \cdots t_{im_i} \end{pmatrix}, z_i = \begin{pmatrix} 1 & 0 \\ 0 & \exp(\gamma s_i) \end{pmatrix} \tag{3}$$

Let $\theta = (\alpha_0, \alpha_1, \gamma, \sigma_0^2, \sigma_1^2, \sigma_\varepsilon^2, \sigma_{01})'$ be all the parameters involved in the model. We only need to solve the likelihood equation to get the estimator $\hat{\theta}$. The logarithm of likelihood function is

$$l(\theta|D) = -\frac{1}{2} \sum_{i=1}^{n} \ln \det(\Sigma_i) - \frac{1}{2} \sum_{i=1}^{n} \left( \begin{pmatrix} X_i \\ Y_i \end{pmatrix} - \mu_i \right)' \Sigma_i^{-1} \left( \begin{pmatrix} X_i \\ Y_i \end{pmatrix} - \mu_i \right) \tag{4}$$

where $\det(\cdot)$ represents the function to calculate determinant. This likelihood equation contains a non-linear exponential function. The analytic expression of parameter estimation $\hat{\theta}$ cannot be obtained by solving the equation. The likelihood equation can be solved by numerical method. The Fisher information matrix is $I(\theta) = E\left(-\partial^2 l(\theta)/\partial\theta\partial\theta'\right)$. The solving process is a bit complicated. Formula (16) in Klein et al. [3] gives the result directly. $I(\theta) = \tilde{I} + I$, $\tilde{I}$ is an additional positive semidefinite matrix which can be seen as the additional information from the initial degradation level of the test units. Li [4] derives the formula of the information matrix

$$
\tilde{I} = \begin{pmatrix}
\frac{n}{\sigma_0^2+\sigma_\varepsilon^2} & 0 & 0 & 0 & 0 & 0 & 0 \\
0 & 0 & 0 & 0 & 0 & 0 & 0 \\
0 & 0 & 0 & 0 & 0 & 0 & 0 \\
0 & 0 & 0 & \frac{n}{2(\sigma_0^2+\sigma_\varepsilon^2)^2} & 0 & \frac{n}{2(\sigma_0^2+\sigma_\varepsilon^2)^2} & 0 \\
0 & 0 & 0 & 0 & 0 & 0 & 0 \\
0 & 0 & 0 & \frac{n}{2(\sigma_0^2+\sigma_\varepsilon^2)^2} & 0 & \frac{n}{2(\sigma_0^2+\sigma_\varepsilon^2)^2} & 0 \\
0 & 0 & 0 & 0 & 0 & 0 & 0
\end{pmatrix} \tag{5}
$$

while $I$ represents the information provided by the degradation process. The $(k,l)$-th entry of the matrix $I$ is

$$
I_\theta(k,l) = \frac{1}{2}\sum_{i=1}^n \mathrm{tr}\left(\Sigma_i^{-1}\frac{\partial\Sigma_i}{\partial\theta_k}\Sigma_i^{-1}\frac{\partial\Sigma_i}{\partial\theta_l}\right) + \sum_{i=1}^n \mathrm{E}\left(\frac{\partial\mu_i'}{\partial\theta_k}\Sigma_i^{-1}\frac{\partial\mu_i}{\partial\theta_l}\right) \tag{6}
$$

where $\partial/\partial\theta_l$ denotes the partial derivative of the $l$-th parameter, $\mathrm{tr}(\cdot)$ solves the trace of square matrix. For any given function $g(\theta)$, the MLE of $g(\theta)$ is $g(\hat{\theta}_{ML})$ by the continuity of maximum likelihood estimation. Then, according to the asymptotic normality of MLE and Delta method, the asymptotic variance of $g(\hat{\theta}_{ML})$ is obtained

$$
\begin{aligned}
\mathrm{Avar}\left(g(\hat{\theta}_{ML})\right) &= \nabla g(\hat{\theta}_{ML})'\mathrm{Avar}\left(\hat{\theta}_{ML}\right)\nabla g(\hat{\theta}_{ML}) \\
&= \nabla g(\hat{\theta}_{ML})'\left[I(\hat{\theta}_{ML})\right]^{-1}\nabla g(\hat{\theta}_{ML})
\end{aligned} \tag{7}
$$

where $\nabla g(\cdot)$ denotes the gradient of function $g(\cdot)$ with respect to parameter $\theta$.

These are the inference methods for stochastic allocation schemes. When we adopt the allocation scheme that takes into account the initial degradation level, the inference process will change a little. The main difference appears in mean $\mu_i$ and covariance matrix $\Sigma_i$ of $(X_i, Y_i)'$. By rewriting the log-likelihood function, the MLE and the Fisher information matrix are similar to the previous ones. Explict expression can be found in Ye et al. [14].

## 2.3   Assessment of Reliability

Let $D_f$ denote a given failure threshold. Under stress level $s$, the failure time $T(s)$ of samples is the first time that the true degradation level $D(t,s) = b_0 + b_1 \exp(\gamma s)t$ reaches $D_f$. Here we assume that the degradation rate is positive, and then we have the following results

$$
\begin{aligned}
F_T(t,s) &= P(T(s)\leqslant t) = P(b_0 + b_1\exp(\gamma s)t \leqslant D_f) \\
&= 1 - \Phi(\kappa)
\end{aligned} \tag{8}
$$

where

$$
\kappa = \frac{D_f - \alpha_0 - \alpha_1\tau(s)}{\sqrt{\sigma_0^2 + \sigma_1^2\tau^2(s) + 2\sigma_{01}\tau(s)}}, \tau(s) = \exp(\gamma s)t
$$

and $\Psi(\cdot)$ represents the cumulative distribution function of the standard normal distribution. In engineering practice, we are generally concerned about the life of the product under normal stress conditions, so we can get the life distribution of the product only by substituting the stress under normal use condition into (8). With the formula of lifetime distribution, the lower confidence limits of reliability for MTTF under a certain level of significance $\alpha$, which is an important performance index in production, can also be obtained. It is given by the formula $F_T(t,s) - z_{1-\alpha} \nabla F_T(t,s)' I^{-1}(\hat{\boldsymbol{\theta}}_{ML}) \nabla F_T(t,s)$, where $z_{1-\alpha}$ is the quantile of standard normal distribution. The Fisher information matrix here also depends on the scheme of allocation and whether the initial state has been taken into consideration. Similar inference process can be obtained for model with negative degradation levels or negative degradation rate.

## 3    Optimal Test Plan for ADT

ADT involves many test variables, including sample size, stress level, measurement time point and so on. The different values of these variables will affect the test duration, cost and the accuracy of the final estimation. Generally, the more samples, the longer the test duration and the more degraded data, the higher the estimation accuracy will be. So we consider the allocation of stress level and sample size at each level under the condition of fixed sample size and measurement times, in order that the subsequent inference will have some excellent properties. This idea was originally introduced by Kiefer [2]. In addition, because of the initial information of the test units, the allocation method may be random or based on the initial state.

In ADT test plan, there are many different criteria to determine the test scheme. For example, in the statistical inference of linear model, the A-optimal criterion, E-optimal criterion and the D-optimal criterion based on design matrix [1]. Here we use the method of minimizing the asymptotic variance of $p$-quantile $t_p$ to obtain the test plan. There are many experimental variables that can be used for decision-making. If we consider them all, it will make the problem very complicated. So we simplify these variables. In addition to the fixed sample size mentioned before, we also assume that the time intervals between measurements are equal. The total number of measurements is fixed in advance. All samples are tested at two different stress levels. Optimal stress levels and the sample size at each stress level are selected to minimize the asymptotic variance $\text{Avar}(\hat{t}_p)$. Since at high stress level, the degradation rate of products will be faster and more life information will be provided, we always assume that $s_H$ is the highest stress level in the degradation model (1), i.e. $s_H = 1$. While the low stress level $s_L$ is the decision variable between intervals $(0,1)$. It can be seen from (7) that the asymptotic variance $\text{Avar}(\hat{t}_p)$ is not related to the specific degradation data in the test process, but only to the decision variables of the test and the selection of the model parameters $\boldsymbol{\theta}$. The model parameter $\boldsymbol{\theta}$ needs to be determined before the test, which is called the plan value $\boldsymbol{\theta}$ of degradation test. It is used to calculate the asymptotic variance and make the calculation result closer to the real situation. We hope that the parameter $\boldsymbol{\theta}$ selected in advance is closer to the real value, so that the test scheme obtained can provide the most reliable results. In fact, we can only estimate the value of parameter $\boldsymbol{\theta}$ after obtaining degenerate data through experiments. There is a contradiction between whether to estimate parameters first or to get test plan first. The plan value of parameters in engineering is usually selected from the data in history. In the absence of prior information, a feasible solution is to conduct a small test to estimate the model parameter $\boldsymbol{\theta}$ before the formal ADT is carried out, and then use these parameters for subsequent test plan. In the next example, we use real degradation data to estimate the parameters and redesign the degradation test scheme with these parameters. Decision-making problems can be written as the following optimization process

$$\min_{(s_L, n_L)} \text{Avar}(\hat{t}_p)$$

It should be noted that the information matrix $I(\theta)$ involved in the formula for calculating the asymptotic variance $\text{Avar}(\hat{t}_p)$ in the random allocation is different from that in the strategic allocation. This is also mentioned in the inference process in Sect. 3.

## 4   Case Study and Simulation Results

This example is a set of transistor data. Lu [5] uses the Wiener process model, and specific data can be found in that appendix. Electronic transistor is a very important component in the electronic equipment. The degradation of these transistors will eventually lead to the failure of the relevant equipment. Since the data in Lu [5] have been processed due to the nature of the patent, we only consider the magnitude of the value, but do not explain its specific physical meanings.

There are 20 samples in this data set, in which there are different 5 temperatures and 2 current levels. Since the method considered in this paper only involves one stress variable, we divide the data into two groups and evaluate the reliability of the data under two kinds of currents. The degradation performance of samples at the same temperature was measured at the same time, and the number of measurements ranged from 6 to 38. The measurement time points under different stress levels are quite different, ranging from 300 h to 15000 h. Lu [5] points out that it may be caused by different degradation mechanism at extreme temperature. So we only adopt the data measured between 25 °C and 75 °C. Figure 1 shows the degradation process with respect to different temperatures at current level 1.

It can be seen that the initial state of different samples varies greatly, ranging from 80 to 110, and the corresponding degradation rate varies slightly. We can see the correlation between the initial state and degradation rate in parameter estimation. The failure threshold varies according to the different application scenarios of transistors. 70 is used as the threshold to illustrate the analysis

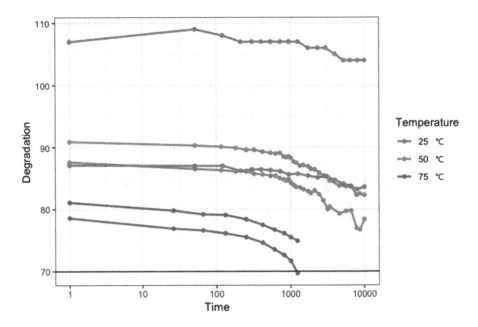

**Fig. 1.** Degradation path under different temperature. Threshold is given by $D_f = 70$.

**Table 1.** MLE of parameters for Transistor Degradation Data. Case 1: including data with random initial state; Case 2: neglecting data at the first inspection time.

|  | Case 1 | Case 2 |
|---|---|---|
| $\hat{\alpha}_0$ | 93.2 | 89.7 |
| $\hat{\alpha}_1$ | $-4.43 \times 10^{-4}$ | $-4.52 \times 10^{-4}$ |
| $\hat{\gamma}$ | 3.17 | 2.88 |
| $\hat{\sigma}_0^2$ | 1.066 | 1.327 |
| $\hat{\sigma}_1^2$ | $1.68 \times 10^{-5}$ | $1.53 \times 10^{-5}$ |
| $\hat{\sigma}_\varepsilon^2$ | 6.82 | 5.13 |
| $\hat{\sigma}_{01}$ | $-0.000389$ | $-0.000392$ |
| $\hat{\rho}$ | $-0.092$ | $-0.087$ |
| Lower bound | 0.9743 | 0.9695 |

results. We calculate the lower confidence limit with respect to 90% significance at 10,000 h at 25 °C. The results of assessment of reliability are as follows.

Table 1 shows the MLE of parameters on transistor degradation data at current level 1. It can be seen that the correlation coefficient between the initial state and the degradation rate is negative. Because it is a decreasing degradation process, the better the initial state, the faster the degradation. We have

calculated the MLE with data at current level 2, which has indicated the same
result.

In order to study the effect of initial information on the evaluation results,
we omit the measurements at time 0 from all the data and re-estimate the
parameters with the remaining data. The results are shown in the third column
in Table 1. The MLEs are closed to that in case 1, however, we get a smaller
lower confidence limit, that is to say, the estimation tends to be conservative,
and the lack of initial information will make the confidence interval wider, which
is consistent with the previous theoretical results. So if we want to have a more
accurate estimate of product reliability, it is benefit to measure the initial state
of the test units and make use of it.

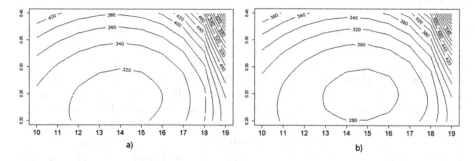

**Fig. 2.** Contour plot of the large-sample variance of $\hat{t}_{0.1}$ using two different allocation
methods. Longitudinal axis represents the stress level after normalization. (a) random
allocation; (b) strategic allocation.

We use the parameters obtained to furtherly optimize the test plan. We measure
with sample size $n = 20$, every 50 h until 1000 h, and design the test by minimizing the asymptotic variance of 0.1-quantile of life at $0\,°C$. Since the optimization
problem is a continuous non-convex problem, if we discretize the stress level at
0.001 intervals, the optimization problem will become a discrete problem. We
only need to get the maximum value at the lattice point. Here we draw the
asymptotic variance contour plots under the two allocation methods as shown
in Fig. 2

**Table 2.** Optimal test plan with two different allocation methods and corresponding
asymptotic variance.

|  | Sample size under $s_L$ | Temperature of $s_L$ | Asymptotic variance |
|---|---|---|---|
| Random allocation | 14 | 18.44 °C | 311.7323 |
| Strategic allocation | 15 | 19.70 °C | 273.8081 |

Table 2 shows that the lowest asymptotic variance 311.73 is achieved when 14 samples are allocated to 18.44 °C at low temperature under the random allocation scheme. The optimal result is 15 samples at low temperature 19.70 °C when applying strategic allocation according to the initial degradation. The corresponding the asymptotic variance is 273.81. That is to say, the variance of the statistics concerned has been reduced to 12.17%, and the effect is significant. Therefore, considering the initial degradation has a positive impact on the test plan. In the other word, to achieve the same variance as that in the strategic allocation scheme, the random allocation need much bigger sample size. This can reduce the cost of the test to a certain extent.

## 5  Conclusion

In this paper, the effect of initial status of product on assessment of reliability and test plan are studied. The general results are obtained through theoretical derivation. In the assessment of reliability, the Fisher information matrix is larger when considering the initial state than neglecting it. The effect on estimating lower confidence limit of reliability is that the estimation under strategic allocation will be more accurate. Analysis on a set of typical electronic product failure data coincide with the theorical result. In terms of testing plan optimization, we compared the traditional random allocation method with the strategic allocation method. Simulation results show that the strategic allocation method can obtain smaller estimation variance under the same sample size and number of measurements, which is conducive to speeding up the degradation test and reducing the test cost.

**Acknowledgements.** The authors are honored to get invitation for contributing a book chapter to celebrate the 80th birthday of Professor Jinhua Cao. The authors are also thankful for the reviewers' comments and suggestions.

## References

1. Chow, S.C.: Advanced Linear Models: Theory and Applications. Routledge, New York (2018)
2. Kiefer, J.: Optimum experimental designs. J. Roy. Stat. Soc.: Ser. B (Methodol.) **21**(2), 272–304 (1959)
3. Klein, A., Mélard, G., Zahaf, T.: Construction of the exact fisher information matrix of gaussian time series models by means of matrix differential rules. Linear Algebra Appl. **321**(1), 209–232 (2000)
4. Li, L.: Design of accelerated degradation test. The university of Chinese Academy of Sciences (2013)
5. Lu, J.: Degradation processes and related reliability models. McGill University Montreal, Canada (1995)
6. Lu, J.C., Park, J., Yang, Q.: Statistical inference of a time-to-failure distribution derived from linear degradation data. Technometrics **39**(4), 391–400 (1997)

7. Marseguerra, M., Zio, E., Cipollone, M.: Designing optimal degradation tests via multi-objective genetic algorithms. Reliab. Eng. Syst. Saf. **79**(1), 87–94 (2003)
8. Meeker, W.Q., Escobar, L.A.: Statistical Methods for Reliability Data. Wiley, New York (2014)
9. Sheng-Tsaing, T., Hong-Fwu, Y.: A termination rule for degradation experiments. IEEE Trans. Reliab. **46**(1), 130–133 (1997)
10. Wang, X., Xu, D.: An inverse gaussian process model for degradation data. Technometrics **52**(2), 188–197 (2010)
11. Weaver, B.P., Meeker, W.Q., Escobar, L.A., Wendelberger, J.: Methods for planning repeated measures degradation studies. Technometrics **55**(2), 122–134 (2013)
12. Whitmore, G.A.: Estimation of wiener diffusion parameters using process measurements subject to error. In: Jewell, N.P., Kimber, A.C., Lee, M.L.T., Whitmore, G.A. (eds.) Lifetime Data: Models in Reliability and Survival Analysis, pp. 363–369. Springer, Heidelberg (1996). https://doi.org/10.1007/978-1-4757-5654-8_47
13. Wu, S.J., Shao, J.: Reliability analysis using the least squares method in nonlinear mixed-effect degradation models. Stat. Sin. **9**(3), 855–877 (1999)
14. Ye, Z.S., Hu, Q., Yu, D.: Strategic allocation of test units in an accelerated degradation test plan. J. Qual. Technol. **51**(1), 64–80 (2019)
15. Yu, H.F., Tseng, S.T.: Designing a degradation experiment. Naval Res. Logistics (NRL) **46**(6), 689–706 (1999)

# Periodic Replacement Policies and Comparisons with Their Extended Policies

Xufeng Zhao[1($\boxtimes$)], Cunhua Qian[2], and Toshio Nakagawa[3]

[1] Nanjing University of Aeronautics and Astronautics, Nanjing 211106, China
xz.cem@nuaa.edu.cn
[2] Nanjing Tech University, Nanjing 211816, China
qch64317@njtech.edu.cn
[3] Aichi Institute of Technology, Toyota 470-0392, Japan
toshi-nakagawa@aitech.ac.jp

**Abstract.** It has been well known that minimal repairs are widely used in planning periodic replacement policies in reliability engineering. In this chapter, we begin with the standard periodic replacement policies that are planned at time $T$ or at failure $K$, respectively, where the cumulative hazard function $H(t)$ is used to count the number of minimal repairs. Next, three extensions of the above standard policies are discussed: (1) When the replacement policies of $T$ and $K$ are planned simultaneously, the approaches of first and last are used to make the best choice. (2) We delay replacement to be done at the first failure over $T$ when it cannot be performed on time $T$. (3) We begin to plan replacement time $T$ once the first failure or the $K$th failure has occurred. We formulate the models of cost rates and give analytical discussions. In addition, comparisons are made for the above policies from point of cost. Finally, numerical examples are illustrated when the failure time has a Weibull distribution.

**Keywords:** Failure rate · Minimal repair · Replacement time · Periodic replacement · Overtime replacement

## 1 Introduction

Manufacturing systems with performance degradation and replacement strategy are commonly encountered in practice. Replacements done after failure and before failure are called corrective replacement and preventive replacement, respectively [1]. However, for a large and complex system, which consists of many kinds of units, minimal repairs that cost less are always took into considerations at failures [2].

Repair models have been studied extensively, such as repairable system subjected to minimal repair [3], age-based replacement with minimal repair [4–6], inspection modeling with minimal repair [7], warranty maintenance with repair

© Springer Nature Singapore Pte Ltd. 2019
Q.-L. Li et al. (Eds.): Cao Festschrift 2019, CCIS 1102, pp. 240–258, 2019.
https://doi.org/10.1007/978-981-15-0864-6_11

time threshold [8], random working models with replacement and minimal repair [9], repairs for multi-sate systems [10], repair process governed by the generalized Polya process [11], and etc.

It has been found that the system operation can be quickly resumed after minimal repairs, even though sometimes the repair is imperfect [12]. In other words, we can make full use of the system under minimal repairs. When minimal repair is taken into account for replacement policies, it is possible to make replacement plans by counting the number of repairs, which is an alternative policy of that planned with the scale of time. Zhao et al. [13] have discussed replacement first and last policies that are planned at time $T$ and at number $N$ of repairs, using the first and last triggering event approaches. The approach of whichever triggering event occurs last, i.e., maintenances based on the assumption of "whichever occurs last" [14] has been proposed as another good choice for the bivariate replacement, not only because it could let the system operate for a longer time, but because it could avoid operational interruptions to complete more running jobs. From this viewpoint, this chapter begins with the standard periodic replacement policies that are planned at time $T$ or at failure $K$, respectively, where the cumulative hazard function $H(t)$ is used to count the number of minimal repairs. Using the approaches of first and last, the replacement first and last policies of $T$ and $K$ are planned simultaneously.

When the system is running some successive jobs without stops, it is better to perform replacement policies after several jobs are completed even though the replacement time has arrived [1]. Replacement policies scheduled at the first completion of some working cycle over a planned time $T$ [15] were modeled. That is, replacement scheduled at continuous times could be modified to be done at discrete applications. We next delay replacement policy to be done at the first failure over $T$ when it cannot be performed on time $T$. Meanwhile, replacement first and last with overtime $T$ and failure $K$ are modeled. In addition, we begin to plan replacement time $T$ once the first failure or the $K$th failure has occurred, as the extensions of the above overtime policy.

In this chapter, we suppose that failures of an operating unit occur at a nonhomogeneous Poisson process with cumulative hazard function $H(t) \equiv \int_0^t h(u)du$, and the failure rate $h(t) \equiv dH(t)/dt = f(t)/\overline{F}(t)$ increases strictly with $t$ from $h(0)$ to $h(\infty) = \infty$, where $f(t)$ is a density function of $F(t) = 1 - e^{-H(t)}$, its mean $\mu \equiv \int_0^\infty \overline{F}(t)dt$, and $\overline{\Phi}(t) \equiv 1 - \Phi(t)$ for any function $\Phi(t)$. Then, the probability that $k$ failures occur in $[0, t]$ is

$$p_k(t) = \frac{H(t)^k}{k!}e^{-H(t)} \quad (k = 0, 1, 2, \cdots).$$

We denote that $P_k(t) = \sum_{j=k}^\infty p_j(t)$ and $\overline{P}_k(t) = 1 - P_k(t)$, then $P_k(0) = 0$, $\overline{P}_k(0) = 1$, $P_k(\infty) = 1$, $\overline{P}_k(\infty) = 0$, $P_0(t) = 1$, $\overline{P}_0(t) = 0$, $\lim_{k \to \infty} P_k(t) = 0$ and $\lim_{k \to \infty} \overline{P}_k(t) = 1$. The above probabilities have the following relations: For $0 < t < \infty$ and $k = 0, 1, 2, \cdots$,

$$P_{k+1}(t) = \int_0^t p_k(u)h(u)du, \qquad \overline{P}_{k+1}(t) = \int_t^\infty p_k(u)h(u)du,$$

$$\int_0^\infty p_k(t)h(t)\mathrm{d}t = 1, \qquad \sum_{k=0}^\infty kp_k(t) = H(t),$$

$$\int_0^\infty H(t)\mathrm{d}P_k(t) = \int_0^\infty \overline{P}_k(t)h(t)\mathrm{d}t = \sum_{j=0}^{k-1}\int_0^\infty p_j(t)h(t)\mathrm{d}t = k.$$

Furthermore, note that

$$Q(T) \equiv \frac{e^{-H(T)}}{\int_T^\infty e^{-H(t)}\mathrm{d}t} = \frac{\overline{F}(T)}{\int_T^\infty \overline{F}(t)\mathrm{d}t} > h(T)$$

increases strictly with $T$ from $1/\mu$ to $h(\infty)$.

For the above assumptions, this chapter formulates the replacement models of cost rates and give analytical discussions. Several comparisons are made from the point of cost. Numerical examples are illustrated when the failure time has a Weibull distribution.

## 2   Standard Policies

We introduce two standard replacement policies in which the unit is replaced preventively at time $T$ and at failure $K$.

### (1) Replacement at $T$

We suppose that the unit undergoes minimal repair at each failure and its failure rate remains undisturbed by repairs. When the unit is replaced at time $T$ ($0 < T < \infty$), the expected cost rate is [1]

$$C(T) = \frac{c_T + c_M H(T)}{T}, \tag{1}$$

where $c_T$ = replacement cost at time $T$ and $c_M$ = cost of minimal repair at each failure. Optimum $T^*$ to minimize $C(T)$ satisfies

$$Th(T) - H(T) = \frac{c_T}{c_M}, \quad i.e., \quad \int_0^T [h(T) - h(t)]\mathrm{d}t = \frac{c_T}{c_M}, \tag{2}$$

and the resulting cost rate is

$$C(T^*) = c_M h(T^*). \tag{3}$$

## (2) Replacement at $K$

When the unit is replaced at failure $K$ $(K = 1, 2, \cdots)$, the expected cost rate is [1]

$$C(K) = \frac{c_K + c_M K}{\int_0^\infty \overline{P}_K(t)\mathrm{d}t},$$ (4)

where $c_K$ = replacement cost at failure $K$. Optimum $K^*$ to minimize $C(K)$ satisfies

$$\frac{1}{\int_0^\infty p_K(t)\mathrm{d}t} \int_0^\infty \overline{P}_K(t)\mathrm{d}t - K \geq \frac{c_K}{c_M}.$$ (5)

Note that $\int_0^\infty p_K(t)\mathrm{d}t$ decreases strictly with $K$ to $1/h(\infty)$.

## (3) Numerical Example

When $H(t) = t^\alpha$, i.e, $h(t) = \alpha t^{\alpha-1}$, Table 1 presents optimum $T^*$ and $K^*$ when $c_T = c_K = 100.0$. Table 1 shows that both $T^*$ and $K^*$ decreases with $\alpha$ and $c_M$. This means that optimum $T^*$ and $K^*$ decrease with the failure rate and the cost of minimal repair.

**Table 1.** Optimum $T^*$ and $K^*$ when $c_T = c_K = 100.0$.

| $c_M$ | $\alpha = 1.2$ | | $\alpha = 2.0$ | |
|---|---|---|---|---|
| | $T^*$ | $K^*$ | $T^*$ | $K^*$ |
| 10 | 26.050 | 51 | 3.162 | 11 |
| 20 | 14.620 | 26 | 2.236 | 6 |
| 30 | 10.428 | 17 | 1.826 | 4 |
| 40 | 8.205 | 13 | 1.581 | 3 |
| 50 | 6.813 | 11 | 1.414 | 3 |
| 60 | 5.853 | 9 | 1.291 | 2 |
| 70 | 5.147 | 8 | 1.195 | 2 |
| 80 | 4.605 | 7 | 1.118 | 2 |
| 90 | 4.175 | 6 | 1.054 | 2 |
| 100 | 3.824 | 6 | 1.000 | 2 |

## 3    Three Extensions

As extended policies, we give the following three policies of replacement first and last, replacement overtime and replacement after failure.

### 3.1   Replacement First and Last

#### (1) Replacement First at $T$ and $K$

When the unit is replaced at time $T$ $(0 < T \le \infty)$ or at failure $K$ $(K = 1, 2, \cdots)$, whichever occurs first, the expected cost rate is

$$C_F(T, K) = \frac{c_T + (c_K - c_T)P_K(T) + c_M \int_0^T \overline{P}_K(t)h(t)\mathrm{d}t}{\int_0^T \overline{P}_K(t)\mathrm{d}t}, \tag{6}$$

which agrees with $C(T)$ in (1) as $K \to \infty$ and $C(K)$ in (4) as $T \to \infty$.

When $c_K = c_T$, we find optimum $T_F^*$ and $K_F^*$ to minimize $C_F(T, K)$. Differentiating $C_F(T, K)$ with respect to $T$ and setting it equal to zero,

$$h(T) \int_0^T \overline{P}_K(t)\mathrm{d}t - \int_0^T \overline{P}_K(t)h(t)\mathrm{d}t = \frac{c_T}{c_M}, \tag{7}$$

whose left-hand side increases strictly with $T$ from 0 to $\infty$. Thus, there exists a finite and unique $T_F^*$ $(0 < T_F^* < \infty)$ which satisfies (7), and the resulting cost rate is

$$C_F(T_F^*, K) = c_M h(T_F^*). \tag{8}$$

Note that the left-hand side of (7) increases strictly with $K$ to that of (2), then $T_F^*$ decreases with $K$ to $T^*$ given in (2).

Forming the inequality $C_F(T, K + 1) - C_F(T, K) \ge 0$,

$$H_1(T, K) \int_0^T \overline{P}_K(t)\mathrm{d}t - \int_0^T \overline{P}_K(t)h(t)\mathrm{d}t \ge \frac{c_T}{c_M}, \tag{9}$$

where

$$H_1(T, K) \equiv \frac{\int_0^T p_K(t)h(t)\mathrm{d}t}{\int_0^T p_K(t)\mathrm{d}t} < h(T),$$

which increases strictly with $K$ to $h(T)$.

Substituting (8) into (9),

$$H_1(T, K) \ge h(T),$$

which does not hold for any $T$. Therefore, optimum policy to minimize $C_F(T, K)$ is $T_F^* = T^*$ and $K_F^* = \infty$, where $T^*$ is given in (2).

Note that the left-hand side of (9) increases with $K$ to that of (2). If $T \le T^*$, then $K_F^* = \infty$, and if $T > T^*$, then there exists a finite and unique minimum $K_F^*$ $(1 \le K_F^* < \infty)$ which satisfies (9).

## (2) Replacement Last at $T$ and $K$

When the unit is replaced at time $T$ $(0 \leq T < \infty)$ or at failure $K$ $(K = 0, 1, 2, \cdots)$, whichever occurs last, the expected cost rate is

$$C_L(T, K) = \frac{c_T + (c_K - c_T) + c_M[H(T) + \int_T^\infty \overline{P}_K(t)h(t)dt]}{T + \int_T^\infty \overline{P}_K(t)dt}, \tag{10}$$

which agrees with $C(T)$ in (1) as $K = 0$ and $C(K)$ in (4) as $T = 0$.

When $c_K = c_T$, we find optimum $T_L^*$ and $K_L^*$ to minimize $C_L(T, K)$. Differentiating $C_L(T, K)$ with respect to $T$ and setting it equal to zero,

$$\int_0^T [h(T) - h(t)]dt - \int_T^\infty \overline{P}_K(t)[h(t) - h(T)]dt = \frac{c_T}{c_M}, \tag{11}$$

whose left-hand side increases strictly with $T$ from $-K$ to $\infty$. Thus, there exists a finite and unique $T_L^*$ $(0 < T_L^* < \infty)$ which satisfies (11), and the resulting cost rate is

$$C_L(T_L^*, K) = c_M h(T_L^*). \tag{12}$$

Note that the left-hand side of (11) decreases strictly with $K$ from that of (2), then $T_L^*$ increases with $K$ from $T^*$ given in (2).

Forming the inequality $C_L(T, K+1) - C_L(T, K) \geq 0$,

$$\int_0^T [H_2(T, K) - h(t)]dt - \int_T^\infty \overline{P}_K(t)[h(t) - H_2(T, K)]dt \geq \frac{c_T}{c_M}, \tag{13}$$

where

$$H_2(T, K) \equiv \frac{\int_T^\infty p_K(t)h(t)dt}{\int_T^\infty p_K(t)dt} \geq h(T),$$

which increases strictly with $K$ from $Q(T) > h(T)$ to $h(\infty)$.

Substituting (11) into (13),

$$H_2(T, K) \geq h(T),$$

which always holds for any $T$. Therefore, optimum policy to minimize $C_L(T, K)$ is $T_L^* = T^*$ and $K_L^* = 0$, where $T^*$ is given in (2).

Note that the left-hand side of (13) increases with $K$ from

$$TQ(T) - H(T) \geq Th(T) - H(T).$$

If $T \geq T^*$, then $K_L^* = 0$, and if $T < T^*$, then there exists a finite and unique minimum $K_L^*$ $(1 \leq K_L^* < \infty)$ which satisfies (13).

In addition, comparing the policies of replacement first and last in (1) and (2), if $T < T^*$, then replacement last is more economical than replacement first; If $T = T^*$, then replacement at time $T^*$ is more economical than both replacement first and last; If $T > T^*$, then replacement first is more economical than replacement last.

**Table 2.** Optimum $T_F^*$ and $T_L^*$ when $c_T = 100.0$ and $\alpha = 1.2$.

| $c_M$ | $K = 5$ | | $K = 10$ | |
|---|---|---|---|---|
| | $T_F^*$ | $T_L^*$ | $T_F^*$ | $T_L^*$ |
| 10 | $\infty$ | 26.050 | 90.672 | 26.050 |
| 20 | 52.621 | 14.620 | 21.517 | 14.620 |
| 30 | 21.147 | 10.428 | 11.940 | 10.431 |
| 40 | 12.487 | 8.206 | 8.624 | 8.233 |
| 50 | 8.844 | 6.815 | 6.947 | 6.916 |
| 60 | 6.926 | 5.862 | 5.901 | 6.072 |
| 70 | 5.761 | 5.171 | 5.165 | 5.500 |
| 80 | 4.978 | 4.651 | 4.613 | 5.094 |
| 90 | 4.413 | 4.249 | 4.178 | 4.791 |
| 100 | 3.982 | 3.933 | 3.825 | 4.559 |

**Table 3.** Optimum $T_F^*$ and $T_L^*$ when $c_T = 100.0$ and $\alpha = 2.0$.

| $c_M$ | $K = 5$ | | $K = 10$ | |
|---|---|---|---|---|
| | $T_F^*$ | $T_L^*$ | $T_F^*$ | $T_L^*$ |
| 10 | 3.439 | 3.163 | 3.185 | 3.179 |
| 20 | 2.270 | 2.259 | 2.236 | 2.401 |
| 30 | 1.833 | 1.902 | 1.826 | 2.135 |
| 40 | 1.583 | 1.716 | 1.581 | 2.001 |
| 50 | 1.415 | 1.603 | 1.414 | 1.921 |
| 60 | 1.291 | 1.527 | 1.291 | 1.868 |
| 70 | 1.195 | 1.473 | 1.195 | 1.830 |
| 80 | 1.118 | 1.432 | 1.118 | 1.801 |
| 90 | 1.054 | 1.401 | 1.054 | 1.779 |
| 100 | 1.000 | 1.375 | 1.000 | 1.761 |

## (3) Numerical Examples

When $H(t) = t^\alpha$, Tables 2 and 3 present optimum $T_F^*$ and $T_L^*$ when $c_T = 100.0$ and $\alpha = 1.2, 2.0$. These two tables show that $T_F^*$ and $T_L^*$ decrease with $c_M$, $T_F^*$ decreases with $K$ and $T_L^*$ increases with $K$. From (8) and (12), we know that if $T_L^* < T_F^*$, replacement last saves more cost than that of replacement first, e.g., when $K = 5$ in Table 2, and so on.

When $H(t) = t^\alpha$, Tables 4 and 5 present optimum $K_F^*$ and $K_L^*$ when $c_T = 100.0$ and $\alpha = 1.2, 2.0$. These two tables show that if finite $K_F^*$ exists, $K_L^* = 0$,

**Table 4.** Optimum $K_F^*$ and $K_L^*$ when $c_T = 100.0$ and $\alpha = 1.2$.

| $c_M$ | $T = 5.0$ | | $T = 10.0$ | | $T^*$ |
|---|---|---|---|---|---|
| | $K_F^*$ | $K_L^*$ | $K_F^*$ | $K_L^*$ | |
| 10 | $\infty$ | 51 | $\infty$ | 51 | 26.050 |
| 20 | $\infty$ | 26 | $\infty$ | 25 | 14.620 |
| 30 | $\infty$ | 17 | $\infty$ | 1 | 10.428 |
| 40 | $\infty$ | 13 | 15 | 0 | 8.205 |
| 50 | $\infty$ | 9 | 11 | 0 | 6.813 |
| 60 | $\infty$ | 5 | 9 | 0 | 5.853 |
| 70 | $\infty$ | 1 | 8 | 0 | 5.147 |
| 80 | 16 | 0 | 7 | 0 | 4.605 |
| 90 | 10 | 0 | 6 | 0 | 4.175 |
| 100 | 7 | 0 | 6 | 0 | 3.824 |

**Table 5.** Optimum $K_F^*$ and $K_L^*$ when $c_T = 100.0$ and $\alpha = 2.0$.

| $c_M$ | $T = 1.0$ | | $T = 2.0$ | | $T^*$ |
|---|---|---|---|---|---|
| | $K_F^*$ | $K_L^*$ | $K_F^*$ | $K_L^*$ | |
| 10 | $\infty$ | 11 | $\infty$ | 10 | 3.162 |
| 20 | $\infty$ | 5 | $\infty$ | 2 | 2.236 |
| 30 | $\infty$ | 4 | 8 | 0 | 1.826 |
| 40 | $\infty$ | 3 | 4 | 0 | 1.581 |
| 50 | $\infty$ | 2 | 3 | 0 | 1.414 |
| 60 | $\infty$ | 1 | 3 | 0 | 1.291 |
| 70 | $\infty$ | 1 | 2 | 0 | 1.195 |
| 80 | $\infty$ | 1 | 2 | 0 | 1.118 |
| 90 | $\infty$ | 1 | 2 | 0 | 1.054 |
| 100 | $\infty$ | 1 | 2 | 0 | 1.000 |

and if finite $K_L^*$ exists, $K_F^* = \infty$, which has been already shown that if $T < T^*$ then $K_F^* = \infty$, and if $T > T^*$ then $K_L^* = 0$ for $T^*$ given in Table 1.

## 3.2  Replacement Over Time

### (1) Replacement Over Time $T$

When the unit is replaced at the first failure over time $T$ ($0 \le T < \infty$), the expected cost rate is [15]

$$C_O(T) = \frac{c_O + c_M[H(T)+1]}{T + \int_T^\infty \overline{F}(t)\mathrm{d}t/\overline{F}(T)}, \tag{14}$$

where $c_O$ = replacement cost over time $T$. Optimum $T_O^*$ to minimize $C_O(T)$ satisfies

$$TQ(T) - H(T) = \frac{c_O}{c_M}, \tag{15}$$

and the resulting cost rate is

$$C(T_O^*) = c_M Q(T_O^*) = \frac{c_O + c_M H(T_O^*)}{T_O^*}, \tag{16}$$

which agrees with (1) as $c_O = c_T$ and $T_O^* = T$.

When $c_O = c_T$, note that $Q(T) > h(T)$, then $T_O^* < T^*$. From (16), $C_O(T_O^*) > C(T^*)$, i.e., replacement at $T$ is more economical than replacement over time $T$.

Table 6 presents optimum $T_O^*$ when $c_O = 100.0$. It shows that $T_O^*$ decreases with $c_M$ and $\alpha$, and $T_O^* < T^*$.

**Table 6.** Optimum $T_O^*$ when $c_O = 100.0$.

| $c_M$ | $\alpha = 1.2$ $T_O^*$ | $\alpha = 2.0$ $T_O^*$ |
|---|---|---|
| 10 | 25.663 | 3.015 |
| 20 | 14.155 | 2.040 |
| 30 | 9.935 | 1.598 |
| 40 | 7.696 | 1.331 |
| 50 | 6.293 | 1.148 |
| 60 | 5.325 | 1.012 |
| 70 | 4.614 | 0.907 |
| 80 | 4.069 | 0.822 |
| 90 | 3.636 | 0.752 |
| 100 | 3.284 | 0.694 |

## (2) Replacement Overtime First with $T$ and $K$

When the unit is replaced at the first failure over time $T$ $(0 \le T < \infty)$ or at failure $K$ $(K = 1, 2, \cdots)$, whichever occurs first, the expected cost rate is

$$C_{OF}(T, K) = \frac{c_O + (c_K - c_O)P_K(T) + c_M \sum_{k=0}^{K-1} P_k(T)}{\int_0^T \overline{P}_K(t)dt + \overline{P}_K(T) \int_T^\infty e^{-H(t)+H(T)}dt}, \qquad (17)$$

which agrees with $C_O(T)$ in (14) as $K \to \infty$ and $C(K)$ in (4) as $T \to \infty$.

When $c_K = c_O$, we find optimum $T_{OF}^*$ and $K_{OF}^*$ to minimize $C_{OF}(T, K)$. Differentiating $C_{OF}(T, K)$ with respect to $T$ and setting it equal to zero,

$$Q(T) \int_0^T \overline{P}_K(t)dt - \int_0^T \overline{P}_K(t)h(t)dt = \frac{c_O}{c_M}, \qquad (18)$$

whose left-hand side increases strictly with $T$ from 0 to $\infty$. Thus, there exists a finite and unique $T_{OF}^*$ $(0 < T_{OF}^* < \infty)$ which satisfies (18), and the resulting cost rate is

$$C_{OF}(T_{OF}^*, K) = c_M Q(T_{OF}^*). \qquad (19)$$

Note that the left-hand side of (18) increases strictly with $K$ to that of (15), then $T_{OF}^*$ decreases with $K$ to $T_O^*$ given in (15).

Forming the inequality $C_{OF}(T, K+1) - C_{OF}(T, K) \ge 0$,

$$H_3(T, K) \left[ \int_0^T \overline{P}_K(t)dt + \overline{P}_K(T) \int_T^\infty e^{-H(t)+H(T)}dt \right] - \sum_{k=0}^{K-1} P_k(T) \ge \frac{c_O}{c_M}, \qquad (20)$$

where

$$H_3(T, K) \equiv \frac{P_K(T)}{\int_0^T [\int_t^\infty e^{-H(u)+H(t)}du]dP_K(t)} < Q(T),$$

which increases strictly with $K$ to $Q(T)$.

Substituting (18) into (20),

$$H_3(T, K) \ge Q(T),$$

which does not hold for any $T$. Therefore, optimum policy to minimize $C_{OF}(T, K)$ is $T_{OF}^* = T_O^*$ and $K_{OF}^* = \infty$, where $T_O^*$ is given in (15).

Note that the left-hand side of (20) increases with $K$ to that of (15). If $T \le T_O^*$, then $K_{OF}^* = \infty$, and if $T > T_O^*$, then there exists a finite and unique minimum $K_{OF}^*$ $(1 \le K_{OF}^* < \infty)$ which satisfies (20).

## (3) Replacement Overtime Last with $T$ and $K$

When the unit is replaced at the first failure over time $T$ $(0 \leq T < \infty)$ or at failure $K$ $(K = 0, 1, 2, \cdots)$, whichever occurs last, the expected cost rate is

$$C_{OL}(T, K) = \frac{c_O + (c_K - c_O)\overline{P}_K(T) + c_M[H(T) + \sum_{k=0}^{K-1} \overline{P}_k(T)]}{T + \int_T^\infty \overline{P}_K(t)dt + P_K(T) \int_T^\infty e^{-H(t)+H(T)}dt}, \qquad (21)$$

which agrees with $C_O(T)$ in (14) as $K = 0$ and $C(K)$ in (4) as $T \to 0$.

When $c_K = c_O$, we find optimum $T_{OL}^*$ and $K_{OL}^*$ to minimize $C_{OL}(T, K)$. Differentiating $C_{OL}(T, K)$ with respect to $T$ and setting it equal to zero,

$$\int_0^T [Q(T) - h(t)]dt + \int_T^\infty \overline{P}_K(t)[Q(T) - h(t)]dt = \frac{c_O}{c_M}, \qquad (22)$$

whose left-hand side increases strictly with $T$ to $\infty$. Thus, there exists a finite and unique $T_{OL}^*$ $(0 \leq T_{OL}^* < \infty)$ which satisfies (22), and the resulting cost rate is

$$C_{OL}(T_{OL}^*, K) = c_M Q(T_{OL}^*). \qquad (23)$$

Note that the left-hand side of (22) increases strictly with $K$ from that of (15), then $T_{OL}^*$ increases with $K$ from $T_O^*$ given in (15).

Forming the inequality $C_{OL}(T, K+1) - C_{OL}(T, K) \geq 0$,

$$H_4(T, K) \left[ T + \int_T^\infty \overline{P}_K(t)dt + P_K(T) \int_T^\infty e^{-H(t)+H(T)}dt \right]$$
$$- \left[ H(T) + \sum_{k=0}^{K-1} \overline{P}_k(T) \right] \geq \frac{c_O}{c_M}, \qquad (24)$$

where

$$H_4(T, K) \equiv \frac{\int_T^\infty p_{K-1}(t)h(t)dt}{\int_T^\infty p_{K-1}h(t)\{\int_t^\infty e^{-H(u)+H(t)}du\}dt} > Q(T),$$

which increases strictly with $K$ from $Q(T)$ to $\infty$.

Substituting (22) into (24),

$$H_4(T, K) > Q(T),$$

which always holds for any $T$. Therefore, optimum policy to minimize $C_{OL}(T, K)$ is $T_{OL}^* = T_O^*$ and $K_{OL}^* = 0$, where $T_O^*$ is given in (15).

Note that the left-hand side of (24) increases with $K$ from that of (15) to $\infty$. If $T \geq T_O^*$, then $K_{OL}^* = 0$, and if $T < T_O^*$, then there exists a finite and unique minimum $K_{OL}^*$ $(1 \leq K_{OL}^* < \infty)$ which satisfies (24).

## (4) Numerical Examples

When $H(t) = t^\alpha$, Tables 7 and 8 present optimum $T^*_{OF}$ and $T^*_{OL}$ when $c_O = 100.0$ and $\alpha = 1.2, 2.0$. From (19) and (23), we know that if $T^*_{OL} < T^*_{OF}$, replacement overtime last saves more cost than that of replacement overtime first, e.g., when $K = 5$ in Table 7, and so on.

**Table 7.** Optimum $T^*_{OF}$ and $T^*_{OL}$ when $c_O = 100.0$ and $\alpha = 1.2$.

| $c_M$ | $K = 5$ | | $K = 10$ | |
|---|---|---|---|---|
| | $T^*_{OF}$ | $T^*_{OL}$ | $T^*_{OF}$ | $T^*_{OL}$ |
| 10 | $\infty$ | 25.663 | 93.317 | 25.663 |
| 20 | 52.686 | 14.155 | 21.092 | 14.155 |
| 30 | 20.720 | 9.935 | 11.447 | 9.937 |
| 40 | 11.999 | 7.696 | 8.105 | 7.720 |
| 50 | 8.324 | 6.294 | 6.419 | 6.385 |
| 60 | 6.386 | 5.331 | 5.367 | 5.527 |
| 70 | 5.210 | 4.630 | 4.630 | 4.945 |
| 80 | 4.423 | 4.102 | 4.074 | 4.531 |
| 90 | 3.856 | 3.692 | 3.638 | 4.223 |
| 100 | 3.425 | 3.368 | 3.285 | 3.986 |

**Table 8.** Optimum $T^*_{OF}$ and $T^*_{OL}$ when $c_O = 100.0$ and $\alpha = 2.0$.

| $c_M$ | $K = 5$ | | $K = 10$ | |
|---|---|---|---|---|
| | $T^*_{OF}$ | $T^*_{OL}$ | $T^*_{OF}$ | $T^*_{OL}$ |
| 10 | 3.299 | 3.015 | 3.037 | 3.030 |
| 20 | 2.072 | 2.058 | 2.040 | 2.207 |
| 30 | 1.604 | 1.667 | 1.598 | 1.919 |
| 40 | 1.332 | 1.458 | 1.331 | 1.774 |
| 50 | 1.148 | 1.330 | 1.148 | 1.686 |
| 60 | 1.012 | 1.243 | 1.012 | 1.626 |
| 70 | 0.907 | 1.180 | 0.907 | 1.584 |
| 80 | 0.822 | 1.133 | 0.822 | 1.552 |
| 90 | 0.752 | 1.096 | 0.752 | 1.527 |
| 100 | 0.694 | 1.066 | 0.694 | 1.508 |

When $H(t) = t^\alpha$, Tables 9 and 10 present optimum $K^*_{OF}$ and $K^*_{OL}$ when $c_O = 100.0$ and $\alpha = 1.2, 2.0$. These two tables show similar tendencies with those in Tables 4 and 5, which has been already shown that if $T < T_O^*$ then $K_{OF}^* = \infty$, and if $T > T_O^*$ then $K_L^* = 0$ for $T_O^*$ given in Table 6.

**Table 9.** Optimum $K_{OF}^*$ and $K_{OL}^*$ when $c_O = 100.0$ and $\alpha = 1.2$.

| $c_M$ | $T = 5.0$ | | $T = 10.0$ | | $T_O^*$ |
|---|---|---|---|---|---|
| | $K_{OF}^*$ | $K_{OL}^*$ | $K_{OF}^*$ | $K_{OL}^*$ | |
| 10 | $\infty$ | 28 | $\infty$ | 28 | 25.663 |
| 20 | $\infty$ | 23 | $\infty$ | 23 | 14.155 |
| 30 | $\infty$ | 17 | 56 | 0 | 9.935 |
| 40 | $\infty$ | 13 | 14 | 0 | 7.696 |
| 50 | $\infty$ | 8 | 11 | 0 | 6.293 |
| 60 | $\infty$ | 1 | 9 | 0 | 5.325 |
| 70 | 17 | 0 | 8 | 0 | 4.614 |
| 80 | 10 | 0 | 7 | 0 | 4.069 |
| 90 | 8 | 0 | 6 | 0 | 3.636 |
| 100 | 6 | 0 | 6 | 0 | 3.284 |

**Table 10.** Optimum $K_{OF}^*$ and $K_{OL}^*$ when $c_O = 100.0$ and $\alpha = 2.0$.

| $c_M$ | $T = 1.0$ | | $T = 2.0$ | | $T_O^*$ |
|---|---|---|---|---|---|
| | $K_{OF}^*$ | $K_{OL}^*$ | $K_{OF}^*$ | $K_{OL}^*$ | |
| 10 | $\infty$ | 11 | $\infty$ | 10 | 3.015 |
| 20 | $\infty$ | 5 | $\infty$ | 1 | 2.040 |
| 30 | $\infty$ | 3 | 5 | 0 | 1.598 |
| 40 | $\infty$ | 2 | 3 | 0 | 1.331 |
| 50 | $\infty$ | 1 | 3 | 0 | 1.148 |
| 60 | $\infty$ | 1 | 2 | 0 | 1.012 |
| 70 | 6 | 0 | 2 | 0 | 0.907 |
| 80 | 4 | 0 | 2 | 0 | 0.822 |
| 90 | 3 | 0 | 2 | 0 | 0.752 |
| 100 | 2 | 0 | 2 | 0 | 0.694 |

## 3.3  Replacement After Failure

### (1) Replacement After the First Failure

When replacement time $T$ is planned after the first failure, i.e., there is no replacement plan before the first failure, the mean time to replacement is

$$\int_0^\infty (t + T)\mathrm{d}F(t) = T + \mu, \tag{25}$$

and the expected number of failures until replacement is

$$\int_0^\infty [1 + H(t + T) - H(t)]\mathrm{d}F(t) = \int_0^\infty H(t + T)\mathrm{d}F(t). \tag{26}$$

Thus, the expected cost rate is

$$C_A(T) = \frac{c_T + c_M \int_0^\infty H(t + T)\mathrm{d}F(t)}{T + \mu}. \tag{27}$$

We find optimum $T_A^*$ to minimize $C_A(T)$. Differentiating $C_A(T)$ with respect to $T$ and setting it equal to zero,

$$(T + \mu)\int_0^\infty h(t + T)\mathrm{d}F(t) - \int_0^\infty H(t + T)\mathrm{d}F(t) = \frac{c_T}{c_M}, \tag{28}$$

whose left-hand side increases strictly with $T$ from $\mu \int_0^\infty h(t)\mathrm{d}F(t) - 1$ to $\infty$. Therefore, there exits a finite and unique $T_A^*$ $(0 < T_A^* < \infty)$ which satisfies (28), and the resulting cost rate is

$$C_A(T_A^*) = c_M \int_0^\infty h(t + T_A^*)\mathrm{d}F(t). \tag{29}$$

Comparing $T_A^*$ in (28) and $T^*$ in (2),

$$L_A(T) \equiv (T + \mu)\int_0^\infty h(t + T)\mathrm{d}F(t) - \int_0^\infty H(t + T)\mathrm{d}F(t) - Th(T) + H(T)$$

$$= \mu \int_0^\infty h(t + T)\mathrm{d}F(t) - \int_0^\infty h(t + T)\overline{F}(t)\mathrm{d}t$$

$$+ T\int_0^\infty [h(t + T) - h(T)]\mathrm{d}F(t).$$

Furthermore,

$$\mu \int_0^\infty h(t + T)\mathrm{d}F(t) - \int_0^\infty h(t + T)\overline{F}(t)\mathrm{d}t$$

$$= \mu \int_0^\infty [h(t + T) - h(T)]\mathrm{d}F(t) - \int_0^\infty [h(t + T) - h(T)]\overline{F}(t)\mathrm{d}t$$

$$= \mu \int_0^\infty \left[\int_0^t \mathrm{d}h(u + T)\right]\mathrm{d}F(t) - \int_0^\infty \left[\int_0^t \mathrm{d}h(u + T)\right]\overline{F}(t)\mathrm{d}t$$

$$= \int_0^\infty \left[\mu\overline{F}(u) - \int_u^\infty \overline{F}(t)\mathrm{d}t\right]\mathrm{d}h(u + T) > 0,$$

as $\int_u^\infty \overline{F}(t)\mathrm{d}t/\overline{F}(u)$ decreases with $u$ from $\mu$ to 0, which follows that $L_A(T) > 0$. This concludes that $T_A^* < T^*$.

Comparing $C_A(T)$ in (27) and $C(T)$ in (1), let $C_A(T) - C(T) = 0$, then,

$$c_T T + c_M T \int_0^\infty H(t+T)\mathrm{d}F(t) - c_T(T+\mu) - c_M(T+\mu)H(T)$$

$$= c_M \int_0^\infty [Th(t+T) - H(T)]\overline{F}(t)\mathrm{d}t = c_T \mu,$$

i.e.,

$$\int_0^\infty [Th(t+T) - H(T)]\overline{F}(t)\mathrm{d}t = \frac{c_T \mu}{c_M}, \tag{30}$$

whose left-hand side increases strictly with $T$ from 0 to $\infty$. Thus, there exists a finite and unique $\widetilde{T}$ which satisfies (30). Therefore, we obtain: If $T \leq \widetilde{T}$, then $C_A(T) \leq C(T)$, and if $T > \widetilde{T}$, then $C_A(T) > C(T)$.

Next, compare $\widetilde{T}$, $T_A^*$ and $T^*$. From (2) and (30),

$$\frac{T}{\mu} \int_0^\infty h(t+T)\overline{F}(t)\mathrm{d}t - H(T) - Th(T) + H(T)$$

$$= \frac{T}{\mu} \int_0^\infty [h(t+T) - h(T)]\overline{F}(t)\mathrm{d}t > 0,$$

which follows that $\widetilde{T} < T^*$.

**Table 11.** Optimum $T_A^*$ and $\widetilde{T}$ when $c_T = 100.0$.

| $c_M$ | $\alpha = 1.2$ | | $\alpha = 2.0$ | |
|---|---|---|---|---|
| | $T_A^*$ | $\widetilde{T}$ | $T_A^*$ | $\widetilde{T}$ |
| 10 | 25.133 | 25.270 | 2.310 | 2.648 |
| 20 | 13.718 | 13.856 | 1.397 | 1.742 |
| 30 | 9.540 | 9.677 | 0.997 | 1.347 |
| 40 | 7.330 | 7.467 | 0.761 | 1.115 |
| 50 | 5.950 | 6.087 | 0.602 | 0.958 |
| 60 | 5.001 | 5.137 | 0.485 | 0.845 |
| 70 | 4.306 | 4.443 | 0.396 | 0.758 |
| 80 | 3.775 | 3.911 | 0.324 | 0.688 |
| 90 | 3.355 | 3.490 | 0.265 | 0.631 |
| 100 | 3.013 | 3.148 | 0.216 | 0.584 |

From (28) and (30),

$$(T + \mu) \int_0^\infty h(t+T)\mathrm{d}F(t) - \int_0^\infty H(t+T)\mathrm{d}F(t)$$

$$- \frac{T}{\mu} \int_0^\infty h(t+T)\overline{F}(t)\mathrm{d}t + H(T)$$

$$= \frac{T+\mu}{\mu} \left[ \mu \int_0^\infty h(t+T)\mathrm{d}F(t) - \int_0^\infty h(t+T)\overline{F}(t)\mathrm{d}t \right] > 0,$$

which follows that $T_A^* < \widetilde{T}$.

When $H(t) = t^\alpha$, Table 11 presents optimum $T_A^*$ and $\widetilde{T}$ when $c_T = 100.0$. This shows that $T_A^* < \widetilde{T} < T^*$, where $T^*$ is given in Table 1. When $\alpha = 1.2$, $\mu = \int_0^\infty \mathrm{e}^{-t^{1.2}}\mathrm{d}t = 0.941$, and when $\alpha = 2.0$, $\mu = \int_0^\infty \mathrm{e}^{-t^{1.2}}\mathrm{d}t = 0.886$, it is easy to shown that $T_A^* < T^* < T_A^* + \mu$, and $T_A^* + \mu$ are almost equal to $T^*$.

## (2) Replacement After the $K$th Failure

When replacement time $T$ is planned after $K$ failures, i.e., there is no replacement plan before the $K$th failure, the mean time to replacement is

$$\int_0^\infty (t+T)\mathrm{d}P_K(t) = T + \int_0^\infty \overline{P}_K(t)\mathrm{d}t, \tag{31}$$

and the expected number of failures until replacement is

$$\int_0^\infty [K + H(t+T) - H(t)]\mathrm{d}P_K(t)$$

$$= K + \int_0^\infty [H(t+T) - H(t)]\mathrm{d}P_K(t) = \int_0^\infty H(t+T)\mathrm{d}P_K(t). \tag{32}$$

Thus, the expected cost rate is

$$C_{AK}(T, K) = \frac{c_T + c_M \int_0^\infty H(t+T)\mathrm{d}P_K(t)}{T + \int_0^\infty \overline{P}_K(t)\mathrm{d}t}, \tag{33}$$

which agrees with (1) when $K = 0$, (4) when $T = 0$, and (27) when $K = 1$.

We find optimum $T_{AK}^*$ and $K_{AK}^*$ to minimize $C_{AK}(T, K)$. Differentiating $C_{AK}(T, K)$ with respect to $T$ and setting it equal to zero,

$$\int_0^\infty h(t+T)\mathrm{d}P_K(t) \left[ T + \int_0^\infty \overline{P}_K(t)\mathrm{d}t \right] - \int_0^\infty H(t+T)\mathrm{d}P_K(t) = \frac{c_T}{c_M}, \tag{34}$$

whose left-hand side increases strictly with $T$ to $\infty$. Thus, there exists a finite and unique $T_{AK}^*$ $(0 \le T_{AK}^* < \infty)$ which satisfies (34), and the resulting cost rate is

$$C_{AK}(T_{AK}^*) = c_M \int_0^\infty h(t+T_{AK}^*)\mathrm{d}P_K(t). \tag{35}$$

**Table 12.** Optimum $T_{AK}^*$ and $K_{AK}^*$ when $c_T = 100.0$ and $\alpha = 1.2$.

| $c_M$ | $K = 5$ or $T = 5.0$ | | $K = 10$ or $T = 10.0$ | |
|---|---|---|---|---|
| | $T_{AK}^*$ | $K_{AK}^*$ | $T_{AK}^*$ | $K_{AK}^*$ |
| 10 | 22.345 | 39 | 19.392 | 28 |
| 20 | 10.967 | 15 | 8.045 | 6 |
| 30 | 6.821 | 8 | 3.927 | 1 |
| 40 | 4.640 | 4 | 1.773 | 1 |
| 50 | 3.288 | 2 | 0.445 | 1 |
| 60 | 2.365 | 1 | 0.000 | 0 |
| 70 | 1.696 | 1 | 0.000 | 0 |
| 80 | 1.188 | 1 | 0.000 | 0 |
| 90 | 0.791 | 1 | 0.000 | 0 |
| 100 | 0.472 | 1 | 0.000 | 0 |

**Table 13.** Optimum $T_{AK}^*$ and $K_{AK}^*$ when $c_T = 100.0$ and $\alpha = 2.0$.

| $c_M$ | $K = 1$ or $T = 1.0$ | | $K = 2$ or $T = 2.0$ | |
|---|---|---|---|---|
| | $T_{AK}^*$ | $K_{AK}^*$ | $T_{AK}^*$ | $K_{AK}^*$ |
| 10 | 2.310 | 5 | 1.870 | 2 |
| 20 | 1.397 | 2 | 0.954 | 1 |
| 30 | 0.997 | 1 | 0.559 | 1 |
| 40 | 0.761 | 1 | 0.324 | 1 |
| 50 | 0.602 | 1 | 0.165 | 1 |
| 60 | 0.485 | 1 | 0.049 | 1 |
| 70 | 0.396 | 1 | 0.000 | 0 |
| 80 | 0.324 | 1 | 0.000 | 0 |
| 90 | 0.265 | 1 | 0.000 | 0 |
| 100 | 0.216 | 1 | 0.000 | 0 |

Note that the left-hand side of (34) increases strictly with $K$ from that of (2), then $T_{AK}^*$ decreases with $K$ from $T^*$ given in (2).

Forming the inequality $C_{AK}(T, K + 1) - C_{AK}(T, K) \geq 0$,

$$H_5(T, K) \left[ T + \int_0^\infty \overline{P}_K(t)\mathrm{d}t \right] - \int_0^\infty H(t + T)\mathrm{d}P_K(t) \geq \frac{c_T}{c_M}, \qquad (36)$$

where

$$H_5(T,K) \equiv \frac{\int_0^\infty p_K(t)h(t+T)dt}{\int_0^\infty p_K(t)dt} \geq \frac{\int_0^\infty \overline{P}_{K+1}(t)h(t+T)dt}{\int_0^\infty \overline{P}_{K+1}(t)dt},$$

which increases strictly with $K$ to $\infty$ and increases strictly with $T$ to $h(\infty)$. Thus, the left-hand side of (36) increases strictly with $K$ from $H_5(T,0)T - H(T)$ to $\infty$. Thus, there exists a finite and unique minimum $K^*_{AK}$ ($0 \leq K^*_{AK} < \infty$) which satisfies (36). Note that the left-hand side of (36) increases with $T$ from that of (5), then $K^*_{AK}$ decreases with $T$ from $K^*$ given in (5).

When $H(t) = t^\alpha$, Tables 12 and 13 presents optimum $T^*_{AK}$ for $K$ and $K^*_{AK}$ for $T$ when $c_T = 100.0$ and $\alpha = 1.2, 2.0$. Tables 12 and 13 show that when given $T$ and $K$ is large or $\alpha$ becomes large, both $T^*_{AK}$ and $K^*_{AK}$ would go to 0.

## 4    Conclusions

This chapter has surveyed the standard replacement policies that are planned at time $T$ and at number $K$ of failures, respectively, and then discussed three extensions of the policies of $T$ and $K$, that is, replacement first and last, replacement overtime first and last, replacement after the first and the $K$th failure. It has been shown that the cumulative hazard function $H(t)$ plays an important role to model the number of failures between replacement policies.

We have also compared the above replacement policies from the point of cost and found the relations of their optimum replacement times or failure numbers for replacement. For example, we have shown that, optimum $T^*_F$ for replacement first decreases with $K$ to $T^*$, optimum $T_O$ for replacement overtime is always less than $T^*$, optimum time $T^*_A < \widetilde{T} < T^*$, where $T^*$ an optimum time for the standard policy. This chapter would provide an interesting work of modelings of replacement policies with minimal repairs for maintainability studies.

**Acknowledgement.** This work is supported by National Natural Science Foundation of China (NO. 71801126), Natural Science Foundation of Jiangsu Province (NO. BK20180412), Aeronautical Science Foundation of China (NO. 2018ZG52080), and Fundamental Research Funds for the Central Universities (NO. NR2018003).

## References

1. Nakagawa, T.: Maintenance Theory of Reliability. Springer, London (2005). https://doi.org/10.1007/1-84628-221-7
2. Barlow, R.E., Proschan, F.: Mathematical Theory of Reliability. Wiley, New York (1965)
3. Pulcini, G.: Mechanical reliability and maintenance models. In: Pham, H. (ed.) Handbook of Reliability Engineering, pp. 317–348. Springer, London (2003). https://doi.org/10.1007/1-85233-841-5_18
4. Chien, Y., Sheu, S.: Extended optimal age-replacement policy with minimal repair of a system subject to shocks. Eur. J. Oper. Res. **174**, 169–181 (2006)

5. Huynh, K., Castro, I., Barros, A., Bérenguer, C.: Modeling age-based maintenance strategies with minimal repairs for systems subject to competing failure modes due to degradation and shocks. Eur. J. Oper. Res. **218**, 140–151 (2012)
6. Lim, J., Qu, J., Zuo, M.: Age replacement policy based on imperfect repair with random probability. Reliab. Eng. Syst. Saf. **149**, 24–33 (2016)
7. Wang, W.: Optimum production and inspection modeling with minimal repair and rework considerations. Appl. Math. Model. **37**, 1618–1626 (2013)
8. Park, M., Jung, K., Park, D.: Optimal post-warranty maintenance policy with repair time threshold for minimal repair. Reliab. Eng. Syst. Saf. **111**, 147–153 (2013)
9. Chang, C.: Optimum preventive maintenance policies for systems subject to random working times, replacement, and minimal repair. Comput. Ind. Eng. **67**, 185–194 (2014)
10. Sheu, S., Chang, C., Chen, Y., Zhang, Z.: Optimal preventive maintenance and repair policies for multi-state systems. Reliab. Eng. Syst. Saf. **140**, 78–87 (2015)
11. Cha, J., Finkelstein, M.: On preventive maintenance under different assumptions on the failure/repair processes. Quality Reliab. Eng. Int. **34**, 66–77 (2018)
12. Yuan, F., Kumar, U.: A general imperfect repair model considering time-dependent repair effectiveness. IEEE Trans. Reliab. **61**, 95–100 (2012)
13. Zhao, X., Al-Khalifa, K.N., Hamouda, A.M.S., Nakagawa, T.: First and last triggering event approaches for replacement with minimal repairs. IEEE Trans. Reliab. **65**, 197–207 (2016)
14. Zhao, X., Nakagawa, T.: Optimization problems of replacement first or last in reliability theory. Eur. J. Oper. Res. **223**, 141–149 (2012)
15. Nakagawa, T., Zhao, X.: Maintenance Overtime Policies in Reliability Theory. Springer, Switzerland (2015). https://doi.org/10.1007/978-3-319-20813-8

# Reliability of a Dual Linear Consecutive System with Three Failure Modes

Rui Peng[1(✉)], Di Wu[2], and Kaiye Gao[3]

[1] School of Economics and Management, Beijing University of Technology,
Beijing, China
pengruisubmit@163.com
[2] School of Management, Xi'an Jiaotong University, Xian, China
13552361171@163.com
[3] School of Economics and Management,
Beijing Information Science & Technology University, Beijing, China
kygao@foxmail.com

**Abstract.** Consecutive systems have applications in the field of telecommunication, transportation, illumination, heating, etc. However, all the existing works just studied the reliability of a single consecutive system. The typical studied consecutive systems include linear/circular consecutive k-out-of-n systems, linear sliding window systems, linear multi-state consecutively connected systems, etc. These models are restricted to the cases where all the system components are arranged on a line or on a circle. In practice, a system may consist of some components arranged on two parallel lines, instead of a single line. An example is the system consisting of road lights at both sides of the highway. In this chapter, a reliability model for system consisting of two linear parallel consecutive subsystems is proposed where three failure modes are considered: (1) the subsystem 1 has at least $k_1$ consecutive failed components; (2) the subsystem 2 has at least $k_2$ consecutive failed components; (3) the system has at least $m$ consecutive failed pairs of components. An iterative approach is proposed to evaluate the reliability of such a system. Numerical examples are presented to illustrate the applications.

**Keywords:** Consecutive systems · Dual system · Failure mode · Iterative approach · Reliability

## 1 Introduction

As an extension of the typical k-out-of-n system [1–9], a consecutive k-out-of-n system fails if consecutive $k$ out of the total $n$ system components fail [10–14]. Depending on whether the components are arranged on a line or on a circle, the system is called linear/circular consecutive system [15]. This kind of consecutive system has applications in telecommunications, pumping systems, heating systems, etc. [16]. For instance, a telecommunication system may consist of several relay stations on a line, with each merely able to send signal to the next $k$ stations. The failure of more than $k$ consecutive relay stations cuts off the signal transmission and fails the system. Due to the practical background of the consecutive system, many generalizations have made by researchers

© Springer Nature Singapore Pte Ltd. 2019
Q.-L. Li et al. (Eds.): Cao Festschrift 2019, CCIS 1102, pp. 259–269, 2019.
https://doi.org/10.1007/978-981-15-0864-6_12

considering different practical factors. Say, some researchers studied the linear multistate consecutively connected system (LMCCS) where each node can provide connection with some following nodes depending on the capacity of the node [17–22]. Except modelling the reliability of such a system, researchers also studied the optimal allocation of components in the system and the optimal maintenance strategy [23]. Some other researchers studied the sliding window systems and its variants [24–26].

However, all the above works are restricted to the case where all the system components are arranged on a line or on a circle. In practice, some system may have components arranged on two parallel lines. Take the highway for examples, it has road lights at both sides. If each side has too many consecutive lights broken, it may affect the traffic. The situation is worse if some consecutive pairs of road lights fail on both sides of the highway. Therefore, this chapter proposes a reliability model for the dual linear consecutive system consisting of two linear consecutive subsystems, each with $n$ components. Figure 1 shows an illustration of such a system. The system is assumed to have three diverse failure modes, i.e., the system fails if: (1) The subsystem 1 has at least $k_1$ consecutive failed components; (2) The subsystem 2 has at least $k_2$ consecutive failed components; (3) The system has $m$ consecutive pairs of failed components, where $m$ is smaller than or equal to $k_1$ and $k_2$.

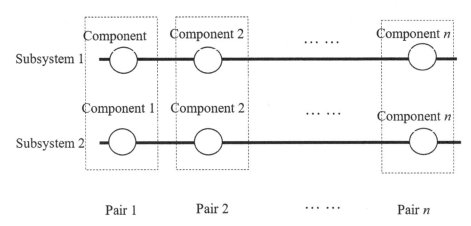

**Fig. 1.** A dual linear consecutive system

The remaining of this chapter is organized as follows. Section 2 proposes an iterative approach to evaluate the reliability of such a system. Section 3 presents a numerical example. Section 4 concludes this chapter and points out the future research.

## 2  The Model

The system consists of two subsystems which are called respectively subsystem 1 and subsystem 2. The subsystem 1 consists of $n$ components and their reliabilities are denoted as $p_1, \ldots, p_n$. Similarly, the subsystem 2 consists of $n$ components and their reliabilities are denoted as $q_1, \ldots, q_n$. The system has three failure modes, i.e., it fails if at least $k_1$ consecutive components fail in subsystem 1, at least $k_2$ consecutive components fail in subsystem 2, or at least $m$ consecutively paired components fail in subsystem 1 and subsystem 2, where $m$ is smaller than both $k_1$ and $k_2$. For the convenience of discussion, the three failure modes are called failure mode 1, failure mode 2, and failure mode 3, respectively. The reason for assuming $m$ to be smaller than both $k_1$ and $k_2$ is that the system has only two failure modes if $m$ is not smaller than either $k_1$ or $k_2$. Say, if $m$ is greater than $k_1$ but smaller than $k_2$, then the failure mode 3 automatically leads to failure mode 1, which makes considering only failure mode 2 and 3 to be sufficient for system reliability evaluation.

The reliability of such a system is denoted as $R(n, k_1, k_2, m, p_1, \ldots, p_n, q_1, \ldots, q_n)$. In order to evaluate the system reliability, an iterative approach is proposed. Use $v_1$ and $v_2$ to denote the indices of the first components in subsystem 1 and subsystem 2 which are working. In order for the system to be reliable, it is easy to see that $v_1$ and $v_2$ must satisfy that $v_1 \le k_1$, $v_2 \le k_2$, and $\min(v_1, v_2) \le m$. In addition, three different cases may happen, that is, $v_1$ is equal to $v_2$, $v_1$ is smaller than $v_2$, and $v_1$ is bigger than $v_2$. These three conditions are discussed as below.

1. In the case where $v_1$ is equal to $v_2$. The conditions that $v_1 \le k_1$, $v_2 \le k_2$ and $\min(v_1, v_2) \le m$ are simplified as $v_1 = v_2 \le m$. Once $v_1 = v_2 \le m$ is satisfied, the conditional system reliability is equal to the reliability of the subsystem consisting of the components of from the $(v_1 + 1)$-th pair to the $n$-th pair of the original system. That is, the conditional system reliability of the system can be obtained by altering the reliability and equals to $R(n - v_1, k_1, k_2, m, p_{v_1+1}, \ldots, p_n, q_{v_1+1}, \ldots, q_n)$.

2. In the case where $v_1$ is bigger than $v_2$. The conditions that $v_1 \le k_1$, $v_2 \le k_2$ and $\min(v_1, v_2) \le m$ are simplified as $v_2 \le m$, $v_2 \le v_1 \le k_1$. The conditional reliability is denoted as $\tilde{R}(n - v_1, n - v_2, k_1, k_2, m, p_{v_1+1}, \ldots, p_n, q_{v_2+1}, \ldots, q_n)$. Note that, here we have introduced $\tilde{R}$, which is different from the system reliability $R$. The notation $\tilde{R}$ is used to describe the reliability of a system consisting of two subsystems of different lengths, whereas $R$ is used to describe the reliability of a system consisting of two subsystems of the same length. In order to use the iterative approach to obtain $R$, it is also needed to build the iterative relationship for $\tilde{R}$. However, to avoid interrupting the reading, this relationship is shown in the later stage.

3. In the case where $v_1$ is smaller than $v_2$. The conditions that $v_1 \le k_1$, $v_2 \le k_2$ and $\min(v_1, v_2) \le m$ are simplified as $v_1 \le m, v_1 \le v_2 \le k_2$. The conditional reliability is denoted as $\tilde{R}(n - v_1, n - v_2, k_1, k_2, m, p_{v_1+1}, \ldots, p_n, q_{v_2+1}, \ldots, q_n)$. Note that the conditional reliability under this case is not the same as the case when $v_1$ is bigger than $v_2$.

Considering the three cases above, it is easy to see that:

$$R(n, k_1, k_2, m, p_1, \ldots, p_n, q_1, \ldots, q_n)$$
$$= \sum_{v_1=1}^{m} (1-p_1)\ldots(1-p_{v_1-1})p_{v_1}(1-q_1)\ldots(1-q_{v_1-1})q_{v_1}$$
$$\times R(n-v_1, k_1, k_2, m, p_{v_1+1}, \ldots, p_n, q_{v_1+1}, \ldots, q_n)$$
$$+ \sum_{v_2=1}^{m} \sum_{v_1=v_2+1}^{k_1} (1-p_1)\ldots(1-p_{v_1-1})p_{v_1}(1-q_1)\ldots(1-q_{v_2-1})q_{v_2}, \qquad (1)$$
$$\times \tilde{R}(n-v_1, n-v_2, k_1, k_2, m, p_{v_1+1}, \ldots, p_n, q_{v_2+1}, \ldots, q_n)$$
$$+ \sum_{v_1=1}^{m} \sum_{v_2=v_1+1}^{k_2} (1-p_1)\ldots(1-p_{v_1-1})p_{v_1}(1-q_1)\ldots(1-q_{v_2-1})q_{v_2}$$
$$\times \tilde{R}(n-v_1, n-v_2, k_1, k_2, m, p_{v_1+1}, \ldots, p_n, q_{v_2+1}, \ldots, q_n)$$

where $(1-p_1)\ldots(1-p_{v_1-1})p_{v_1}(1-q_1)\ldots(1-q_{v_2-1})q_{v_2}$ is the probability that the first working component in subsystem 1 is $v_1$ and the first working component in subsystem 2 is $v_2$. From Eq. (1), it can be seen that the system reliability is decomposed into the reliabilities of systems of smaller size. However, since the right hand side of Eq. (1) contains not only $R$, but also $\tilde{R}$, the iterative relationship of $\tilde{R}$ also needs to be provided. Similarly, two different cases need to be considered, i.e., $v_1 < v_2$ and $v_2 < v_1$. Since the two cases are symmetric, it is only needed to know how to decompose $\tilde{R}$ without loss of generality when $v_1 < v_2$. Use $v_3$ to denote the second working component in subsystem 1. It is easy to see that there are four different cases which are $v_3 < v_2$, $v_3 = v_2$, $n \geq v_3 > v_2$, and $v_3$ does not exist. Note that the last case corresponds to the situation that all the components after component $v_1$ in subsystem 1 are failed. Therefore, the conditional reliability of the system $\tilde{R}$ can be decomposed as

$$\tilde{R}(n-v_1, n-v_2, k_1, k_2, m, p_{v_1+1}, \ldots, p_n, q_{v_2+1}, \ldots, q_n | v_1 < v_2)$$
$$= \sum_{v_3=v_1+1}^{v_2-1} 1(v_3 - v_1 - 1 < m)(1-p_{v_1+1})\ldots(1-p_{v_3-1})p_{v_3}$$
$$\times \tilde{R}(n-v_3, n-v_2, k_1, k_2, m, p_{v_3+1}, \ldots, p_n, q_{v_2+1}, \ldots, q_n)$$
$$+ 1(v_2 - v_1 - 1 < m)(1-p_{v_1+1})\ldots(1-p_{v_2-1})p_{v_2}R(n-v_2, k_1, k_2, m, p_{v_2+1}, \ldots p_n, q_{v_2+1}, \ldots, q_n),$$
$$+ \sum_{v_3=v_2+1}^{n} 1(v_3 - v_1 - 1 < k_1, v_2 - v_1 - 1 < m)(1-p_{v_1+1})\ldots(1-p_{v_3-1})p_{v_3}$$
$$\times \tilde{R}(n-v_3, n-v_2, k_1, k_2, m, p_{v_3+1}, \ldots, p_n, q_{v_2+1}, \ldots, q_n)$$
$$+ 1(n-v_1 < k_1, v_2 - v_1 - 1 < m)(1-p_{v_1+1})\ldots(1-p_{n-1})(1-p_n)RR(n-v_2, m, q_{v_2+1}, \ldots, q_n)$$
$$(2)$$

where $1()$ is the unity function that $1(True) = 1$ and $1(False) = 0$, and $RR(n - v_2, m, q_{v_2+1}, \ldots, q_n)$ is the reliability of a consecutive $m$-out-of-$n$-$v$-$2$ system with reliabilities of components equaling to $q_{v_2+1}, \ldots, q_n$. Note that $\tilde{R}$ is the summation of the probabilities of four different cases mentioned.

From Eqs. (1) and (2), it is easy to see that the $R$ can be finally decomposed to the reliabilities of single linear consecutive systems. For any single consecutive $kk$-out-of-$nn$ system, the system reliability can be readily calculated from another iterative approach. For the convenience of discussion, we use $RRR(kk, nn, h_1, \ldots, h_{nn})$ to represent the reliability of a consecutive $kk$-out-of-$nn$ system with components reliabilities equaling to $h_1, \ldots, h_{nn}$. Thus, the following relationship holds as

$$
\begin{aligned}
RRR(kk, nn, h_1, \ldots, h_{nn}) &= \sum_{v_4=1}^{kk} (1 - h_1) \ldots (1 - h_{v_4-1}) h_{v_4} \\
&\times RRR(kk, nn - kk, h_{v_4+1}, \ldots, h_{nn})
\end{aligned}
\tag{3}
$$

where $v_4$ is the first working component in the consecutive $kk$-out-of-$nn$ system.

## 3 Illustrative Example

To illustrate the calculating process of the proposed model, we present a brief example in this section. Here consider a dual linear consecutive system where each of its two subsystems consists of 4 components. The reliabilities of the components in subsystem 1 and 2 are denoted as $p_1, p_2, p_3, p_4$ and $q_1, q_2, q_3, q_4$ respectively. The system fails if any 3 or more consecutive components fail at any one subsystems, or any 2 or more consecutively paired components fail. Thus, according to the definition of the proposed model, $n = 4$, $k_1 = k_2 = 3$, $m = 2$.

According to Eq. (1), we have

$$
\begin{aligned}
&R(4, 3, 3, 2, p_1, p_2, p_3, p_4, q_1, q_2, q_3, q_4) \\
&= \sum_{v_1=1}^{2} (1 - p_1) \ldots (1 - p_{v_1-1}) p_{v_1} (1 - q_1) \ldots (1 - q_{v_1-1}) q_{v_1} \\
&\quad \times R(4 - v_1, 3, 3, 2, p_{v_1+1}, \ldots, p_4, q_{v_1+1}, \ldots, q_4) \\
&\quad + \sum_{v_2=1}^{2} \sum_{v_1=v_2+1}^{3} (1 - p_1) \ldots (1 - p_{v_1-1}) p_{v_1} (1 - q_1) \ldots (1 - q_{v_2-1}) q_{v_2}, \\
&\quad \times \tilde{R}(4 - v_1, 4 - v_2, 3, 3, 2, p_{v_1+1}, \ldots, p_4, q_{v_2+1}, \ldots, q_4) \\
&\quad + \sum_{v_1=1}^{2} \sum_{v_2=v_1+1}^{3} (1 - p_1) \ldots (1 - p_{v_1-1}) p_{v_1} (1 - q_1) \ldots (1 - q_{v_2-1}) q_{v_2} \\
&\quad \times \tilde{R}(4 - v_1, 4 - v_2, 3, 3, 2, p_{v_1+1}, \ldots, p_4, q_{v_2+1}, \ldots, q_4) \\
&= I_1 + I_2 + I_3
\end{aligned}
\tag{4}
$$

where $I_1, I_2, I_3$ indicate the first, second and third items respectively.

According to Eqs. (1) and (2), we can obtain the following derivation process:

First, $I_1$ can be further decomposed as:

$$
\begin{aligned}
I_1 &= p_1 q_1 R(3,3,3,2,p_2,p_3,p_4,q_2,q_3,q_4) \\
&+ (1-p_1)p_2(1-q_1)q_2 R(2,3,3,2,p_3,p_4,q_3,q_4) \\
&= p_1 q_1 \sum_{v_1=1}^{2} (1-p_2)\ldots(1-p_{1+v_1-1})p_{1+v_1}(1-q_2)\ldots(1-q_{1+v_1-1})q_{1+v_1} \\
&\times R(3-v_1,3,3,2,p_{1+v_1+1},\ldots,p_4,q_{1+v_1+1},\ldots,q_4) \\
&+ p_1 q_1 \sum_{v_2=1}^{2}\sum_{v_1=v_2+1}^{3} (1-p_2)\ldots(1-p_{1+v_1-1})p_{1+v_1}(1-q_2)\ldots(1-q_{1+v_2-1})q_{1+v_2} \\
&\times \tilde{R}(3-v_1,3-v_2,3,3,2,p_{v_1+1},\ldots,p_4,q_{v_2+1},\ldots,q_4) \\
&+ p_1 q_1 \sum_{v_1=1}^{2}\sum_{v_2=v_1+1}^{3} (1-p_2)\ldots(1-p_{1+v_1-1})p_{1+v_1}(1-q_2)\ldots(1-q_{1+v_2-1})q_{1+v_2} \\
&\times \tilde{R}(3-v_1,3-v_2,3,3,2,p_{v_1+1},\ldots,p_4,q_{v_2+1},\ldots,q_4) \\
&+ (1-p_1)p_2(1-q_1)q_2 R(2,3,3,2,p_3,p_4,q_3,q_4) \\
&= I_{1,1}+I_{1,2}+I_{1,3}+I_{1,4}
\end{aligned}
\tag{5}
$$

where $I_{1,1}, I_{1,2}, I_{1,3}, I_{1,4}$ indicate the first to the fourth items respectively. Specifically,

$$
\begin{aligned}
I_{1,1} &= p_1 p_2 q_1 q_2 R(3-1,3,3,2,p_3,p_4,q_3,q_4) \\
&+ p_1(1-p_2)p_3 q_1(1-q_2)q_3 R(3-2,3,3,2,p_4,q_4) \\
&= p_1 p_2 q_1 q_2[1-(1-p_3)(1-p_4)(1-q_3)(1-q_4)]+p_1(1-p_2)p_3 q_1(1-q_2)q_3
\end{aligned}
\tag{6}
$$

$$
\begin{aligned}
I_{1,2} &= p_1(1-p_2)p_3 q_1 q_2 \tilde{R}(3-2,3-1,3,3,2,p_4,q_3,q_4) \\
&+ p_1(1-p_2)(1-p_3)p_4 q_1 q_2 \tilde{R}(3-3,3-1,3,3,2,q_3,q_4) \\
&+ p_1(1-p_2)(1-p_3)p_4 q_1(1-q_2)q_3 \tilde{R}(3-3,3-2,3,3,2,q_4) , \\
&= p_1(1-p_2)p_3 q_1 q_2 + p_1(1-p_2)(1-p_3)p_4 q_1 q_2 \\
&+ p_1(1-p_2)(1-p_3)p_4 q_1(1-q_2)q_3
\end{aligned}
\tag{7}
$$

$$
\begin{aligned}
I_{1,3} &= p_1 p_2 q_1(1-q_2)q_3 \tilde{R}(3-1,3-2,3,3,2,p_3,p_4,q_4) \\
&+ p_1 p_2 q_1(1-q_2)(1-q_3)q_4 \tilde{R}(3-1,3-3,3,3,2,p_3,p_4) \\
&+ p_1(1-p_2)p_3 q_1(1-q_2)(1-q_3)q_4 \tilde{R}(3-2,3-3,3,3,2,p_4), \\
&= p_1 p_2 q_1(1-q_2)q_3 + p_1 p_2 q_1(1-q_2)(1-q_3)q_4 \\
&+ p_1(1-p_2)p_3 q_1(1-q_2)(1-q_3)p_4
\end{aligned}
\tag{8}
$$

$$
\begin{aligned}
I_{1,4} &= (1-p_1)p_2(1-q_1)q_2 R(2,3,3,2,p_3,p_4,q_3,q_4) \\
&= (1-p_1)p_2(1-q_1)q_2[1-(1-p_3)(1-p_4)(1-q_3)(1-q_4)] .
\end{aligned}
\tag{9}
$$

Second, for $I_2$

$$
\begin{aligned}
I_2 &= (1-p_1)p_2 q_1 \tilde{R}(4-2,4-1,3,3,2,p_3,p_4,q_2,q_3,q_4) \\
&+ (1-p_1)(1-p_2)p_3 q_1 \tilde{R}(4-3,4-1,3,3,2,p_4,q_2,q_3,q_4) \\
&+ (1-p_1)(1-p_2)p_3(1-q_1)q_2 \tilde{R}(4-3,4-2,3,3,2,p_4,q_3,q_4) , \\
&= I_{2,1}+I_{2,2}+I_{2,3}
\end{aligned}
\tag{10}
$$

where $I_{2,1}, I_{2,2}, I_{2,3}$ indicate the first, second and third items respectively. Specifically,

$$
\begin{aligned}
I_{2,1} &= (1-p_1)p_2q_1\tilde{R}(4-2,4-1,3,3,2,p_3,p_4,q_2,q_3,q_4)\\
&= (1-p_1)p_2q_1\\
&\times\{1(2-1-1<2)q_2R(4-2,3,3,2,p_3,p_4,q_3,q_4)\\
&+1(3-1-1<3,2-1-1<2)(1-q_2)q_3R(4-2,4-3,3,3,2,p_3,p_4,q_4)\\
&+1(4-1-1<3,2-1-1<2)(1-q_2)(1-q_3)q_4R(4-2,4-4,3,3,2,p_3,p_4),\\
&+1(4-1<3,2-1-1<2)(1-q_2)(1-q_3)(1-q_4)RR(4-2,2,p_3,p_4)\}\\
&= (1-p_1)p_2q_1\\
&\times\{q_2[1-(1-p_3)(1-p_4)(1-q_3)(1-q_4)]\\
&+(1-q_2)q_3+(1-q_2)(1-q_3)q_4+0\}
\end{aligned} \tag{11}
$$

$$
\begin{aligned}
I_{2,2} &= (1-p_1)(1-p_2)p_3q_1\tilde{R}(4-3,4-1,3,3,2,p_4,q_2,q_3,q_4)\\
&= (1-p_1)(1-p_2)p_3q_1[1-(1-q_2)(1-q_3)(1-q_4)]
\end{aligned}, \tag{12}
$$

$$
\begin{aligned}
I_{2,3} &= (1-p_1)(1-p_2)p_3(1-q_1)q_2\tilde{R}(4-3,4-2,3,3,2,p_4,q_3,q_4)\\
&= (1-p_1)(1-p_2)p_3(1-q_1)q_2\times 1
\end{aligned} \tag{13}
$$

Third, for $I_3$

$$
\begin{aligned}
I_3 &= p_1(1-q_1)q_2\tilde{R}(4-1,4-2,3,3,2,p_2,p_3,p_4,q_3,q_4)\\
&+p_1(1-q_1)(1-q_2)q_3\tilde{R}(4-1,4-3,3,3,2,p_2,p_3,p_4,q_4)\\
&+(1-p_1)p_2(1-q_1)(1-q_2)q_3\tilde{R}(4-2,4-3,3,3,2,p_3,p_4,q_4),\\
&= I_{3,1}+I_{3,2}+I_{3,3}
\end{aligned} \tag{14}
$$

where $I_{3,1}, I_{3,2}, I_{3,3}$ indicate the first, second and third items respectively. Specifically,

$$
\begin{aligned}
I_{3,1} &= p_1(1-q_1)q_2\tilde{R}(4-1,4-2,3,3,2,p_2,p_3,p_4,q_3,q_4)\\
&= p_1(1-q_1)q_2\{\\
&1(2-1-1<2)p_2R(4-2,3,3,2,p_3,p_4,q_3,q_4)\\
&+1(3-1-1<3,2-1-1<2)(1-p_2)p_3\tilde{R}(4-3,4-2,3,3,2,p_4,q_3,q_4)\\
&+1(4-1-1<3,2-1-1<2)(1-p_2)(1-p_3)p_4\tilde{R}(4-4,4-2,3,3,2,q_3,q_4),\\
&+1(4-1<3,2-1-1<2)(1-p_2)(1-p_3)(1-p_4)RR(4-2,2,q_3,q_4)\\
&=p_1(1-q_1)q_2\{p_2[1-(1-p_3)(1-p_4)(1-q_3)(1-q_4)]\\
&+(1-p_2)p_3+(1-p_2)(1-p_3)p_4+0\}
\end{aligned} \tag{15}
$$

$$
\begin{aligned}
I_{3,2} &= p_1(1-q_1)(1-q_2)q_3\tilde{R}(4-1,4-3,3,3,2,p_2,p_3,p_4,q_4)\\
&= p_1(1-q_1)(1-q_2)q_3[1-(1-p_2)(1-p_3)(1-p_4)]
\end{aligned}, \tag{16}
$$

$$
\begin{aligned}
I_{3,3} &= (1-p_1)p_2(1-q_1)(1-q_2)q_3\tilde{R}(4-2,4-3,3,3,2,p_3,p_4,q_4)\\
&= (1-p_1)p_2(1-q_1)(1-q_2)q_3\times 1
\end{aligned}. \tag{17}
$$

Finally, by adding up all the above probabilities leading to system success, we can obtain the function of the system reliability, which is

$$
\begin{aligned}
R(4,3,3,2,p_1,p_2,p_3,p_4,q_1,q_2,q_3,q_4) &= I_1 + I_2 + I_3 \\
&= I_{1,1} + I_{1,2} + I_{1,3} + I_{1,4} + I_{2,1} + I_{2,2} + I_{2,3} + I_{3,1} + I_{3,2} + I_{3,3} \\
&= \{p_1 p_2 q_1 q_2 [1 - (1-p_3)(1-p_4)(1-q_3)(1-q_4)] \\
&\quad + p_1(1-p_2)p_3 q_1(1-q_2)q_3\} \\
&\quad + \{p_1(1-p_2)p_3 q_1 q_2 + p_1(1-p_2)(1-p_3)p_4 q_1 q_2 \\
&\quad + p_1(1-p_2)(1-p_3)p_4 q_1(1-q_2)q_3\} \\
&\quad + \{p_1 p_2 q_1(1-q_2)q_3 + p_1 p_2 q_1(1-q_2)(1-q_3)q_4 \\
&\quad + p_1(1-p_2)p_3 q_1(1-q_2)(1-q_3)q_4\} \\
&\quad + \{(1-p_1)p_2(1-q_1)q_2[1-(1-p_3)(1-p_4)(1-q_3)(1-q_4)]\} \\
&\quad + \{(1-p_1)p_2 q_1\{q_2[1-(1-p_3)(1-p_4)(1-q_3)(1-q_4)] \\
&\quad + (1-q_2)q_3 + (1-q_2)(1-q_3)q_4 + 0\}\} \\
&\quad + \{(1-p_1)(1-p_2)p_3 q_1[1-(1-q_2)(1-q_3)(1-q_4)]\} \\
&\quad + \{(1-p_1)(1-p_2)p_3(1-q_1)q_2 \times 1\} \\
&\quad + \{p_1(1-q_1)q_2\{p_2[1-(1-p_3)(1-p_4)(1-q_3)(1-q_4)] \\
&\quad + (1-p_2)p_3 + (1-p_2)(1-p_3)p_4 + 0\}\} \\
&\quad + \{p_1(1-q_1)(1-q_2)q_3[1-(1-p_2)(1-p_3)(1-p_4)]\} \\
&\quad + \{(1-p_1)p_2(1-q_1)(1-q_2)q_3 \times 1\}
\end{aligned}
$$

$\hspace{11cm}(18)$

Then we can calculate the reliability of the system by substituting the numerical values of $p_1, p_2, p_3, p_4$ and $q_1, q_2, q_3, q_4$ into Eq. (18). To present the variation trend of reliability, we assume that the basic reliabilities of these components are 0.5 and then vary the reliability of the concerned variable in two ways: (1) varying the reliability of only one component at a time; (2) varying the reliability of a pair of components at a time.

Figure 2 shows the first case. The solid line illustrates the variation of system reliability by the variation of $p_1$ from 0 to 1 for $p_2 = p_3 = p_4 = q_1 = q_2 = q_3 = q_4 = 0.5$. The dashed line illustrates the variation of system reliability by the variation of $p_2$ from 0 to 1 for $p_1 = p_3 = p_4 = q_1 = q_2 = q_3 = q_4 = 0.5$. As the system is symmetric for the first half components (components 1 and 2 in both two subsystems) and the other half components (components 3 and 4 in both two subsystems), the variation of system reliability by the variation of $p_3$ and $p_4$ is the same as the variation of system reliability by the variation of $p_2$ and $p_1$ respectively, so we do not display the other two curves in this figure.

It can be seen from Fig. 2 that the reliability of the system increases with the reliabilities of its components, which is consistent with common sense that the system is more reliable if its components are more reliable. In addition, it should be noted that $p_2$ grows faster than $p_1$ which makes the surpass for the curve of $p_2$ to the curve of $p_1$ when they are 0.5. This phenomenon represents that the component deployed at the middle influences the reliability of the system more than the component deployed at the beginning and the end. The two curves intersect at the point of $p_1 = p_2 = 0.5$ where two curves represent the same case and $p_1 = p_2 = p_3 = p_4 = q_1 = q_2 = q_3 = q_4 = 0.5$.

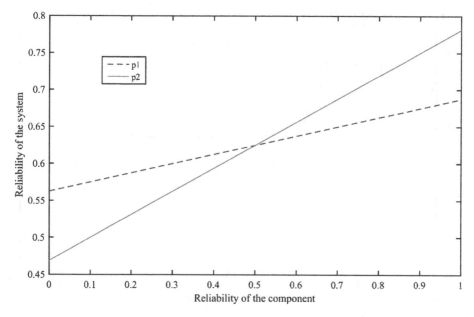

**Fig. 2.** System reliability as functions of $p_1$ and $p_2$.

Figure 3 shows the second case. The solid line illustrates the variation of system reliability by the synchronous variations of $p_1$ and $q_1$ from 0 to 1 for $p_2 = p_3 = p_4 = q_2 = q_3 = q_4 = 0.5$. The dashed line illustrates the variation of system's reliability by the synchronous variations of $p_2$ and $q_2$ from 0 to 1 for $p_1 = p_3 = p_4 = q_1 = q_3 = q_4 = 0.5$. Besides, the curves for $p_3$, $q_3$ and $p_4$, $q_4$ are also not plotted in this figure due to the symmetry of the system.

It can be seen from Fig. 3 that the reliability of the system increases with the reliability of its components. Compared with the curves in Fig. 2, the ranges of two curves in Fig. 3 are larger. For example, the system reliability for $(p_1, q_1)$ in Fig. 3 starts from about 0.3 and ends over 0.9, while the system reliability for $p_1$ in Fig. 2 starts from about 0.45 and ends under 0.8. This is because that a pair of components is more important than just one single component to the duel system. Just like Fig. 2, the two curves in Fig. 3 meet at the point where $p_1 = p_2 = p_3 = p_4 = q_1 = q_2 = q_3 = q_4 = 0.5$. Besides, the growth rate of the curve for $(p_2, q_2)$ is faster than the growth rate of the curve for $(p_1, q_1)$, which is consistent with Fig. 2, representing that the paired components deployed at the middle is more important to the system reliability compared with the paired components deployed at the beginning and the end.

With an example of small scale system, we illustrated the feasibility of the proposed reliability model and the process of the proposed iterative approach to evaluate the reliability. In practice, the model and the iterative approach proposed in this chapter can be straightforwardly applied to large scale systems easily using computer programming for the complete calculation.

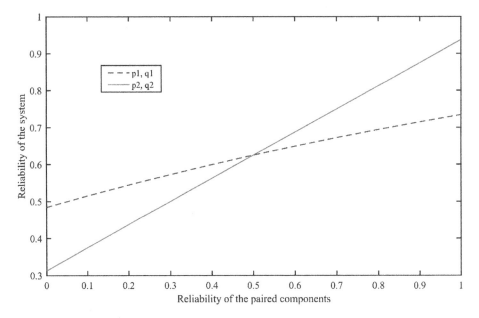

**Fig. 3.** The curves of system's reliability as functions of paired components' reliability: $(p_1, q_1)$ and $(p_2, q_2)$.

## 4    Conclusions

In this chapter, we proposed a model to evaluate the reliability of dual linear consecutive system. The dual linear consecutive system consists of two subsystems where components are arranged on two parallel lines, which has been widely applied in practice. Three failure modes are considered: (1) at least $k_1$ consecutive components fail in subsystem 1; (2) at least $k_2$ consecutive components fail in subsystem 2; (3) at least $m$ consecutively paired components fail in subsystem 1 and subsystem 2, where $m$ is smaller than both $k_1$ and $k_2$. This study developed an iterative approach to construct the evaluation model. A brief example is presented to illustrate the calculating process of the proposed mode. Future work can be devoted to the optimal positioning of the components in the system.

## References

1. Zhu, X., Boushaba, M., Coit, D.W., Benyahia, A.: Reliability and importance measures for m-consecutive-k, l-out-of-n system with non-homogeneous Markov-dependent components. Reliab. Eng. Syst. Saf. **167**, 1–9 (2017)
2. Qiu, S., Sallak, M., Schön, W., Ming, H.X.G.: Extended LK heuristics for the optimization of linear consecutive-k-out-of-n: F systems considering parametric uncertainty and model uncertainty. Reliab. Eng. Syst. Saf. **175**, 51–61 (2018)

3. Mohammadi, F., Sáenz-de-Cabezón, E., Wynn, H.P.: Efficient multi-cut enumeration of k-out-of-n: F and consecutive k-out-of-n: F systems. Pattern Recogn. Lett. **102**, 82–88 (2018)
4. Yuan, L., Cui, Z.D.: Reliability analysis for the consecutive-k-out-of-n: F system with repairmen taking multiple vacations. Appl. Math. Model. **37**(7), 4685–4697 (2013)
5. Salehi, E.T., Asadi, M., Eryılmaz, S.: Reliability analysis of consecutive k-out-of-n system with non-identical components life times. J. Stat. Plan. Infer. **141**(8), 2920–2932 (2011)
6. Yamamoto, H., Zuo, M.J., Akiba, T., Tian, Z.: Recursive formulas for the reliability of multi-state consecutive k-out-of-n: G systems. IEEE Trans. Reliab. **55**(1), 98–104 (2006)
7. Shen, J., Cui, L., Du, S.: Birnbaum importance for linear consecutive-k-out-of-n systems with sparse d. IEEE Trans. Reliab. **64**(1), 359–375 (2015)
8. Cui, L., Lin, C., Du, S.: m-consecutive-k, l-out-of-n system. IEEE Trans. Reliab. **64**(1), 386–393 (2015)
9. Lin, C., Cui, L., Coit, D.W., Lv, M.: Reliability modeling on consecutive-k(r)-out-of-n(r): F linear zigzag structure and circular polygon structure. IEEE Trans. Reliab. **65**(3), 1509–1521 (2016)
10. Dafnis, S.D., Makri, F.S., Philippou, A.N.: The reliability of a generalized consecutive system. Appl. Math. Comput. **359**, 186–193 (2019)
11. Cui, L.: The IFR property for consecutive-k-out-of-n: F systems. Stat. Probab. Lett. **59**(4), 405–414 (2002)
12. Eryilmaz, S.: Reliability properties of consecutive k-out-of-n systems of arbitrarily dependent components. Reliab. Eng. Syst. Saf. **94**(2), 350–356 (2009)
13. Mo, Y.C., Xing, L.D., Cui, L.R., Si, S.B.: MDD-based performability analysis of multi-state linear consecutive-k-out-of-n: F systems. Reliab. Eng. Syst. Saf. **166**, 124–131 (2017)
14. Dui, H.Y., Si, S.B., Yam, R.C.M.: Importance measures for optimal structure in linear consecutive-k-out-of-n systems. Reliab. Eng. Syst. Saf. **169**, 339–350 (2018)
15. Zuo, M.J., Liang, M.: Reliability of multistate consecutively-connected systems. Reliab. Eng. Syst. Saf. **44**(2), 173–176 (1994)
16. Peng, R., Xiao, H.: Reliability of linear consecutive-k-out-of-n systems with two change points. IEEE Trans. Reliab. **67**(3), 1019–1029 (2018)
17. Malinowski, J., Preuss, W.: Reliability a 2-way linear consecutively connected system with multistate components. Microelectron. Reliab. **36**(10), 1483–1488 (1996)
18. Levitin, G.: Optimal allocation of multistate elements in a linear consecutively-connected system. IEEE Trans. Reliab. **52**(2), 192–199 (2003)
19. Levitin, G., Yeh, W.C., Dai, Y.: Minimizing bypass transportation expenses in linear multistate consecutively-connected systems. IEEE Trans. Reliab. **63**(1), 230–238 (2014)
20. Levitin, G., Xing, L., Benhaim, H., Dai, Y.: M/NCCS: linear consecutively connected systems subject to combined gap constraints. Int. J. Gen. Syst. **44**, 833–848 (2015)
21. Yu, H., Yang, J., Peng, R., Zhao, Y.: Reliability evaluation of linear multi-state consecutively-connected systems constrained by m consecutive and n total gaps. Reliab. Eng. Syst. Saf. **150**, 35–43 (2014)
22. Yu, H., Yang, J., Peng, R., Zhao, Y.: Linear multi-state consecutively-connected systems constrained by m consecutive and n total gaps. Reliab. Eng. Syst. Saf. **150**, 35–43 (2016)
23. Peng, R., Xie, M., Ng, S.H., Levitin, G.: Element maintenance and allocation for linear consecutively connected systems. IIE Trans. **44**(11), 964–973 (2012)
24. Levitin, G., Ben-Haim, H.: Consecutive sliding window systems. Reliab. Eng. Syst. Saf. **96**(10), 1367–1374 (2011)
25. Xiang, Y.P., Levitin, G.: Combined m-consecutive and k-out-of-n sliding window systems. Eur. J. Oper. Res. **219**(1), 105–113 (2012)
26. Xiao, H., Peng, R., Wang, W.B., Zhao, F.: Optimal element loading for linear sliding window systems. J. Risk Reliab. **230**(1), 75–84 (2016)

# Component Importance-Driven Reliability Optimization for Linear Consecutive k out of n Systems

Hongyan Dui[1] and Shubin Si[2(✉)]

[1] School of Management Engineering,
Zhengzhou University, Zhengzhou, China
duihongyan@zzu.edu.cn
[2] School of Mechanical Engineering,
Northwestern Polytechnical University, Xi'an, China
sisb@nwpu.edu.cn

**Abstract.** Importance measures in reliability systems are used to identify weak single component or multiple components in contributing to proper functioning of the system. They can be used to analyze the structure optimization with component reliabilities for reconfigurable systems, and solve the system optimization problem for a kind of component assignment problem for the consecutive k out of n systems. In reconfigurable systems, the system structure can be changed by reordering the components. The optimal structure can be determined for any specific combination of system components. For the change of the reliability of specific components, the importance of the corresponding components and the optimal system structure can be found. The relationships between reliability allocation and the importance are studied by building the reliability of the optimal system structure as a function of component reliability. At last, a numerical example is used to demonstrate the proposed method.

**Keywords:** Reliability systems · Importance measures · Optimization · Structure

## 1 Introduction

Importance measures are widely used for identifying weaknesses components and supporting the system reliability improvement. Many works have evaluated the system reliability [1–3]. Importance measures were originally proposed by Birnbaum. Then a lot of importance measures have been introduced to apply in consecutive k-out-of-n systems [4–9]. The Birnbaum importance measures have also been used to solve the component assignment problem and study their optimal configuration for consecutive-k-out-of-n system in work [10–13]. Papastavridis [14] analyzed the computation of the Birnbaum importance in consecutive-k-out-of-n: F systems. Lin et al. [15] obtained the component ranks based on the structure importance. Zuo [16] analyzed the relationships of Birnbaum importance between consecutive-k-out-of-n: G and F systems. Kuo et al. [17] gave the relationships between the consecutive k-out-of-n: F and G systems

© Springer Nature Singapore Pte Ltd. 2019
Q.-L. Li et al. (Eds.): Cao Festschrift 2019, CCIS 1102, pp. 270–284, 2019.
https://doi.org/10.1007/978-981-15-0864-6_13

for the optimal configuration. Dui et al. [18, 19] analyzed the changes of different importance measures when the optimal system reliability is changed.

A consecutive-k-out-of-n system is composed of n ordered components, in which the system fails or works only when at least k consecutive components fail or work [20]. In real life, the linear consecutive-k-out-of-n systems have many applications in engineering area, such as oil pipeline, street-lighting systems, long-distance telecommunication systems, and relay stations [21]. In actual practice, the system optimal structure may change with some component reliabilities change. The problem of the structure optimization of a linear consecutive-k-out-of-n system is to assign the components to n positions on the line to maximize the reliability of the system [22]. The optimal permutation is to assign component reliabilities in a certain order to system components. An optimal assignment is called invariant if it depends only on the ordering of the component reliability, but not on their actual values. When the optimal arrangement is not invariant, a heuristic algorithm is used to determine the approximate optimal assignment.

In this paper, the possible changes of the optimal structure and the lifetime are considered with importance measures. These can describe the component importance changes when with the component orders changes in optimal systems.

The rest of the paper is organized as follows. Section 2 analyzes the importance measures for system optimization. A numerical example is used to demonstrate the system change based on the component importance values in Sect. 3. Section 4 gives the conclusions.

## 2 Importance Measures for System Optimization

In this section, we discuss consecutive $k$-out-of-$n$: F systems of performance level $K$. For a multi-state system of performance level $K$, the system (component) fails when the state of system (component) is less than $K$, and works when the state of system (component) is more than or equal to $K$. So the multi-state consecutive $k$-out-of-$n$: F system of performance level $K$ is equivalent to: $\Phi(X(t)) < K$ if and only if at least $k_K$ consecutive components have $X_i(t) < K$.

The Birnbaum importance of component $i$ in multi-state systems with the performance level $K$ is as in (1),

$$I(BM)_i^t = \Pr\{\Phi(X(t)) \geq K | X_i(t) \geq K\} - \Pr\{\Phi(X(t)) \geq K | X_i(t) < K\}, \qquad (1)$$

where $\Phi(X(t))$ is the system structure function, $X(t)$ state vector of the components $(X_1(t), X_2(t), \ldots, X_n(t))$ and $X_i(t)$ the state of component $i$, $X_i(t) = 0, 1, 2, \ldots, M$.

The integrated importance of component $i$ for multi-state system reliability of performance level $K$ is defined as in (2),

$$I(IIM)_i^t = \Pr\{X_i(t) \geq K\} \cdot \lambda_i(t) I(BM)_i^t, \qquad (2)$$

where $\lambda_i(t)$ is the failure rate of component $i$ at time $t$.

The Mean Absolute Deviation [23] evaluates the effect of all states of a component on the system reliability as in (3),

$$I(\text{MAD})_i^t = \sum_l P(X_i(t) = l) |P(\Phi(l_i, X(t)) \geq d) - P(\Phi(X(t)) \geq d)|, \qquad (3)$$

where $d$ represents the system constant demand.

For one multi-state system of $K$ level, the system constant demand $d$ represents $K$ level, and component $i$ has two classes of states, $X_i(t) \geq K$ or $X_i(t) < K$. Because $\text{Pr}\{\Phi(X(t)) \geq K\} = \text{Pr}(X_i(t) \geq K)\,\text{Pr}\{\Phi(X(t)) \geq K | X_i(t) \geq K\} + \text{Pr}(X_i(t) < K)\,\text{Pr}\{\Phi(X(t)) \geq K | X_i(t) < K\}$, we have $I(\text{MAD})_i^t = 2\,\text{Pr}(X_i(t) \geq K)\,\text{Pr}(X_i(t) < K)I(BM)_i^t$.

Then Mean Absolute Deviation for multi-state systems of performance level $K$ is as in (4),

$$I(\text{MAD})_i^t = 2\,\text{Pr}(X_i(t) \geq K)\,\text{Pr}(X_i(t) < K)I(BM)_i^t. \qquad (4)$$

Let $R^t(j)$ is the reliability of multi-state consecutive $k$-out-of-$j$: F subsystem consisting of components $1, 2, \ldots, j$ at time $t$, $R^{\prime t}(j)$ the reliability of multi-state consecutive $k$-out-of-$j$: F subsystem consisting of components $(n - j + 1), (n - j + 2), \ldots, (n - 1), n$.

Then the Birnbaum importance for the consecutive $k$-out-of-$n$: F system of performance level $K$ is as in (5),

$$I(BM)_i^t = \frac{\partial R^t(n)}{\partial P_{i,K}(t)} = \frac{R^t(i-1)R^{\prime t}(n-i) - R^t(n)}{1 - P_{i,K}(t)}, \qquad (5)$$

where $P_{i,K}(t) = \text{Pr}\{X_i(t) \geq K\}$.

The integrated importance for the consecutive $k$-out-of-$n$: F system in multi-state systems of performance level $K$ is as in (6),

$$I(IIM)_i^t = \text{Pr}\{X_i(t) \geq K\} \cdot \lambda_i(t)I(BM)_i^t = P_{i,K}(t) \cdot \lambda_i(t)\frac{R^t(i-1)R^{\prime t}(n-i) - R^t(n)}{1 - P_{i,K}(t)}. \qquad (6)$$

The Mean Absolute Deviation for the consecutive $k$-out-of-$n$: F system in multi-state systems of performance level $K$ is as in (7),

$$I(\text{MAD})_i^t = 2P_{i,K}(t)[R^t(i-1)R^{\prime t}(n-i) - R^t(n)]. \qquad (7)$$

If a component reliability changes, the optimal system structure may also change. Then, the importance of the corresponding component will change with the change of the optimal system structure. Suppose that $R^t_{opt}(i-1)$, $R'^t_{opt}(n-i)$, $R^t_{opt}(n)$ are the system or subsystem reliabilities obtained for the optimal configurations at time $t$. For the optimal structure of the consecutive $k$-out-of-$n$: F system, the Birnbaum importance of component $i$ is $I(BM)^t_i = \left(R^t_{opt}(i-1)R'^t_{opt}(n-i) - R^t_{opt}(n)\right)/\left(1 - P_{i,K}(t)\right)$, the integrated importance of component $i$ in consecutive k-out-of-n: F system can be represented that $I(IIM)^t_i = P_{i,K}(t) \cdot \lambda_i(t) \cdot \left(R^t_{opt}(i-1)R'^t_{opt}(n-i) - R^t_{opt}(n)\right)/\left(1 - P_{i,K}(t)\right)$, and Mean Absolute Deviation of component $i$ is $I(MAD)^t_i = 2P_{i,K}(t) \left[R^t_{opt}(i-1)R'^t_{opt}(n-i) - R^t_{opt}(n)\right]$.

## 3 Numerical Examples

In this section, component $n$ follows the two-parameter Weibull distribution, and other components follow exponential distribution. In exponential distributions, the failure rate of component $i$ is $\lambda_i = 0.2 + 0.2(i-1)$. For component $n$, the Weibull distribution is $F(t) = 1 - \exp[-(t/\beta)^\alpha]$, where $\beta$ and $\alpha$ are the scale and shape parameters, respectively, and $\alpha > 0$, $\beta > 0$. When $\alpha = 1$, the Weibull distribution becomes into the exponential distributions. Let $\alpha = 0.1$, $\alpha = 3$ to represent the decreasing and increasing failure rate. We use the consecutive-$k$-out-of-5: F and G systems to illustrate the value change of the proposed measures with the optimal system structure for $k = 3, 4$.

For consecutive-3-out-of-5: F and G systems, the reliability changes of system, target component (component 5), and other components (components 1, 2, 3, 4) are as in Fig. 1. If the failure rate is smaller, then the component reliability decreases slower. So the reliability curves from top to bottom are the ones of components 1, 2, 3, 4, respectively, in Fig. 1. When shape parameter is 0.1 or 3 for component 5, the reliability curve of component 5 intersects with that of components 1, 2, 3, 4.

Figures 2 and 3 show different component importance measures for consecutive-3-out-of-5: F and G systems when shape parameter is 0.1.

Figures 4 and 5 show different component importance measures for consecutive-3-out-of-5: F and G systems when shape parameter is 3.

Figures 6 and 7 show different component importance measures for consecutive-4-out-of-5: F and G systems when shape parameter is 0.1.

Figures 8 and 9 show different component importance measures for consecutive-4-out-of-5: F and G systems when shape parameter is 3.

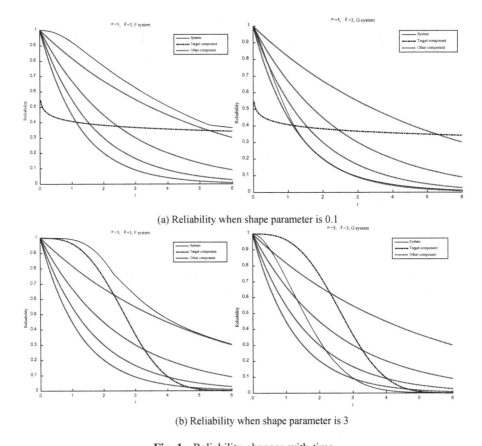

(a) Reliability when shape parameter is 0.1

(b) Reliability when shape parameter is 3

**Fig. 1.** Reliability changes with time

From Figs. 2 to 9, when different curves interact, the change of system optimal structure occurs. Besides, importance values of different components in optimal structure change with time. The original permutation may not be optimal after the change in the reliability of a specific component. The importance index of a specific component obtained with respect to the possible system reconfiguration indicates that the most promising component reliability change can be attained when the system structure can be easily modified. The optimal system structure varies with the change in component reliability. Thus, component importance is related to the optimal system structure. When the reliability curve of the target component intersects with one of the other components, the increase in the importance index corresponds to the change in the system's optimal configuration.

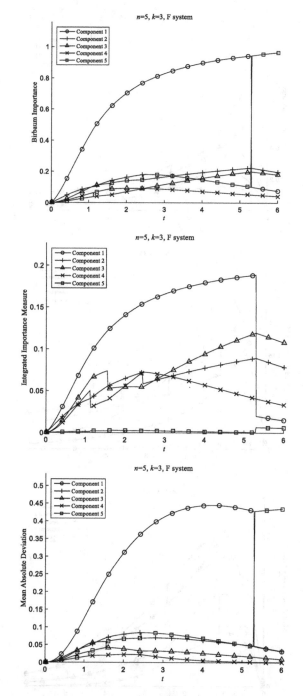

**Fig. 2.** All component importance values for consecutive-3-out-of-5: F systems when shape parameter is 0.1

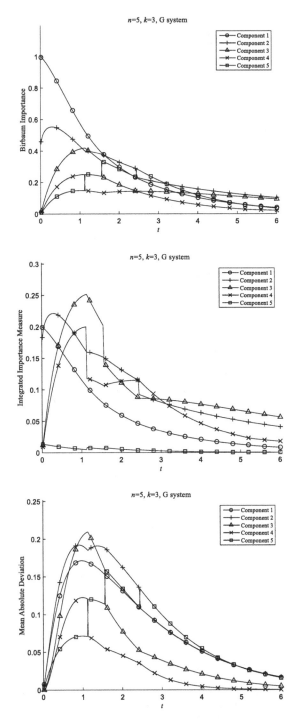

**Fig. 3.** All component importance values for consecutive-3-out-of-5: G systems when shape parameter is 0.1

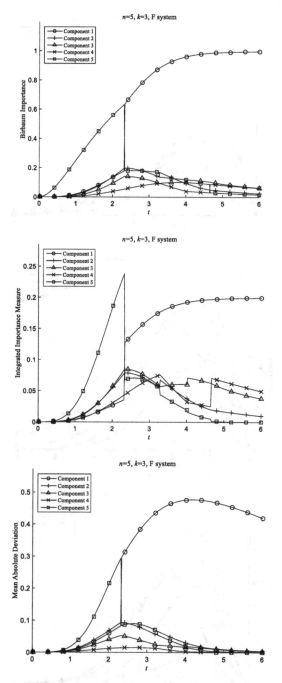

**Fig. 4.** All component importance values for consecutive-3-out-of-5: F systems when shape parameter is 3

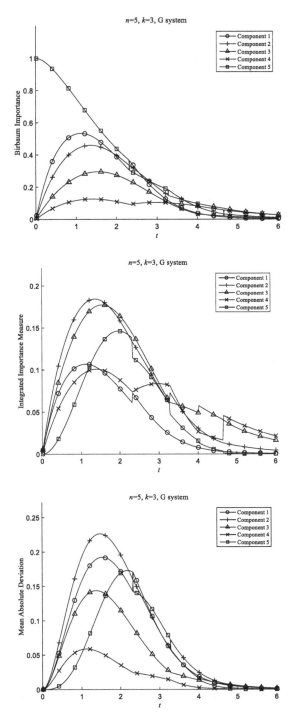

**Fig. 5.** All component importance values for consecutive-3-out-of-5: G systems when shape parameter is 3

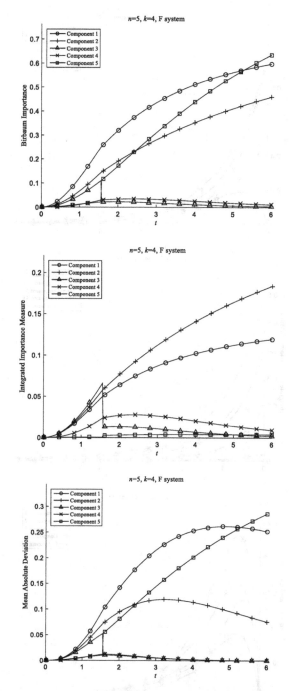

**Fig. 6.** All component importance values for consecutive-4-out-of-5: F systems when shape parameter is 0.1

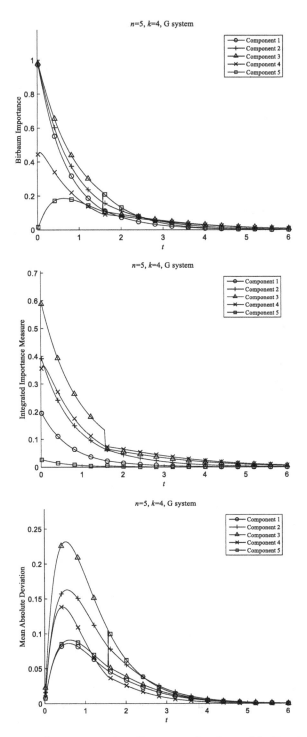

**Fig. 7.** All component importance values for consecutive-4-out-of-5: G systems when shape parameter is 0.1

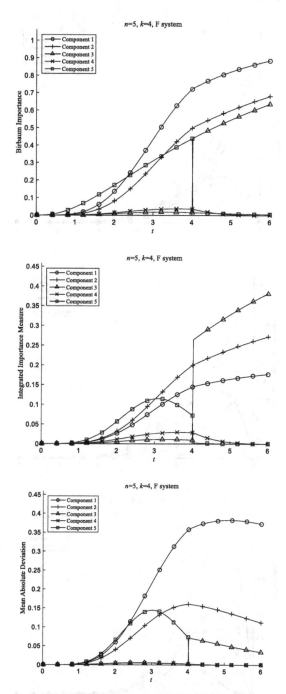

**Fig. 8.** All component importance values for consecutive-4-out-of-5: F systems when shape parameter is 3

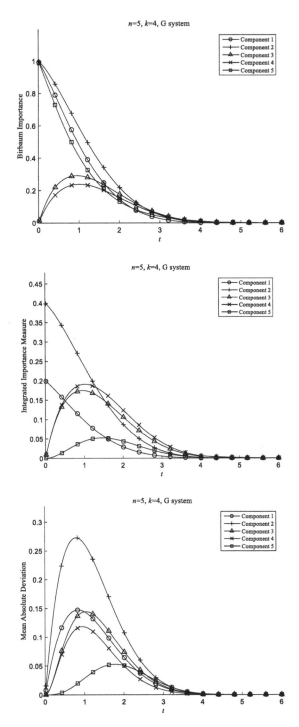

**Fig. 9.** All component importance values for consecutive-4-out-of-5: G systems when shape parameter is 3

## 4    Concluding Remarks

This paper studied the optimization methods of component importance in reliability systems. Based on the important measure, we analyze the relationships between the component reliability with the degradation characteristics and the optimal dispatching system to solve the reliability optimization problem of consecutive k out of n systems.

In future work, the system optimization involves minimizing the system cost subject to reliability requirement or maximizing the system reliability under resource constraints, such as cost, weight. It has attracted many researchers due to the high requirements of performance for various kinds of systems. Besides, the states of components and systems in research works could be classified into binary state system, multi-state system and continuous state system. We also study the improvement methods of system structure to increase component reliability, add redundant components, reassign interchangeable components and optimize the combinations of above methods.

**Acknowledgments.** The authors gratefully acknowledge the financial supports for this research from the National Natural Science Foundation of China (Nos. 71771186, 71501173).

## References

1. Cui, L., Li, H.: Analytical method for reliability and MTTF assessment of coherent systems with dependent components. Reliab. Eng. Syst. Saf. **92**(3), 300–307 (2007)
2. Cui, L., Xu, Y., Zhao, X.: Developments and applications of the finite Markov chain imbedding approach in reliability. IEEE Trans. Reliab. **59**(4), 685–690 (2010)
3. Zhao, X., Cui, L.: On the accelerated scan finite Markov chain imbedding approach. IEEE Trans. Reliab. **58**(2), 383–388 (2009)
4. Levitin, G., Podofillini, L., Zio, E.: Generalized importance measures for multi-state elements based on performance level restrictions. Reliab. Eng. Syst. Saf. **82**(3), 287–298 (2003)
5. Dui, H., Li, S., Xing, L., Liu, H.: System performance-based joint importance analysis guided maintenance for repairable systems. Reliab. Eng. Syst. Saf. **186**, 162–175 (2019)
6. Dui, H., Si, S., Yam, R.C.M.: A cost-based integrated importance measure of system components for preventive maintenance. Reliab. Eng. Syst. Saf. **168**, 98–104 (2017)
7. Dui, H., Si, S., Wu, S., Yam, R.C.M.: An importance measure for multistate systems with external factors. Reliab. Eng. Syst. Saf. **167**, 49–57 (2017)
8. Dui, H., Chen, L., Wu, S.: Generalized integrated importance measure for system performance evaluation: application to a propeller plane system. Eksploatacja i Niezawodnosc-Maintenance and Reliability **19**(2), 279–286 (2017)
9. Dui, H., Si, S., Zuo, M., Sun, S.: Semi-Markov process-based integrated importance measure for multi-state systems. IEEE Trans. Reliab. **64**(2), 754–765 (2015)
10. Zio, E., Podofillini, L.: Monte-Carlo simulation analysis of the effects on different system performance levels on the importance on multi-state components. Reliab. Eng. Syst. Saf. **82**(1), 63–73 (2003)
11. Kuo, W., Zhu, X.: Importance Measures in Reliability, Risk, and Optimization: Principles and Applications. Wiley, UK (2012)

12. Kuo, W., Zhu, X.: Some recent advances on importance measures in reliability. IEEE Trans. Reliab. **61**(2), 344–360 (2012)
13. Wu, S., Coolen, F.P.A.: A cost-based importance measure for system components: an extension of the Birnbaum importance. Eur. J. Oper. Res. **225**(1), 189–195 (2013)
14. Papastavridis, S.: The most important component in a consecutive-k-out-of-n: F system. IEEE Trans. Reliab. **36**(2), 266–268 (1987)
15. Lin, F.H., Kuo, W., Hwang, F.K.: Structure importance of consecutive-k-out-of-n systems. Oper. Res. Lett. **25**(2), 101–107 (1999)
16. Zuo, M.J.: Reliability and component importance of a consecutive-k-out-of-n system. Microelectron. Reliab. **33**(2), 243–258 (1993)
17. Kuo, W., Zhang, W.X., Zuo, M.: A consecutive-k-out-of-n: G system: the mirror image of a consecutive-k-out-of-n: F system. IEEE Trans. Reliab. **39**(2), 244–253 (1990)
18. Dui, H., Si, S., Yam, R.C.M.: Importance measures for optimal structure in linear consecutive-k-out-of-n systems. Reliab. Eng. Syst. Saf. **169**, 339–350 (2018)
19. Si, S., Levitin, G., Dui, H., Sun, S.: Importance analysis for reconfigurable systems. Reliab. Eng. Syst. Saf. **126**, 72–80 (2014)
20. Zuo, M.J., Kuo, W.: Design and performance analysis of consecutive-k-out-of-n structure. Naval Res. Logistics **37**, 203–230 (1990)
21. Levitin, G.: Optimal allocation of multistate elements in a linear consecutively-connected system. IEEE Trans. Reliab. **52**, 192–199 (2003)
22. Cui, L.R., Hawkes, A.G.: A note on the proof for the optimal consecutive-k-out-of-n: G line for n ≦ 2k. J. Stat. Plan. Infer. **138**, 1516–1520 (2008)
23. Ramirez-Marquez, J.E., Coit, D.W.: Composite importance measures for multi-state systems with multi-state components. IEEE Trans. Reliab. **54**(3), 517–529 (2005)

# On the Equivalence of the Coefficient of Variation Ordering and the Lorenz Ordering Within Two-Parameter Families

Yugu Xiao[1(✉)] and Jing Yao[2]

[1] Center for Applied Statistics, School of Statistics, Renmin University of China, Beijing, People's Republic of China
yuguxiao@ruc.edu.cn
[2] Maxwell Institute for Mathematical Sciences and School of Mathematical and Computer Sciences, Heriot-Watt University, Edinburgh EH14 4AS, UK
j.yao@hw.ac.uk

**Abstract.** It is well-known that the Lorenz ordering, which is widely used to rank the inequality of income, will lead to the ordering of coefficient of variation. This paper finds that these two stochastic orders are equivalent within several common two-parameter families of distributions including the location-scale family, some scale and shape parameter family. Our finding manifests that once the compared life distributions or income distributions belong to a two-parameter family discussed above, rankings by the Lorenz curve and by the coefficient of variation for inequality generate the same order. Furthermore, a simple general sufficient condition without limiting within two-parameter families for this property is provided. These results could extend application of coefficient of variation, which can be regarded as a proxy of Lorenz curve in many cases for an inequality ranking or orderings of life distributions, even if the life has asymmetric heavy-tail distribution.

**Keywords:** Lorenz order · Coefficient of variation · Location-scale family · Scale and shape parameter family

## 1 Introduction

The Lorenz curve, which is a very popular and powerful tool to measure the inequality of income, has been extensively studied in the literature. See, e.g., [1–3]. To make the compared income distributions sortable in the Lorenz sense, various conditions have already been introduced in many studies such as [4–7].

Another relatively simple and fairly common inequality measure is the coefficient of variation (CV), which is formally defined by $CV_X = \sigma_X/\mu_X$ for a random variable $X$ with the mean $\mu_X$ and the standard deviation $\sigma_X$. It is well known that the Lorenz order is a very strict order and implies the inequality ordering under many indices, including the CV, the Gini index [8], the Pietra index [9] and the Amato index [10]. It is noteworthy that they are not equivalent

© Springer Nature Singapore Pte Ltd. 2019
Q.-L. Li et al. (Eds.): Cao Festschrift 2019, CCIS 1102, pp. 285–294, 2019.
https://doi.org/10.1007/978-981-15-0864-6_14

in general. More specifically, as pointed out by [11], the CV order can not imply the Lorenz order.

However, the consistency of the mean-standard deviation decision and the expected utility maximization is established by [12] and [13] under the assumption that the compared risks belong to the location-scale (LS) family. Within the LS family of distributions, [14] also show that any reward-to-risk performance measure with positive homogeneity and functional translation invariance is an increasing function of the Sharpe ratio. From these successful attempts, it seems workable to investigate the equivalence of the CV order and the Lorenz order by restricting the compared risks to some family of distributions. However, to the best of our knowledge, few papers have been devoted to this interesting problem. It is thus the objective of this paper to shed some light on this topic.

In this paper, we find that the CV order and the Lorenz order are indeed equivalent within a large number of two-parameter families of distributions for which the CV exists. The first family is the LS family including many widely used distributions such as the exponential, extreme value distribution of type I, Laplace, logistic, half-logistic, normal, half-normal, and the uniform distributions. We refer to [15] for more properties of LS distributions. The second distribution family we consider is the scale and shape parameter family, which includes the scale-power parameter family and the scale-convolution parameter family [5]. This family is quite large and encompasses the gamma distribution, lognormal distribution, Pareto III (also known as Fisk distribution), the Weibull distribution and so on. In addition to these two families of distributions, a general sufficient condition for this property is provided, inverse Weibull distribution and inverse Gaussian distribution given as an example.

## 2   Equivalence of the CV Order and the Lorenz Order

First, we provide the definitions of Lorenz curve and Lorenz order.

**Definition 1 (Lorenz curve and Lorenz order).** *The Lorenz curve of a random variable $X$ with cumulative distribution function(c.d.f.) $F_X(\cdot)$ and a positive finite mean $\mu_X$ is given by*

$$L_X(p) = \frac{\int_0^p F_X^{-1}(s)ds}{\mu_X}, \quad p \in [0,1], \tag{1}$$

*where $F_X^{-1}(s) = sup\{x : F_X(x) \le s\}$ for $0 < s < 1$ is the right-continuous inverse function of $F_X(\cdot)$.*

*Based upon the Lorenz curve, $X$ is said to be smaller than another random variable $Y$ with a positive finite mean in the Lorenz order(denoted $X \le_{Lorenz} Y$) if $L_X(p) \ge L_Y(p)$ for all $p \in (0,1)$.*

[7] points out that $X \le_{Lorenz} Y$ if and only if $X/\mu_X \le_{cx} Y/\mu_Y$, where the convex order is formally defined below:

**Definition 2 (Convex order).** *The random variable $X$ is said to be smaller than $Y$ in the convex order (denoted by $X \leq_{cx} Y$) if $E[\phi(X)] \leq E[\phi(Y)]$ holds for all convex functions $\phi(\cdot)$ provided that the expectations exist.*

Based upon the above definition of convex order, now we have

$$X \leq_{Lorenz} Y \Longleftrightarrow X/\mu_X \leq_{cx} Y/\mu_Y \Longrightarrow CV_X \leq CV_Y, \tag{2}$$

assuming that $X$ and $Y$ have a finite variance. It is noteworthy that the reverse is generally incorrect as emphasized by [11]. However, if the compared risks are restricted to follow some two-parameter distributions, the CV order can imply the Lorenz order, as discussed below.

## 2.1   The LS Family of Distributions

In this subsection, we assume that the compared risks satisfy the LS property defined below:

**Definition 3 (Location-scale family).** *A random variable $X$ belongs to the LS family if its c.d.f. is in the form of*

$$F_X(x|a,b) = H\left(\frac{x-a}{b}\right), \quad x \in R, \ a \in R, \ b > 0,$$

*where $a$ is a location parameter, $b$ is a scale parameter, and $H(\cdot)$ is a c.d.f. independent of $a$ and $b$. [15].*

It is necessary to point out that the location parameter and the scale parameter need not be the mean and standard deviation of a random variable, respectively. Furthermore, the LS distribution may have infinite mean or variance, and Cauchy distribution is a special example. The LS family of distributions is quite a large family and encompasses many widely used statistical distributions listed in Sect. 1.2 of [15].

**Theorem 1.** *Suppose that a random variable $X$ with positive finite mean and variance belongs to the LS family. Then the Lorenz curve $L_X(p)$ of $X$ is decreasing in $CV_X$ for all $p \in (0,1)$.*

*Proof.* Note that the inverse function $F_X^{-1}(\cdot)$ is positively homogeneous and translation invariant. We have

$$L_X(p) = \frac{\int_0^p F_X^{-1}(s)ds}{\mu_X} = \frac{\int_0^p (\sigma_X F_Z^{-1}(s) + \mu_X)ds}{\mu_X} = CV_X \times \int_0^p F_Z^{-1}(s)ds + p, \quad p \in (0,1), \tag{3}$$

where $Z = \dfrac{X - \mu_X}{\sigma_X}$.

Since it is assumed that $X$ follows the LS distribution with a location parameter $a$ and a scale parameter $b$, we let $W = (X - a)/b$ such that it has the c.d.f. $H(\cdot)$. With the help of $W$, the random variable $Z$ could be rewritten by

$$Z = \frac{X - (a + b\mu_W)}{b\sigma_W} = \frac{W}{\sigma_W} - \frac{\mu_W}{\sigma_W},$$

which in turn implies that the distribution of $Z$ relies only upon the c.d.f. $H(\cdot)$.

The residual task is to show $\int_0^p F_Z^{-1}(s)ds \leq 0$ for all $p \in (0,1)$. More specifically, noting that $E[Z] = 0$ and that $F_Z^{-1}(\cdot)$ is a right-continuous increasing function, the result is trivial if $F_Z^{-1}(p) \leq 0$. Otherwise, if $F_Z^{-1}(p) > 0$, we have

$$\int_0^p F_Z^{-1}(s)ds = \int_0^1 F_Z^{-1}(s)ds - \int_p^1 F_Z^{-1}(s)ds = E[Z] - \int_p^1 F_Z^{-1}(s)ds < 0.$$

As a consequence, it follows from (3) that the Lorenz curve $L_X(p)$ is decreasing in $CV_X$ for each $p \in (0,1)$. The proof is finally completed.

An explanation on how Theorem 1 implies the equivalence between the CV order and the Lorenz order can be provided. Actually, the decreasing property of $L_X(p)$ with respect to $CV_X$ is not enough to the equivalence. However, the formula (3) shows that $L_X(p)$ linearly depends on $CV_X$ through a coefficient determined by the baseline distribution. This property is the reason why the CV order and the Lorenz order are equivalent.

While LS family is quite large, it is impossible to contain all the two-parameter distributions. Thus, we proceed to analyze this equivalence over another fairly common class of distributions in the next subsection.

## 2.2   The Scale and Shape Parameter Family of Distributions

In this subsection, we restrict the compared risks to the scale and shape parameter family which includes the scale-power parameter family and the scale-convolution parameter family as special cases. We define the scale and shape parameter family following [16].

**Definition 4 (Scale and shape parameter family).** *Suppose that in terms of the distribution function $F(\cdot|\beta)$, a two-parameter distribution function $F(\cdot|\alpha, \beta)$ is defined as*

$$F(x|\alpha, \beta) = F(\alpha x|\beta),$$

*where $\alpha > 0$ is a scale parameter and $\beta \in I$ is a shape parameter. Then $\{F(\cdot|\alpha, \beta) : \alpha > 0, \beta \in I\}$ is called a scale and shape parameter family with the underlying distribution $F(\cdot|\beta)$.*

**Theorem 2.** *Suppose that $\{F(\cdot|\alpha, \beta) : \alpha > 0, \beta \in I\}$ is a scale and shape parameter family of distributions with $F(0|\alpha, \beta) = 0$ and a finite variance. If $F(\cdot|\beta)$ is monotonic in the shape parameter $\beta$ in the sense of Lorenz order, then the CV order implies the Lorenz order.*

*Proof.* [16] claims that the CV of a random variable belonging to the scale and shape parameter family is a function of the shape parameter $\beta$ and irrelevant with the scale parameter $\alpha$. If $F(\cdot|\beta)$ is monotonic in $\beta$ in the sense of Lorenz order, then Eq. (2) implies that the CV must be a monotonic function of $\beta$. If the CV is strictly monotonic in $\beta$, then it follows from Eq. (2) that the CV order is consistent with the Lorenz order.

Otherwise, if the CV of $F(\cdot|\beta_1)$ is equal to that of $F(\cdot|\beta_2)$ with $\beta_1 < \beta_2$, then we can show that the Lorenz curves are equal for these two distributions. More specifically, we denote by $Y_1$ and $Y_2$ the random variables corresponding to these two distributions. Since it is assumed that the Lorenz curve is monotonic in $\beta$, we can see from (2) that $Y_1/\mu_{Y_1}$ and $Y_2/\mu_{Y_2}$ are comparable in convex order. Noting that $CV_{Y_1} = CV_{Y_2}$, it follows from Theorem 3.A.42 in [6] that $Y_1/\mu_{Y_1}$ and $Y_2/\mu_{Y_2}$ are equal in distribution. Noting that $L_Y(p) = L_{tY}(p)$ for all $p \in (0,1)$ and $t > 0$, we get that $Y_1$ and $Y_2$ are equal in the Lorenz order. The proof is thus completed.

In the above theorem, the assumption of the monotonicity of $F(\cdot|\beta)$ with respect to the shape parameter in the sense of Lorenz order at first seems very strict and uneasy to be verified. Exactly, this assumption holds true for many scale-shape distributions such as the scale-power parameter family and the scale-convolution parameter family which are defined below.

**Definition 5 (Scale-power parameter family and scale-convolution parameter family).** *For a scale and shape parameter family $\{F(\cdot|\alpha, \beta) : \alpha > 0, \beta \in I, F(0|\alpha, \beta) = 0\}$ with the underlying distribution $F(\cdot|\beta)$, if the underlying distribution is in the form of*

$$F(x|\beta) = F(x^\beta), \ \beta > 0,$$

*then this family and $\beta$ are called the scale-power parameter family and the power parameter, respectively.*

*On the other hand, if the index set $I$ is closed under addition and the underlying distribution $F(\cdot|\beta)$ satisfies the following property*

$$F(\cdot|\beta_1 + \beta_2) = F(\cdot|\beta_1) * F(\cdot|\beta_2) \text{ for all } \beta_1, \beta_2 \in I,$$

*then $\{F(\cdot|\alpha, \beta) : \alpha > 0, \beta \in I, F(0|\alpha, \beta) = 0\}$ and the shape parameter $\beta$ are said to be a scale-convolution parameter family and the convolution parameter, respectively.*

The above assumptions of the underlying distribution $F(\cdot|\beta)$ are weak such that these two subclasses of distributions include many commonly used distribution functions. Just to name a few, the lognormal, Pareto III and Weibull distributions belong to the scale-power parameter family, and the scale-convolution parameter family includes the Gamma distribution. For more detailed discussion of this two subclasses of distributions, we refer to [5].

**Corollary 1.** *Suppose that $\{F(\cdot|\alpha, \beta) : \alpha > 0, \beta \in I, F(0|\alpha, \beta) = 0\}$ is a scale and shape parameter family of distributions with a finite CV. If it is a scale-power parameter family or a scale-convolution parameter family, then the CV order implies Lorenz order.*

*Proof.* It follows from Proposition 7.D.5 and Proposition 7.J.6. in [5] that the Lorenz order is monotonic in the shape parameter $\beta$ if it is a power parameter or a convolution parameter. The final result can be easily derived by using Theorem 2.

## 2.3   A Sufficient Condition for the Equivalence Property

We have shown that the CV order and the Lorenz order are equivalent when the compared risks are restricted to the LS family or some scale and shape parameter family. It is noteworthy that the assumption of these two families is sufficient but not necessary to investigate the equivalence of these two stochastic orders. To extend this, a simple general sufficient condition without limiting within two-parameter families for the equivalence property is provided.

**Theorem 3.** *Assume that nonnegative random variables $X$ and $Y$ have distribution functions $F_X$ and $F_Y$, expectations $\mu_X$ and $\mu_Y$, and $CV_X$ and $CV_Y$, respectively. If $F_X(\mu_X x)$ crosses $F_Y(\mu_Y x)$ at most once, then $CV_X \leq CV_Y$ implies $X \leq_{Lorenz} Y$.*

*Proof.* Let $R = X/\mu_X$ and $V = Y/\mu_Y$ with distributions $F_R$ and $F_V$. Noting that $E(R) = E(V) = 1$, the Karlin-Novikoff cut criterion will imply that $R$ and $V$ are comparable in convex order, then it follows from (2) that $X$ and $Y$ can be ranked in the Lorenz order. Thus, we only have to show the CV order can imply the convex order by taking the advantage of the up-crossing property of distribution functions.

In fact, $F_R$ crosses $F_V$ or $F_X(\mu_X x)$ crosses $F_Y(\mu_Y x)$ exactly once, denoting the crossing point as $x_0$. Since

$$E(R) - E(V) = \int_0^{+\infty} (\overline{F}_R(x) - \overline{F}_V(x))dx = 0,$$

where $\overline{F}$ denotes survival function, we have

$$(CV_X)^2 - (CV_Y)^2 = Var(R) - Var(V) = E(R^2) - E(V^2)$$

$$= 2\int_0^{+\infty} x(\overline{F}_R(x) - \overline{F}_V(x))dx$$

$$= 2\int_0^{+\infty} (x - x_0)(\overline{F}_R(x) - \overline{F}_V(x))dx \leq 0. \tag{4}$$

It follows that $\overline{F}_R$ crosses $\overline{F}_V$ from above since the two factors of the integrand in Eq. (4) must have opposite signs for all $x$. Applying Theorem 3.A.44. in [6], $R$ is less than $V$ in convex order. The proof is thus completed.

The key function of this condition is a handy to avoid identifying which families the compared life distributions belong to.

*Example 1 (Inverse Weibull distribution).* Considering the inverse Weibull distribution, also known as Fréchet distribution, i.e. the distribution function

$$F_X(x) = exp\{-(x/\alpha)^{-\beta}\}, \quad \alpha > 0, \quad \beta > 0, \quad x > 0.$$

Its expectation is $\alpha\Gamma(1 - \frac{1}{\beta})$ for $\beta > 1$, and its variance is $\alpha^2(\Gamma(1 - \frac{2}{\beta}) - (\Gamma(1 - \frac{1}{\beta}))^2)$ for $\beta > 2$.

Assume that nonnegative random variables $X$ and $Y$ have inverse Weibull distributions $F_X$ and $F_Y$ with parameters $(\alpha_1, \beta_1)$ and $(\alpha_2, \beta_2)$, $\beta_i > 2, i = 1, 2$, respectively. Consider $F_X(\mu_X x) = F_Y(\mu_Y x)$, i.e.

$$exp\{-(\Gamma(1 - \frac{1}{\beta_1})x)^{-\beta_1}\} = exp\{-(\Gamma(1 - \frac{1}{\beta_2})x)^{-\beta_2}\}, \quad \beta_1, \; \beta_2 > 2, \quad x > 0.$$

Then $F_X(\mu_X x) - F_Y(\mu_Y x) = 0$ has one solution $x = exp\{\frac{\beta_2 A_2 - \beta_1 A_1}{\beta_1 - \beta_2}\}$ for $\beta_1 \neq \beta_2$ in $(0, \infty)$, where $A_i = ln(\Gamma(1 - \frac{1}{\beta_i}))$, $i = 1, 2$. It follows that $CV_X \leq (\geq)CV_Y$ implies $X \leq_{Lorenz} (\geq_{Lorenz})Y$ by Theorem 3. For the special case $\beta_1 = \beta_2$, i.e. $F_X(\mu_X x) = F_Y(\mu_Y x)$ for all $x$, and hence $X$ and $Y$ are equal in the Lorenz order.

A specific example for the distribution function and the Lorenz curve of Inverse Weibull distribution is shown in Fig. 1.

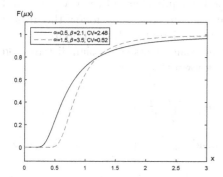 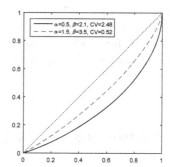

**Fig. 1.** A specific example for the distribution function (left figure) and the Lorenz curve (right figure) of Inverse Weibull distribution

*Example 2 (Inverse Gaussian distribution).* The inverse Gaussian distribution, which is also known as the Wald distribution, is a two-parameter family of continuous probability distributions with the density function

$$f_X(x) = \frac{\sqrt{\theta}}{\sqrt{2\pi x^3}} exp\left\{-\frac{\theta(x - m)^2}{2m^2 x}\right\}, \quad \theta, m, x > 0.$$

Its c.d.f. can be given by

$$F_X(x|m, \theta) = \Phi\left(\sqrt{\frac{\theta}{x}}\left(\frac{x}{m} - 1\right)\right) + exp\left\{\frac{2\theta}{m}\right\}\Phi\left(-\sqrt{\frac{\theta}{x}}\left(\frac{x}{m} + 1\right)\right),$$

where $\Phi(\cdot)$ is the c.d.f. of standard normal distribution.

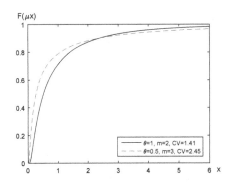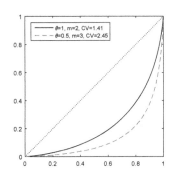

**Fig. 2.** A specific example for the distribution function (left figure) and the Lorenz curve (right figure) of Inverse Gaussian distribution

Obviously, it belongs neither to the LS family nor the scale and shape parameter family. However, we can help establish the equivalence of the CV order and the Lorenz order by Theorem 3. That is if the compared risks follow the inverse Gaussian distribution, then the CV order is consistent with the Lorenz order.

In fact, for random variables $Y_i$ that follow the inverse Gaussian distribution with parameters $(\theta_i, m_i)$ for $i = 1, 2$, it is known that

$$\mu_{Y_i} = m_i \quad \text{and} \quad \sigma_{Y_i} = \sqrt{\frac{m_i^3}{\theta_i}},$$

which in turn imply $CV_{Y_i} = \sqrt{\frac{m_i}{\theta_i}}$. Now consider $\mu_{Y_1} f_{Y_1}(\mu_{Y_1} x) = \mu_{Y_2} f_{Y_2}(\mu_{Y_2} x)$, that is,

$$\frac{\sqrt{\theta_1}}{\sqrt{2\pi m_1 x^3}} exp\left\{-\frac{\theta_1(x-1)^2}{2m_1 x}\right\} = \frac{\sqrt{\theta_2}}{\sqrt{2\pi m_2 x^3}} exp\left\{-\frac{\theta_2(x-1)^2}{2m_2 x}\right\} \tag{5}$$

for $\theta_1, \theta_2, m_1, m_2, x > 0$, which is equivalent to

$$\left(\frac{1}{CV_{Y_2}^2} - \frac{1}{CV_{Y_1}^2}\right)(x-1)^2 + 2ln\left(\frac{CV_{Y_2}}{CV_{Y_1}}\right)(x-1) + 2ln\left(\frac{CV_{Y_2}}{CV_{Y_1}}\right) = 0.$$

It is trivial that the above equation has at most two real solutions, and hence the distribution functions $F_{\frac{Y_i}{\mu_{Y_i}}}(\cdot|m_i, \theta_i)$ for $i = 1, 2$ intersect at most once. By Theorem 3 again, that $Y_1$ and $Y_2$ can be ranked in the Lorenz order.

And if $CV_{Y_1} = CV_{Y_2}$, we can see from (5) that $\{Y_i/\mu_{Y_i}\}_{i=1}$ are equal in distribution and hence $Y_i$ for $i = 1, 2$ are equal in the Lorenz order.

A specific example for the distribution function and the Lorenz curve of Inverse Gaussian distribution is shown in Fig. 2.

## 3  Conclusion

It is well known that the Lorenz order is a very strict order and implies the inequality ordering under many indices, and the CV order can not imply the

Lorenz order in general. However, this paper shows that these two stochastic orders are equivalent if the compared risks are restricted to some two-parameter family of distributions. Due to the easy calculation of CV, our result can help ease the difficulty in ranking the income inequality once the income distributions belong to the two-parameter family considered in this paper.

In the literature, the CV measure is often criticized for not incorporating the higher moments especially of the risks with asymmetric heavy tails, and hence is usually replaced by Gini index [8] to measure the inequality. It is noteworthy that the ordering under Gini index can also be implied by the Lorenz order. As a consequence, our result manifests that Gini index and the CV measure exactly generate the identical ranking order when the compared risks belong to some special two-parameter family.

**Acknowledgments.** The authors would like to thank Professor Yichun Chi at the Central University of Finance and Economics, Professor Qihe Tang at the University of Iowa and Professor Xiaojun Shi at the Remnin University of China for helpful comments. The first author was supported by the Major Project of the National Social Science Fund of China (16ZDA052), and by the MOE(China) National Key Research Bases for Humanities and Social Sciences (16JJD910001).

# References

1. Atkinson, A.B.: On the measurement of inequality. J. Econ. Theory. **5**, 244–263 (1970)
2. Rothschild, M., Stiglitz, J.E.: Some further results on the measurement of in equality. J. Econ. Theory. **6**, 188–204 (1973)
3. Cowell, F.: Measuring Inequality, 3rd edn. Oxford University Press, New York (2011)
4. Foster, J.E., Shorrocks, A.F.: Inequality and poverty orderings. Eur. Econ. Rev. **32**, 654–661 (1988)
5. Marshall, A.W., Olkin, I.: Life Distributions. Springer, New York (2007). https://doi.org/10.1007/978-0-387-68477-2
6. Shaked, M., Shanthikumar, J.G.: Stochastic Orders. Springer, New York (2007). https://doi.org/10.1007/978-0-387-34675-5
7. Marshall, A.W., Olkin, I., Arnold, B.: Inequalities: Theory of Majorization and Its Applications, 2nd edn. Springer, New York (2010). https://doi.org/10.1007/978-0-387-68276-1
8. Gini, C.: Variabilità e mutabilità. Studi Economico-Giuridici dell'Università di Cagliari, vol. 3, pp. 1–158 (1912)
9. Pietra, G.: Delle relazioni fra indici di variabilita, Note I e II. Atti del Reale Instituto Veneto di Scienze, Lettere ed Arti, vol. 74, pp. 775–804 (1915)
10. Amato, V.: Metodologia Statistica Strutturale. Cacucci, Bari (1968)
11. Arnold, B.C.: Majorization and the Lorenz Order: A Brief Introduction. Lecture Notes in Statistics, vol. 43. Springer, Heidelberg (1987). https://doi.org/10.1007/978-1-4615-7379-1
12. Meyer, J.: Two-moment decision models and expected utility maximization. Am. Econ. Rev. **77**, 421–430 (1987)

13. Levy, H.: Two-moment decision models and expected utility maximization: comment. Am. Econ. Rev. **79**, 597–600 (1989)
14. Schuhmacher, F., Eling, M.: A decision-theoretic foundation for reward-to-risk performance measures. J. Bank. Financ. **36**, 2077–2082 (2012)
15. Rinne, H.: Location-Scale Distributions. In: Lovric, M. (ed.) International Encyclopedia of Statistical Science, pp. 752–754. Springer, Heidelberg (2011). https://doi.org/10.1007/978-3-642-04898-2
16. Newby, M.J.: The properties of moment estimators for the Weibull distribution based on the sample coefficient of variation. Technometrics **22**, 187–194 (1980)

# Statistical Inference of the Competing Risks Model with Modified Weibull Distribution Under Adaptive Type-II Progressive Hybrid Censoring

Wang Yan[1,2(✉)] and Shi Yimin[1]

[1] Northwestern Polytechnical University, Xi'an 710072, China
wywzyf@126.com
[2] Xi'an Polytechnic University, Xi'an 710048, China

**Abstract.** This paper considers the statistical inference of competing risks model with modified Weibull distribution which not only covers the increasing and decreasing hazard rate function, and also represents a bathtub-shaped hazard rate behavior. Based on the adaptive Type-II progressive hybrid censored data, the maximum likelihood estimations of the unknown parameters are obtained, and then the Bayes approach, combined Gibbs sampling method, is also considered with gamma priors of the scale parameters and vague priors for the shape parameters. Finally, two data sets with a real data set and a Monte Carlo simulate data set are analyzed to investigate the performance of the purposed methods.

**Keywords:** Competing risks model · Modified Weibull distribution · Adaptive Type-II progressive hybrid censoring · Maximum likelihood estimator · Bayes estimator · Gibbs sample

## 1 Introduction

In reliability and survival analysis, many units have the "bathtub-shaped" hazard rates. This hazard rate initially decreases because of the burn-in effect, and then it reaches a stable value and finally the rate increases as the units wear out or aging. Common life distributions, including Exponential, Weibull, Gamma and ect, generally do not have a bathtub-shaped hazard rate and are not suitable to model this life data. So, recent years many models by generalizing or extending the Weibull distribution have been proposed to deal with this problem. For example, the exponentiated Weibull distribution (EW) distribution [1], modified Weibull (MW) distribution [2], extended flexible Weibull distribution [3], generalized modified Weibull distribution [4], the beta modified Weibull distribution [5] and Kumaraswamy modified Weibull distribution [6]. Among them, the three parameters MW distribution, due to the less parameters and closed cumulative distribution function (CDF), has been considered by many scholars. Jiang et al. [7] estimated the parameters of MW distribution by using maximum-likelihood method and Markov chain Monte Carlo method based on complete data. Updhyay and Gupta [8] considered Bayse estimations (BEs) of the parameters for MW

© Springer Nature Singapore Pte Ltd. 2019
Q.-L. Li et al. (Eds.): Cao Festschrift 2019, CCIS 1102, pp. 295–312, 2019.
https://doi.org/10.1007/978-981-15-0864-6_15

distribution. Jiang et al. [9] derived maximum-likelihood estimations (MLEs) of the parameters based on complete data and Type-II censored data as well as proved the existence and uniqueness of the MLEs. Peng and Yan [10] acquired the MLEs and BEs of the MW distribution under Type-II progressive censored scheme. Vidović [11] obtained the MLEs of MW distribution in the presence of upper $k$th record values and confirmed the existence and uniqueness of MLEs. EL-Sagheer et al. [12] got the point and interval estimators of MW distribution in constant-stress partially accelerated life test based on Type-II progressive censored sample.

Due to the complexity of the external environment and internal structure, the failure of unit would be caused by more than one failure causes, and every cause possible leads to the final failure of the unit. So, the failure of unit can be seen as the competing result of failure causes. The competing risks model occurres in many application fields such as life test, engineering, medical statistics and reliability. The statistical analysis for competing risks model has been considered by many authors. For example, Bhattacharya et al. [13] considered the competing risk model with Weibull distribution. Anmadi et al. [14] analyzed competing risks model with exponential distribution under middle censored. Feizjavadian and Hashemi [15] studied the dependent competing risks by using Marshall-Olkin bivariate Weibull (MOBW) distribution in progressively hybrid censoring. Wu et al. [16] discussed the dependent competing risks by using copula function under progressively hybrid censoring in accelerated life test.

Censoring is a very common phenomenon in life test and many authors made statistical inference of the life test under different censorings. The most common censoring schemes are Type-I, Type-II, progressive censoring scheme (PCS) and progressively hybrid censoring scheme (PHCS) and the details of these censorings can be seen in Epstein [17], Childs et al. [18], Kundu and Joarder [19] and Childs et al. [20]. Compared to Type-I, Type-II, PCS, the PHCS provides the flexibility to remove surviving testing units and terminate the life test. However, as referred in [20], the PHCS also has the shortcomings. In the Type-I PHCS, the effective sample size is random and it may be a very small number which will result in low efficiency of the statistical inference. In the Type-II PHCS, there is enough effective sample size but the termination time of the test is difficult to predict.

In order to increase the efficiency of statistical inference as well as save the total test time, Ng et al. [21] suggested the adaptive Type-II progressive hybrid scheme (AT-II PHCS). This scheme works as follows. Suppose $n$ identical units are put on life test. The observed number of failures $m(m < n)$ is fixed in advance and the test time is allowed to run over the time $T$ which is given beforehand. The progressive censoring scheme $(R_1, R_2, \ldots, R_m)$ is specified but the values of some of the $R_i$ may change accordingly during the life test. During the life test, upon the $i$th failure observed, $R_i$ units are randomly removed from the test. We denote the $m$ completely observed lifetimes by $X_{i:m:n}, i = 1, 2, \ldots, m$. If the $m$th failure time occurs before time $T$ (i.e. $X_{m:m:n} < T$), the test stops at time $X_{m:m:n}$ with the unchanged progressive censoring scheme $(R_1, R_2, \ldots, R_m)$ where $R_m = n - m - \sum_{i=1}^{m} R_i$. Otherwise, if the $J$th failure occurs before time $T$, i.e. $X_{J:m:n} < T < X_{J+1:m:n}$, where $J < m$, then we adapt the number of units progressively removed from test upon failures by setting $R_{J+1} = R_{J+2} = \ldots = R_{m-1} = 0$ and at the time $X_{m:m:n}$ all remaining units $R_m$ are removed, where $R_m = n - m - \sum_{i=1}^{J} R_i$. Thus, in this case, the effectively applied progressive censored scheme is $\left( R_1, \ldots, R_J, 0, \ldots, 0, n - m - \sum_{i=1}^{J} R_i \right)$. For convenience, let $X_i = X_{i:m:n}, i = 1, 2, \ldots, m$. After the above test is carried out, we can get the following observation data.

Case 1: $(X_1, R_1), \ldots, (X_m, R_m)$, if $X_m < T$, where $R_m = n - m - \sum_{i=1}^{m} R_i$.

Case 2: $(X_1, R_1), \ldots, (X_J, R_J)(X_{J+1}, 0), \ldots, (X_{m-1}, 0), (X_m, R_m)$, if $X_J < T < X_{J+1}$, $J < m$, where $R_m = n - m - \sum_{i=1}^{J} R_i$.

The main advantage of this scheme is to speed up the test when the test duration exceeds predetermined time $T$ and assure us to obtain effective failure numbers for the statistical inference.

Currently, there are few literatures related to competing risks model with a bathtub-shaped lifetime distribution. Therefore, the purpose of this paper is to make statistical inference of the competing risks model with MW distribution under AT-II PHCS. The main contributions of this paper can be expounded from two aspects. (1) Construct the competing risks model of the product which lifetime variables follow the MW distribution. The MW distribution has the bathtub-shaped failure rate which is common in practical application and many studies show that the MW distribution can better fit the real life data of a part of the product. (2) Investigate the statistical inference problem of the competing risks model with MW distribution under AT-II PHCS. In the AT-II PHCS, the experimenter can obtain the fixed failures and also terminate the test as soon as possible after exceeding the ideal test time.

The remainder of this paper is organized as follows. In Sect. 2, we describe the MW distribution and property of its hazard rate function (HRF). The competing risks model under AT-II PHCS is proposed and the MLEs of unknown parameters are derived in Sect. 3. The BEs and highest posterior density (HPD) credible intervals (CIs) of the parameters based on gamma priors for the scale parameters and vague priors for the shape parameters are presented in Sect. 4. The simulation results and data analysis are showed in Sect. 5. Some conclusions are provided in Sect. 6.

## 2 The Modified Weibull Distribution and Property of Its Hazard Rate Function

In this section, we make a description on the MW distribution and it's HRF.

The CDF and probability density function (PDF) of the MW distribution can be represented as follows, respectively.

$$F(x; \alpha, \lambda, \beta) = 1 - \exp\left(-\beta x^\alpha e^{\lambda x}\right) \tag{1}$$

$$f(x; \alpha, \lambda, \beta) = \beta(\alpha + \lambda x)x^{\alpha-1}e^{\lambda x}\exp\left(-\beta x^\alpha e^{\lambda x}\right), x \geq 0, \alpha, \lambda > 0, \beta \geq 0 \tag{2}$$

where $\alpha$ and $\lambda$ are the shape parameters of the distribution, and $\beta$ is the scale parameter, respectively.

The HRF of the MW distribution is

$$H(x; \theta) = \frac{f(x; \theta)}{1 - F(x; \theta)} = \beta(\alpha + \lambda x)x^{\alpha-1}e^{\lambda x} \tag{3}$$

The first derivation of the HRF with respect to $x$ is

$$H'(x; \theta) = \beta x^{\alpha-2}e^{\lambda x}\left(\lambda^2 x^2 + 2\alpha\lambda x + \alpha^2 - \alpha\right) \tag{4}$$

From (4), we can see that the shape of the HRF mainly depends on the parameter $\alpha$. At the same time, the parameter $\lambda$ also plays an indirect role in assessing the shape of the HRF.

(1) If $\alpha \geq 1$, the function $H'(x; \theta) > 0$ for all $x > 0$, which implying HRF is an increasing function.

(2) If $0 < \alpha < 1$, the function $H'(x; \theta)$ has a change point where the curve of the HRF changes its behavior. Based on the quadratic expression $\lambda^2 x^2 + 2\alpha\lambda x + \alpha^2 - \alpha$, we can obtain the change point which is $x = (\sqrt{\alpha} - \alpha)/\lambda$. When $0 < x < (\sqrt{\alpha} - \alpha)/\lambda$, the $H'(x; \theta) < 0$ which implying the HRF is a decreasing function. If the $x > (\sqrt{\alpha} - \alpha)/\lambda$, the $H'(x; \theta) > 0$ which implying the HRF is an increasing function. In summary, HRF is decreasing initially and then increasing, which implying the HRF is a bathtub-shape function. The figure of the HRF of MW distribution is seen in Fig. 1.

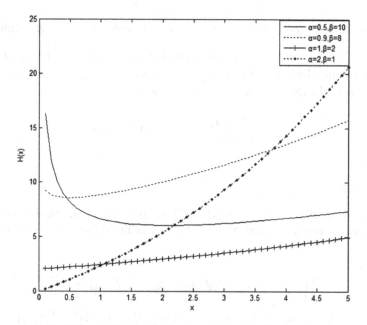

**Fig. 1.** HRF of MW distribution for different $(\alpha, \beta)$ when $\lambda = 0.1$

## 3  The Competing Risks Model Description and MLEs

This section is developed to describe the competing risks model by using MW distribution and consider the point estimations of the unknown parameters by employing maximum likelihood method based on the competing risks data under AT-II PHCS.

### 3.1  The Competing Risks Model Description

Suppose $n$ independent and identical units are put on life test and each unit has two independent failure causes. The lifetimes of $n$ units are denoted by $X_1, X_2, \ldots, X_n$ and for $i$th unit, the lifetime is $X_i = \min(X_{i1}, X_{i2}), i = 1, 2, \ldots, n$ where $X_{ij}, i = 1, 2, \ldots, n, j = 1, 2$ represents the latent failure time of the $i$th unit under the $j$th failure cause. Suppose that the latent failure time $X_{ij}, i = 1, 2, \ldots, n, j = 1, 2$ follows the MW distribution. Let $\theta = (\alpha, \lambda, \beta_1, \beta_2)$, the CDF and PDF of the MW distribution can be expressed as follows, respectively.

$$F_j(x; \theta) = 1 - \exp\left(-\beta_j x^\alpha e^{\lambda x}\right) \tag{5}$$

$$f_j(x; \theta) = \beta_j(\alpha + \lambda x)x^{\alpha-1}e^{\lambda x}\exp\left(-\beta_j x^\alpha e^{\lambda x}\right), x \geq 0, \lambda > 0, \beta_j > 0, j = 1, 2 \tag{6}$$

Under AT-II PHCS, the failure number $m$, the progressive censoring scheme $(R_1, R_2, \ldots, R_m)$ and the terminal time $T$ where $T \in (0, +\infty)$ are predetermined. For the time $T$, if $X_J \leq T \leq X_{J+1}$, where $J < m$, the dependent censoring scheme is expressed as $R_i^* = R_i I(X_J < T), i = 1, 2, \ldots, m - 1$ and $R_m^* = \sum\limits_{k=j+1}^{m} R_k$. Otherwise, $R_i^* = R_i, i = 1, 2, \ldots, m$. Then, we can obtain the adaptive Type-II progressive censored competing risks data $(X_1, c_1, R_1^*), (X_2, c_2, R_2^*), \ldots, (X_m, c_m, R_m^*)$, where $c_i$ is failure cause of the $i$th unit and is defined as follows.

$$c_i = \begin{cases} 0, \text{ if } X_{i1} < X_{i2} \\ 1, \text{ if } X_{i1} > X_{i2} \end{cases}, i = 1, 2, \ldots, m \tag{7}$$

## 3.2   The MLEs

Under the AT-II PHCS, let $J = \max\{i : X_i \leq T\}$ be denoted the number of failures before $T$. Based on the competing risks data, the likelihood function can be written as follows.

$$\begin{aligned} L(\theta) &\propto \prod_{i=1}^{m} [f_1(x_i)\bar{F}_2(x_i)]^{\delta_1(c_i)} [f_2(x_i)\bar{F}_1(x_i)]^{\delta_2(c_i)} [\bar{F}_1(x_i)\bar{F}_2(x_i)]^{R_i^*} \\ &\propto \beta_1^{m_1} \beta_2^{m_2} \prod_{i=1}^{m} (\alpha + \lambda x_i) x_i^{\alpha-1} e^{\lambda x_i} \exp\left[-(\beta_1 + \beta_2)x_i^\alpha e^{\lambda x_i}(R_i^* + 1)\right] \end{aligned} \tag{8}$$

Where $\delta_j(\cdot), j = 1, 2$ is the indicator function, $\delta_j(c_i) = 1$, if $c_i = j$; $\delta_j(c_i) = 0$, if $c_i \neq j$; $m_j = \sum\limits_{i=1}^{m} \delta_j(c_i), j = 1, 2$ is the failure number which is affected by the $j$th failure cause.

On the basis of (8), the log-likelihood function can be transformed into the following result.

$$\begin{aligned} \ln L(\theta) &= m_1 \ln \beta_1 + m_2 \ln \beta_2 + \sum_{i=1}^{m} \ln(\alpha + \lambda x_i) + \sum_{i=1}^{m} \lambda x_i \\ &+ (\alpha - 1)\sum_{i=1}^{m} \ln x_i - (\beta_1 + \beta_2)\sum_{i=1}^{m} x_i^\alpha e^{\lambda x_i}(R_i^* + 1) \end{aligned} \tag{9}$$

By setting the first partial derivative of $\ln L$ with respect to $\alpha$, $\lambda$ and $\beta_j, j = 1, 2$ to zero, we can obtain the following results.

$$\frac{\partial \ln L(\theta)}{\partial \alpha} = \sum_{i=1}^{m} \left(\frac{1}{\alpha + \lambda x_i} + \ln x_i\right) - (\beta_1 + \beta_2)\sum_{i=1}^{m} x_i^\alpha e^{\lambda x_i} \ln x_i(R_i^* + 1) = 0 \tag{10}$$

$$\frac{\partial \ln L(\theta)}{\partial \lambda} = \sum_{i=1}^{m} \left(\frac{x_i}{\alpha + \lambda x_i} + x_i\right) - (\beta_1 + \beta_2)\sum_{i=1}^{m} x_i^{\alpha+1} e^{\lambda x_i}(R_i^* + 1) = 0 \tag{11}$$

$$\frac{\partial \ln L(\theta)}{\partial \beta_j} = \frac{m_j}{\beta_j} - \sum_{i=1}^{m} x_i^\alpha e^{\lambda x_i}(R_i^* + 1) = 0, j = 1, 2 \tag{12}$$

Since the MLEs of the $\alpha$, $\lambda$ and $\beta_j, j = 1, 2$ cannot be solved analytically from (10) to (12), we can use the Newton-Raphson iteration method to obtain the MLEs of the parameters. The specific steps of method can be expresses as follows.

Step 1: Give the initial values of $\alpha$, $\lambda$ and $\beta_j, j = 1, 2$, say $\alpha^{(0)}$, $\lambda^{(0)}$ and $\beta_j^{(0)}, j = 1, 2$, respectively.

Step 2: In the $k$th iteration, calculate $\left( \frac{\partial \ln L}{\partial \alpha}, \frac{\partial \ln L}{\partial \lambda}, \frac{\partial \ln L}{\partial \beta_1}, \frac{\partial \ln L}{\partial \beta_2} \right) \Big|_{\left( \alpha^k, \lambda^k, \beta_1^k, \beta_2^k \right)}$ and the observed Fisher information matrix $I\left( \alpha^k, \lambda^k, \beta_1^k, \beta_2^k \right)$. Where

$$I(\alpha, \lambda, \beta_1, \beta_2) = \begin{bmatrix} I_{11} & I_{12} & I_{13} & I_{14} \\ I_{21} & I_{22} & I_{23} & I_{24} \\ I_{31} & I_{32} & I_{33} & I_{34} \\ I_{41} & I_{42} & I_{43} & I_{44} \end{bmatrix}$$

where

$$I_{11} = -E\left( \frac{\partial^2 \ln L}{\partial \alpha^2} \right), I_{22} = -E\left( \frac{\partial^2 \ln L}{\partial \lambda^2} \right), I_{(j+2)(j+2)} = -E\left( \frac{\partial^2 \ln L}{\partial \beta_j \partial \beta_j} \right)$$

$$I_{12} = I_{21} = -E\left( \frac{\partial^2 \ln L}{\partial \alpha \partial \lambda} \right), I_{1(j+2)} = I_{(j+2)1} = -E\left( \frac{\partial^2 \ln L}{\partial \alpha \partial \beta_j} \right), I_{2(j+2)} = I_{(j+2)2} = -E\left( \frac{\partial^2 \ln L}{\partial \lambda \partial \beta_j} \right) j = 1, 2.$$

Step 3: Update $(\alpha, \lambda, \beta_1, \beta_2)$ by

$$\left( \alpha^{(k+1)}, \lambda^{(k+1)}, \beta_1^{(k+1)}, \beta_2^{(k+1)} \right) =$$
$$\left( \alpha^k, \lambda^k, \beta_1^k, \beta_2^k \right) + \left( \frac{\partial \ln L}{\partial \alpha}, \frac{\partial \ln L}{\partial \lambda}, \frac{\partial \ln L}{\partial \beta_1}, \frac{\partial \ln L}{\partial \beta_2} \right) \Big|_{\left( \alpha^k, \lambda^k, \beta_1^k, \beta_2^k \right)} \times I^{-1}\left( \alpha^k, \lambda^k, \beta_1^k, \beta_2^k \right).$$

Step 4: Setting $k = k + 1$ the MLEs of the parameters (denoted by $\hat{\alpha}, \hat{\lambda}$ and $\hat{\beta}_j, j = 1, 2$) can be obtained by repeating the step 2–3 until

$$\left| \left( \alpha^{(k+1)}, \lambda^{(k+1)}, \beta_1^{(k+1)}, \beta_2^{(k+1)} \right) - \left( \alpha^k, \lambda^k, \beta_1^k, \beta_2^k \right) \right| < \varepsilon$$

where $\varepsilon$ is a threshold value and fixed in advance.

# 4 Bayes Analysis

Bayes method is an effective alternative to traditional statistics and has been widely used in statistical inference. Since the Bayes analysis considers the prior information, this may result in a moderate effective on the final inference. In this section, we provide the Bayes inference for the unknown parameters, which including the BEs and HPD CIs.

We assume that $\alpha$, $\lambda$ and $\beta_j, j = 1, 2$ are independent with each other. For the parameter $\beta_j, j = 1, 2$, the gamma distribution with the parameters $(a_j, b_j), j = 1, 2$ is selected as the prior distribution, because it is the conjugate distribution when $\alpha$ and $\lambda$ are known, and the PDF of $\beta_j$ is written as follows.

$$\pi(\beta_j) = \frac{b_j^{a_j}}{\Gamma(a_j)} \beta_j^{a_j-1} e^{-b_j \beta_j}, j = 1, 2$$

For the parameters $\alpha$ and $\lambda$, we select the independent vague priors and the PDF are written as follows.

$$\pi(\alpha) = \frac{1}{c_1}, \pi(\lambda) = \frac{1}{c_2},$$

where the hyper parameters $a_j, b_j, c_j, j = 1, 2$ are known and non-negative.

Thus, the joint prior function of the parameters can be written as follows.

$$\pi(\alpha, \lambda, \beta_1, \beta_2) = \frac{b_1^{a_1} b_2^{a_2}}{c_1 c_2 \Gamma(a_1) \Gamma(a_2)} \beta_1^{a_1-1} e^{-b_1 \beta_1} \beta_2^{a_2-1} e^{-b_2 \beta_2} \tag{13}$$

According to (8), the posterior density function is obtained as follows.

$$\pi(\alpha, \lambda, \beta_1, \beta_2 | x)$$

$$= \frac{L(x; \alpha, \lambda, \beta_1, \beta_2) \pi(\alpha, \lambda, \beta_1, \beta_2)}{\int_0^{+\infty} \int_0^{+\infty} \int_0^{+\infty} \int_0^{+\infty} L(x; \alpha, \lambda, \beta_1, \beta_2) \pi(\alpha, \lambda, \beta_1, \beta_2) d\alpha d\lambda d\beta_1 d\beta_2}$$

$$\propto \beta_1^{m_1 + a_1 - 1} \beta_2^{m_2 + a_2 - 1} \prod_{i=1}^{m} (\alpha + \lambda x_i) x_i^{\alpha-1} e^{\lambda x_i}$$

$$\times \exp\left\{ -\beta_1 \left( \sum_{i=1}^{m} x_i^\alpha e^{\lambda x_i} (R_i^* + 1) + b_1 \right) \right\} \exp\left\{ -\beta_2 \left( \sum_{i=1}^{m} x_i^\alpha e^{\lambda x_i} (R_i^* + 1) + b_2 \right) \right\} \tag{14}$$

The BE of any function of parameters, say $\varpi(\alpha, \lambda, \beta_1, \beta_2)$, under squared error loss function can be expressed as follows.

$$\hat{\varpi}(\alpha, \lambda, \beta_1, \beta_2) = E(\varpi(\alpha, \lambda, \beta_1, \beta_2) | x)$$
$$= \int_0^{+\infty} \int_0^{+\infty} \int_0^{+\infty} \int_0^{+\infty} \varpi(\alpha, \lambda, \beta_1, \beta_2) \pi(\alpha, \lambda, \beta_1, \beta_2 | x) d\alpha d\lambda d\beta_1 d\beta_2 \tag{15}$$

Because the BEs of the parameters cannot be got analytically, we can adapt the Gibbs sample method to calculate the BEs.

From (14), the full posterior conditional distributions of the parameters $\alpha$, $\lambda$ and $\beta_j, j = 1, 2$ can be expressed as follows.

$$\pi(\beta_j | \alpha, \lambda, x) \propto \beta_j^{m_j + a_j - 1} \exp\left\{ -\beta_j \left( \sum_{i=1}^{m} x_i^\alpha e^{\lambda x_i} (R_i^* + 1) + b_j \right) \right\}, j = 1, 2 \tag{16}$$

$$\pi(\alpha|\lambda,\beta_1,\beta_2,x) \propto \prod_{i=1}^{m}(\alpha+\lambda x_i)x_i^{\alpha-1}\exp\left[-(\beta_1+\beta_2)x_i^{\alpha}e^{\lambda x_i}(R_i^*+1)\right] \qquad (17)$$

$$\pi(\lambda|\alpha,\beta_1,\beta_2,x) \propto \prod_{i=1}^{m}(\alpha+\lambda x_i)e^{\lambda x_i}\exp\left[-(\beta_1+\beta_2)x_i^{\alpha}e^{\lambda x_i}(R_i^*+1)\right] \qquad (18)$$

From (16), we can easily get that it is a gamma density function, and it can be written as follows.

$$\beta_j|\alpha,\lambda,x \sim Gamma\left(m_j+a_j,\left(\sum_{i=1}^{m}x_i^{\alpha}e^{\lambda x_i}(R_i^*+1)+b_j\right)\right), j=1,2$$

For the parameters $\alpha$ and $\lambda$, the full posterior conditions are both log-concave functions. The following lemma is used for further development.

**Lemma 1:** $\pi(\alpha|\lambda,\beta_1,\beta_2,x)$ and $\pi(\lambda|\alpha,\beta_1,\beta_2,x)$ are both the log-concave functions.
  Proof:

$$\ln\pi(\alpha|\lambda,\beta_1,\beta_2,x) \propto \sum_{i=1}^{m}l(\alpha+\lambda x_i)+(a-1)\sum_{i=1}^{m}\ln x_i-(\beta_1+\beta_2)\sum_{i=1}^{m}x_i^{\alpha}e^{\lambda x_i}(R_i^*+1)$$

$$\ln\pi(\lambda|\alpha,\beta_1,\beta_2,x) \propto \sum_{i=1}^{m}\ln(\alpha+\lambda x_i)+\lambda\sum_{i=1}^{m}x_i-(\beta_1+\beta_2)\sum_{i=1}^{m}x_i^{\alpha}e^{\lambda x_i}(R_i^*+1)$$

Let $h(\alpha)=\ln\pi(\alpha|\lambda,\beta_1,\beta_2,x)$, $g(\lambda)=\ln\pi(\lambda|\alpha,\beta_1,\beta_2,x)$, then

$$h'(\alpha) \propto \sum_{i=1}^{m}\frac{1}{(\alpha+\lambda x_i)}+\sum_{i=1}^{m}\log x_i-(\beta_1+\beta_2)\sum_{i=1}^{m}x_i^{\alpha}e^{\lambda x_i}(R_i^*+1)(\log x_i)$$

$$h''(\alpha) \propto -\sum_{i=1}^{m}\frac{1}{(\alpha+\lambda x_i)^2}-(\beta_1+\beta_2)\sum_{i=1}^{m}x_i^{\alpha}e^{\lambda x_i}(R_i^*+1)(\log x_i)^2<0$$

$$g'(\lambda) \propto \sum_{i=1}^{m}\frac{x_i}{(\alpha+\lambda x_i)}+\sum_{i=1}^{m}x_i-(\beta_1+\beta_2)\sum_{i=1}^{m}x_i^{\alpha}e^{\lambda x_i}(R_i^*+1)x_i$$

$$g''(\lambda) \propto -\sum_{i=1}^{m}\frac{x_i^2}{(\alpha+\lambda x_i)^2}-(\beta_1+\beta_2)\sum_{i=1}^{m}x_i^{\alpha}e^{\lambda x_i}(R_i^*+1)x_i^2<0$$

Since $h''(\alpha)<0,g''(\lambda)<0$, $h(\alpha),g(\lambda)$ are log-concave functions.
  To obtain the BEs of the parameters, we use the Gibbs sample method, and the specific steps expressed as follows.

Step 1: Initial values $(\alpha,\lambda,\beta_1,\beta_2)=\left(\alpha^{(0)},\lambda^{(0)},\beta_1^{(0)},\beta_2^{(0)}\right)$.

Step 2: Generate $\beta_j^{(1)}$ from

$$Gamma\left(m_j+a_j,\left(\sum_{i=1}^{m}x_i^{\alpha^{(0)}}e^{\lambda^{(0)}x_i}(R_i^*+1)+b_j\right)\right), j=1,2.$$

Step 3: Generate $\alpha^{(1)}$ from (17) by using the adaptive rejection sampling method [22].

Step 4: Generate $\lambda^{(1)}$ from (18) by using the adaptive rejection sampling method.

Step 5: Repeat step 2–4 for $N$ times, and we obtained the

$$\left(\alpha^{(k)}, \lambda^{(k)}, \beta_1^{(k)}, \beta_2^{(k)}\right), k = 1, 2, \ldots \ldots, N.$$

Step 6: Under the squared error loss function, the BE of $\varpi(\alpha, \lambda, \beta_1, \beta_2)$ is

$$\hat{\varpi}(\alpha, \lambda, \beta_1, \beta_2) = E[\varpi(\alpha, \lambda, \beta_1, \beta_2)|x]$$
$$= \sum_{k=1}^{N} \varpi\left(\alpha^{(k)}, \lambda^{(k)}, \beta_1^{(k)}, \beta_2^{(k)}\right)/N.$$

For given $\gamma$, the HPD CI of $\varpi(\alpha, \lambda, \beta_1, \beta_2)$ can be obtained by the following three steps.

Step 1: Arrange $\varpi_i = \varpi\left(\alpha^{(i)}, \lambda^{(i)}, \beta_1^{(i)}, \beta_2^{(i)}\right), i = 1, 2, \ldots, N$ in ascending order, referred to as $\varpi_{(1)}, \varpi_{(2)}, \ldots, \varpi_{(N)}$.

Step 2: Consider the $100(1 - \gamma)\%$ CIs as $\left[\varpi_{(l)}, \varpi_{(l+N(1-\gamma))}\right], l = 1, \ldots, N - (1 - \alpha)N$.

Step 3: Consider the shortest interval from $\left[\varpi_{(l)}, \varpi_{(l+N(1-\gamma))}\right], l = 1, \ldots, N - (1 - \alpha)N$ as the HPD CI.

## 5  Data Analysis

In this section, we consider two numerical examples based on a real data set and the simulate data to investigate the performance of the proposed methods.

### 5.1  Real Data Set

In this section, we consider the real data from Aarset [23] which is given in Table 1. This real data set has been considered by Lai et al. [2], Ng [24] and Jiang et al. [7] and proved with a bathtub-shaped failure rate property. The [2] indicated that the set fit by MW distribution is better than the other extended Weibull distribution.

Now we create an adaptive Type-II progressive hybrid data set with $m = 35$, $T = 90$ and $R_4 = 4, R_{12} = 4, R_{25} = 3, R_{33} = 4, R_i = 0, i \neq 4, 12, 25, 33$. The data set is presented in Table 2. For the simple, the cause 1 data is denoted by $T_1$ and the cause 2 data denoted by $T_2$.

**Table 1.** The competing risks data from Aarset with two failure causes

| Cause1 | 0.1 | 0.2 | 1 | 1 | 1 | 3 | 6 | 11 | 12 | 18 | 18 | 18 | 32 | 36 | 40 |
|--------|-----|-----|----|----|----|----|----|----|----|----|----|----|----|----|----|
|        | 45  | 47  | 50 | 63 | 67 | 67 | 75 | 79 | 83 | 84 | 85 | 85 | 85 |    |    |
| Cause2 | 1   | 1   | 2  | 7  | 18 | 18 | 21 | 45 | 55 | 60 | 63 | 67 | 67 | 72 | 82 |
|        | 82  | 84  | 84 | 85 | 85 | 86 | 86 |    |    |    |    |    |    |    |    |

**Table 2.** The adaptive Type-II progressive hybrid data set

| Cause1 | 0.1 | 0.2 | 1 | 1 | 3 | 6 | 11 | 12 | 18 | 32 | 36 | 40 | 47 | 50 | 63 |
|--------|-----|-----|----|----|----|----|----|----|----|----|----|----|----|----|----|
|        | 67  | 75  | 79 | 84 | 85 |    |    |    |    |    |    |    |    |    |    |
| Cause2 | 1   | 1   | 2  | 7  | 18 | 21 | 45 | 55 | 60 | 67 | 72 | 82 | 84 | 85 | 86 |

In competing risks model, the life data is from MW distribution with different parameters. Before the data is analyzed, we examine that the assumption of the equality of the scale parameters of MW distribution is whether reasonable or not. It is assumed that $T_1$ and $T_2$ follow the MW distribution with the parameters $(\alpha_1, \lambda_1, \beta_1)$ and $(\alpha_2, \lambda_2, \beta_2)$, respectively. The likelihood rate test (LRS) for following hypothesis is discussed.

$$H_0 : \alpha_1 = \alpha_2, \lambda_1 = \lambda_2 \text{V.S} H_1 : H_0 \text{ is not true}$$

The LR statistic $-2(L_0 - L_1) = 4.7541$, where $L_0$ and $L_1$ is the maximum log-likelihood values under $H_0$ and $H_1$, respectively and the associated value $p$ is greater than 0.05. Hence, we should accept the null hypothesis.

For the BEs and HPD CIs of the parameters, we consider the prior information with $a_j = 0.0001$ and $b_j = 0.0001, j = 1, 2$, the point and interval estimations of the parameters are shown in Table 3 and the trace plot of 2000 iterations of the parameters listed in Fig. 2.

**Table 3.** The point and interval estimations of the parameters

| Estimator | $\alpha$ | $\lambda$ | $\beta_1$ | $\beta_2$ |
|-----------|----------|-----------|-----------|-----------|
| MLE   | 0.3115 | 0.0101 | 0.1913 | 0.0789 |
| Bayes | 0.3356 | 0.0154 | 0.1536 | 0.1067 |
| HPD   | (0.2825,0.3811) | (0.0115,0.00183) | (0.1082,0.1987) | (0.0513,0.1438) |

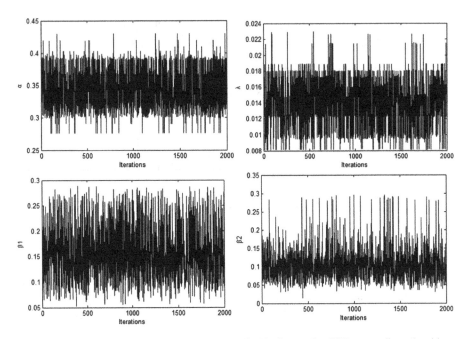

**Fig. 2.** Trace plot for the parameters of MW distribution under Gibbs sampling algorithm

## 5.2 Simulation Data Set

Without loss of generality, we consider the completing risks model has two independent failure causes and the lifetime of each failure cause follows the MW distribution. The initial values of the parameters are $(\alpha, \lambda, \beta_1, \beta_2) = (0.75, 2, 1.5, 1.8)$ and the sample size is $n = 30, 40, 60$. The MLEs, BEs and HPD CIs of the unknown parameters are obtained under different values $m$ and $T$. Three different progressive censoring schemes are used, which are denoted as follows.

> Scheme 1: $R_1 = R_2 = \ldots = R_{m-1} = 1$ and $R_m = n - 2m + 1$;
> Scheme 2: $R_1 = R_2 = \ldots = R_{m-1} = 0$ and $R_m = n - m$;
> Scheme 3: $R_1 = n - m$ and $R_2 = \ldots = R_m = 0$.

To compute the BEs and HPD CIs of the parameters, the hyper parameters of the prior are given by $a_j = 0.0001$ and $b_j = 0.0001, j = 1, 2$. Now, the same process is repeated 1000 times, and then the mean squared error (MSE) are computed, which is considered by $MSE = \sum_{i=1}^{100} (\phi_i - \hat{\phi}_i)^2 / 1000$, where $\phi_i$ and $\hat{\phi}_i$ denote the initial values and the estimation values of $\alpha, \lambda$ and $\beta_j, j = 1, 2$, respectively. Under different progressive censoring schemes, the point estimations of the unknown parameters are shown in Tables 4 and 5. And the HPD CIs with the given significant level (0.05) is also computed and the results are shown in Tables 6 and 7.

**Table 4.** The MSEs of the parameters when $T = 0.5$

| $n$ | $m$ | Scheme | Estimator | $\alpha$ | $\lambda$ | $\beta_1$ | $\beta_2$ |
|---|---|---|---|---|---|---|---|
| 30 | 10 | 1 | MLE | 0.234 | 0.4393 | 0.4115 | 0.4182 |
| | | | Bayes | 0.1454 | 0.3678 | 0.3626 | 0.3611 |
| | | 2 | MLE | 0.2365 | 0.4446 | 0.4199 | 0.4261 |
| | | | Bayes | 0.1457 | 0.3598 | 0.3612 | 0.3596 |
| | | 3 | MLE | 0.2432 | 0.4524 | 0.4205 | 0.4253 |
| | | | Bayes | 0.1484 | 0.3679 | 0.3635 | 0.3652 |
| | 15 | 1 | MLE | 0.216 | 0.4259 | 0.3927 | 0.3957 |
| | | | Bayes | 0.1326 | 0.3492 | 0.3406 | 0.3481 |
| | | 2 | MLE | 0.2197 | 0.4273 | 0.4044 | 0.4108 |
| | | | Bayes | 0.1374 | 0.3358 | 0.3428 | 0.3351 |
| | | 3 | MLE | 0.2189 | 0.4167 | 0.4123 | 0.4087 |
| | | | Bayes | 0.1365 | 0.3485 | 0.3417 | 0.3412 |
| 40 | 15 | 1 | MLE | 0.1969 | 0.3704 | 0.3489 | 0.3709 |
| | | | Bayes | 0.1156 | 0.3111 | 0.3143 | 0.3123 |
| | | 2 | MLE | 0.2059 | 0.3766 | 0.3421 | 0.3826 |
| | | | Bayes | 0.1204 | 0.3095 | 0.3192 | 0.3126 |
| | | 3 | MLE | 0.2103 | 0.3748 | 0.3446 | 0.3875 |
| | | | Bayes | 0.1191 | 0.3020 | 0.3148 | 0.3167 |
| | 20 | 1 | MLE | 0.1758 | 0.3517 | 0.3228 | 0.3438 |
| | | | Bayes | 0.1024 | 0.2902 | 0.2772 | 0.2963 |
| | | 2 | MLE | 0.1803 | 0.3573 | 0.3147 | 0.3583 |
| | | | Bayes | 0.1103 | 0.2852 | 0.2812 | 0.2926 |
| | | 3 | MLE | 0.1816 | 0.3609 | 0.3232 | 0.3528 |
| | | | Bayes | 0.1112 | 0.2913 | 0.2814 | 0.2913 |
| 60 | 20 | 1 | MLE | 0.1555 | 0.3262 | 0.3018 | 0.3146 |
| | | | Bayes | 0.0954 | 0.2632 | 0.2478 | 0.2722 |
| | | 2 | MLE | 0.1632 | 0.3171 | 0.2957 | 0.3182 |
| | | | Bayes | 0.1001 | 0.2597 | 0.2519 | 0.2702 |
| | | 3 | MLE | 0.1613 | 0.3227 | 0.2988 | 0.3261 |
| | | | Bayes | 0.0979 | 0.2586 | 0.2553 | 0.2735 |
| | 30 | 1 | MLE | 0.1278 | 0.2823 | 0.2708 | 0.2839 |
| | | | Bayes | 0.0723 | 0.2354 | 0.2036 | 0.2337 |
| | | 2 | MLE | 0.1373 | 0.2937 | 0.2752 | 0.2864 |
| | | | Bayes | 0.0804 | 0.2283 | 0.2132 | 0.2371 |
| | | 3 | MLE | 0.1369 | 0.2921 | 0.2861 | 0.2896 |
| | | | Bayes | 0.0841 | 0.2221 | 0.2119 | 0.2387 |

**Table 5.** The MSEs of the parameters when $T = 1$

| $n$ | $m$ | Scheme | Estimator | $\alpha$ | $\lambda$ | $\beta_1$ | $\beta_2$ |
|---|---|---|---|---|---|---|---|
| 30 | 10 | 1 | MLE | 0.2352 | 0.4351 | 0.4252 | 0.4211 |
| | | | Bayes | 0.1507 | 0.3589 | 0.3612 | 0.3595 |
| | | 2 | MLE | 0.2277 | 0.4343 | 0.4173 | 0.4145 |
| | | | Bayes | 0.1496 | 0.3602 | 0.3649 | 0.3556 |
| | | 3 | MLE | 0.2318 | 0.4376 | 0.4225 | 0.4168 |
| | | | Bayes | 0.1483 | 0.3611 | 0.3689 | 0.3574 |
| | 15 | 1 | MLE | 0.2182 | 0.4164 | 0.3999 | 0.3972 |
| | | | Bayes | 0.1355 | 0.3343 | 0.3407 | 0.3357 |
| | | 2 | MLE | 0.2194 | 0.4236 | 0.3853 | 0.3831 |
| | | | Bayes | 0.1314 | 0.3301 | 0.3346 | 0.3409 |
| | | 3 | MLE | 0.2219 | 0.4213 | 0.3921 | 0.3879 |
| | | | Bayes | 0.1297 | 0.3319 | 0.3367 | 0.3416 |
| 40 | 15 | 1 | MLE | 0.1961 | 0.3773 | 0.3567 | 0.3658 |
| | | | Bayes | 0.1123 | 0.3102 | 0.3021 | 0.3109 |
| | | 2 | MLE | 0.2064 | 0.3814 | 0.3571 | 0.3672 |
| | | | Bayes | 0.1157 | 0.2992 | 0.3031 | 0.3118 |
| | | 3 | MLE | 0.2035 | 0.3843 | 0.3572 | 0.3711 |
| | | | Bayes | 0.1146 | 0.2989 | 0.3108 | 0.3146 |
| | 20 | 1 | MLE | 0.1731 | 0.3432 | 0.3278 | 0.3395 |
| | | | Bayes | 0.1012 | 0.2892 | 0.2777 | 0.2955 |
| | | 2 | MLE | 0.1728 | 0.3374 | 0.3256 | 0.3302 |
| | | | Bayes | 0.1053 | 0.2781 | 0.2765 | 0.2889 |
| | | 3 | MLE | 0.1707 | 0.3473 | 0.3309 | 0.3415 |
| | | | Bayes | 0.1064 | 0.2875 | 0.2798 | 0.2912 |
| 60 | 20 | 1 | MLE | 0.1571 | 0.3114 | 0.305 | 0.3092 |
| | | | Bayes | 0.0913 | 0.2652 | 0.2448 | 0.2657 |
| | | 2 | MLE | 0.1534 | 0.3182 | 0.3013 | 0.3085 |
| | | | Bayes | 0.0956 | 0.2537 | 0.2479 | 0.2648 |
| | | 3 | MLE | 0.1569 | 0.3119 | 0.3117 | 0.3145 |
| | | | Bayes | 0.0973 | 0.2618 | 0.2503 | 0.2687 |
| | 30 | 1 | MLE | 0.1284 | 0.2757 | 0.2639 | 0.2743 |
| | | | Bayes | 0.0759 | 0.2302 | 0.2135 | 0.2324 |
| | | 2 | MLE | 0.1311 | 0.2836 | 0.2632 | 0.2791 |
| | | | Bayes | 0.0813 | 0.2313 | 0.2131 | 0.2352 |
| | | 3 | MLE | 0.1313 | 0.2856 | 0.2709 | 0.2784 |
| | | | Bayes | 0.0897 | 0.2357 | 0.2157 | 0.2388 |

From the Tables 4 and 5, it can be seen that: with the sample size $n$ increasing, both the MSEs of the MLEs and BEs for the unknown parameters decrease. Furthermore, as the sample size $n$ is fixed, the MSEs of the MLEs and BEs for the unknown parameters decrease when the effective sample size $m$ increase. The BEs of the unknown parameters is better than MLEs and it's MSEs is always less than the MLE's. The reason is that the Bayes approach considers the effective prior information which including historical experience and information. From Tables 6 and 7, it can be observed that: the lengths of the HPD CIs are decreased with the sample size $n$ or the effective sample size $m$ increase.

**Table 6.** The HPD CIs of the parameters when $T = 0.5$

| $n$ | $m$ | Scheme | Estimator | $\alpha$ | $\lambda$ | $\beta_1$ | $\beta_2$ |
|---|---|---|---|---|---|---|---|
| 30 | 10 | 1 | HPD | (0.6187,0.9173) | (1.7915,2.3066) | (1.1912,1.9254) | (1.4722,2.2714) |
| | | | Length | 0.2986 | 0.5151 | 0.7342 | 0.7992 |
| | | 2 | HPD | (0.616,0.9198) | (1.7899,2.3016) | (1.1894,1.9279) | (1.4014,2.2157) |
| | | | Length | 0.3038 | 0.5118 | 0.7385 | 0.8143 |
| | | 3 | HPD | (0.6178,0.9254) | (1.7896,2.3092) | (1.1861,1.9254) | (1.4199,2.2681) |
| | | | Length | 0.3076 | 0.5196 | 0.7393 | 0.8482 |
| | 15 | 1 | HPD | (0.638,0.8865) | (1.8467,2.2792) | (1.2153,1.8546) | (1.4837,2.1843) |
| | | | Length | 0.2485 | 0.4325 | 0.6393 | 0.7005 |
| | | 2 | HPD | (0.6237,0.8768) | (1.8557,2.2862) | (1.2184,1.8497) | (1.4992,2.1925) |
| | | | Length | 0.2532 | 0.4305 | 0.6313 | 0.6933 |
| | | 3 | HPD | (0.6381,0.8959) | (1.847,2.2812) | (1.2218,1.8712) | (1.4911,2.1954) |
| | | | Length | 0.2578 | 0.4342 | 0.6494 | 0.7043 |
| 40 | 15 | 1 | HPD | (0.6496,0.8876) | (1.86,2.2662) | (1.2389,1.8297) | (1.5588,2.1943) |
| | | | Length | 0.2379 | 0.4062 | 0.5908 | 0.6354 |
| | | 2 | HPD | (0.6447,0.8783) | (1.8573,2.2652) | (1.2244,1.8136) | (1.5511,2.1863) |
| | | | Length | 0.2336 | 0.4079 | 0.5892 | 0.6352 |
| | | 3 | HPD | (0.6459,0.8844) | (1.8653,2.2745) | (1.2473,1.8392) | (1.5519,2.1882) |
| | | | Length | 0.2385 | 0.4092 | 0.5919 | 0.6363 |
| | 20 | 1 | HPD | (0.6591,0.865) | (1.8673,2.1879) | (1.2749,1.7838) | (1.5863,2.1154) |
| | | | Length | 0.2058 | 0.3206 | 0.5089 | 0.5291 |
| | | 2 | HPD | (0.6623,0.8652) | (1.8652,2.1891) | (1.2895,1.7986) | (1.5831,2.1162) |
| | | | Length | 0.203 | 0.3239 | 0.5091 | 0.5331 |
| | | 3 | HPD | (0.6556,0.8655) | (1.8629,2.1857) | (1.2546,1.7659) | (1.5826,2.1184) |
| | | | Length | 0.2099 | 0.3228 | 0.5113 | 0.5357 |
| 60 | 20 | 1 | HPD | (0.6698,0.8539) | (1.8752,2.1456) | (1.2912,1.7454) | (1.6224,2.0741) |
| | | | Length | 0.1841 | 0.2704 | 0.4542 | 0.4518 |
| | | 2 | HPD | (0.6686,0.8512) | (1.8952,2.1633) | (1.2938,1.7469) | (1.6139,2.0745) |
| | | | Length | 0.1826 | 0.2681 | 0.4531 | 0.4606 |
| | | 3 | HPD | (0.6756,0.8612) | (1.8672,2.143) | (1.2861,1.7438) | (1.6099,2.0681) |
| | | | Length | 0.1856 | 0.2758 | 0.4577 | 0.4582 |
| | 30 | 1 | HPD | (0.6931,0.8418) | (1.9473,2.0968) | (1.3588,1.7118) | (1.6649,2.0295) |
| | | | Length | 0.1487 | 0.1495 | 0.353 | 0.3646 |
| | | 2 | HPD | (0.6911,0.8325) | (1.9366,2.0811) | (1.3491,1.6897) | (1.6597,2.016) |
| | | | Length | 0.1414 | 0.1445 | 0.3406 | 0.3563 |
| | | 3 | HPD | (0.6868,0.8319) | (1.9207,2.0729) | (1.3418,1.705) | (1.6475,2.0139) |
| | | | Length | 0.1451 | 0.1522 | 0.3632 | 0.3664 |

**Table 7.** The HPD CIs of the parameters when $T = 1$

| $n$ | $m$ | Scheme | Estimator | $\alpha$ | $\lambda$ | $\beta_1$ | $\beta_2$ |
|---|---|---|---|---|---|---|---|
| 30 | 10 | 1 | HPD | (0.6125,0.9095) | (1.7878,2.3053) | (1.1958,1.9317) | (1.4331,2.2377) |
| | | | Length | 0.297 | 0.5174 | 0.7358 | 0.8045 |
| | | 2 | HPD | (0.6147,0.9146) | (1.7937,2.3058) | (1.1929,1.9367) | (1.4488,2.2392) |
| | | | Length | 0.2999 | 0.5121 | 0.7438 | 0.7904 |
| | | 3 | HPD | (0.6171,0.9153) | (1.7789,2.3071) | (1.1792,1.9151) | (1.4187,2.2716) |
| | | | Length | 0.2982 | 0.5282 | 0.7358 | 0.8529 |
| | 15 | 1 | HPD | (0.6397,0.8857) | (1.8511,2.2808) | (1.2282,1.8676) | (1.5017,2.2061) |
| | | | Length | 0.2459 | 0.4296 | 0.6394 | 0.7044 |
| | | 2 | HPD | (0.6305,0.8848) | (1.8349,2.2674) | (1.2153,1.8492) | (1.5011,2.2043) |
| | | | Length | 0.2542 | 0.4326 | 0.6339 | 0.7032 |
| | | 3 | HPD | (0.6376,0.8939) | (1.8497,2.2801) | (1.2178,1.8641) | (1.4827,2.1854) |
| | | | Length | 0.2563 | 0.4305 | 0.6463 | 0.7027 |
| 40 | 15 | 1 | HPD | (0.6447,0.8825) | (1.8653,2.2714) | (1.2412,1.8255) | (1.5649,2.1954) |
| | | | Length | 0.2378 | 0.4061 | 0.5843 | 0.6305 |
| | | 2 | HPD | (0.6453,0.8782) | (1.848,2.2528) | (1.2288,1.8143) | (1.5599,2.1962) |
| | | | Length | 0.2329 | 0.4048 | 0.5856 | 0.6363 |
| | | 3 | HPD | (0.6473,0.8844) | (1.8683,2.2749) | (1.2416,1.8337) | (1.5526,2.1925) |
| | | | Length | 0.2371 | 0.4066 | 0.5921 | 0.6399 |
| | 20 | 1 | HPD | (0.6618,0.8666) | (1.8639,2.1925) | (1.2662,1.7792) | (1.5814,2.1096) |
| | | | Length | 0.2048 | 0.3286 | 0.513 | 0.5282 |
| | | 2 | HPD | (0.6571,0.8614) | (1.8772,2.1977) | (1.2799,1.7813) | (1.5785,2.1176) |
| | | | Length | 0.2043 | 0.3205 | 0.5015 | 0.5392 |
| | | 3 | HPD | (0.6577,0.8671) | (1.8731,2.1986) | (1.2826,1.7975) | (1.5858,2.1244) |
| | | | Length | 0.2094 | 0.3255 | 0.5149 | 0.5386 |
| 60 | 20 | 1 | HPD | (0.6702,0.8547) | (1.8733,2.1497) | (1.2958,1.7512) | (1.6331,2.0845) |
| | | | Length | 0.1845 | 0.2764 | 0.4554 | 0.4514 |
| | | 2 | HPD | (0.6602,0.845) | (1.8921,2.1539) | (1.3029,1.7567) | (1.6188,2.0828) |
| | | | Length | 0.1848 | 0.2618 | 0.4538 | 0.464 |
| | | 3 | HPD | (0.6726,0.8624) | (1.8834,2.1547) | (1.2924,1.7584) | (1.6087,2.0696) |
| | | | Length | 0.1898 | 0.2713 | 0.466 | 0.4609 |
| | 30 | 1 | HPD | (0.6976,0.8401) | (1.9342,2.0795) | (1.3614,1.7196) | (1.6706,2.0259) |
| | | | Length | 0.1425 | 0.1454 | 0.3582 | 0.3553 |
| | | 2 | HPD | (0.6864,0.8312) | (1.9387,2.0867) | (1.3517,1.6936) | (1.6694,2.0256) |
| | | | Length | 0.1448 | 0.148 | 0.3418 | 0.3562 |
| | | 3 | HPD | (0.6821,0.8362) | (1.9116,2.0727) | (1.3274,1.6927) | (1.6753,2.0338) |
| | | | Length | 0.1541 | 0.1611 | 0.3653 | 0.3585 |

# 6   Conclusion

Under AT-II PHCS, we have studied the statistical inference of the competing risks model by using the MW distribution, which not only coverers the increasing and decreasing hazard rate function, and also has a bathtub-shaped behavior. The MLEs, BEs and HPD CIs of the unknown parameters are obtained. Finally, two data sets (real data set and simulation data set) are provided to evaluate the performance of all procedures, and the results show that when the sample size or the effective sample size are large, both the MSEs of MLEs and BEs for the parameters are have the better performance and meanwhile, the HPD CIs for parameters have the shorter lengths. The BEs for parameters give the better performance than that of MLEs in term of MSE in most case.

An assumption is made in this paper that the competing risks are statistically independent. The case, however, where the competing risks are dependent, is very common in practice and the related statistical inference with dependent competing risks model is possible future work.

**Acknowledgments.** This works was supported by the National Natural Science Foundation of China (71571144, 71401134, 11501433, 71171164, 70471057) and the Program of International Cooperation and Exchanges in Science and Technology Funded by Shaanxi Provience (2016KW-033).

# References

1. Mudholkar, G.S., Srivastava, D.K.: Exponentiated Weibull family for analyzing bathtub failure rate data. IEEE Trans. Reliab. **42**(2), 299–302 (1993)
2. Lai, C.D., Xie, M., Murthy, D.N.P.: A modified Weibull distribution. IEEE Trans. Reliab. **52**(1), 33–37 (2003)
3. Bebbington, M., Lai, C.D., Zitikis, R.: A flexible Weibull extension. Reliab. Eng. Syst. Saf. **92**(3), 719–726 (2007)
4. Carrasco, M., Ortega, E.M.M., Cordeiro, G.M.: A generalized modified Weibull distribution for lifetime modeling. Comput. Stat. Data Anal. **53**(2), 450–462 (2005)
5. Silva, G.O., Ortega, E.M.M., Cordeiro, G.M.: The beta modified Weibull distribution. Lifetime Data Anal. **16**(3), 409–430 (2010)
6. Cordeiro, G.M., Ortega, E.M.M., Silva, G.O.: The Kumaraswamy modified Weibull distribution: theory and applications. J. Stat. Comput. Simul. **84**(7), 1387–1411 (2014)
7. Jiang, H., Xie, M., Tang, L.C.: Markov chain Monte Carlo methods for parameters estimation of the modified Weibull distribution. J. Appl. Stat. **35**(6), 647–658 (2008)
8. Upadhyay, S.K., Gupta, A.: A Bayes analysis of modified Weibull distribution via Markov chain Monte Carlo simulation. J. Stat. Comput. Simul. **80**(3), 241–254 (2010)
9. Jiang, H., Xie, M., Tang, L.C.: On MLEs of the parameters of a modified Weibull distribution for progressively type-II censored samples. J. Appl. Stat. **37**(4), 617–627 (2010)
10. Peng, X.Y., Yan, Z.Z.: Estimation and application for a new extended Weibull distribution. Reliab. Eng. Sys. Safety **121**(7), 34–42 (2014)
11. Vidović, Z.: On MLEs of the parameters of a modified Weibull distribution based on record values. J. Appl. Stat. **18**, 1–10 (2018)
12. EL-Sagheer, R.M., Mohamed, N.M.: On estimation of modified Weibull parameters inpresence of accelerated life test. J. Stat. Theory Pract. **12**(3), 542–560 (2018)
13. Bhattacharya, S., Pradhan, B., Kundu, D.: Analysis of hybrid censored competing risks data. Statistics **48**(5), 1138–1154 (2014)
14. Ahmadi, K., Rezaei, M., Yousefzadeh, F.: Statistical analysis of middle censored competing risks data with exponential distribution. J. Stat. Comput. Simul. **87**(16), 3082–3110 (2017)

15. Feizjavadian, S.H., Hasheme, R.: Analysis of dependent competing risks in the presence of progressive hybrid censoring using Marshall-Olkin bivariate Weibull distribution. Comput. Stat. Data Anal. **82**(2), 19–34 (2015)

16. Wu, M., Shi, Y.M., Zhang, C.F.: Statistical analysis of dependent competing risks model in accelerated life testing under progressively hybrid censoring using copula function. Commun. Stat. Simul. Comput. **6**(5), 4004–4017 (2017)

17. Epstein, B.: Truncated life-tests in the exponential case. Ann. Math. Statist. **25**, 555–564 (1954)

18. Childs, A., Chandrasekar, B., Balakrishnan, N., et al.: Exact likelihood inference based on Type-I and Type-II hybrid censored samples from the exponential distribution. Ann. Inst. Statist. Math. **55**(2), 319–330 (2003)

19. Kundu, D., Joarder, A.: Analysis of type-II progressively hybrid censored data. Comput. Stat. Data Anal. **50**(10), 2509–2528 (2006)

20. Childs, A., Chandrasekar, B., Balakrishnan, N.: Exact likelihood inference for an exponential parameter under progressive hybrid censoring schemes. In: Vonta, E., Nikulin, M., Limnios, N., Huber-Carol, C. (eds.) Statistical Models and Methods for Biomedical and Technical Systems, pp. 319–330. Birkhauser Boston, Boston (2008)

21. Ng, H.K.T., Kundu, D., Chan, P.S.: Statistical analysis of exponential lifetimes under an adaptive Type-II progressively censoring scheme. Wiley Online Lib. **56**, 687–698 (2009)

22. Gilks, W.R., Wild, P.: Adaptive rejection sampling for Gibbs sampling. Appl. Stat. **41**, 337–348 (1992)

23. Aarset, M.V.: How to identify bathtub hazard rate. IEEE Trans. Reliab. **36**(1), 106–108 (1987)

24. Ng, H.K.T.: Parameter estimation for a modified Weibull distribution for progressively type-II censored samples. IEEE Trans. Reliab. **54**(3), 374–380 (2005)

# Queueing Theory

# Functional Law of the Iterated Logarithm for Multiclass Queues with Preemptive Priority Service Discipline: The Overloaded Case

Yongjiang Guo[✉] and Xiyang Hou

School of Science, Beijing University of Posts and Telecommunications,
Beijing 100876, People's Republic of China
yongerguo@bupt.edu.cn

**Abstract.** A functional *law of the iterated logarithm* (LIL) and its corresponding LIL are established for an overloaded multiclass queueing model with preemptive priority service discipline. The functional LIL and the LIL limits quantify the magnitude of asymptotic stochastic fluctuations of the stochastic processes compensated by their deterministic fluid limits in two forms: the functional and numerical, respectively. We establish the functional LIL and LIL limits for five performance measures: queue length, workload, busy time, idle time and number of departures. By the primitive data of the first and second moments of the interarrival and service times, all the functional LILs are expressed into some compact sets of continuous functions and their corresponding LILs are some analytic functions. The proofs are based on the fluid approximation and the strong approximation of the queueing system, with the fluid approximation characterizing the expected values of the performance functions and the strong approximation approximating discrete performance processes with reflected Brownian motions.

**Keywords:** Multiclass queue · The functional law of the iterated logarithm · The law of the iterated logarithm · Strong approximation · Brownian motion

## 1 Introduction

We study one type of functional *law of the iterated logarithm* (LIL) and two types of LIL limits for the overloaded multiclass $(GI/GI)^K/1/PPSD$ queueing system, which consists of one server, $K$ classes of customers, a *preemptive priority service discipline* (PPSD) with class $i$ having preemptive priority over class $j$ for $1 \leq i < j \leq K$, class-dependent and exogenous renewal arrival processes (the first $GI$) and *independent and identically distributed* (i.i.d.) general service times (the second $GI$).

© Springer Nature Singapore Pte Ltd. 2019
Q.-L. Li et al. (Eds.): Cao Festschrift 2019, CCIS 1102, pp. 315–343, 2019.
https://doi.org/10.1007/978-981-15-0864-6_16

In the queueing literature, researcher put their attention on the multi-class priority queues heavily because the models are relevant to many real applications. For example, patients are treated sequently based on their severity levels in emergency rooms; the VIP customers experience less waiting time in service systems such as call centers; customers taking "fastpasses" can jump over the long regular waiting lines in entertainment parks such as Disneyland. Many asymptotic theories have been established for queueing models with priority policy including diffusion approximations [1,7,31], strong approximations [3,5,11,38] and heavy-traffic weak convergence results [1,20,35]. Since strong approximations are important building blocks for our analysis, we note that the most relevant work to the current paper is [5] among the literature on priority queues, which developed the strong approximations for the performance measures for the $(GI/GI)^K/1/$PPSD queueing system. Another related paper is [15], which studied some similar results on the functional LIL and LIL for the underloaded and critical loaded for the $(GI/GI)^K/1/$PPSD queueing system and complement the current work.

The limits of the functional LIL and the LIL, for a stochastic process such as renewal process, can quantify its magnitude of asymptotic stochastic fluctuations by its fluid limit (mean value) in two forms: the functional and numerical, respectively. In literature, one type of the functional LIL and two types of the LIL limits for stochastic processes are captured by the researcher generally. The functional LIL is usually thought of as being developed by Strassen [33] for Brownian motion firstly, and the two LIL limits here are referred to the Lévy's LIL [25,26] and its later generalized LIL by Csörgő and Révész [8], respectively. For the overloaded $(GI/GI)^K/1/PPSD$ system, we develop the functional LIL and the two LILs in Lévy's type for performance measures: queue length, workload, busy time, idle time and number of departures. Although based on a common $(GI/GI)^K/1/PPSD$ queueing model with [18], we aim to develop the functional LIL and the LIL in Lévy's type in this paper, and authors mainly focused on the LIL in Csörgő and Révész's type in [18]. Besides the different issues, the analysis for the LILs are also different: in this paper the LIL limits are obtained through finding the supremum and the infimum of the obtained compact set of the functional LIL limits, in [18] the LILs are got through analyzing the fluid and strong approximations.

The functional LIL above is firstly obtained by Strassen [33] for standard Brownian motion. It qualifies the asymptotic rate of the increasing variability around the mean zero through continuous functions. Assume $W(t)$ is a one-dimensional standard Brownian motion or Wiener process, and define its associated functional LIL–scaled process: $W^n(t) = W(nt)/\sqrt{2n \log \log n}$ for all $t \in [0,1]$. Strassen's functional LIL tells us that the sequence $\{W^n(t), n = 3, 4, \dots\}$ is *relatively compact*, which means its every subsequence has a convergent subsubsequence, and that all the limits of the convergent subsequences are in a compact set: the space of absolutely continuous functions $x$ with $x(0) = 0$ and $\int_0^1 [\dot{x}(t)]^2 \, dt \leq 1$, where $\dot{x}$ is the derivative of $x$. This compact set of the limit points is

$$\left\{ x \in \mathbb{C}^1[0,1] : \ x(0) = 0, \ \int_0^1 [\dot{x}(t)]^2 \, \mathrm{d}t \le 1 \right\},$$

where $\mathbb{C}^1[0,1]$ is the functional space of the one-dimensional continuous functions defined on $[0,1]$. One can see [8,33] for the functional LIL of the multi–dimensional Brownian motion. Strassen's result on the functional LIL has many applications besides in queueing networks, such as, diffusion process [2], Markov chain [34].

The two LILs: the Lévy's LIL and the generalization by Csörgő and Révész, are firstly established for standard Brownian motion too. Different with the functional LIL, the LIL use the numerical value to qualify the asymptotic rate of the increasing variability. For the defined $W(t)$ above, the Lévy's LIL in [25, 26] is to find the superior and inferior limits: *with probability one* (w.p.1),

$$\limsup_{L \to \infty} \frac{W(L)}{\sqrt{2L \log \log L}} = -\liminf_{L \to \infty} \frac{W(L)}{\sqrt{2L \log \log L}} = 1, \tag{1.1}$$

and the generalized LIL by Csörgő and Révész [8] is: w.p.1,

$$\limsup_{L \to \infty} \frac{\sup_{0 \le t \le L} |W(t)|}{\sqrt{2L \log \log L}} = 1. \tag{1.2}$$

There is a weaker form of LIL in the literture:

$$\sup_{0 \le t \le L} |W(t)| = O(\sqrt{L \log \log L}) \quad \text{w.p.1}.$$

The functional $O$ means that $f(t) = O(g(t))$ as $t \to \infty$ if $\limsup_{t \to \infty} |f(t)/g(t)| \le M$ for some $M > 0$. See [4–6] for more details. The LIL seems more numerical because it provides an explicit value (the "1" in (1.1) and (1.2)) to characterize the asymptotic rate of the increasing variability, however the functional LIL can be thought of as another mathematical issue and tells us the fluctuation in some compact sets. Mathematically, the LIL can be thought of as the supremum or the infimum of the functional LIL set supported on $[0,1]$.

In the literature of queueing networks, three methods exist for the functional LIL and LIL limits to our best knowledge. Based on the functional LIL by Strassen, Iglehart [24] reconstructed a process through renewal processes to develop the functional LIL and the LIL in Lévy's type for queue length, departure and waiting time of the multiple channel queueing systems. Minkevičius etc. [29, 30, 32] used a probability inequality method to obtain the LIL in Lévy's type for multiphase queue, single-server and multi-server open queueing networks in strictly heavy traffic. By the method based on fluid and strong approximations, Guo, Liu and Pei [19] studied the functional LIL and the LIL in Lévy's type for a single-server queue with batch arrival and feedbacks; Guo and Li [16] obtained the functional LIL and the LIL in Lévy's type for the performance functions of a two-stage tandem queue, whose generalized LIL in Csörgő and Révész's type is obtained in [17]; Guo and Liu [18] got the LIL in Csörgő and Révész's type for the $(GI/GI)^K/1/PPSD$ queueing system.

We summarize our contributions below. First, we provide a whole analysis for the functional LIL and LIL limits: from how to define them by renewal process, fluid and strong approximations to how to prove or compute them with the help of Brownian motion. Second, we obtain the functional LIL and the LIL in Lévy's type for five key performance measures for the overloaded $(GI/GI)^K/1/PPSD$ queueing system: the queue length, workload, busy time, idle time and departure processes (see Sect. 2 for their definitions). Third, the functional LIL and LIL in Lévy's type obtained for overloaded $(GI/GI)^K/1/PPSD$ queueing system cover three types of situations: the traffic intensity less than, equivalent to and more than one. Forth, we identify the functional LIL of performance measures as compact sets of analytic functions in terms of the primitive model parameters, and the LIL as analytic functions, where the model parameters are the first and second moments of the interarrival and service times. Fifth, the functional LIL and LIL limits provide interesting and sometimes counterintuitive observations. For instance, the functional LIL and the LIL for the workload and busy time are almost the same for classes with traffic intensity no more than one; the function LIL– and LIL–versions of the Little's law, which identifies the relationship between the queue length and workload processes, are preserved well for classes with traffic intensity no more than one and fail for classes with traffic intensity more than one.

An approach based on strong approximation (or the strong approximation approach) are used to find all the functional LIL and LIL limits. Next we simply introduce the strong approximation and the strong approximation approach through a renewal process. Define $\{N(t), t \geq 0\}$ to be a renewal process with rate $\lambda > 0$ and interarrival time variance $\sigma^2 < \infty$. Let $\bar{N}(t) = \lambda t$ and $\tilde{N}(t) = \lambda t + \lambda^{3/2}\sigma W(t)$ with $W$ being a one-dimensional standard Brownian motion. We note here that $\bar{N}(t)$ and $\tilde{N}(t)$ are the fluid and strong approximations of $N(t)$, respectively. Assume that the $r$th moment of the interarrival times exists, $r > 2$ is a positive number, then, by [21, 22],

$$\sup_{0 \leq t \leq L} \left| N(t) - \tilde{N}(t) \right| = o(L^{1/r}), \quad \text{w.p.1}, \tag{1.3}$$

where the function "$o(\cdot)$" says that $f(t) = o(g(t))$ as $t \to \infty$ if $\lim_{t\to\infty} |f(t)/g(t)| = 0$. Equation (1.3) tells us that the renewal process $N(t)$ can be approximated by a Brownian motion with a positive drift $\lambda$ and the approximating error is in the order of $o(L^{1/r})$. The functional LIL of $N(t)$ is the limit of the scaled sequence: for all $t \in [0, 1]$, as $n \to \infty$, w.p.1,

$$\frac{N(nt) - \bar{N}(nt)}{\sqrt{2n \log \log n}} = \frac{N(nt) - \tilde{N}(nt)}{\sqrt{2n \log \log n}} + \frac{\tilde{N}(nt) - \bar{N}(nt)}{\sqrt{2n \log \log n}} = \frac{\lambda^{3/2}\sigma W(nt)}{\sqrt{2n \log \log n}} + o(1), \tag{1.4}$$

where the second equality is from (1.3). From above, the functional LIL of renewal process is changed into a problem of Brownian motion. Similar result holds for the LIL, because, as (1.4), w.p.1,

$$\frac{N(t) - \bar{N}(t)}{\sqrt{2t \log \log t}} = \frac{N(t) - \widetilde{N}(t)}{\sqrt{2t \log \log t}} + \frac{\widetilde{N}(t) - \bar{N}(t)}{\sqrt{2t \log \log t}} = \frac{\lambda^{3/2} \sigma W(t)}{\sqrt{2t \log \log t}} + o(1), \quad t \to \infty. \quad (1.5)$$

However, it is a pity to note that the functional LIL scaled process (1.4) is relatively compact and then has no unique limit. As a result, what we can do is (i) to find all the limits of convergent subsequences which is the functional LIL issue, (ii) to find the superior and inferior limits of the scaled process in (1.5) which is exactly the LIL. For the application of strong approximation, see random walks [8–10]. For applications in queueing models, see [6] for $GI/GI/1$ queue, [13] for $GI/GI/\infty$ queue, [37] for multiple channel queue, [14] for tandem-queue network, [4,23,36] for generalized Jackson network, [38] for non-preemptive priority queue, [27,28] for time-dependent Markovian network queues and [5,6] for feedforward queueing networks.

The strong approximation approach follows four steps: (i) To establish the fluid and strong approximations for the performance measures of interest (e.g., the queue length and workload processes). (ii) To associate the functional LILs of performance measures of interest with their corresponding strong approximations, which are usually the continuous functions of Brownian motions. (iii) To find the closed-form functional LIL of Brownian functions, which is generally a compact set of continuous functions. (iv) To compute the LIL through analyzing the functional LIL set. Since there exist many properties of Brownian motion, the strong approximation approach DOES take advantage of these existing functional LIL results and helps us to find the functional LIL for the queueing model. However, two big difficulties exist for the strong approximation approach: to develop the fluid and strong approximations for given queuing systems. In addition, obtaining the LIL from a given functional LIL set may not be straightforward. The previous methods for the functional LIL and LIL include the continuous mapping method [5,6,36,37] and probability inequality method [23,38].

The rest paper is organized as follows. In Sect. 2, we formalize the $(GI/GI)^K/1/PPSD$ queueing model, define the key performance measures, and introduce their fluid limits for applications below. In Sect. 3, we define the functional LIL and its corresponding LIL limits for the processes of interest and introduce our main results, that is, Theorems 3.1–3.3. We also given some remarks to provide insights of these results. In Sect. 4, we prove the main result Theorems 3.1–3.3, whose bases: Strassen's functional LIL and the strong approximation, are also introduced in this section, we also develop some results on the relatively compact. Finally, in Sect. 5 we draw conclusions.

We close the introduction by summarizing all notations. All random variables and processes are assumed to be defined on a common probability space $(\Omega, \mathcal{F}, P)$. Let $E(X)$ and $Var(X)$ be the mean and variance for random variable $X$, respectively. Write $X =_d Y$ if $X$ and $Y$ are identically distributed.

For positive integer $k$, let $\mathbb{R}^k$ and $\mathbb{R}_+^k$ be the sets of the $k$-dimensional and nonnegative real numbers, respectively. Define $\mathbb{R} = \mathbb{R}^1$ and $\mathbb{R}_+ = \mathbb{R}_+^1$. Let prime be the transpose of a vector or matrix. Given $a = (a_1, a_1, \ldots, a_k)' \in \mathbb{R}^k$, $|a| = (a_1^2 + \cdots + a_k^2)^{1/2}$ denotes the Euclidean norm. Let "$\equiv$" denote a definition. $[a]^+ \equiv \max\{a, 0\}$ for $a \in \mathbb{R}$. For $a, b \in \mathbb{R}$, define $a \vee b = \max\{a, b\}$, $a \wedge b = \min\{a, b\}$. Let $\mathbb{D}^k[a, b]$ be the space of $k$-dimensional right continuous functions on $[a, b)$ having left limits on $(a, b]$, endowed the Skorohod topology, see [12]. Assume that $\mathbb{C}^k[a, b]$ is the subset of continuous functions in $\mathbb{D}^k[a, b]$. Let $\mathbb{D} \equiv \mathbb{D}^1$ and $\mathbb{C} \equiv \mathbb{C}^1$. Define $\mathbb{D}_0 \equiv \{x \in \mathbb{D} : x(0) \geq 0\}$. Given two functions $f$ and $g$, define $f \circ g(t) = f(g(t))$. It says that $f^n \rightrightarrows \mathcal{K}_f$ w.p.1 if $\{f^n, n \geq 1\}$ is relatively compact and the set of all limit points are in the compact set $\mathcal{K}_f$. Define the uniform norm $||f||_L \equiv \sup_{0 \leq t \leq L} |f(t)|$ for function $f$. We say that $f^n \to f$ *uniformly on compact set* (u.o.c.) if $||f^n - f||_L \to 0$, as $n \to \infty$. Let $\varphi(t) \equiv \sqrt{2t \log \log t}$ for all $t$ bigger than Eular constant. Let $e(\cdot)$ be the identity mapping $e(t) \equiv t$ for all $t$, $\eta(\cdot)$ be the zero mapping $\eta(t) = 0$ for all $t$, and $\mathbf{1}_C(\cdot)$ be the indicator function of some given set $C$, i.e., $\mathbf{1}_C(s) = 1$ if $s \in C$ and $0$ otherwise.

## 2    The Model

We consider the model consisting of a single server and $K$ queues, $K \geq 2$. Each queue $k$ is fed by an external class-$k$ arrival process, $1 \leq k \leq K$. In each queue, customers are served in the order of arrival. A PPSD policy is enforced among $K$ classes: If a customer of higher priority arrives, the low-priority customer that is currently being served (if any) will be immediately bumped out of service and placed at the head of line of its own queue; after all customers of higher priorities leave the system, the server will resume serving that preempted customer until its service is completed or another interruption by a customer of higher priority. We label these classes from 1 to $K$ with class 1 takes the highest priority while class $K$ the lowest.

For each class $k$, let $u_k(n)$ and $v_k(n)$ be the interarrival time (time between two consecutive arrivals) and service time of the $n^{\text{th}}$ customer. Let $u_k = \{u_k(n), n = 1, 2, \ldots\}$ and $v_k = \{v_k(n), n = 1, 2, \ldots\}$ be two independent i.i.d. sequence of non-negative random variables, having means $\mathsf{E}[u_k(1)] \equiv 1/\lambda_k$ and $\mathsf{E}[v_k(1)] \equiv 1/\mu_k$, variances $Var[v_k(1)]$ and $Var[v_k(1)]$, and the squared coefficients of variation $c_{a,k}^2 \equiv Var[u_k(1)]/(\mathsf{E}[u_k(1)])^2$ and $c_{s,k}^2 \equiv Var[v_k(1)]/(\mathsf{E}[v_k(1)])^2$, respectively. Define the partial sums

$$U_k(n) \equiv \sum_{i=1}^{n} u_k(i) \quad \text{and} \quad V_k(n) \equiv \sum_{i=1}^{n} v_k(i), \quad n = 1, 2, \ldots,$$

and their corresponding renewal processes

$$A_k(t) \equiv \max\{n \geq 0 : U_k(n) \leq t\} \quad \text{and} \quad S_k(t) \equiv \max\{n \geq 0 : V_k(n) \leq t\},$$

where $A_k(t)$ counts the total number of arrivals for class $k$ customers in the time interval $(0, t]$ and $S_k(t)$ counts the number of class $k$ customers the server can potentially serve in $(0, t]$ if there are no class $i$ customers with $i < k$.

Define the overall traffic intensity

$$\rho \equiv \sum_{k=1}^{K} \rho_k \quad \text{with} \quad \rho_k \equiv \frac{\lambda_k}{\mu_k}, \quad k = 1, 2, \ldots, K.$$

We say the system is underloaded when $\rho < 1$, critically loaded when $\rho = 1$ and overloaded when $\rho > 1$. Let $c_k^2 \equiv c_{a,k}^2 + c_{s,k}^2$ be the variability coefficient for class $k$ (capturing the variabilities of both the arrival and service distributions), and let $c_k > 0$. Let

$$\sigma_k^2 \equiv \sum_{j=1}^{k} \rho_j\, w_j \quad \text{with} \quad w_j \equiv \frac{c_j^2}{\mu_j}.$$

Here $\sigma_k^2$ can be understood as the (weighted) cumulative utilization of service capacity by the first $k$ classes.

Let $Q_k(t)$ be the total number of class-$k$ customers in the system at time $t$, let $Z_k(t)$ be the workload for class $k$ at time $t$, that is the total amount of time required to process all class $k$ customers assuming no future arrivals and no class $i < k$ customers after time $t$. Let $T_k(t)$ be the total amount of time the server is busy serving class $k$ customers in $[0, t]$, that is $T_1(t) = \int_0^t \mathbf{1}_{\{Q_1(s)>0\}} ds$ and

$$T_k(t) = \int_0^t \mathbf{1}_{\{Q_k(s)>0, Q_i(s)=0, i<k\}} ds \quad \text{for} \quad 2 \leq k \leq K.$$

Let $I_k(t)$ be the residual time in $[0, t]$ available to serve classes $k+1, \ldots, K$ after serving the first $k$ classes, i.e., $I_k(t) = t - \sum_{i=1}^{k} T_i(t)$. Define $Y_k(t) = \mu_k I_k(t)$ for $k = 1, 2, \ldots, K$. Let $D_k(t) = S_k(T_k(t))$ count the total number of class $k$ customers that complete service by time $t$. Let $Q \equiv (Q_1, \ldots, Q_K)'$ be the vector of the queue length processes, also let $Z$, $B$, $I$ and $D$ be the vectors of the workload, busy time, idle time and departure processes in the same token.

We have the dynamical equations:

$$Q_k(t) = A_k(t) - D_k(t) \geq 0, \quad Z_k(t) = V_k(A_k(t)) - T_k(t), \quad \int_0^t Q_k(t) dI_k(t) = 0,$$

where the first equation holds by flow conservation, the second holds because $V_k(A_k(t))$ represents the total amount of work (measured in time units) of class-$k$ arrivals in $[0, t]$, and the third holds because the idle process $I_k(t)$ increases only when $Q_k(t) = 0$.

Next, we give the fluid limit for the performance measures defined above. The fluid limit is based on the functional strong law of large numbers of stochastic process. Define the corresponding scaled processes as

$$\bar{Q}^{(n)}(t) = \frac{1}{n} Q(nt), \quad \bar{Z}^{(n)}(t) = \frac{1}{n} Z(nt), \quad \bar{T}^{(n)}(t) = \frac{1}{n} T(nt),$$

$$\bar{I}^{(n)}(t) = \frac{1}{n} I(nt), \quad \bar{D}^{(n)}(t) = \frac{1}{n} D(nt).$$

We summarize the fluid limits [11] in the next lemma, also see [3, 18] for details and proofs.

**Lemma 2.1** (Fluid limits for the $(GI/GI)^K/1/PPSD$ queue). *Assume the system is initially empty. If* $\mathsf{E}[u_k(1)] < \infty$ *and* $\mathsf{E}[v_k(1)] < \infty$, *then*

$$\left(\bar{Q}^{(n)}, \bar{Z}^{(n)}, \bar{T}^{(n)}, \bar{I}^{(n)}, \bar{D}^{(n)}\right) \to (\bar{Q}, \bar{Z}, \bar{T}, \bar{I}, \bar{D}) \equiv \bar{\mathbb{X}}, \quad u.o.c., \ w.p.1, \quad as \ n \to \infty,$$

*where* $\bar{\mathbb{X}} \equiv (\bar{\mathbb{X}}_1, \ldots, \bar{\mathbb{X}}_K)$ *with the kth element* $\bar{\mathbb{X}}_k = (\bar{Q}_k, \bar{Z}_k, \bar{T}_k, \bar{I}_k, \bar{D}_k)$ *satisfying*

$$\bar{Q}_k(t) \equiv \lambda_k t - \bar{D}_k(t) = \bar{X}_k(t) + \bar{Y}_k(t) \geq 0, \quad \bar{X}_k(t) \equiv (\lambda_k - \mu_k)t + \mu_k \sum_{l=1}^{k-1} \bar{T}_l(t),$$

$$\bar{Y}_k(t) \equiv \Psi(\bar{X}_k)(t), \quad \bar{T}_k(t) \equiv t - \sum_{l=1}^{k-1} \bar{T}_l(t) - \bar{I}_k(t), \quad \bar{I}_k(t) \equiv \frac{\bar{Y}_k(t)}{\mu_k},$$

$$\bar{D}_k(t) \equiv \mu_k \bar{T}_k(t), \quad \bar{Z}_k(t) \equiv \frac{\bar{Q}_k(t)}{\mu_k}, \quad k = 1, \ldots, K, \tag{2.1}$$

*and functions* $\Phi$ *and* $\Psi$ *are defined for* $x \in \mathbb{D}_0$ *as*

$$\Psi(x)(t) \equiv \sup_{0 \leq s \leq t} \{-x(s)\}^+ \quad and \quad \Phi(x)(t) \equiv x(t) + \sup_{0 \leq s \leq t} \{-x(s)\}^+. \tag{2.2}$$

**Remark 2.1** (Oblique reflection mapping). *The mapping* $(\Psi, \Phi)$ *is known as the one dimensional oblique reflection mapping, and is Lipschitz continuous in uniform norm, see [6] for detailed discussions. Alternative representation for* $(\Psi, \Phi)$ *is given below.*

**Definition 2.1** (Definition of oblique reflection mapping). *For any function* $x \in \mathbb{D}_0$, *if there exists a unique pair of functions* $z, y \in \mathbb{D}_0$ *satisfying*

(i) $z(t) = x(t) + y(t) \geq 0$;
(ii) $y$ *is nondecreasing and* $y(0) = 0$;
(iii) $\int_0^\infty z(t)dy(t) = 0$,

*then* $(z, y) \equiv (\Phi, \Psi)(x)$ *is called the one dimensional oblique reflection mapping.*

The objective of the rest of the paper is to establish the functional LIL and the LIL limits for performance functions $(Q_k, Z_k, B_k, I_k, D_k, 1 \leq k \leq K)$ and identify the LIL limits as simple functions of the primitive model data

$$\mathcal{D} \equiv \left(\lambda_k, \mu_k, c_{a,k}^2, c_{s,k}^2, c_k^2, 1 \leq k \leq K\right), \tag{2.3}$$

and identify the functional LIL limits as functional set in term of data in (2.3) together with some compact sets of continuous functions.

# 3    Main Results

In this section, we give the funcitional LIL and its corresponding LILin Lévy's type for the performance measures in the $(GI/GI)^K/1/PPSD$ queueing system. We firstly formalize the LIL and the functional LIL limits into mathematical problems and then present main results in Theorems 3.1–3.3.

## 3.1    The LIL and the Functional LIL Scalings

Now we are ready to define the LIL and the functional LIL scalings based on the fluid limit given in Lemma 2.1.

**The LIL Scalings and Limits.** We define the LIL-scaled processes for $Q_k$:

$$Q_{k,sup}^* \equiv \limsup_{t \to \infty} \frac{Q_k(t) - \bar{Q}_k(t)}{\varphi(t)}, \quad Q_{k,inf}^* \equiv \liminf_{t \to \infty} \frac{Q_k(t) - \bar{Q}_k(t)}{\varphi(t)}, \quad k = 1, 2, \ldots, K. \quad (3.1)$$

Similarly, we define the following LIL-scaled notations in the same token of (3.1): for $k = 1, 2, \ldots, K$,

$$Z_{k,sup}^*, Z_{k,inf}^*; \quad T_{k,sup}^*, T_{k,inf}^*; \quad I_{k,sup}^*, I_{k,inf}^*; \quad D_{k,sup}^*, D_{k,inf}^*. \quad (3.2)$$

Let

$$\mathcal{X}_{k,sup}^* \equiv \left( Q_{k,sup}^*, Z_{k,sup}^*, T_{k,sup}^*, I_{k,sup}^*, D_{k,sup}^* \right),$$
$$\mathcal{X}_{k,inf}^* \equiv \left( Q_{k,inf}^*, Z_{k,inf}^*, T_{k,inf}^*, I_{k,inf}^*, D_{k,inf}^* \right), \quad k = 1, 2, \ldots, K. \quad (3.3)$$

We will express all LIL limits in (3.3) as functions of the primitive data (2.3).

**The Functional LIL Scalings and Limits.** For all $t \in [0,1]$ and $n = 3, 4, \ldots$, define the functional LIL-scaled process for $Q_k$:

$$Q_k^n(t) \equiv \frac{Q_k(nt) - \bar{Q}_k(nt)}{\varphi(n)}, \quad k = 1, 2, \ldots, K. \quad (3.4)$$

Similarly we define the functional LIL-scaled processes: $Z_k^n(t)$, $T_k^n(t)$, $I_k^n(t)$, $Y_k^n(t)$, $D_k^n(t)$ in the same token of (3.4), $k = 1, 2, \ldots, K$. We will develop all the functional LIL results by showing that

$$(Q_k^n, Z_k^n, T_k^n, I_k^n, D_k^n) \Rightarrow (\mathcal{K}_{Q_k}, \mathcal{K}_{Z_k}, \mathcal{K}_{T_k}, \mathcal{K}_{I_k}, \mathcal{K}_{D_k}) \equiv \mathcal{K}_k^*, \quad \text{w.p.1}, \quad k = 1, 2, \ldots, K, \quad (3.5)$$

which are expressed in terms of the input data (2.3) and the compact set $\mathcal{G}_k$ defined as

$$\mathcal{G}_k(\delta) \equiv \left\{ x \in \mathbb{C}^k[0,1] : x(0) = 0, \int_0^1 [\dot{x}(t)]^2 \, dt \leq \delta^2 \right\}, \quad \delta > 0, \quad (3.6)$$

where the square denotes inner product, and $\dot{x}(t)$ denotes the derivative of $x(t)$ which exists almost everywhere with respect to Lebesgue measure. We simply denote $\mathcal{G}_1$ by $\mathcal{G}$. Strassen [33] proved that $\mathcal{G}_k(\delta)$ is a compact set in $\mathbb{C}^k[0,1]$ for any $\delta > 0$, and that for $x \in \mathcal{G}_k(\delta)$ and $0 \leq a \leq b \leq 1$, $|x(b) - x(a)| \leq \delta(b-a)^{1/2}$.

**Remark 3.1** (Understanding the functional LIL and LIL). *From the defini-tions of the LIL* (3.1) *and functional LIL* (3.4), *it is easy to see that both of them refine the fluid limit of renewal process because they are not only cen-tered by the fluid limit, but also show us the asymptotic deviations around the fluid limit. Meanwhile we note that the mathematical forms presented by the LIL and the functional LIL are different: the LIL is numerical and the functional LIL is functional. Especially, the functional set of the functional LIL consist of univariate or multivariate functions. Generally, the sup-LIL and inf-LIL are the supremum and the infimum of some function in its corresponding functional LIL limit, respectively. When the functional LIL is a set of the univariate functions, it is easy to find the LIL limits, however, when the functional LIL is a set of multivariate functions, it is difficult to analyze the functional set and find the LIL limits.*

## 3.2    The LIL and the Functional LIL Limits

We now give our main results: the functional LIL and its corresponding LIL in Lévy's type for the queue length, workload, busy time, idle time and departure processes in the overload regime. All proofs are given in Sect. 4.

Throughout the rest of the paper, we suppose that, for all $k = 1, \ldots, K$,

$$\mathsf{E}\left[u_k(1)^r\right] < \infty \quad \text{and} \quad \mathsf{E}\left[v_k(1)^r\right] < \infty \quad \text{for some } r > 2. \tag{3.7}$$

For applications, we define the continuous mapping $G_1 : \mathbb{C} \times \mathbb{C} \to \mathbb{C}$ by

$$G_1(x, y)(t) = \inf_{0 \le s \le t} \left[x(s) - y(s)\right]^+ + y(t). \tag{3.8}$$

Define the continuous mapping $G_2 : \mathbb{C} \times \mathbb{C} \to \mathbb{C}$ by

$$G_2(x, y)(t) = \inf_{0 \le s \le t} \left[x(s)\right]^+ + y(t). \tag{3.9}$$

Suppose that the $(GI/GI)^K/1/PPSD$ queueing system is in the overloaded regime: $\rho > 1$, there are three sub-cases categorized by the values of $\rho_1, \ldots, \rho_K$:

**Case 1.** There exists a $k_0 : 1 \le k_0 < K$ such that $\sum_{j=1}^{k_0} \rho_j = 1$ and $\sum_{j=1}^{k_0+1} \rho_j > 1$;

**Case 2.** There exists a $k_0 : 1 \le k_0 < K$ such that $\sum_{j=1}^{k_0} \rho_j < 1$ and $\sum_{j=1}^{k_0+1} \rho_j > 1$;

**Case 3.** $\rho_1 > 1$.

Next, we give our main results according to three cases defined above.

**Theorem 3.1** (The limits in Case 1). *Suppose that the $(GI/GI)^K/1/PPSD$ queueing system is in* **Case 1** *of the overloaded regime, the functional LIL limits for class 1 to class $k_0 - 1$ satisfy, w.p.1,*

$$\mathcal{K}_k^* = \left\{ \left(\eta, \eta, \frac{\lambda_k^{1/2} c_k}{\mu_k} x, -\sigma_k x, \lambda_k^{1/2} c_{a,k} x\right) : x \in \mathcal{G}(1) \right\}, \quad k = 1, 2, \ldots, k_0 - 1, \tag{3.10}$$

*The functional LIL limits for class $k_0$ satisfy, w.p.1,*

$$\mathcal{K}_{k_0}^* = \left\{ (\Phi(\mu_{k_0} x), \Phi(x), G_1(y), \Psi(x), G_1(\mu_K y)) : x \in \mathcal{G}(\sigma_{k_0}), y \in \mathcal{G}_2(\sigma_{k_0}) \right\}, \quad (3.11)$$

*For class $k_0 + 1$, the functional LIL, w.p.1,*

$$\mathcal{K}_{k_0+1} = \left\{ (G_2(q_{k_0+1} x), G_2(z_{k_0+1} x), \Psi(\sigma_{k_0} y), \eta, \mu_{k_0+1} \Psi(\sigma_{k_0} y)) : x \in \mathcal{G}_2(1), y \in \mathcal{G}(1) \right\}, \quad (3.12)$$

*where*

$$q_{k_0+1} = \sqrt{\lambda_{k_0+1} c_{a,k_0+1}^2 + \mu_{k_0+1}^2 \sigma_{k_0}^2}, \quad z_{k_0+1} = \sqrt{\sigma_{k_0}^2 + \frac{\lambda_{k_0+1} c_{k_0+1}^2}{\mu_{k_0+1}^2}}, \quad (3.13)$$

*and for $k = k_0 + 2, k_0 + 3, \ldots, K$ if $k_0 + 1 < K$, the functional LIL limits*

$$\mathcal{K}_k = \left\{ \left( \lambda_k^{1/2} c_{a,k} x, \frac{c_k \sqrt{\lambda_k}}{\mu_k} x, \eta, \eta, \eta \right) : x \in \mathcal{G}(1) \right\}. \quad (3.14)$$

*For the LIL limits, the classes from 1 to $k_0 - 1$ satisfy, w.p.1,*

$$\mathcal{X}_{k,sup}^* = -\mathcal{X}_{k,inf}^* = \left( 0, 0, \frac{\lambda_k^{1/2} c_k}{\mu_k}, \sigma_k, \lambda_k^{1/2} c_{a,k} \right), \quad k = 1, 2, \ldots, k_0 - 1. \quad (3.15)$$

*The class $k_0$ satisfies, w.p.1,*

$$Q_{k_0,sup}^* = \mu_{k_0} Z_{k_0,sup}^* = \mu_{k_0} I_{k_0,sup}^* = \sigma_{k_0}, \quad Q_{k_0,inf}^* = Z_{k_0,inf}^* = I_{k_0,inf}^* = 0, \quad (3.16)$$

*The class $k_0 + 1$ satisfies, w.p.1,*

$$T_{k_0+1,sup}^* = \sigma_{k_0+1}, \quad D_{k_0+1,sup}^* = \mu_{k_0+1} \sigma_{k_0+1},$$
$$T_{k_0+1,inf}^* = D_{k_0+1,inf}^* = I_{k_0+1,sup}^* = I_{k_0+1,inf}^* = 0, \quad (3.17)$$

*the classes from $k_0 + 2, k_0 + 3, \ldots, K$ satisfy, w.p.1,*

$$Q_{k,sup}^* = -Q_{k,inf}^* = \lambda_k^{1/2} c_{a,k}, \quad Z_{k,sup}^* = -Z_{k,inf}^* = \frac{c_k \sqrt{\lambda_k}}{\mu_k}, \quad T_{k,sup}^* = T_{k,inf}^* = I_{k,sup}^*$$

$$= I_{k,inf}^* = D_{k,sup}^* = D_{k,inf}^* = 0, \quad k = k_0 + 2, k_0 + 3, \ldots, K. \quad (3.18)$$

**Remark 3.2** (Understanding the limits in Case 1). *The Case 1 in the overloaded regime is more complicated than the underloaded and critically loaded cases* [15]*, the reasons is that class $k_0$ is a critically loaded state: $\sum_{k=1}^{k_0} \rho_k = 1$. Since the first $k_0$ classes form an underloaded $(GI/GI)^{k_0}/1/PPSD$ queueing system, readers can refer to* [15] *for insights, we only go to understand the classes $k_0 + 1, \ldots, K$, which characterize the overloaded regime. When $k_0$ is the critical critically loaded class, the asymptotic stochastic fluctuations, for class $k_0 + 1$*

and classes $k_0 + 2, k_0 + 3, \ldots, K$, are very different. This is so because class $k_0 + 1$ is still influenced by the performance of the first $k$ classes (although the first $k_0$ classes have utilized all service capacity, it is still possible to serve some (perhaps very little) class $k_0 + 1$ customers), however the customers of classes $k_0 + 2, k_0 + 3, \ldots, K$ are almost never served. This difference is well embodied in the parameters in (3.12)–(3.13) and (3.14). For example, (i) for the queue length, the second term $\mu_{k_0+1}^2 \sigma_{k_0}^2$ of $q_{k_0+1}$ in (3.13) represents the influence from the first $k_0$ classes, and the corresponding parameter $\lambda_k^{1/2} c_{a,k}$ for $k > k_0 + 1$ in (3.14) is from its own arrival process and is independent of the first $k_0$ classes, $k > k_0 + 1$; (ii) for busy time, the asymptotic fluctuation of the $T_{k_0+1}$ is in fact identical with $I_{k_0}$ because $T_{k_0+1}(t) = I_{k_0}(t) - I_{k_0+1}(t)$ and the deviation from $I_{k_0+1}(t)$ is almost negligible, however the asymptotic fluctuation of the $T_k(t) = I_{k-1}(t) - I_k(t)$ is almost zero with the same reason that the deviation of $I_k(t)$ is almost negligible for $k > k_0 + 1$; (iii) for departure, since the departure is heavily dependent on the busy time, it follows that the asymptotic fluctuation of $D_{k_0+1}$ is embodied by $\sigma_{k_0}$, the fluctuation parameter of the first $k$ classes, nevertheless the asymptotic fluctuation for classes $k > k_0 + 1$ is zero because the corresponding asymptotic fluctuation for busy time is almost negligible.

**Theorem 3.2** (The limits in Case 2). *Suppose that the $(GI/GI)^K/1/PPSD$ queueing system is in **Case 2** of the overloaded regime, the functional LIL limits for class 1 to class $k_0 - 1$ satisfy (3.10) with $k = 1, 2, \ldots, k_0$; The functional LIL limits for classes $k_0 + 2$ to $K$ satisfy (3.14) with $k = k_0 + 2, k_0 + 3, \ldots, K$. For class $k_0 + 1$,*

$$\mathcal{K}_{k_0+1} = \left\{ \left( q_{k_0+1}^* x, \sigma_{k_0+1} x, -\sigma_{k_0} x, \eta, d_{k_0+1}^* x \right) : x \in \mathcal{G}(1) \right\}; \qquad (3.19)$$

*where*

$$q_{k_0+1}^* = \sqrt{ \mu_{k_0+1}^2 \sigma_{k_0}^2 + \lambda_{k_0+1} c_{a,k_0+1}^2 + \mu_{k_0+1} c_{s,k_0+1}^2 \sqrt{ 1 - \sum_{i=1}^{k_0} \rho_i } },$$

$$d_{k_0+1}^* = \sqrt{ \mu_{k_0+1}^2 \sigma_{k_0}^2 + \mu_{k_0+1} c_{s,k_0+1}^2 \sqrt{ 1 - \sum_{i=1}^{k_0} \rho_i } }. \qquad (3.20)$$

*For the LIL limits, the classes from 1 to $k_0$ satisfy (3.15) with $k = 1, 2, \ldots, k_0$, the class $k_0 + 1$ satisfies*

$$\mathcal{X}_{k_0+1,sup}^* = -\mathcal{X}_{k_0+1,inf}^* = \left( q_{k_0+1}^*, \sigma_{k_0+1}, \sigma_{k_0}, 0, d_{k_0+1}^* \right), \quad w.p.1, \qquad (3.21)$$

*the classes from $k_0 + 2$ to $K$ satisfy (3.18) with $k = k_0 + 2, k_0 + 3, \ldots, K$.*

**Remark 3.3** (Understanding the limits in Case 2). *When $k_0$ is the last underloaded class and $k_0 + 1$ is the first overloaded class, that is, the system is in the Case 2, class $k_0 + 1$ can utilize the $(1 - \sum_{i=1}^{k_0} \rho_i)$ parts of capacity of the server,* classes $k_0 + 2, k_0 + 3, \ldots, K$ are almost never served. The performance for class $k(> k_0)$ can refer to classes $k_0 + 2, k_0 + 3 \ldots, K$ in Case 1. For class $k_0 + 1$, the deviation parameter for queue length $q^*_{k_0+1}$ consists of three parts: $\mu^2_{k_0+1}\sigma^2_{k_0}$, $\lambda_{k_0+1}c^2_{a,k_0+1}$ and $\mu_{k_0+1}c^2_{s,k_0+1}\sqrt{1 - \sum_{i=1}^{k_0}\rho_i}$, representing three asymptotic fluctuations from the first $k_0$ classes, the arrival and the service of class $k_0 + 1$, respectively. For workload processes of classes $k_0$ and $k_0 + 1$, their deviation parameters have similar structure: $\sigma_{k_0}$ and $\sigma_{k_0+1}$, because, for class $k_0 + 1$, the workload process keeps track of the total amount of unfinished service times, their unfinished service variability will still make an impact to the total workload. For busy time $T_{k_0+1}$, its deviation parameter $\sigma_{k_0}$ is the same as the idle time $I_{k_0}$, this is so because, $I_k(t)$, the remaining service capacity available for low-priority classes $k > k_0$, will asymptotically all be devoted to class $k_0 + 1$. For departure $D_{k_0+1}$, the deviation parameter $d^*_{k_0+1}$ in (3.20) includes the influence $\mu^2_{k_0+1}\sigma^2_{k_0}$ from the first $k_0$ class, and the influence $\mu_{k_0+1}c^2_{s,k_0+1}\sqrt{1 - \sum_{i=1}^{k_0}\rho_i}$ from its own service, however is short of its own arrival influence because the queue length of class $k_0 + 1$ will go to infinity, and then its own arrival has no impact on it.

**Theorem 3.3** (The limits in Case 3). *Suppose that the $(GI/GI)^K/1/PPSD$ queueing system is in* **Case 3** *of the overloaded regime, the functional LIL limits (3.14) holds for classes $k = 2, 3, \ldots, K$. For class 1, we have the functional LIL limits:*

$$\mathcal{K}_1 = \left\{\left(\sqrt{\lambda_1 c^2_{a,1} + \mu_1 c^2_{s,1}}\, x, \sigma_1 x, \eta, \eta, \mu_1^{1/2}c_{s,1}x\right) : x \in \mathcal{G}(1)\right\}. \qquad (3.22)$$

*For the LIL limits, the class 1 satisfies*

$$\mathcal{X}^*_{1,sup} = -\mathcal{X}^*_{1,inf} = \left(\sqrt{\lambda_1 c^2_{a,1} + \mu_1 c^2_{s,1}}, \sigma_1, 0, 0, \mu_1^{1/2}c_{s,1}\right), \quad w.p.1, \quad (3.23)$$

*the classes from 2 to $K$ satisfy (3.18) with $k = 2, 3, \ldots, K$.*

**Remark 3.4** (The Little's law). *Together with Theorems 3.1–3.3, it says that the Little's law between the queue length and the workload processes holds for classes with traffic intensity no more than one and fails in classes with traffic intensity more than one, that is,*

$$\mathcal{K}_{Q_k} = \mu_k \mathcal{K}_{Z_k}, \quad Q^*_{k,sup} = \mu_k Z^*_{k,sup}, \quad Q^*_{k,inf} = \mu_k Z^*_{k,inf} \quad for \quad k : \sum_{i=1}^{k}\rho_i \leq 1,$$

$$\mathcal{K}_{Q_k} \neq \mu_k \mathcal{K}_{Z_k}, \quad Q^*_{k,sup} \neq \mu_k Z^*_{k,sup}, \quad Q^*_{k,inf} \neq \mu_k Z^*_{k,inf} \quad for \quad k : \sum_{i=1}^{k}\rho_i > 1.$$

*For the class $k: \sum_{i=1}^{k}\rho_i > 1$, its queue length $Q_k$ will go to infinity with time increasing. The unfinished class-k customers play no role on the queue length because the queue length counts the number of them. However, the unfinished class-k customers maybe make an big impact on the workload if their service is high variable.*

**Remark 3.5** (The functional LIL sets consisting of binary functions). *From Theorems 3.1–3.3, we can find that the functional LIL sets, for classes $k, k+1$ with $\rho_k = 1$, consist of some continuous binary functions, such as, $G_1(x, y)$ in (3.8) and $G_2(x, y)$ in (3.9). All other functional LIL sets are composed of continuous unary functions. Through finding the supremum and infimum of the functional set, we can get the superior and inferior LIL limits, such as, $Q_{k,sup}^*$ and $Q_{k,inf}^*$ with $\rho_k < 1$. Nevertheless, for the functional LIL sets consisting of binary functions, such as, $\mathcal{K}_{Q_k} = G_2(\mathcal{G}_2(q_{k_0+1}))$ in (3.12), we do not have good idea to analyze their the supremum and infimum, as a result, we do not show readers their corresponding LIL limits. Mathematically, it is interesting to develop new methods to analyze such binary functional sets and find the corresponding the supremum and infimum or the LIL limits.*

# 4 Proofs

In this section, we prove our main results: Theorems 3.1–3.3. We will prove them by the strong approximation approach. We first give some basis for proof in Sect. 4.1, and then prove Theorems 3.1–3.3 in Sect. 4.2, respectively.

## 4.1 The Primitive Basis for Proofs

We give some basis for applications in the proof, including the strong approximation of the performance measures which are some equations based on Brownian motions, the functional LIL of Brownian motions and its corresponding continuous mapping theorem given by Strassen in [33] and some results on the relatively compact limits.

The idea of the strong approximation is to approximate a discrete process, such as the queue length $Q$, by the sum of two continuous functions: (i) the deterministic fluid function $\bar{Q}$ and (ii) standard Brownian motions, with $\bar{Q}$ characterizing the mean value and the Brownian motions quantifying the stochastic fluctuations around that mean value. We next introduce the strong approximations for the $(GI/GI)^K/1/PPSD$ system, and see Lemma 1 and Corollaries 1 and 3 in [18] for details.

**Lemma 4.1** (Strong approximations for $(GI/GI)^K/1/PPSD$). *If (3.7) holds, then, w.p.1,*

$$\left\| Q_k - \tilde{Q}_k \right\|_L = o(L^{1/r}), \quad \left\| Z_k - \tilde{Z}_k \right\|_L = o(L^{1/r}), \quad \left\| T_k - \tilde{T}_k \right\|_L = o(L^{1/r}),$$

$$\left\| I_k - \tilde{I}_k \right\|_L = o(L^{1/r}), \quad \left\| D_k - \tilde{D}_k \right\|_L = o(L^{1/r}), \quad k = 1, 2, \ldots, K, \qquad (4.1)$$

*where*

$$\widetilde{Q}_k(t) \equiv \widetilde{X}_k(t) + \widetilde{Y}_k(t) = \Phi(\widetilde{X}_k)(t), \quad \widetilde{Y}_k(t) \equiv \Psi(\widetilde{X}_k)(t),$$

$$\widetilde{X}_k(t) \equiv (\lambda_k - \mu_k)t + \mu_k \sum_{l=1}^{k-1} \widetilde{T}_l(t) + \widetilde{W}_k(t),$$

$$\widetilde{T}_k(t) \equiv t - \sum_{j=1}^{k-1} \widetilde{T}_j(t) - \widetilde{I}_k(t), \quad \widetilde{I}_k(t) \equiv \frac{1}{\mu_k}\widetilde{Y}_k(t),$$

$$\widetilde{Z}_k(t) \equiv \frac{1}{\mu_k}\widetilde{Q}_k(t) + \frac{1}{\mu_k}\left[\mu_k^{1/2}c_{s,k}W_{s,k}(\bar{T}_k(t)) - \mu_k^{1/2}c_{s,k}W_{s,k}(\rho_k t)\right],$$

$$\widetilde{D}_k(t) \equiv \mu_k\widetilde{T}_k(t) + \mu_k^{1/2}c_{s,k}W_{s,k}(\bar{T}_k(t)),$$

$$\widetilde{W}_k(t) \equiv \lambda_k^{1/2}c_{a,k}W_{a,k}(t) - \mu_k^{1/2}c_{s,k}W_{s,k}(\bar{T}_k(t)), \tag{4.2}$$

$W_{a,k}$ and $W_{s,k}$ are independent standard Brownian motions associated with the arrival and service processes of class $k$, respectively, and $\Psi$ and $\Phi$ are defined in (2.2). Let $W_k(t) \equiv \mu_k \sum_{l=1}^{k} \widetilde{W}_l(t)/\mu_l$, $k = 1, 2, \ldots, K$, the functional measures $\widetilde{X}_k(t)$ and $\widetilde{B}_k(t)$ in (4.2) satisfy

$$\widetilde{X}_k(t) - \bar{X}_k(t) = -\sum_{l=1}^{k-1} \frac{\mu_k}{\mu_l}\left[\widetilde{Q}_l(t) - \bar{Q}_l(t)\right] + W_k(t), \tag{4.3}$$

$$\bar{T}_k(t) - \widetilde{T}_k(t) = \frac{1}{\mu_k}\left[\widetilde{Q}_k(t) - \bar{Q}_k(t)\right] - \frac{1}{\mu_k}\widetilde{W}_k(t). \tag{4.4}$$

*The approximated queue length* $\widetilde{Q}_k(t)$ *satisfy: If* $\rho_k < 1$, $k = 1, 2, \ldots, K$, *then* $\left\|\widetilde{Q}_k\right\|_L = O(\log L)$ *w.p.1 as* $L \to \infty$.

Lemma 4.1 provides a basis for our strong approximation approach. By the strong approximation above, we can transform the problem $\mathcal{K}_k^*$ into a Brownian motion associated problem. Now we defined the Brownian motion scaling processes: The functional LIL scaled processes for $\widetilde{Q}_k$:

$$\widetilde{Q}_k^n(t) = \frac{\widetilde{Q}_k(nt) - \bar{Q}_k(nt)}{\varphi(n)}, \quad k = 1, 2, \ldots, K. \tag{4.5}$$

Similarly, we define the scaled processes $\widetilde{Z}_k^n(t), \widetilde{T}_k^n(t), \widetilde{I}_k^n(t), \widetilde{Y}_k^n(t), \widetilde{D}_k^n(t)$ in the same token of (4.5) for $k = 1, 2, \ldots, K$, respectively. Let

$$\left(\widetilde{Q}_k^n, \widetilde{Z}_k^n, \widetilde{T}_k^n, \widetilde{I}_k^n, \widetilde{D}_k^n\right) \rightrightarrows \left(\mathcal{K}_{\widetilde{Q}_k}, \mathcal{K}_{\widetilde{Z}_k}, \mathcal{K}_{\widetilde{T}_k}, \mathcal{K}_{\widetilde{I}_k}, \mathcal{K}_{\widetilde{D}_k}\right), \quad \text{w.p.1}, \quad k = 1, 2, \ldots, K,$$

if the limits on the right exist. It similarly follows from Lemma 4.3 in [16] that

$$\mathcal{K}_k^* = \left(\mathcal{K}_{\widetilde{Q}_k}, \mathcal{K}_{\widetilde{Z}_k}, \mathcal{K}_{\widetilde{T}_k}, \mathcal{K}_{\widetilde{I}_k}, \mathcal{K}_{\widetilde{D}_k}\right), \quad k = 1, 2, \ldots, K. \tag{4.6}$$

In words, if we need to find the functional LILs $\mathcal{K}_k^*$ in (3.5), we only go to compute the corresponding Brownian motion problems in (4.6).

Strassen [33] firstly developed the functional LIL for Brownian motion as follows.

**Lemma 4.2** (Strassen's functional LIL result). *If $W_a, W_b$ are two mutually independent one-dimensional standard Brownian motions, $\sigma_a \neq 0, \sigma_b \neq 0$ are two constants, then, for any $t \in [0.1]$, w.p.1,*

$$\frac{W_a(nt)}{\varphi(n)} \rightrightarrows \mathcal{G}(1) \quad and \quad \left(\frac{\sigma_a W_a(nt)}{\varphi(n)}, \frac{\sigma_b W_b(nt)}{\varphi(n)}\right) \rightrightarrows \mathcal{G}_2\left(\sqrt{\sigma_a^2 + \sigma_b^2}\right).$$

The following Lemma 4.3 is a Corollary of Theorem 3 in [33], called the continuous mapping theorem for the relatively compact.

**Lemma 4.3** (Strassen's continuous mapping theorem). *Let $\{x_n : n \geq 1\}$ be a relatively compact sequence in $\mathbb{C}^k[0, 1]$ endowed with the uniform norm and with the compact set $\mathcal{G}_k$ as its set of limit points. If $f$ is a continuous function on $\mathbb{C}^k[0, 1]$ into some metric space $\mathbb{S}$ with Borel sets $\psi$, then the sequence $\{f(x_n) : n \geq 1\}$ is relatively compact in $(\mathbb{S}, \psi)$ and the set of its limit points coincides with $f(\mathcal{G}_k)$, a compact set.*

The following Lemma 4.4 is mainly used in Theorems 3.1 to deal with the relatively compact of the sum of two or more functions, where one function sequence converges to single zero-point set.

**Lemma 4.4** (Relatively compact for sum function). *Consider three sequences of relatively compact functions: $\{f_i^n(t)\} \subset \mathbb{C}[0,1]$, satisfying that $f_i^n(t) \rightrightarrows \mathcal{K}_i$ for $i = 1, 2, 3$, $(f_i^n(t), f_j^n(t)) \rightrightarrows \mathcal{K}_{ij}$, $i \neq j$, $t \in [0, 1]$. Suppose $\mathcal{K}_1 = \{\eta(t), t \in [0,1]\} \equiv \{0\}$, we have (i) $f_1^n(t) + f_2^n(t) \rightrightarrows \mathcal{K}_2$, (ii) $(f_2^n(t), f_1^n(t) + f_3^n(t)) \rightrightarrows \mathcal{K}_{23}$ as $n \to \infty$, where $\mathcal{K}_2$ is a compact set of univariate functions, $\mathcal{K}_{23}$ is a compact set of binary functions.*

*Proof.* We first note that (i) is a special case of (ii), it remains to prove (ii). By the definition of relatively compact, since $(f_2^n(t), f_3^n(t)) \rightrightarrows \mathcal{K}_{23}$, then for any subsequence $\{(f_2^{n_k}(t), f_3^{n_k}(t)), k = 1, 2, \dots\}$ of the given sequence $\{(f_2^n(t), f_3^n(t)), n = 1, 2, \dots\}$, there exists a convergent subsubsequence $\{(f_2^{n_{k_l}}(t), f_3^{n_{k_l}}(t)), l = 1, 2, \dots\}$ with its convergence limit, say a binary function $x_{23}(t)$, in $\mathcal{K}_{23}$ for all $t \in [0, 1]$, that is,

$$(f_2^{n_{k_l}}(t), f_3^{n_{k_l}}(t)) \to x_{23}(t) \quad for\ all \quad t \in [0, 1] \quad as \quad l \to \infty.$$

Notice that $\mathcal{K}_1 = \{\eta(t), t \in [0, 1]\}$, the sequence $\{f_1^n(t)\}$ is in fact a convergent sequence: $f_1^n(t) \to \eta(t) \equiv 0$ for all $t \in [0, 1]$ as $n \to \infty$. For its subsubsequence $\{(f_1^{n_{k_l}}(t), n = 1, 2, \dots\}$, we also have $f_1^{n_{k_l}}(t) \to 0$ for all $t \in [0, 1]$ as $l \to \infty$. So, these follow that

$$(f_2^{n_{k_l}}(t), f_1^{n_{k_l}}(t) + f_3^{n_{k_l}}(t)) \to x_{23}(t) \quad for\ all \quad t \in [0, 1] \quad as \quad l \to \infty.$$

Hence, $(f_2^n(t), f_1^n(t) + f_3^n(t)) \rightrightarrows \mathcal{K}_{23}$ for all $t \in [0, 1]$ as $n \to \infty$.

The next Lemma 4.5 is mainly used in Theorem 3.1 to deal with the relatively compact of some compound functions whose some function sequences are convergent with limit zero.

**Lemma 4.5** (Relatively compact for the compound). *Given three functions* $g_i(t) \in \mathbb{C}[0,1]$, $i = 1, 2, 3$ *and let* $g_i^n(t) = g_i(nt)/\varphi(n)$. *Suppose that* $g_1$ *is continuous under the uniform norm,* $g_2^n(t) \to \{\eta(t), t \in [0,1]\} \equiv \{0\}$ *for all* $t \in [0,1]$, $g_3^n(t)$ *is relatively compact and satisfies that* $g_3^n(t) \rightrightarrows \mathcal{K}_{g_3}$ *for all* $t \in [0,1]$, *where* $\mathcal{K}_{g_3} \subset \mathbb{C}[0,1]$ *is a compact set. Then, for all* $t \in [0,1]$, *both* $g_1 \circ g_3^n(t)$ *and* $g_1 \circ (g_2^n + g_3^n)(t)$ *are relatively compact with the identical limit set* $g_1(\mathcal{K}_{g_3})$, *that is, as* $n \to \infty$,

$$g_1 \circ g_3^n(t) \rightrightarrows g_1(\mathcal{K}_{g_3}) \quad and \quad g_1 \circ (g_2^n + g_3^n)(t) \rightrightarrows g_1(\mathcal{K}_{g_3}).$$

*Proof.* Because $g_1$ is a continuous mapping under the uniform norm, $g_3^n(t) \rightrightarrows \mathcal{K}_{g_3}$ for all $t \in [0,1]$, it follows from Lemma 4.3 that $g_1 \circ g_3^n(t) \rightrightarrows g_1(\mathcal{K}_{g_3})$ as $n \to \infty$. By the definition of relatively compact, since $g_3^n(t) \rightrightarrows \mathcal{K}_{g_3}$ for all $t \in [0,1]$, then for any subsequence $\{(g_3^{n_k}(t), k = 1, 2, \dots\}$ of the given sequence $\{(g_3^n(t)), n = 1, 2, \dots\}$, there exists a convergent subsubsequence $\{g_3^{n_{k_l}}(t), l = 1, 2, \dots\}$, which converges to a limit in $\mathcal{K}_{g_3}$, say $x_3(t) \in \mathcal{K}_{g_3}$ for all $t \in [0,1]$, that is,

$$g_3^{n_{k_l}}(t) \to x_3(t) \quad \text{for all} \quad t \in [0,1] \quad \text{as} \quad l \to \infty.$$

Since $g_2^n(t) \to 0$ for all $t \in [0,1]$, the subsbusequence $g_2^{n_{k_l}}(t) \to 0$ for all $t \in [0,1]$ as $l \to \infty$. As a result, for the sequence $\{g_2^n + g_3^n, n = 1, 2, \dots\}$, its subsbusequence

$$(g_2^{n_{k_l}} + g_3^{n_{k_l}})(t) = g_2^{n_{k_l}}(t) + g_3^{n_{k_l}}(t) \to x_3(t) \quad \text{for all} \quad t \in [0,1] \quad \text{as} \quad l \to \infty.$$

That is, $(g_2^n + g_3^n)(t) \rightrightarrows \mathcal{K}_{g_3}$. This follows that $g_1 \circ (g_2^n + g_3^n)(t) \rightrightarrows g_1(\mathcal{K}_{g_3})$.

## 4.2    Proofs of Theorems 3.1–3.3

In this section, we go to prove Theorems 3.1–3.3. Before proving, we first present the following Lemma 4.6 for application, which helps us to deal with the process $Q_k$ easily. Reader can refer to Lemma 6 in [18] for its proof.

**Lemma 4.6.** *Suppose that the overloaded* $(GI/GI)^K/1/PPSD$ *is in Case 1, that is, there exists a* $k_0: 1 \le k_0 < K$ *such that* $\sum_{j=1}^{k_0} \rho_j = 1$ *and* $\sum_{j=1}^{k_0+1} \rho_j > 1$, *then for* $i = 1, 2, \dots, K - k_0$,

$$\widetilde{X}_{k_0+i}(t) - \bar{X}_{k_0+i}(t) = \widetilde{W}_{k_0+i}(t) - \mu_{k_0+i}\widetilde{I}_{k_0+i-1}(t), \tag{4.7}$$

$$\widetilde{Q}_{k_0+i}(t) - \bar{Q}_{k_0+i}(t) = \widetilde{W}_{k_0+i}(t) - \mu_{k_0+i}\left[\widetilde{I}_{k_0+i-1}(t) - \widetilde{I}_{k_0+i}(t)\right]. \tag{4.8}$$

**Proof of Theorem** 3.1. We firstly note that the queueing system is in **Case** 1: There exists a $k_0: 1 \le k_0 < K$ such that $\sum_{j=1}^{k_0} \rho_j = 1$ and $\sum_{j=1}^{k_0+1} \rho_j > 1$. Since the first $k_0$ classes form a critically loaded $(GI/GI)^{k_0}/1/PPSD$ queueing system, the functional LIL (3.10) and the LIL (3.15) of classes $1, 2, \dots, k_0 - 1$

can be obtained similarly with Theorem 3.1 in [15], and the functional LIL (3.11) and the LIL (3.16) are similar with Theorem 3.2 in [15]. It remains to prove the functional LILs (3.12), (3.14) and the LILs (3.17), (3.18) for classes $k_0 + 1, k_0 + 2, \ldots, K$.

For applications, we give the fluid solution to (2.1):

$$
\bar{\mathbb{X}}_k(t) = \begin{cases} \left(0, 0, \rho_k t, (1 - \sum_{i=1}^{k} \rho_i)t, \lambda_k t\right), & k = 1, 2, \ldots, k_0, \\ (\lambda_k t, \rho_k t, 0, 0, 0), & k = k_0 + 1, k_0 + 2, \ldots, K. \end{cases} \tag{4.9}
$$

**The Functional LIL for Class $k_0 + 1$.** Now we are ready to deal with the functional LIL for class $k_0 + 1$. By (4.3),

$$
\tilde{X}_{k_0+1}(t) = \bar{X}_{k_0+1}(t) - \sum_{l=1}^{k_0} \frac{\mu_{k_0+1}}{\mu_l} \tilde{Q}_l(t) + \sum_{l=1}^{k_0} \frac{\mu_{k_0+1}}{\mu_l} \widetilde{W}_l(t) + \lambda_{k_0+1}^{1/2} c_{a,k_0+1} W_{a,k_0+1}(t). \tag{4.10}
$$

Notice that, in the above equality, the last two terms are driftless Brownian motions, the second term satisfies, for all $t \in [0, 1]$,

$$
\sum_{l=1}^{k_0-1} \frac{\mu_{k_0+1}}{\mu_l} \tilde{Q}_l^n(t) = \sum_{l=1}^{k_0-1} \frac{\mu_{k_0+1}}{\mu_l} \frac{\tilde{Q}_l^n(nt)}{\varphi(n)} \rightrightarrows \{0\}, \quad \frac{\tilde{Q}_{k_0}^n(nt)}{\varphi(n)} \rightrightarrows \Phi(\mathcal{G}(\mu_{k_0}\sigma_{k_0})), \quad \text{w.p.1},
$$

as $n \to \infty$, and then

$$
\sum_{l=1}^{k_0-1} \frac{\mu_{k_0+1}}{\mu_l} \frac{\tilde{Q}_l(t)}{t} \to 0, \quad \frac{\tilde{Q}_{k_0}(t)}{t} \to 0, \quad \text{w.p.1, as} \quad t \to \infty.
$$

This follows that $\lim_{t \to \infty} \tilde{X}_{k_0+1}(t)/t = \lim_{t \to \infty} \bar{X}_{k_0+1}(t)/t = \lambda_{k_0+1}$ w.p.1, or equivalently $\lim_{t \to \infty} \tilde{X}_{k_0+1}(t) = +\infty$, w.p.1. So, this, together with the definition of the mapping $\Psi$ in (2.2), implies that $\sup_{t \geq 0} \tilde{Y}_{k_0+1}(t) < \infty$ w.p.1, and as a result, for all $t \in [0, 1]$,

$$
\tilde{I}_{k_0+1}^n(t) = \frac{\tilde{Y}_{k_0+1}^n(t)}{\mu_{k_0+1}} = \frac{\tilde{Y}_{k_0+1}(nt)}{\mu_{k_0+1}\varphi(n)} \to 0, \quad \text{w.p.1}.
$$

For $T_{k_0+1}$, by (4.2) and (4.9),

$$
\tilde{T}_{k_0+1}^n(t) = \frac{\tilde{T}_{k_0+1}(nt)}{\varphi(n)} = \frac{\tilde{I}_{k_0}(nt) - \tilde{I}_{k_0+1}(nt)}{\varphi(n)} = \tilde{I}_{k_0}^n(t) - \tilde{I}_{k_0+1}^n(t) \rightrightarrows \Psi(\mathcal{G}(\sigma_{k_0})), \quad \text{w.p.1}
$$

for all $t \in [0, t]$, where the relatively compact limits follows from the functional LIL of $\tilde{I}_{k_0}^n(t)$. For $Q_{k_0+1}$, it follows from (4.2) and (4.9) that, for all $t \in [0, 1]$, w.p.1,

$$
\tilde{D}_{k_0+1}^n(t) = \frac{\tilde{D}_{k_0+1}(nt) - \bar{D}_{k_0+1}(nt)}{\varphi(n)} = \frac{\mu_{k_0+1}\tilde{T}_{k_0+1}(nt)}{\varphi(n)} = \mu_{k_0+1}\tilde{T}_{k_0+1}^n(t) \rightrightarrows \mu_{k_0+1}\Psi(\mathcal{G}(\sigma_{k_0})).
$$

For $Q_{k_0+1}$, it follows from (4.4), (4.8) and (4.9) that

$$\tilde{Q}_{k_0+1}^n(t) = \frac{\tilde{Q}_{k_0+1}(nt) - \bar{Q}_{k_0+1}(nt)}{\varphi(n)} = \frac{\widetilde{W}_{k_0+1}(nt)}{\varphi(n)} - \mu_{k_0+1}\left(\tilde{I}_{k_0}^n(t) - \tilde{I}_{k_0+1}^n(t)\right)$$

$$= \frac{\lambda_{k_0+1}^{1/2} c_{a,k_0+1} W_{a,k_0+1}(nt)}{\varphi(n)} - \frac{\mu_{k_0+1}\tilde{I}_{k_0}(nt)}{\varphi(n)} + \mu_{k_0+1}\tilde{I}_{k_0+1}^n(t)$$

$$= \frac{\lambda_{k_0+1}^{1/2} c_{a,k_0+1} W_{a,k_0+1}(nt)}{\varphi(n)} - \frac{\mu_{k_0+1}}{\mu_{k_0}\varphi(n)} \sup_{0\le s\le nt}\left[-\tilde{X}_{k_0}(s)\right]^+ + \mu_{k_0+1}\tilde{I}_{k_0+1}^n(t) \qquad (4.11)$$

$$= \frac{\lambda_{k_0+1}^{1/2} c_{a,k_0+1} W_{a,k_0+1}(nt)}{\varphi(n)} + \mu_{k_0+1}\tilde{I}_{k_0+1}^n(t)$$

$$- \frac{\mu_{k_0+1}}{\mu_{k_0}\varphi(n)} \sup_{0\le s\le nt}\left[\sum_{l=1}^{k_0-1} \frac{\mu_{k_0}}{\mu_l}\tilde{Q}_l(s) - W_{k_0}(s)\right]^+$$

$$= \frac{\lambda_{k_0+1}^{1/2} c_{a,k_0+1} W_{a,k_0+1}(nt)}{\varphi(n)} + \mu_{k_0+1}\tilde{I}_{k_0+1}^n(t)$$

$$- \sup_{0\le s\le nt}\left[\sum_{l=1}^{k_0-1} \frac{\mu_{k_0+1}\tilde{Q}_l(s)}{\mu_l\varphi(n)} - \frac{\mu_{k_0+1}W_{k_0}(s)}{\mu_{k_0}\varphi(n)}\right]^+$$

$$= \frac{\lambda_{k_0+1}^{1/2} c_{a,k_0+1} W_{a,k_0+1}(nt)}{\varphi(n)} + \mu_{k_0+1}\tilde{I}_{k_0+1}^n(t) - \Psi(-\sum_{l=1}^{k_0-1} \frac{\mu_{k_0+1}\tilde{Q}_l}{\mu_l\varphi(n)} + \frac{\mu_{k_0+1}}{\mu_{k_0}}\frac{W_{k_0}}{\varphi(n)})(nt).$$

Notice that, for all $t \in [0,1]$, $\tilde{I}_{k_0+1}^n(t) \to 0$ and $\sum_{l=1}^{k_0-1} \tilde{Q}_l^n(t) \to 0$ as $n \to \infty$. In order to find the functional LIL limit of $\tilde{Q}_{k_0+1}^n(t)$, it suffices to compute the functional LIL limit of

$$\frac{\lambda_{k_0+1}^{1/2} c_{a,k_0+1} W_{a,k_0+1}(nt)}{\varphi(n)} - \Psi(\frac{\mu_{k_0+1}}{\mu_{k_0}}\frac{W_{k_0}}{\varphi(n)})(nt)$$

$$= \frac{\lambda_{k_0+1}^{1/2} c_{a,k_0+1} W_{a,k_0+1}(nt)}{\varphi(n)} + \inf_{0\le s\le nt}\left[\frac{\mu_{k_0+1}}{\mu_{k_0}}\frac{W_{k_0}}{\varphi(n)}(s)\right]^+ \qquad (4.12)$$

$$= G_2(\frac{\mu_{k_0+1}}{\mu_{k_0}}\frac{W_{k_0}}{\varphi(n)}, \frac{\lambda_{k_0+1}^{1/2} c_{a,k_0+1} W_{a,k_0+1}}{\varphi(n)})(nt),$$

where $G_2$ is defined in (3.9). Because $W_{k_0}$ and $W_{a,k_0+1}$ are independent Brownian motions, and

$$Var\left(\lambda_{k_0+1}^{1/2} c_{a,k_0+1} W_{a,k_0+1}(t) + \frac{\mu_{k_0+1}}{\mu_{k_0}} W_{k_0}(t)\right) = q_{k_0+1}^2 t,$$

where $q_{k_0+1}$ is defined in (3.13), together with Lemmas 4.2 and 4.3 we get that, for all $t \in [0,1]$, w.p.1,

$$G_2(\frac{\mu_{k_0+1}}{\mu_{k_0}}\frac{W_{k_0}}{\varphi(n)}, \frac{\lambda_{k_0+1}^{1/2} c_{a,k_0+1} W_{a,k_0+1}}{\varphi(n)})(nt) \rightrightarrows G_2(\mathcal{G}_2(q_{k_0+1})).$$

So, by Lemma 4.5, for all $t \in [0,1]$, $\tilde{Q}_{k_0+1}^n(t) \rightrightarrows G_2(\mathcal{G}_2(q_{k_0+1}))$ w.p.1.

We next go to find the functional LIL limit for $\widetilde{Z}_{k_0+1}^n(t)$. By (4.2), (4.3) and (4.9),

$$
\begin{aligned}
\widetilde{Z}_{k_0+1}^n(t) &= \frac{\widetilde{Z}_{k_0+1}(nt) - \bar{Z}_{k_0+1}(nt)}{\varphi(n)} \\
&= \frac{1}{\mu_{k_0+1}} \frac{\widetilde{Q}_{k_0+1}(nt) - \mu_{k_0+1}^{1/2} c_{s,k_0+1} W_{s,k_0+1}(\rho_{k_0+1} nt) - \bar{Q}_{k_0+1}(nt)}{\varphi(n)} \\
&= \frac{1}{\mu_{k_0+1}} \frac{\widetilde{W}_{k_0+1}(nt) - \mu_{k_0+1}^{1/2} c_{s,k_0+1} W_{s,k_0+1}(\rho_{k_0+1} nt)}{\varphi(n)} - \left( \widetilde{I}_{k_0}^n(t) - \widetilde{I}_{k_0+1}^n(t) \right) \\
&= \frac{1}{\mu_{k_0+1}} \frac{\lambda_{k_0+1}^{1/2} c_{a,k_0+1} W_{a,k_0+1}(nt) - \mu_{k_0+1}^{1/2} c_{s,k_0+1} W_{s,k_0+1}(\rho_{k_0+1} nt)}{\varphi(n)} \\
&\quad - \frac{1}{\mu_{k_0}} \widetilde{Y}_{k_0}^n(t) + \widetilde{I}_{k_0+1}^n(t) \\
&= \frac{1}{\mu_{k_0+1}} \frac{\lambda_{k_0+1}^{1/2} c_{a,k_0+1} W_{a,k_0+1}(nt) - \mu_{k_0+1}^{1/2} c_{s,k_0+1} W_{s,k_0+1}(\rho_{k_0+1} nt)}{\varphi(n)} \\
&\quad - \frac{1}{\mu_{k_0} \varphi(n)} \Psi(\widetilde{X}_{k_0})(nt) + \widetilde{I}_{k_0+1}^n(t) \\
&= \frac{1}{\mu_{k_0+1}} \frac{\lambda_{k_0+1}^{1/2} c_{a,k_0+1} W_{a,k_0+1}(nt) - \mu_{k_0+1}^{1/2} c_{s,k_0+1} W_{s,k_0+1}(\rho_{k_0+1} nt)}{\varphi(n)} \\
&\quad - \Psi\left(- \sum_{l=1}^{k_0-1} \frac{\widetilde{Q}_l}{\mu_l \varphi(n)} + \frac{W_{k_0}}{\mu_{k_0} \varphi(n)}\right)(nt) + \widetilde{I}_{k_0+1}^n(t),
\end{aligned}
$$

where the third equality holds similarly with the third equality for $\widetilde{Q}_{k_0+1}^n(t)$ above, the forth equality holds because $\bar{T}_{k_0+1}(t) = 0$ and $\widetilde{I}_{k_0}^n(t) = \Psi(\widetilde{X}_{k_0})(nt)/(\mu_{k_0}\varphi(n))$, the sixth equality holds similarly with (4.11). As analysis for $Q_{k_0+1}$, if we compute the functional LIL for $Z_{k_0+1}$, by Lemma 4.5 it suffices to consider the following functional LIL, as (4.12),

$$
\begin{aligned}
&\frac{1}{\mu_{k_0+1}} \frac{\lambda_{k_0+1}^{1/2} c_{a,k_0+1} W_{a,k_0+1}(nt) - \mu_{k_0+1}^{1/2} c_{s,k_0+1} W_{s,k_0+1}(\rho_{k_0+1} nt)}{\varphi(n)} - \Psi\left(\frac{W_{k_0}}{\mu_{k_0}\varphi(n)}\right)(nt) \\
&= \frac{\lambda_{k_0+1}^{1/2} c_{a,k_0+1} W_{a,k_0+1}(nt) - \mu_{k_0+1}^{1/2} c_{s,k_0+1} W_{s,k_0+1}(\rho_{k_0+1} nt)}{\mu_{k_0+1}\varphi(n)} + \inf_{0 \le s \le nt}\left[\frac{W_{k_0}(s)}{\mu_{k_0}\varphi(n)}\right]^+ \\
&= G_2\left(\frac{W_{k_0}}{\mu_{k_0}\varphi(n)}, \frac{\lambda_{k_0+1}^{1/2} c_{a,k_0+1} W_{a,k_0+1} - \mu_{k_0+1}^{1/2} c_{s,k_0+1} W_{s,k_0+1}(\rho_{k_0+1}\cdot)}{\mu_{k_0+1}\varphi(n)}\right)(nt),
\end{aligned}
$$

where $G_2$ is defined in (3.9). Because $W_{k_0}$, $W_{a,k_0+1}$ and $W_{s,k_0+1}$ are independent Brownian motions, and

$$
Var\left(\frac{W_{k_0}(t)}{\mu_{k_0}} + \frac{\lambda_{k_0+1}^{1/2} c_{a,k_0+1} W_{a,k_0+1}(t) - \mu_{k_0+1}^{1/2} c_{s,k_0+1} W_{s,k_0+1}(\rho_{k_0+1} t)}{\mu_{k_0+1}}\right) = z_{k_0+1}^2 t,
$$

where $z_{k_0+1}$ is defined in (3.13), it follows from Lemmas 4.2 and 4.3 that, for all $t \in [0,1]$, w.p.1,

$$
G_2\left(\frac{W_{k_0}}{\mu_{k_0}\varphi(n)}, \frac{\lambda_{k_0+1}^{1/2} c_{a,k_0+1} W_{a,k_0+1} - \mu_{k_0+1}^{1/2} c_{s,k_0+1} W_{s,k_0+1}(\rho_{k_0+1}\cdot)}{\mu_{k_0+1}\varphi(n)}\right)(nt) \rightrightarrows G_2(\mathcal{G}_2(z_{k_0+1})),
$$

then $\widetilde{Z}^n_{k_0+1}(t) \Rightarrow G_2(\mathcal{G}_2(z_{k_0+1}))$ w.p.1 for all $t \in [0, 1]$.

**The Functional LIL for Classes $k_0 + 2$ to $K$.** We next consider the functional LIL limits for classes $k_0 + 2, k_0 + 3, \ldots, K$. We first prove $\widetilde{I}_k(t) < \infty$ w.p.1 for all $t$ and $k = k_0 + 2, k_0 + 3, \ldots, K$ by induction. Since $\bar{T}_{k_0+2}(t) = 0$, rewriting (4.7) yields that

$$\widetilde{X}_{k_0+2}(t) - \bar{X}_{k_0+2}(t) = \lambda^{1/2}_{k_0+2} c_{a,k_0+2} W_{a,k_0+2}(t) - \mu_{k_0+2}\widetilde{I}_{k_0+1}(t).$$

Notice that $\widetilde{I}^n_{k_0+1}(t) \to 0$ w.p.1 for all $t \in [0,1]$, then $\widetilde{I}_{k_0+1}(t)/t \to 0$ w.p.1 as $t \to \infty$. So,

$$\lim_{t\to\infty} \frac{\widetilde{X}_{k_0+2}(t)}{t} = \lim_{t\to\infty} \frac{\bar{X}_{k_0+2}(t)}{t} = \lambda_{k_0+2} > 0 \quad \text{w.p.1,}$$

and $\lim_{t\to\infty} \widetilde{X}_{k_0+2}(t) = \infty$ w.p.1. Hence, $\widetilde{I}_{k_0+2}(t) = \sup_{0 \le s \le t} \left(-\widetilde{X}_{k_0+2}(s)\right)^+ < +\infty$ w.p.1.

Next, we suppose that, for all $t$, $\widetilde{I}_k(t) < \infty$ w.p.1 $k = k_0 + 2, k_0 + 3, \ldots, k_0 + i$ with $i < K - k_0$. Since $\bar{T}_{k_0+i+1}(t) = 0$, (4.7) implies that

$$\widetilde{X}_{k_0+i+1}(t) - \bar{X}_{k_0+i+1}(t) = \lambda^{1/2}_{k_0+i+1} c_{a,k_0+i+1} W_{a,k_0+i+1}(t) - \mu_{k_0+i+1}\widetilde{I}_{k_0+i}(t).$$

By the induction hypothesis, $\widetilde{I}_{k_0+i}(t) < \infty$ w.p.1 for all $t$, then $\widetilde{I}_{k_0+i}(t)/t \to 0$ w.p.1 as $t \to \infty$. With similar analysis of $\widetilde{I}_{k_0+2}$, we have $\lim_{t\to\infty} \widetilde{X}_{k_0+i+1}(t) = \infty$ w.p.1, and as a result, $\widetilde{I}_{k_0+i+1}(t) = \sup_{0 \le s \le t} \left(-\widetilde{X}_{k_0+i+1}(s)\right)^+ < +\infty$ w.p.1. Hence, we proved that $\widetilde{I}_k(t) < \infty$ w.p.1 for all $t$ and $k = k_0 + 2, k_0 + 3, \ldots, K$. This follows that, for all $t \in [0, 1]$ and $i = 1, 2, \ldots, K - k_0$,

$$\widetilde{I}^n_{k_0+i}(t) = \frac{\widetilde{I}_{k_0+i}(nt) - \bar{I}_{k_0+i}(nt)}{\varphi(n)} = \frac{\widetilde{I}_{k_0+i}(nt)}{\varphi(n)} \to 0, \quad \text{w.p.1,}$$

and then $\widetilde{I}^n_{k_0+i}(t) \Rightarrow \{0\}$ w.p.1 for all $t \in [0, 1]$ and $i = 1, 2, \ldots, K - k_0$.

We now go to find the functional LIL for $T_k$ and $D_k$, $k = k_0+2, k_0+3, \ldots, K$. Notice that $\widetilde{I}^n_k(t) \Rightarrow \{0\}$ w.p.1 for all $t \in [0, 1]$ and $k = k_0 + 2, k_0 + 3, \ldots, K$, we have, by (4.2) and (4.9), $k = k_0 + 2, k_0 + 3, \ldots, K$,

$$\widetilde{T}^n_{k_0+i}(t) = \frac{\widetilde{T}_{k_0+i}(nt)}{\varphi(n)} = \frac{\widetilde{I}_{k_0+i-1}(nt) - \widetilde{I}_{k_0+i}(nt)}{\varphi(n)} = \widetilde{I}^n_{k_0+i-1}(t) - \widetilde{I}^n_{k_0+i}(t) \Rightarrow 0, \quad \text{w.p.1,}$$

$$\widetilde{D}^n_{k_0+i}(t) = \frac{\widetilde{D}_{k_0+i}(nt)}{\varphi(n)} = \frac{\mu_{k_0+i}\widetilde{T}_{k_0+i}(nt)}{\varphi(n)} = \mu_{k_0+i}\widetilde{T}^n_{k_0+i}(t) \Rightarrow 0, \quad \text{w.p.1.}$$

For $Q_k$, $k = k_0 + 2, k_0 + 3, \ldots, K$, since $\bar{T}_k(t) = 0$ for all $k = k_0 + 2, k_0 + 3, \ldots, K$, it follows from (4.8) that, for all $t \in [0, 1]$ and $k = k_0+2, k_0+3, \ldots, K$,

$$\widetilde{Q}^n_k(t) = \frac{\widetilde{Q}_k(nt) - \bar{Q}_k(nt)}{\varphi(n)} = \frac{\lambda^{1/2}_k c_{a,k} W_k(nt)}{\varphi(n)} - \mu_k \left[\widetilde{I}^n_{k-1}(t) - \widetilde{I}^n_k(t)\right] \Rightarrow \mathcal{G}(\lambda^{1/2}_k c_{a,k}), \quad \text{w.p.1,}$$

where the relatively compact limit holds because $\widetilde{I}_k^n(t) \rightrightarrows \{0\}$ w.p.1 for all $t \in [0, 1]$ and $k = k_0 + 2, k_0 + 3, \ldots, K$.

For $Z_k$, $k = k_0 + 2, k_0 + 3, \ldots, K$, by (4.2) and (4.9), for all $t \in [0, 1]$,

$$
\begin{aligned}
\widetilde{Z}_k^n(t) &= \frac{\widetilde{Z}_k(nt) - \bar{Z}_k(nt)}{\varphi(n)} = \frac{\widetilde{Q}_k(nt) - \mu_k^{1/2} c_{s,k} W_{s,k}(n\rho_k t) - \bar{Q}_k(nt)}{\mu_k \varphi(n)} \\
&= \frac{1}{\mu_k \varphi(n)} \left[ \lambda_k^{1/2} c_{a,k} W_{a,k}(nt) - \mu_k^{1/2} c_{s,k} W_{s,k}(n\rho_k t) \right] - \widetilde{I}_{k-1}^n(t) + \widetilde{I}_k^n(t) \\
&\rightrightarrows \mathcal{G}\left( \frac{c_k \sqrt{\lambda_k}}{\mu_k} \right), \quad \text{w.p.1,}
\end{aligned}
$$

where the relatively compact limit holds because the variance of the sum of $\lambda_k^{1/2} c_{a,k} W_{a,k}(t)$ and $\mu_k^{1/2} c_{s,k} W_{s,k}(\rho_k t)$ is $\lambda_k c_k^2 t$, and $\widetilde{I}_k^n(t) \rightrightarrows \{0\}$ w.p.1 for all $t \in [0, 1]$ and $k = k_0 + 2, k_0 + 3, \ldots, K$.

**The LIL.** We now go to find the LIL (3.17) for class $k_0 + 1$ and the LIL (3.18) for classes $k_0 + 2, \ldots, K$. Notice that, for $\delta > 0$ given above, by (2.1) on page 169 in [24], $|y(b)| \leq \delta\sqrt{b} \leq \delta$ for any $y \in \mathcal{G}(\delta)$ and $0 \leq b \leq 1$, then, for any $\delta > 0$,

$$
\sup_{x \in \Psi(\mathcal{G}(\delta))} x(1) = \sup_{y \in \mathcal{G}(\delta)} \sup_{0 \leq s \leq 1} \{-y(s)\} = \delta, \quad \inf_{x \in \Psi(\mathcal{G}(\delta))} x(1) = \inf_{y \in \mathcal{K}(\delta)} \sup_{0 \leq s \leq 1} \{-y(s)\} = 0, \quad (4.13)
$$

where the supremum and infimum are attained for the functions $y(s) = -s$ and $y(s) = 0$ respectively. We get the LIL limits for $T_{k_0+1,sup}^*, T_{k_0+1,inf}^*, D_{k_0+1,sup}^*$ and $D_{k_0+1,inf}^*$. The LIL limits $I_{k_0+1,sup}^* = I_{k_0+1,inf}^* = 0$ follows from $\mathcal{K}_{I_{k_0+1}} = \{0\}$. That is, (3.17) holds. For classes $k_0 + 2, k_0 + 3, \ldots, K$, their LIL limits can be obtained similarly with Theorem 3.1 in [15], which follows (3.18). □

**Proof of Theorem** 3.2. We firstly note that there exists a $k_0$: $1 \leq k_0 < K$ such that $\sum_{j=1}^{k_0} \rho_j < 1$ and $\sum_{j=1}^{k_0+1} \rho_j > 1$. As Theorem 3.1, we first go to find the functional LIL and then the LIL limits. Notice that the first $k_0$ classes form a underloaded $(GI/GI)^{k_0}/1/PPSD$ queueing system, we have the functional LIL and the LIL limits the same as (3.10) and (3.15) with $k = 1, 2, \ldots, k_0$, respectively, the functional LIL and the LIL limits for classes $k_0 + 2, k_0 + 3, \ldots, K$ are the same as (3.14) and (3.18) respectively. So, we only need to consider class $k_0 + 1$, that is, the functional LIL (3.19) and the LIL (3.21).

For applications, we give the following fluid solution to (2.1) for class $k = k_0 + 1$:

$$
\bar{\mathbb{X}}_k(t) = \left( \mu_k (\sum_{l=1}^{k} \rho_l - 1)t, (\sum_{l=1}^{k} \rho_l - 1)t, (1 - \sum_{l=1}^{k-1} \rho_l)t, 0, \mu_k (1 - \sum_{l=1}^{k-1} \rho_l)t \right), k = k_0 + 1. \quad (4.14)
$$

**The Functional LIL for Class $k_0 + 1$.** We now consider the functional LIL limits for class $k_0 + 1$. As the analysis for (4.10), by (4.3) we have

$$
\widetilde{X}_{k_0+1}(t) = \bar{X}_{k_0+1}(t) - \sum_{l=1}^{k_0} \frac{\mu_{k_0+1}}{\mu_l} \widetilde{Q}_l(t) + \sum_{l=1}^{k_0+1} \frac{\mu_{k_0+1}}{\mu_l} \widetilde{W}_l(t). \quad (4.15)
$$

where $\bar{X}_{k_0+1}(t) = \mu_{k_0+1}\left(\sum_{l=1}^{k_0+1}\rho_l - 1\right)t = \bar{Q}_{k_0+1}(t)$ by (2.1). Since the first $k_0$ classes form a underloaded $(GI/GI)^{k_0}/1/\text{PPSD}$ queueing system, we have, for all $t \in [0, 1]$,

$$\frac{1}{\varphi(n)}\sum_{l=1}^{k_0}\frac{\mu_{k_0+1}}{\mu_l}\widetilde{Q}_l(nt) \Rightarrow \{0\}, \quad \text{w.p.1}, \tag{4.16}$$

and then,

$$\frac{1}{t}\sum_{l=1}^{k_0}\frac{\mu_{k_0+1}}{\mu_l}\widetilde{Q}_l(t) \to 0, \quad \text{w.p.1, as} \quad t \to \infty.$$

Notice that $\widetilde{W}_l(t)$ is a driftless Brownian motion, we have $\lim_{t\to\infty}\tilde{X}_{k_0+1}(t) = +\infty$ w.p.1, as a result, $\tilde{I}_{k_0+1}(t) = \sup_{0\leq s\leq t}\left[-\tilde{X}_{k_0+1}(s)\right]^+/\mu_{k_0+1} < \infty$ w.p.1. So, it follows that, for all $t \in [0, 1]$, $\tilde{I}_{k_0+1}^n(t) = \tilde{I}_{k_0+1}(nt)/\varphi(n) \Rightarrow \{0\}$ w.p.1.

For the functional LIL of $T_{k_0+1}$, since $\tilde{T}_{k_0+1}(t) = \tilde{I}_{k_0}(t) - \tilde{I}_{k_0+1}(t)$ and $\bar{T}_{k_0+1}(t) = \bar{I}_{k_0}(t) - \bar{I}_{k_0+1}(t)$, we have, for all $t \in [0, 1]$,

$$\tilde{T}_{k_0+1}^n(t) = \frac{\tilde{T}_{k_0+1}(nt) - \bar{T}_{k_0+1}(nt)}{\varphi(n)} = \frac{\tilde{I}_{k_0}(nt) - \bar{I}_{k_0}(nt)}{\varphi(n)} - \frac{\tilde{I}_{k_0+1}(nt) - \bar{I}_{k_0+1}(nt)}{\varphi(n)}$$

$$= \tilde{I}_{k_0}^n(t) - \tilde{I}_{k_0+1}^n(t) \Rightarrow -\mathcal{G}(\sigma_{k_0}), \quad \text{w.p.1},$$

because, for all $t \in [0, 1]$, w.p.1, $\tilde{I}_{k_0}^n(t) \Rightarrow -\mathcal{G}(\sigma_{k_0})$, see Theorem 3.1 in [15], and $\tilde{I}_{k_0+1}^n(t) \Rightarrow \{0\}$ as above.

For the functional LIL of $Q_{k_0+1}$, by (4.2), (4.3) and (4.14),

$$\tilde{Q}_{k_0+1}^n(t) = \frac{\tilde{Q}_{k_0+1}(nt) - \bar{Q}_{k_0+1}(nt)}{\varphi(n)} = \frac{\tilde{X}_{k_0+1}(t) - \bar{X}_{k_0+1}(nt) + \tilde{Y}_{k_0+1}(nt)}{\varphi(n)}$$

$$= \frac{1}{\varphi(n)}\left(-\sum_{l=1}^{k_0}\frac{\mu_{k_0+1}}{\mu_l}\widetilde{Q}_l(nt) + W_{k_0+1}(nt)\right) + \mu_{k_0+1}\tilde{I}_{k_0+1}^n(t).$$

Notice that (4.16) holds, $\tilde{I}_{k_0+1}^n(t) \Rightarrow \{0\}$ for all $t \in [0, 1]$ w.p.1, and by Eqs. (4.1) and (4.14),

$$W_{k_0+1}(t) = \sum_{l=1}^{k_0+1}\frac{\mu_{k_0+1}}{\mu_l}\widetilde{W}_l(t) = \sum_{l=1}^{k_0}\frac{\mu_{k_0+1}}{\mu_l}\widetilde{W}_l(t) + \lambda_{k_0+1}^{1/2}c_{a,k_0+1}W_{a,k_0+1}(t)$$

$$- \mu_{k_0+1}^{1/2}c_{s,k_0+1}W_{s,k_0+1}\left(\left(1 - \sum_{l=1}^{k_0}\rho_l\right)t\right) \tag{4.17}$$

is a driftless Brownian motion with variance parameter $q_{k_0+1}^*$ which is defined in (3.20), we have, for all $t \in [0, 1]$, $\tilde{Q}_{k_0+1}^n(t) \Rightarrow \mathcal{G}(q_{k_0+1}^*)$ w.p.1.

For the functional LIL of $Z_{k_0+1}$, by (4.3),

$$\widetilde{X}_{k_0+1}(t) - \bar{X}_{k_0+1}(t) = -\sum_{l=1}^{k_0} \frac{\mu_{k_0+1}}{\mu_l} \widetilde{Q}_l(t) + W_{k_0+1}(t),$$

then, with (4.2), (4.14) and (4.17),

$$\begin{aligned}
\widetilde{Z}_{k_0+1}^n(t) &= \frac{\widetilde{Z}_{k_0+1}(nt) - \bar{Z}_{k_0+1}(nt)}{\varphi(n)} = \frac{1}{\mu_{k_0+1}\varphi(n)} \left[ \widetilde{X}_{k_0+1}(nt) - \bar{X}_{k_0+1}(nt) \right] + \widetilde{I}_{k_0+1}^n(t) \\
&\quad + \frac{1}{\mu_{k_0+1}\varphi(n)} \left[ \mu_{k_0+1}^{1/2} c_{s,k_0+1} W_{s,k_0+1}(\bar{T}_{k_0+1}(nt)) - \mu_{k_0+1}^{1/2} c_{s,k_0+1} W_{s,k_0+1}(n\rho_{k_0+1}t) \right] \\
&= \widetilde{I}_{k_0+1}^n(t) - \sum_{l=1}^{k_0} \frac{1}{\mu_l} \widetilde{Q}_l^n(t) + \sum_{l=1}^{k_0+1} \frac{1}{\mu_l \varphi(n)} \left[ \lambda_l^{1/2} c_{a,l} W_{a,l}(nt) - \mu_l^{1/2} c_{s,l} W_{s,l}(n\rho_l t) \right].
\end{aligned}$$

Similarly, for all $t \in [0,1]$,

$$\widetilde{I}_{k_0+1}^n(t) \rightrightarrows \{0\}, \quad \sum_{l=1}^{k_0} \frac{1}{\mu_l} \widetilde{Q}_l^n(t) \rightrightarrows \{0\} \quad \text{w.p.1}, \tag{4.18}$$

and by Lemma 4.2, for all $t \in [0,1]$,

$$\sum_{l=1}^{k_0+1} \frac{1}{\mu_l \varphi(n)} \left[ \lambda_l^{1/2} c_{a,l} W_{a,l}(nt) - \mu_l^{1/2} c_{s,l} W_{s,l}(n\rho_l t) \right] \rightrightarrows \mathcal{G}(\sigma_{k_0+1}) \quad \text{w.p.1},$$

So, for all $t \in [0,1]$, $\widetilde{Z}_{k_0+1}^n(t) \rightrightarrows \mathcal{G}(\sigma_{k_0+1})$ w.p.1.

Finally, we go to find the functional LIL for $D_{k_0+1}$. By (4.2), (4.3), (4.4), (4.14) and (4.17), we have

$$\begin{aligned}
\widetilde{D}_{k_0+1}^n(t) &= \frac{\widetilde{D}_{k_0+1}(nt) - \bar{D}_{k_0+1}(nt)}{\varphi(n)} \\
&= \frac{\mu_{k_0+1}}{\varphi(n)} \left[ \widetilde{T}_{k_0+1}(nt) - \bar{T}_{k_0+1}(nt) \right] + \frac{\mu_{k_0+1}^{1/2} c_{s,k_0+1} W_{s,k_0+1}(\bar{T}_{k_0+1}(nt))}{\varphi(n)} \\
&= -\frac{1}{\varphi(n)} \left[ \widetilde{Q}_{k_0+1}(nt) - \bar{Q}_{k_0+1}(nt) \right] + \frac{\lambda_{k_0+1}^{1/2} c_{a,k_0+1} W_{a,k_0+1}(nt)}{\varphi(n)} \\
&= -\frac{1}{\varphi(n)} \left[ \widetilde{X}_{k_0+1}(nt) - \bar{X}_{k_0+1}(nt) \right] - \mu_{k_0+1} \widetilde{I}_{k_0+1}^n(t) + \frac{\lambda_{k_0+1}^{1/2} c_{a,k_0+1} W_{a,k_0+1}(nt)}{\varphi(n)} \\
&= \sum_{l=1}^{k_0} \frac{\mu_{k_0+1}}{\mu_l} \widetilde{Q}_l^n(t) - \mu_{k_0+1} \widetilde{I}_{k_0+1}^n(t) - \sum_{l=1}^{k_0} \frac{\mu_{k_0+1} \widetilde{W}_l(nt)}{\mu_l \varphi(n)} \\
&\quad + \frac{\mu_{k_0+1}^{1/2} c_{s,k_0+1} W_{s,k_0+1}(\bar{T}_{k_0+1}(nt))}{\varphi(n)},
\end{aligned}$$

because, as (4.18), for all $t \in [0,1]$, the first two terms

$$\sum_{l=1}^{k_0} \frac{\mu_{k_0+1}}{\mu_l} \widetilde{Q}_l^n(t) \rightrightarrows \{0\}, \quad \mu_{k_0+1} \widetilde{I}_{k_0+1}^n(t) \rightrightarrows \{0\}, \quad \text{w.p.1},$$

and for the last two terms, notice that, for all $t \geq 0$, $\bar{T}_{k_0+1}(t) = (1 - \sum_{l=1}^{k_0} \rho_l)t$,

$$
Var\left(\sum_{l=1}^{k_0} \frac{\mu_{k_0+1}\widetilde{W}_l(t)}{\mu_l} + \mu_{k_0+1}^{1/2}c_{s,k_0+1}W_{s,k_0+1}(\bar{T}_{k_0+1}(t))\right) = (d_{k_0+1}^*)^2 t,
$$

where $d_{k_0+1}^*$ is defined in (3.20), then, for all $t \in [0, 1]$, by Lemma 4.2,

$$
-\sum_{l=1}^{k_0} \frac{\mu_{k_0+1}\widetilde{W}_l(nt)}{\mu_l\varphi(n)} + \frac{\mu_{k_0+1}^{1/2}c_{s,k_0+1}W_{s,k_0+1}(\bar{T}_{k_0+1}(nt))}{\varphi(n)} \rightrightarrows \mathcal{G}(d_{k_0+1}^*), \quad \text{w.p.1,}
$$

we have, together with Lemma 4.4, $\widetilde{D}_{k_0+1}^n(t) \rightrightarrows \mathcal{G}(d_{k_0+1}^*)$ w.p.1 for all $t \in [0, 1]$.

**The LIL.** The LIL limits (3.21) for classes $k_0+1, \ldots, K$ can be obtained similarly with the part of **the LIL** of Theorem 3.1 in [15]. □

**Proof of Theorem 3.3.** We first note that the traffic intensity $\rho_1 > 1$. Since the $(GI/GI)^1/1/PPSD$ is a overloaded queueing system, the classes from 2 to $K$ operate similarly with classes $k_0 + 2, k_0 + 3, \ldots, K$ in Theorem 3.2, it follows the functional LIL and the LIL limits for class 2 to $K$ satisfy (3.14) and (3.18) with $k = 2, 3, \ldots, K$, respectively. It suffices to find the functional LIL and the LIL for class 1. For applications, we first give the class-1's fluid solution to (2.1): $\bar{\mathbb{X}}_1(t) = ((\lambda_1 - \mu_1)t, (1 - \rho_1)t, t, 0, \mu_1 t)$ for all $t \geq 0$.

**The Functional LIL for Class 1.** Now, we are ready to compute the functional LIL limits for class 1. We first deal with $I_1$. Since $\rho_1 > 1$ and

$$
\widetilde{X}_1(t) = \mu_1(\rho_1 - 1)t + \lambda_1^{1/2}c_{a,1}W_{a,1}(t) - \mu_1^{1/2}c_{s,1}W_{s,1}(t),
$$

we have $\lim_{t\to\infty} \widetilde{X}_1(t)/t = \mu_1(\rho_1 - 1) > 0$ w.p.1, as a result, $\lim_{t\to\infty} \widetilde{X}_1(t) = \infty$ w.p.1. So,

$$
\Psi(\widetilde{X}_1)(nt) = \sup_{0 \leq s \leq nt}\left[-\widetilde{X}_1(s)\right]^+ < \infty \quad \text{and} \quad \widetilde{I}_1^n(t) = \frac{1}{\mu_1\varphi(n)}\Psi(\widetilde{X}_1)(nt) \rightrightarrows \{0\}, \quad \text{w.p.1,}
$$

for all $t \in [0, 1]$. As a result, for all $t \in [0, 1]$,

$$
\widetilde{T}_1^n(t) = \frac{\bar{T}_1(nt) - \tilde{T}_1(nt)}{\varphi(n)} = -\frac{\tilde{I}_1(nt)}{\varphi(n)} = -\widetilde{I}_1^n(t) \rightrightarrows \{0\}, \quad \text{w.p.1,}
$$

For $Q_1$, since $\bar{Q}_1(t) = \bar{X}_1(t)$ for all $t \geq 0$, by Lemma 4.2,

$$
\widetilde{Q}_1^n(t) = \frac{\widetilde{Q}_1(nt) - \bar{Q}_1(nt)}{\varphi(n)} = \frac{1}{\varphi(n)}\left[\widetilde{X}_1(nt) + \mu_1\tilde{I}_1(nt) - \bar{X}_1(nt)\right]
$$

$$
= \frac{1}{\varphi(n)}\left[\lambda_1^{1/2}c_{a,1}W_{a,1}(nt) - \mu_1^{1/2}c_{s,1}W_{s,1}(nt)\right] + \mu_1\widetilde{I}_1^n(t)
$$

$$
\rightrightarrows \mathcal{G}(\sqrt{\lambda_1 c_{a,1}^2 + \mu_1 c_{s,1}^2}), \quad \text{w.p.1}
$$

for all $t \in [0,1]$, because $\widetilde{I}_1^n(t) \Rightarrow \{0\}$ w.p.1 for all $t \in [0,1]$ and the variance of the Brownian motion $\lambda_1^{1/2} c_{a,1} W_{a,1}(t) - \mu_1^{1/2} c_{s,1} W_{s,1}(t)$ is $(\lambda_1 c_{a,1}^2 + \mu_1 c_{s,1}^2)t$. For the functional LIL of $Z_1$, by (2.1) and (4.2),

$$\bar{Z}_1(t) = \frac{\bar{Q}_1(t)}{\mu_1} \quad \text{and} \quad \widetilde{Z}_1(t) = \frac{\widetilde{Q}_1(t)}{\mu_1} + \frac{\mu_1^{1/2} c_{s,1} W_{s,1}(t) - \mu_1^{1/2} c_{s,1} W_{s,1}(\rho_1 t)}{\mu_1},$$

and then, by Lemma 4.2, for all $t \in [0,1]$,

$$\begin{aligned}
\widetilde{Z}_1^n(t) &= \frac{\widetilde{Z}_1(nt) - \bar{Z}_1(nt)}{\varphi(n)} = \frac{\widetilde{Q}_1^n(t)}{\mu_1} + \frac{\mu_1^{1/2} c_{s,1} W_{s,1}(nt) - \mu_1^{1/2} c_{s,1} W_{s,1}(n\rho_1 t)}{\mu_1 \varphi(n)} \\
&= \widetilde{I}_1^n(t) + \frac{\widetilde{W}_1(nt)}{\mu_1 \varphi(n)} + \frac{\mu_1^{1/2} c_{s,1} W_{s,1}(nt) - \mu_1^{1/2} c_{s,1} W_{s,1}(n\rho_1 t)}{\mu_1 \varphi(n)} \\
&= \widetilde{I}_1^n(t) + \frac{\lambda_1^{1/2} c_{a,1} W_{a,1}(nt) - \mu_1^{1/2} c_{s,1} W_{s,1}(n\rho_1 t)}{\mu_1 \varphi(n)} \Rightarrow \mathcal{G}(\sigma_1), \quad \text{w.p.1,}
\end{aligned}$$

because the variance of Brownian motion $(\lambda_1^{1/2} c_{a,1} W_{a,1}(t) - \mu_1^{1/2} c_{s,1} W_{s,1}(\rho_1 t))$ is $\lambda(c_{a,1}^2 + c_{s,1}^2)t = \mu_1^2 \sigma_1^2 t$. For the functional LIL of $D_1$, since, by (2.1) and (4.2), for all $t \in [0,1]$,

$$\widetilde{D}_1^n(t) = \frac{\widetilde{D}_1(nt) - \bar{D}_1(nt)}{\varphi(n)} = \mu_1 \widetilde{T}_1^n(t) + \frac{\mu_1^{1/2} c_{s,1} W_{s,1}(nt)}{\varphi(n)} \Rightarrow \mathcal{G}(\mu_1^{1/2} c_{s,1}), \quad \text{w.p.1,}$$

because, for all $t \in [0,1]$, $\widetilde{T}_1^n(t) \Rightarrow \{0\}$ w.p.1.

**The LIL.** When $\rho_1 > 1$, the LIL limits can be obtained in two cases: class 1 and classes from 2 to $K$. For class 1, we can get the LIL limits (3.23) as the analysis in part: **The LIL** in Theorem 3.1 in [15]. For classes 2 to $K$, its proof is similar too. □

## 5    Conclusion

In this paper, we introduced a model of an overloaded multi-class single-server queue under a priority service discipline, called $(GI/GI)^K/1/PPSD$ queueing system. Specifically, the $(GI/GI)^K/1/PPSD$ model consists of a single server and $K$ classes of customers, with class 1 taking the highest priority, class 2 the second highest priority, and so on, class $K$ the lowest priority. The service discipline is assumed to be preemptive.

We developed the functional LIL and its corresponding LIL in Lévy's type (1.1) for the overloaded $(GI/GI)^K/1/PPSD$ queueing system by focusing on five key performance measures: queue length, workload, busy time, idle time, and departure processes. The functional LIL and LIL limits refine the previous work on the functional strong law of large numbers (or fluid limit) and show us more insight on the asymptotic fluctuations around the fluid limits in the functional and numerical form, respectively. All the functional LIL and LIL limits are identified respectively to be the functional compact set and functions by the first and second moments of primitive data: the arrival and service. Our main results, Theorems 3.1–3.3, give us a whole analysis on the asymptotic deviation covering all classes with traffic intensity less than, equivalent to and more than one.

The approach here to analysis for main results: Theorems 3.1–3.3, are based on the fluid and strong approximations, called the strong approximation approach, which operates in four steps: find the fluid solution and strong approximation, transferring the discrete problems on the functional LIL into their corresponding strong approximation problems, finding the functional LIL limits and finding the LIL in Lévy's type from some compact set given by the functional LIL limits.

It is also interesting and important to find the asymptotic fluctuations around the fluid limit for some other queueing systems. From the main results: Theorems 3.1–3.3, it says that more random parameters, such that arrival and service times, makes the system fluctuation around the mean value bigger. However, this changing fluctuation does not embody in other approximations, such as the fluid, diffusion and strong approximations. The functional LIL and LIL limits can help us to find more insight on the asymptotic fluctuations in numerical and functional forms.

**Acknowledgements.** This work is supported by NSFC grants 11871116 and 11971074.

# Appendix

To improve the paper's readability, we summarize the notations of the performance functions used in this paper. In the following two tables, the first one is for the fluid approximation, strong approximation, the functional LIL and its corresponding LIL in Lévy's type, for $k = 1, 2, \ldots, K$,

| Performance functions | Queue length | Workload | Busy time | Idle time | Departure |
|---|---|---|---|---|---|
| Original notation | $Q_k(t)$ | $Z_k(t)$ | $B_k(t)$ | $I_k(t)$ | $D_k(t)$ |
| Fluid approximation | $\bar{Q}_k(t)$ | $\bar{Z}_k(t)$ | $\bar{B}_k(t)$ | $\bar{I}_k(t)$ | $\bar{D}_k(t)$ |
| Strong approximation | $\widetilde{Q}_k(t)$ | $\widetilde{Z}_k(t)$ | $\widetilde{B}_k(t)$ | $\widetilde{I}_k(t)$ | $\widetilde{D}_k(t)$ |
| Functional LIL | $\mathcal{K}_{Q_k}$ | $\mathcal{K}_{Z_k}$ | $\mathcal{K}_{B_k}$ | $\mathcal{K}_{I_k}$ | $\mathcal{K}_{D_k}$ |
| And for strong approximation | $\mathcal{K}_{\widetilde{Q}_k}$ | $\mathcal{K}_{\widetilde{Z}_k}$ | $\mathcal{K}_{\widetilde{B}_k}$ | $\mathcal{K}_{\widetilde{I}_k}$ | $\mathcal{K}_{\widetilde{D}_k}$ |
| Lévy's LIL of superior limit | $Q^*_{k,sup}(t)$ | $Z^*_{k,sup}(t)$ | $B^*_{k,sup}(t)$ | $I^*_{k,sup}(t)$ | $D^*_{k,sup}(t)$ |
| Lévy's LIL of inferior limit | $Q^*_{k,inf}(t)$ | $Z^*_{k,inf}(t)$ | $B^*_{k,inf}(t)$ | $I^*_{k,inf}(t)$ | $D^*_{k,inf}(t)$ |

The second is for the scaled process of the LIL in Lévy's type, the functional LIL and the fluid limits, for $k = 1, 2, \ldots, K$,

| Scalings | of Lévy's LIL | of functional LIL | and of strong approximation | of fluid |
|---|---|---|---|---|
| for $Q_k$ | $\frac{Q_k(t)-\bar{Q}_k(t)}{\varphi(t)}$ | $Q_k^n(t) \equiv \frac{Q_k(nt)-\bar{Q}_k(nt)}{\varphi(n)}$ | $\tilde{Q}_k^n(t) \equiv \frac{\bar{Q}_k(nt)-\bar{Q}_k(nt)}{\varphi(n)}$ | $\bar{Q}_k^{(n)}(t) \equiv \frac{1}{n}Q^n(nt)$ |
| for $Z_k$ | $\frac{Z_k(t)-\bar{Z}_k(t)}{\varphi(t)}$ | $Z_k^n(t) \equiv \frac{Z_k(nt)-\bar{Z}_k(nt)}{\varphi(n)}$ | $\tilde{Z}_k^n(t) \equiv \frac{\bar{Z}_k(nt)-\bar{Z}_k(nt)}{\varphi(n)}$ | $\bar{Z}_k^{(n)}(t) \equiv \frac{1}{n}Z^n(nt)$ |
| for $B_k$ | $\frac{B_k(t)-\bar{B}_k(t)}{\varphi(t)}$ | $B_k^n(t) \equiv \frac{B_k(nt)-\bar{B}_k(nt)}{\varphi(n)}$ | $\tilde{B}_k^n(t) \equiv \frac{\bar{B}_k(nt)-\bar{B}_k(nt)}{\varphi(n)}$ | $\bar{B}_k^{(n)}(t) \equiv \frac{1}{n}B^n(nt)$ |
| for $I_k$ | $\frac{I_k(t)-\bar{I}_k(t)}{\varphi(t)}$ | $I_k^n(t) \equiv \frac{I_k(nt)-\bar{I}_k(nt)}{\varphi(n)}$ | $\tilde{I}_k^n(t) \equiv \frac{\bar{I}_k(nt)-\bar{I}_k(nt)}{\varphi(n)}$ | $\bar{I}_k^{(n)}(t) \equiv \frac{1}{n}I^n(nt)$ |
| for $D_k$ | $\frac{D_k(t)-\bar{D}_k(t)}{\varphi(t)}$ | $D_k^n(t) \equiv \frac{D_k(nt)-\bar{D}_k(nt)}{\varphi(n)}$ | $\tilde{D}_k^n(t) \equiv \frac{\bar{D}_k(nt)-\bar{D}_k(nt)}{\varphi(n)}$ | $\bar{D}_k^{(n)}(t) \equiv \frac{1}{n}D^n(nt)$ |

# References

1. Bramson, M., Dai, J.G.: Heavy traffic limits for some queueing networks. Ann. Appl. Probab. **11**(1), 49–90 (2001)
2. Caramellino, L.: Strassen's law of the iterated logarithm for diffusion processes for small time. Stochast. Process. Appl. **74**(1), 1–19 (1998)
3. Chen, H.: Fluid approximations and stability of multiclass queueing networks I: wrok-conserving disciplines. Ann. Appl. Probab. **5**(3), 637–665 (1995)
4. Chen, H., Mandelbaum, A.: Hierarchical modeling of stochastic networks, Part I: fluid models. In: Yao, D.D. (ed.) Stochastic Modeling and Analysis of Manufacturing Systems, pp. 47–105. Springer, New York (1994). https://doi.org/10.1007/978-1-4612-2670-3_2
5. Chen, H., Shen, X.: Strong approximations for multiclass feedforward queueing networks. Ann. Appl. Probab. **10**(3), 828–876 (2000)
6. Chen, H., Yao, D.D.: Fundamentals of Queueing Networks. Springer, New York (2001). https://doi.org/10.1007/978-1-4757-5301-1
7. Chen, H., Zhang, H.: A sufficient condition and a necessary condition for the diffusion approximations of multiclass queueing networks under priority service disciplines. Queueing Syst. **34**(1–4), 237–268 (2000)
8. Csörgő, M., Révész, P.: Strong Approximations in Probability and Statistics. Academic Press, New York (1981)
9. Csörgő, M., Horváth, L.: Weighted Approximations in Probability and Statistics. Wiley, New York (1993)
10. Csörgő, M., Deheuvels, P., Horváth, L.: An approximation of stopped sums with applications in queueing theory. Adv. Appl. Probab. **19**(3), 674–690 (1987)
11. Dai, J.G.: On the positive Harris recurrence for multiclass queueing networks: a unified approach via fluid limit models. Ann. Appl. Probab. **5**(1), 49–77 (1995)
12. Ethier, S.N., Kurtz, T.G.: Markov Processes: Characterization and Convergence. Wiley, New York (1986)
13. Glynn, P.W., Whitt, W.: A new view of the heavy-traffic limit for infinite-server queues. Adv. Appl. Probab. **23**(1), 188–209 (1991)
14. Glynn, P.W., Whitt, W.: Departures from many queues in series. Ann. Appl. Probab. **1**(4), 546–572 (1991)
15. Guo, Y., Hou X.: Functional law of the iterated logarithm for multiclass queues with preemptive priority service discipline: the underloaded and critically loaded cases (2018)
16. Guo, Y., Li, Z.: Asymptotic variability analysis for a two-stage tandem queue, part I: the functional law of the iterated logarithm. J. Math. Anal. Appl. **450**, 1479–1509 (2017)
17. Guo, Y., Li, Z.: Asymptotic variability analysis for a two-stage tandem queue, part II: the law of the iterated logarithm. J. Math. Anal. Appl. **450**, 1510–1534 (2017)

18. Guo, Y., Liu, Y.: A law of iterated logarithm for multiclass queues with preemptive priority service discipline. Queueing Syst. **79**(3), 251–291 (2015)
19. Guo, Y., Liu, Y., Pei, R.: Functional law of iterated logarithm for multi-server queues with batch arrivals and customer feedback. Ann. Oper. Res. **264**, 157–191 (2018)
20. Harrison, M.: A limit theorem for priority queues in heavy traffic. J. Appl. Probab. **10**(4), 907–912 (1973)
21. Horváth, L.: Strong approximation of renewal processes. Stochast. Process. Appl. **18**(1), 127–138 (1984)
22. Horváth, L.: Strong approximation of extended renewal processes. Ann. Probab. **12**(4), 1149–1166 (1984)
23. Horváth, L.: Strong approximations of open queueing networks. Math. Oper. Res. **17**(2), 487–508 (1992)
24. Iglehart, G.L.: Multiple channel queues in heavy traffic: IV. Law of the iterated logarithm. Z.Wahrscheinlichkeitstheorie verw. Geb. **17**, 168–180 (1971)
25. Lévy, P.: Théorie de l'addition des variables aléatories. Gauthier-Villars, Paris (1937)
26. Lévy, P.: Procesus stochastique et mouvement Brownien. Gauthier-Villars, Paris (1948)
27. Mandelbaum, A., Massey, W.A.: Strong approximations for time-dependent queues. Math. Oper. Res. **20**(1), 33–64 (1995)
28. Mandelbaum, A., Massey, W.A., Reiman, M.: Strong approximations for Markovian service networks. Queueing Syst. **30**, 149–201 (1998)
29. Minkevičius, S., Steišūnas, S.: A law of the iterated logarithm for global values of waiting time in multiphase queues. Stat. Probab. Lett. **61**(4), 359–371 (2003)
30. Minkevičius, S.: On the law of the iterated logarithm in multiserver open queueing networks. Stochastics **86**(1), 46–59 (2014)
31. Peterson, W.P.: A heavy traffic limit theorem for networks of queues with multiple customer types. Math. Oper. Res. **16**(1), 90–118 (1991)
32. Sakalauskas, L.L., Minkevičius, S.: On the law of the iterated logarithm in open queueing networks. Eur. J. Oper. Res. **20**(3), 632–640 (2000)
33. Strassen, V.: An invariance principle for the law of the iterated lorgarith. Z. Wahrscheinlichkeitstheorie verw. Geb. **3**(3), 211–226 (1964)
34. Tsai, T.H.: Empirical law of the iterated logarithm for Markov chains with a countable state space. Stat. Probab. Lett. **89**(2), 175–191 (2000)
35. Whitt, W.: Weak convergence theorems for priority queues: preemptive-resume discipline. J. Appl. Probab. **8**(1), 74–94 (1971)
36. Zhang, H.: Strong approximations of irreducible closed queueing networks. Advances in Appl. Probab. **29**(2), 498–522 (1997)
37. Zhang, H., Hsu, G.X., Wang, R.X.: Strong approximations for multiple channels in heavy traffic. J. Appl. Prob. **27**(3), 658–670 (1990)
38. Zhang, H., Hsu, G.X.: Strong approximations for priority queues: head-of-the-line-first discipline. Queueing Syst. **10**(3), 213–234 (1992)

# Functional Law of the Iterated Logarithm for Multiclass Queues with Preemptive Priority Service Discipline: The Underloaded and Critically Loaded Cases

Yongjiang Guo$^{(\boxtimes)}$ and Xiyang Hou

School of Science, Beijing University of Posts and Telecommunications,
Beijing 100876, People's Republic of China
yongerguo@bupt.edu.cn

**Abstract.** A functional *law of the iterated logarithm* (LIL) and its corresponding LIL are established for a underloaded and critically loaded multiclass queueing system with preemptive priority service discipline, covering five performance measures: queue length, workload, busy time, idle time and number of departures. All the functional LIL and the LIL limits quantify the magnitude of asymptotic stochastic fluctuations of the performance compensated by their deterministic fluid limits in two forms: the functional and numerical, respectively. By the primitive data of the first and second moments of the interarrival and service times, all the functional LILs are expressed into some compact sets of continuous functions and all the LILs are some analytic functions. The proofs are based on the fluid approximation and the strong approximation of the queueing system, with the fluid approximation characterizing the expected values of the performance functions and the strong approximation approximating discrete performance processes with reflected Brownian motions.

**Keywords:** Multiclass queue · The functional law of the iterated logarithm · The law of the iterated logarithm · Strong approximation · Brownian motion

## 1 Introduction

This paper can be thought of as a companion of [10]. Both of papers focus on a common multiclass $(GI/GI)^K/1/PPSD$ queueing system, which consists of one server and $K$ customer classes, all customers arrive at the system from outside following class-dependent renewal processes (the first $GI$), and are served by a *preemptive priority service discipline* (PPSD) with class 1 taking the highest priority, class 2 taking the second highest priority, and so on, class $K$ taking the lowest priority, the service times are assumed to be *independent and identically distributed* (i.i.d.) non-exponential random variables (the second $GI$). However, this paper and [10] have different research issues. Guo [10] aims to the overloaded regime of the system, defined by the traffic intensity $\rho > 1$, however this

© Springer Nature Singapore Pte Ltd. 2019
Q.-L. Li et al. (Eds.): Cao Festschrift 2019, CCIS 1102, pp. 344–360, 2019.
https://doi.org/10.1007/978-981-15-0864-6_17

paper considers the underloaded and critically loaded regimes with the traffic intensity $\rho < 1$ and $\rho = 1$, respectively. We develop, for the underloaded and critically loaded regimes, a functional *law of the iterated logarithm* (LIL) and its corresponding LIL in Lévy's type for five key performance measures: queue length, workload, busy time, idle time and the number of departures.

Among literature on queueing studies, it is emphasized that the previous works of [1, 3, 6, 10] and [13] are the most revalant to the current paper. In [1, 6], the authors obtained the fluid approximation for queueing networks under static buffer priority service policy, which helps us understand the deterministic dynamic performance for the $(GI/GI)^K/1/PPSD$ queueing system. Chen and Shen [3] developed the strong approximation for a feedforward queueing networks, which follows the strong approximation for the $(GI/GI)^K/1/PPSD$ queueing system. Guo etc. [10, 13] considered the $(GI/GI)^K/1/PPSD$ queueing system. Guo [10] studied the overloaded regime for the $(GI/GI)^K/1/PPSD$ queueing system, and obtained the functional LIL and the LIL in Lévy's type. Guo and Liu [13] established the LIL limits in Csörgő and Révész's type. For the underloaded and critically loaded regimes of the $(GI/GI)^K/1/PPSD$ queueing system, we establish the functional LIL and its associated LIL in Lévy's type for the five performance measures. Other literature on multiclass priority queueing systems can refer to [2, 4] for fluid approximation, [8, 9, 22–24] for strong approximations, [11, 14, 15] for functional LIL and [12, 18–20] for LIL.

Both functional LIL and LIL limits quantize the asymptotic stochastic fluctuations of stochastic processes around their fluid limits in functional and numerical forms, respectively. To our best knowledge, there are one type of the functional LIL and two types of the LIL limits for stochastic processes in literature. The functional LIL generally originates to Strassen [21], and the two types of LIL limits are the Lévy's LIL [16, 17] and its later generalized LIL by Csörgő and Révész [5]. The functional LIL and the LIL limits are both firstly established for the standard Brownian motion. Let $W(t)$ be a one-dimensional standard Brownian motion, and $W^n(t) = W(nt)/\sqrt{2n \log \log n}$ be the corresponding functional LIL–scaled process for $t \in [0, 1]$ and $n = 3, 4, \ldots$. Strassen's functional LIL tells us that, (i) the scaled sequence $\{W^n(t), n = 3, 4, \ldots\}$ is relatively compact, which means that every subsequence has a convergent subsubsequence, (ii) all the limits of the convergent subsequences form a compact set in the space of absolutely continuous functions $x$, which satisfies that $x(0) = 0$, $\int_0^1 [\dot{x}(t)]^2 \, dt \leq 1$, with $\dot{x}$ being the derivative of $x$ and the square brackets being the inner product. Mathematically, the compact set has the following form:

$$\left\{ x \in \mathbb{C}^1[0,1] : \ x(0) = 0, \ \int_0^1 [\dot{x}(t)]^2 \, dt \leq 1 \right\},$$

where $\mathbb{C}^1[0,1]$ is the space of the one-dimensional continuous functions on $[0,1]$. The Lévy's LIL is to consider the superior and inferior limits: *with probability one* (w.p.1),

$$\limsup_{L \to \infty} \frac{W(L)}{\sqrt{2L \log \log L}} = -\liminf_{L \to \infty} \frac{W(L)}{\sqrt{2L \log \log L}} = 1, \qquad (1.1)$$

and the generalized LIL by Csörgő and Révész [5] is: w.p.1,

$$\limsup_{L \to \infty} \frac{\sup_{0 \le t \le L} |W(t)|}{\sqrt{2L \log \log L}} = 1. \tag{1.2}$$

Besides the above LILs in (1.1) and (1.2), there is a weaker type of LIL: for Brwonian motion $W$ defined above, $\sup_{0 \le t \le L} |W(t)| = O(\sqrt{L \log \log L})$ w.p.1, where we say $f(t) = O(g(t))$ as $t \to \infty$ if $\limsup_{t \to \infty} |f(t)/g(t)| \le N$ for some $N > 0$. Readers can refer to [2–4] for more details.

The contributions of this paper are summarized below. As a supplement to [10], we study the underloaded and critically loaded regimes, and develop the functional LIL and the LIL in Lévy's type for five key performance measures of the $(GI/GI)^K/1/PPSD$ queueing system, including the queue length, workload, busy time, idle time and departure processes (see Sect. 2 for their definitions). All the functional LIL limits of the above performance measures are identified to be as compact sets of analytic functions and the corresponding LIL in Lévy's type as analytic functions, where all the functions are constructed in terms of the primitive model parameters, and the model parameters are the first and second moments of the interarrival and service times. The main results, Theorems 3.1 and 3.2, are proved by an approach based on strong approximation (or strong approximation approach), which is different from the previous approach based on probabilistic inequality or renewal process construction. Some interesting observations from Theorems 3.1 and 3.2 are given, for example, the functional LIL– and LIL–versions of the Little's law between the queue length and workload processes are preserved well in the underloaded and critically loaded regimes, the LIL of the queue length, as the function of traffic intensity $\rho$, is continuous when $\rho < 1$ and has a big upper skip at $\rho = 1$.

The presentation of the paper will proceed as follows. First, the $(GI/GI)^K/1/ PPSD$ queueing model, the associated key performance functions, and their corresponding fluid limits are in Sect. 2. Then, in Sect. 3, the functional LIL and its corresponding LIL limits for the processes of interest, that is, Theorems 3.1 and 3.2, are presented. For reader's convenience, some insights of these results are given too. The proofs of Theorems 3.1 and 3.2 are given in Sect. 4, in which we also present the proof's bases before proving: Strassen's functional LIL, the strong approximation, and some results on the relatively compact. Finally, we draw conclusions in Sect. 5.

We conclude the introduction with an account of the notations and conventions used throughout the paper. Let $(\Omega, \mathcal{F}, \mathsf{P})$ a common probability space on which all random variables and processes are defined, $\mathsf{E}(\cdot)$ and $\mathrm{Var}(\cdot)$ be the corresponding mean and variance, respectively. " $=_d$ " means equality in distribution. Given any positive integer $k$, $\mathbb{R}^k$ and $\mathbb{R}^k_+$ are the sets of the $k$-dimensional real numbers and nonnegative real numbers, especially, $\mathbb{R} = \mathbb{R}^1$ and $\mathbb{R}_+ = \mathbb{R}^1_+$. Let prime be the transpose of a vector or matrix. For $x = (x_1, x_1, \ldots, a_k)' \in \mathbb{R}^k$, $|x| = (x_1^2 + \cdots + x_k^2)^{1/2}$ is the Euclidean norm. For $x \in \mathbb{R}$, $[x]^+ \equiv \max\{x, 0\}$, where "$\equiv$" denotes a definition. Given $x, y \in \mathbb{R}$, let $x \vee y = \max\{x, y\}$, and $x \wedge y = \min\{x, y\}$. Suppose $\mathbb{D}^k[x, y]$ is the space of $k$-dimensional right continuous functions on $[x, y)$ with left limits on $(x, y]$, endowed the Skorohod topology [7]. Suppose that $\mathbb{C}^k[x, y]$ is the subset of continuous functions in $\mathbb{D}^k[x, y]$. Especially $\mathbb{D} \equiv \mathbb{D}^1$, $\mathbb{C} \equiv \mathbb{C}^1$ and $\mathbb{D}_0 \equiv \{x \in \mathbb{D} : x(0) \geq 0\}$. Given functions $f$ and $g$, $f \circ g(t) = f(g(t))$, $\dot{f}(t)$ denotes the derivative of $f(t)$. If $\{f^n, n \geq 1\}$ is relatively compact and the set of all limit points form a compact set $\mathcal{K}_f$, we say that $f^n \rightrightarrows \mathcal{K}_f$ w.p.1. Let $||f||_L \equiv \sup_{0 \leq t \leq L} |f(t)|$ be the uniform norm of $f$. We say that $f^n \to f$ uniformly on compact set (u.o.c.) if $||f^n - f||_L \to 0$, as $n \to \infty$. Let $\varphi(t) \equiv \sqrt{2t \log\log t}$ for all $t$ bigger than Eular constant. Define the identity mapping $e$ with $e(t) \equiv t$, the zero mapping $\eta$ with $\eta(t) = 0$ for all $t$, and the indicator function $1_C$ with $1_C(s) = 1$ if $s \in C$ and 0 otherwise.

## 2  The $(GI/GI)^K/1/PPSD$ queueing system

As described in the first paragraph in Sect. 1, the $(GI/GI)^K/1/PPSD$ queueing system has one single-server station serving $K$ classes of customers, $K \geq 2$. Each class has its own exogenous arrival process and its own priority. We assume that class-1 has the highest priority, class-2 has the second highest priority, and so on, class-$K$ has the lowest priority. A PPSD policy is operated among all classes. In words, customers with identical priority are severed accordingly to the order of arrivals, and customers with different priorities are served accordingly to their priorities: A newly arrival of class-$k$ can preempt the service of class-$i$ with $i > k$, and the interrupted class-$i$ customer resumes its service only when all customers with higher priority than class-$i$ complete service.

**The Primitive Data.** Suppose that $u_k \equiv \{u_k(n), n = 1, 2, \dots\}$ and $v_k \equiv \{v_k(n), n = 1, 2, \dots\}$, $k = 1, 2, \dots, K$, are $2K$ sequences of mutually independent i.i.d. random variables, denote the interarrival and service time sequences. More specially, $u_k(n)$ denotes the class-$k$ interarrival time between the $(n-1)st$ and the $n$th arrivals, and $v_k(n)$ denotes the service time of the $n$th served the class-$k$ customer. Assume that the sequences $u_k$ and $v_k$ have the following moments: means $\mathsf{E}[u_k(1)] \equiv 1/\lambda_k$ and $\mathsf{E}[v_k(1)] \equiv 1/\mu_k$, variances $Var[v_k(1)]$ and $Var[v_k(1)]$, respectively. Define the corresponding squared coefficients of variation: $c_{a,k}^2 \equiv Var[u_k(1)]/(\mathsf{E}[u_k(1)])^2$ and $c_{s,k}^2 \equiv Var[v_k(1)]/(\mathsf{E}[v_k(1)])^2$. Define the arrival process:

$$A_k(t) \equiv \max\{n \geq 0 : U_k(n) \leq t\} \quad \text{with} \quad U_k(n) \equiv \sum_{i=1}^{n} u_k(i),$$

where $A_k(t)$ denotes the total number of arrivals for class-$k$ customers in $(0, t]$, the service process:

$$S_k(t) \equiv \max\{n \geq 0 : V_k(n) \leq t\} \quad \text{with} \quad V_k(n) \equiv \sum_{i=1}^{n} v_k(i), \quad n = 1, 2, \dots.$$

where $S_k(t)$ is the number of class-$k$ customers the server serves in $t$ units of time.

**The Dynamical Equations.** We define the following performance measures. For class $k, k = 1, 2, \ldots, K$,

- $Q_k(t)$ is the queue length process and denotes the number of class-$k$ customers in the system at time $t$;
- $Z_k(t)$ is the workload process and denotes the time that the server needs to finish the current customers of class $k$ in system at time $t$;
- $T_k(t)$ is the busy time process and is the total amount of time the server is busy serving class $k$ customers in $[0, t]$;
- $I_k(t) = t - \sum_{i=1}^{k} T_i(t)$ is the residual time in $[0, t]$ and denotes the time available for the server to serve classes $k + 1, \ldots, K$ after serving the first $k$ classes;
- $D_k(t)$ is the departure process and counts the completed customers of class $k$ in $[0, t]$.

The dynamical equations are given:

$$Q_k(t) = A_k(t) - D_k(t) \geq 0, \quad D_k(t) = S_k(T_k(t)),$$
$$T_k(t) = \int_0^t \mathbf{1}_{\{Q_k(s)>0, Q_i(s)=0, i<k\}} ds, \quad T_1(t) = \int_0^t \mathbf{1}_{\{Q_1(s)>0\}} ds,$$
$$Z_k(t) = V_k(A_k(t)) - T_k(t),$$
$$\int_0^t Q_k(t) dI_k(t) = 0,$$

where the first equation follows from the flow conservation, the forth holds because $V_k(A_k(t))$ denotes the amount of work of class-$k$ arrivals by time $t$, and the last is from the fact that $I_k(t)$ increases only when $Q_k(t) = 0$.

We define the traffic intensity

$$\rho \equiv \sum_{k=1}^{K} \rho_k \quad \text{with} \quad \rho_k \equiv \frac{\lambda_k}{\mu_k}, \quad k = 1, 2, \ldots, K.$$

We say the system is underloaded when $\rho < 1$, critically loaded when $\rho = 1$ and overloaded when $\rho > 1$.

**The Fluid Limit.** We now give the fluid limit for applications next. First, we define the fluid scaled process:

$$\bar{Q}^{(n)}(t) = \frac{1}{n} Q(nt), \quad \bar{Z}^{(n)}(t) = \frac{1}{n} Z(nt), \quad \bar{T}^{(n)}(t) = \frac{1}{n} T(nt),$$
$$\bar{I}^{(n)}(t) = \frac{1}{n} I(nt), \quad \bar{D}^{(n)}(t) = \frac{1}{n} D(nt).$$

The following fluid limits is from [6].

**Lemma 2.1** (Fluid limits [6]). *Suppose that the system is initially empty, that is $Q_k(0) = 0, k = 1, 2, \ldots, K$. If $\mathsf{E}[u_k(1)] < \infty$ and $\mathsf{E}[v_k(1)] < \infty$, we have*

$$\left(\bar{Q}^{(n)}, \bar{Z}^{(n)}, \bar{T}^{(n)}, \bar{I}^{(n)}, \bar{D}^{(n)}\right) \to (\bar{Q}, \bar{Z}, \bar{T}, \bar{I}, \bar{D}) \equiv \bar{\mathbb{X}}, \quad u.o.c., \quad w.p.1, \quad as \ n \to \infty,$$

*where the limit*

$$\bar{\mathbb{X}} \equiv (\bar{\mathbb{X}}_1, \ldots, \bar{\mathbb{X}}_K), \quad having\ the\ element \quad \bar{\mathbb{X}}_k = \left(\bar{Q}_k, \bar{Z}_k, \bar{T}_k, \bar{I}_k, \bar{D}_k\right), \quad k = 1, \ldots, K,$$

*satisfies*

$$\bar{Q}_k(t) \equiv \mu_k \bar{Z}_k(t) \equiv \lambda_k t - \bar{D}_k(t) = \bar{X}_k(t) + \bar{Y}_k(t) \geq 0,$$

$$\bar{X}_k(t) \equiv (\lambda_k - \mu_k)t + \mu_k \sum_{l=1}^{k-1} \bar{T}_l(t),$$

$$\bar{T}_k(t) \equiv t - \sum_{l=1}^{k-1} \bar{T}_l(t) - \bar{I}_k(t),$$

$$\bar{I}_k(t) \equiv \frac{\bar{Y}_k(t)}{\mu_k}, \quad \bar{Y}_k(t) \equiv \Psi(\bar{X}_k)(t),$$

$$\bar{D}_k(t) \equiv \mu_k \bar{T}_k(t), \tag{2.1}$$

*the functions $\Phi$ and $\Psi$ are defined for $x \in \mathbb{D}_0$ as*

$$\Psi(x)(t) \equiv \sup_{0 \leq s \leq t} \{-x(s)\}^+, \quad \Phi(x)(t) \equiv x(t) + \sup_{0 \leq s \leq t} \{-x(s)\}^+. \tag{2.2}$$

The mapping $(\Psi, \Phi)$ in (2.2) is known as the one dimensional oblique reflection mapping, and is Lipschitz continuous in uniform norm, see [4] for more details.

**The Objective.** As a supplement to [10], we go to find the LIL in Lévy's type and functional LIL for the five main performance measures $(Q_k, Z_k, B_k, I_k, D_k, 1 \leq k \leq K)$ in the underloaded and critically loaded regimes. The LIL will present us some analytic function of the primitive moment data

$$\mathcal{D} \equiv \left(\lambda_k, \mu_k, c_{a,k}^2, c_{s,k}^2, c_k^2, 1 \leq k \leq K\right), \tag{2.3}$$

the functional LIL will show us some compact set of continuous functions of the data (2.3).

# 3  Main Results

In this section, we establish the LIL in Lévy's type and functional LIL for the five main performance measures $(Q_k, Z_k, B_k, I_k, D_k, 1 \leq k \leq K)$ in the underloaded and critically loaded regimes. We first define the LIL and functional LIL in Sect. 3.1 and give the main results in Sect. 3.2.

## 3.1   The Primitive Scalings

We define the primitive scalings firstly for the LIL and secondly the functional LIL.

**The LIL Scalings.** We define the LIL-scaled processes: for $k = 1, 2, \ldots, K$,

$$Q^*_{k,sup} \equiv \limsup_{t \to \infty} \frac{Q_k(t) - \bar{Q}_k(t)}{\varphi(t)}, \quad Q^*_{k,inf} \equiv \liminf_{t \to \infty} \frac{Q_k(t) - \bar{Q}_k(t)}{\varphi(t)},$$

$$Z^*_{k,sup} \equiv \limsup_{t \to \infty} \frac{Z_k(t) - \bar{Z}_k(t)}{\varphi(t)}, \quad Z^*_{k,inf} \equiv \liminf_{t \to \infty} \frac{Z_k(t) - \bar{Z}_k(t)}{\varphi(t)},$$

$$T^*_{k,sup} \equiv \limsup_{t \to \infty} \frac{T_k(t) - \bar{T}_k(t)}{\varphi(t)}, \quad T^*_{k,inf} \equiv \liminf_{t \to \infty} \frac{T_k(t) - \bar{T}_k(t)}{\varphi(t)},$$

$$I^*_{k,sup} \equiv \limsup_{t \to \infty} \frac{I_k(t) - \bar{I}_k(t)}{\varphi(t)}, \quad I^*_{k,inf} \equiv \liminf_{t \to \infty} \frac{I_k(t) - \bar{I}_k(t)}{\varphi(t)},$$

$$D^*_{k,sup} \equiv \limsup_{t \to \infty} \frac{D_k(t) - \bar{D}_k(t)}{\varphi(t)}, \quad D^*_{k,inf} \equiv \liminf_{t \to \infty} \frac{D_k(t) - \bar{D}_k(t)}{\varphi(t)}. \quad (3.1)$$

For $k = 1, 2 \ldots, K$, let

$$\mathcal{X}^*_{k,sup} \equiv \left( Q^*_{k,sup}, Z^*_{k,sup}, T^*_{k,sup}, I^*_{k,sup}, D^*_{k,sup} \right),$$
$$\mathcal{X}^*_{k,inf} \equiv \left( Q^*_{k,inf}, Z^*_{k,inf}, T^*_{k,inf}, I^*_{k,inf}, D^*_{k,inf} \right). \quad (3.2)$$

We go to identify all the LIL limits in (3.2) as analytic functions based on (2.3).

**The Functional LIL Scalings.** We define the functional LIL-scaled processes: for $t \in [0,1]$ and $n = 3, 4, \ldots,,$

$$Q^n_k(t) \equiv \frac{Q_k(nt) - \bar{Q}_k(nt)}{\varphi(n)}, \quad Z^n_k(t) \equiv \frac{Z_k(nt) - \bar{Z}_k(nt)}{\varphi(n)}, \quad T^n_k(t) \equiv \frac{T_k(nt) - \bar{T}_k(nt)}{\varphi(n)},$$

$$I^n_k(t) \equiv \frac{I_k(nt) - \bar{I}_k(nt)}{\varphi(n)}, \quad D^n_k(t) \equiv \frac{D_k(nt) - \bar{D}_k(nt)}{\varphi(n)}, \quad \text{for} \quad k = 1, 2, \ldots, K. \quad (3.3)$$

We go to establish the following result:

$$(Q^n_k, Z^n_k, T^n_k, I^n_k, D^n_k) \Rightarrow (\mathcal{K}_{Q_k}, \mathcal{K}_{Z_k}, \mathcal{K}_{T_k}, \mathcal{K}_{I_k}, \mathcal{K}_{D_k}) \equiv \mathcal{K}^*_k, \quad \text{w.p.1}, \quad k = 1, 2, \ldots, K, (3.4)$$

which are identified by the data (2.3) and the compact set $\mathcal{G}_k$:

$$\mathcal{G}_k(\delta) \equiv \left\{ x \in \mathbb{C}^k[0,1] : x(0) = 0, \int_0^1 |\dot{x}(t)|^2 \, dt \leq \delta^2 \right\}, \quad \delta > 0. \quad (3.5)$$

In [21] Strassen told us that $\mathcal{G}_k(\delta)$ is a compact set in $\mathbb{C}^k[0,1]$ for any $\delta > 0$, and that for $x \in \mathcal{G}_k(\delta)$ and $0 \leq a \leq b \leq 1$, $|x(b) - x(a)| \leq \delta(b-a)^{1/2}$.

## 3.2   The LIL and the Functional LIL Limits

We now give our main results: Theorems 3.1 and 3.2. All proofs are given in Sect. 4. Throughout the rest of the paper, we suppose that, for all $k = 1, \ldots, K$,

$$\mathsf{E}\left[u_k(1)^r\right] < \infty \quad \text{and} \quad \mathsf{E}\left[v_k(1)^r\right] < \infty \quad \text{for some } r > 2. \tag{3.6}$$

**Theorem 3.1** (Limits in the underloaded regime). *Suppose* (3.6) *holds. If* $\rho < 1$, *then, w.p.1, the functional LIL*

$$\mathcal{K}_k^* = \left\{ \left( \eta, \eta, \frac{\lambda_k^{1/2} c_k}{\mu_k} x, -\sigma_k x, \lambda_k^{1/2} c_{a,k} x \right) : x \in \mathcal{G}(1) \right\}, \quad k = 1, 2, \ldots, K, \tag{3.7}$$

*and the LIL*

$$\mathcal{X}_{k,sup}^* = -\mathcal{X}_{k,inf}^* = \left( 0, 0, \frac{\lambda_k^{1/2} c_k}{\mu_k}, \sigma_k, \lambda_k^{1/2} c_{a,k} \right), \quad k = 1, 2, \ldots, K, \tag{3.8}$$

*where*

$$\sigma_k^2 \equiv \sum_{j=1}^k \rho_j \frac{c_j^2}{\mu_j} \quad \text{with} \quad c_j^2 \equiv c_{a,j}^2 + c_{s,j}^2,$$

*and* $\sigma_k^2$ *can be understood as the (weighted) cumulative utilization of service capacity by the first* $k$ *classes.*

**Remark 3.1** (Insight from the underloaded regime). *If the* $(GI/GI)^K/1/ PPSD$ *queueing system is in the underloaded regime, then all classes are in light traffic, and all queue lengths are stochastic bounded (not growing with time* $t$*), and this follows the zero functional LIL- and LIL-limits. Since the fluid and strong approximations satisfy the Little's law in the underloaded regime, the similar zero functional LIL- and LIL-limits are followed for the workload process. In the sense of the functional LIL and LIL, all the workload (or waiting time) process are asymptotically negligible in the underloaded regime, so all customers are quickly served upon arrival. This show us that the departure process performs similarly with its corresponding arrival and is almost independent of the service. This gives us the reason that the class-k departure's LIL and the functional LIL limits are only expressed by the arrival parameter* $\lambda^{1/2} c_{a,k}$*. Because* $\sum_{i=1}^k T_k(t) + I_k(t) = t$ *for all* $t \geq 0$*, the stochastic deviations, from the first* $k$ *classes's busy time* $\sum_{i=1}^k T_k(t)$ *and the idle time* $I_k(t)$*, are mutually supplement each other. As a result, the deviation parameter of the idle time* $I_k(t)$ *is embodied by* $\sigma_k$*:* $\sigma_k^2 = \sum_{i=1}^k (\lambda_k^{1/2} c_k/\mu_k)^2$*, the parameter for busy time from classes 1 to k, see* (3.7) *and* (3.8).

**Theorem 3.2** (Limits in the critically loaded regimes). *Suppose (3.6) holds. If* $\rho = 1$, *then, the functional LILs* $\mathcal{K}_k^*$ *satisfy (3.7) for* $k = 1, 2, \ldots, K - 1$ *w.p.1, and for class* $K$,

$$\mathcal{K}_K^* = \{(\Phi(\mu_K x), \Phi(x), G_1(y), \Psi(x), G_1(\mu_K y)) : x \in \mathcal{G}(\sigma_K), y \in \mathcal{G}_2(\sigma_K)\}, \quad w.p.1, \ (3.9)$$

*where* $G_1 : \mathbb{C} \times \mathbb{C} \to \mathbb{C}$ *is a continuous mapping defined by*

$$G_1(x, y)(t) = \inf_{0 \leq s \leq t} [x(s) - y(s)] + y(t). \tag{3.10}$$

*For the LIL limits,* $\mathcal{X}_{k,sup}^*$ *and* $\mathcal{X}_{k,inf}^*$ *satisfy (3.8) for classes* $k = 1, 2, \ldots, K-1$, *and for class* $K$,

$$Q_{K,sup}^* = \mu_K Z_{K,sup}^* = \mu_K I_{K,sup}^* = \sigma_K, \quad Q_{K,inf}^* = Z_{K,inf}^* = I_{K,inf}^* = 0, \quad w.p.1. \ (3.11)$$

**Remark 3.2** (Insight from the critically loaded regime). *If the* $(GI/GI)^K/1/$ *PPSD is in the critically loaded regime, then classes* $1, 2, \ldots, K - 1$ *are in light traffic, class* $K$ *is a critical loaded state. Under the PPSD service discipline, the performance of class* $K$ *is impacted by the classes* $1, 2, \ldots, K - 1$ *with higher priority, this is why the deviation parameter* $\sigma_K$ *is in (3.9) and (3.11). The zero limits of the LIL in (3.11) comes from their definitions which can be understood together with the oblique reflection mapping (2.2) and the Little's law:* $Q_{K,sup}^* = \mu_K Z_{K,sup}^*$ *and* $\mathcal{K}_{Q_K} = \mu_K \mathcal{K}_{Q_K}$. *Finally we note that, the LIL of the queue length, as the function of traffic intensity* $\rho$, *is continuous when* $\rho < 1$ *and has a big upper skip at* $\rho = 1$.

## 4    Proofs

In this section, we prove Theorems 3.1 and 3.2 by the strong approximation approach. We first give some basis for proof in Sect. 4.1, and then prove them in Sect. 4.2.

### 4.1    Strong Approximation and Associated Reslults

We give the basis for the proof including the strong approximation and Strassen's functional LIL. The strong approximation for $(GI/GI)^K/1/PPSD$ queueing is a Brownian motion system associated with the fluid approximation. Strassen's functional LIL helps us get the functional LIL of the main performance measures based on the Brownian motion system.

**Lemma 4.1** (Strong approximations) . *Suppose that (3.6) holds, we have, w.p.1,*

$$\left\|Q_k - \widetilde{Q}_k\right\|_L = o(L^{1/r}), \quad \left\|Z_k - \widetilde{Z}_k\right\|_L = o(L^{1/r}), \quad \left\|T_k - \widetilde{T}_k\right\|_L = o(L^{1/r}),$$

$$\left\|I_k - \widetilde{I}_k\right\|_L = o(L^{1/r}), \quad \left\|D_k - \widetilde{D}_k\right\|_L = o(L^{1/r}), \quad k = 1, 2, \ldots, K, \tag{4.1}$$

*where*

$$\widetilde{Q}_k(t) \equiv \widetilde{X}_k(t) + \widetilde{Y}_k(t) = \Phi(\widetilde{X}_k)(t), \quad \widetilde{Y}_k(t) \equiv \Psi(\widetilde{X}_k)(t),$$

$$\widetilde{X}_k(t) \equiv (\lambda_k - \mu_k)t + \mu_k \sum_{l=1}^{k-1} \widetilde{T}_l(t) + \widetilde{W}_k(t),$$

$$\widetilde{T}_k(t) \equiv t - \sum_{j=1}^{k-1} \widetilde{T}_j(t) - \widetilde{I}_k(t), \quad \widetilde{I}_k(t) \equiv \frac{1}{\mu_k} \widetilde{Y}_k(t),$$

$$\widetilde{Z}_k(t) \equiv \frac{1}{\mu_k} \widetilde{Q}_k(t) + \frac{1}{\mu_k} \left[ \mu_k^{1/2} c_{s,k} W_{s,k}(\bar{T}_k(t)) - \mu_k^{1/2} c_{s,k} W_{s,k}(\rho_k t) \right],$$

$$\widetilde{D}_k(t) \equiv \mu_k \widetilde{T}_k(t) + \mu_k^{1/2} c_{s,k} W_{s,k}(\bar{T}_k(t)),$$

$$\widetilde{W}_k(t) \equiv \lambda_k^{1/2} c_{a,k} W_{a,k}(t) - \mu_k^{1/2} c_{s,k} W_{s,k}(\bar{T}_k(t)), \tag{4.2}$$

$\Psi$ *and* $\Phi$ *are defined in (2.2),* $W_{a,k}$ *and* $W_{s,k}$ *are independent standard Brownian motions which can be looked as from the arrival and service processes of class* $k$, *respectively. Define* $W_k(t) \equiv \mu_k \sum_{l=1}^{k} \widetilde{W}_l(t)/\mu_l$, $k = 1, 2, \ldots, K$, $\bar{X}_k(t)$ *and* $\bar{B}_k(t)$ *satisfy*

$$\widetilde{X}_k(t) - \bar{X}_k(t) = -\sum_{l=1}^{k-1} \frac{\mu_k}{\mu_l} \left[ \widetilde{Q}_l(t) - \bar{Q}_l(t) \right] + W_k(t), \tag{4.3}$$

$$\bar{T}_k(t) - \widetilde{T}_k(t) = \frac{1}{\mu_k} \left[ \widetilde{Q}_k(t) - \bar{Q}_k(t) \right] - \frac{1}{\mu_k} \widetilde{W}_k(t). \tag{4.4}$$

*In addition, If* $\rho_k < 1$, $k = 1, 2, \ldots, K$, *then* $\left\| \widetilde{Q}_k \right\|_L = O(\log L)$ *w.p.1 as* $L \to \infty$.

Lemma 4.1 approximates the discrete $(GI/GI)^K/1/PPSD$ queueing system into a continuous Brownian motion system. In fact, based on the strong approximation given by Lemma 4.1 the discrete function LIL problem 3.4 can be transformed into its corresponding continuous problem of Brownian motion. To this end, we define the following scaled processes based on the strong approximation.

$$\widetilde{Q}_k^n(t) = \frac{\widetilde{Q}_k(nt) - \bar{Q}_k(nt)}{\varphi(n)}, \quad \widetilde{Z}_k^n(t) \equiv \frac{\widetilde{Z}_k(nt) - \bar{Z}_k(nt)}{\varphi(n)}, \quad \widetilde{T}_k^n(t) \equiv \frac{\widetilde{T}_k(nt) - \bar{T}_k(nt)}{\varphi(n)},$$

$$\widetilde{I}_k^n(t) \equiv \frac{\widetilde{I}_k(nt) - \bar{I}_k(nt)}{\varphi(n)}, \quad \widetilde{D}_k^n(t) \equiv \frac{\widetilde{D}_k(nt) - \bar{D}_k(nt)}{\varphi(n)}, \quad \text{for} \quad k = 1, 2, \ldots, K. \tag{4.5}$$

Let

$$\left( \widetilde{Q}_k^n, \widetilde{Z}_k^n, \widetilde{T}_k^n, \widetilde{I}_k^n, \widetilde{D}_k^n \right) \Rightarrow \left( \mathcal{K}_{\widetilde{Q}_k}, \mathcal{K}_{\widetilde{Z}_k}, \mathcal{K}_{\widetilde{T}_k}, \mathcal{K}_{\widetilde{I}_k}, \mathcal{K}_{\widetilde{D}_k} \right), \quad \text{w.p.1}, \quad k = 1, 2, \ldots, K,$$

if the relatively compact limits on the right exist. Similarly with Lemma 4.3 in [11] we conclude that

$$\mathcal{K}_k^* = \left( \mathcal{K}_{\widetilde{Q}_k}, \mathcal{K}_{\widetilde{Z}_k}, \mathcal{K}_{\widetilde{T}_k}, \mathcal{K}_{\widetilde{I}_k}, \mathcal{K}_{\widetilde{D}_k} \right), \quad k = 1, 2, \ldots, K. \tag{4.6}$$

It means in 4.6 that it suffices to find the functional LIL in (4.6) if we are interested in the functional LIL in (3.4).

The following result is Strassen's functional LIL for BM [21].

**Lemma 4.2** (Strassen's functional LIL). *Suppose that $W_a, W_b$ are two mutually independent one-dimensional standard Brownian motions, $\sigma_a, \sigma_b$ are two nonzero constants, we have, for any $t \in [0.1]$, w.p.1,*

$$\frac{W_a(nt)}{\varphi(n)} \rightrightarrows \mathcal{G}(1) \quad and \quad \left(\frac{\sigma_a W_a(nt)}{\varphi(n)}, \frac{\sigma_b W_b(nt)}{\varphi(n)}\right) \rightrightarrows \mathcal{G}_2\left(\sqrt{\sigma_a^2 + \sigma_b^2}\right).$$

### 4.2   Proof of Theorems 3.1 and 3.2

We first prove Theorem 3.1 and then Theorem 3.2.

**Proof of Theorem** 3.1. We first go to find the functional LIL and then the LIL in Lévy's type. For applications, we give the following fluid solution to (2.1) in this underloaded regime:

$$\bar{\mathbb{X}}_k(t) = (0, 0, \rho_k t, (1 - \sum_{i=1}^{k} \rho_i)t, \lambda_k t) \quad \text{for} \quad k = 1, 2, \ldots, K, \quad t \geq 0.$$

**The Functional LIL for the Underloaded Regime.** By Lemma 4.1, $\left\|\widetilde{Q}_k\right\|_L = O(\log L)$ w.p.1 as $L \to \infty$. So, with (4.5) $\widetilde{Q}_k^n(t) \rightrightarrows \eta(t) \equiv 0$ w.p.1 for all $t \in [0, 1]$. For $Z_k$, it follows from (2.1) and (4.2) that $\bar{Q}_k(t) = \mu_k \bar{Z}_k(t)$ and $\widetilde{Q}_k(t) = \mu_k \widetilde{Z}_k(t)$, and than $\widetilde{Z}_k^n(t) \rightrightarrows \eta(t) \equiv 0$ w.p.1 for all $t \in [0, 1]$. We now go to find the functional LIL limits for $T_k$ and $I_k$. By Lemmas 4.2 and (4.4), we have, $k = 1, 2, \ldots, K$,

$$\widetilde{T}_k^n(t) = \frac{\bar{T}_k(nt) - \tilde{T}_k(nt)}{\varphi(n)} = -\frac{1}{\mu_k}\widetilde{Q}_k^n(t) + \frac{1}{\mu_k}\frac{\widetilde{W}_k(nt)}{\varphi(n)} \rightrightarrows \frac{1}{\mu_k}\mathcal{G}(\lambda_k^{1/2}c_k), \quad \text{w.p.1,}$$

where, in the underloaded regime, $\widetilde{W}_k(t) = \lambda_k^{1/2}c_{a,k}W_{a,k}(t) - \mu_k^{1/2}c_{s,k}W_{s,k}(\rho_k t)$ is a driftless Brownian motion with variance parameter $\lambda_k^{1/2}c_k$. For $I_k$,

$$\widetilde{I}_k^n(t) = \frac{\bar{I}_k(nt) - \tilde{I}_k(nt)}{\varphi(n)} = \frac{\sum_{l=1}^{k}\left(\bar{T}_l(nt) - \widetilde{T}_l(nt)\right)}{\varphi(n)}$$

$$= \sum_{l=1}^{k}\frac{1}{\mu_l}\widetilde{Q}_l^n(t) - \frac{\sum_{l=1}^{k}\frac{1}{\mu_l}\widetilde{W}_l(nt)}{\varphi(n)} \rightrightarrows -\mathcal{G}(\sigma_k), \quad \text{w.p.1,} \qquad (4.7)$$

because $\widetilde{Q}_l^n(t) \rightrightarrows \eta(t) \equiv 0$ w.p.1 and $\sum_{l=1}^{k}\widetilde{W}_l(t)/\mu_l$ is a driftless Brownian motion with variance parameter $\sigma_k$. For $D_k$,

$$\widetilde{D}_k^n(t) = \frac{\bar{D}_k(nt) - \bar{D}_k(nt)}{\varphi(n)} = \frac{\mu_k^{1/2} c_{s,k} W_{s,k}(\bar{T}_k(nt)) + \mu_k(\widetilde{T}_k(nt) - \bar{T}_k(nt))}{\varphi(n)}$$

$$= \frac{\mu_k^{1/2} c_{s,k} W_{s,k}(\bar{T}_k(nt)) - \left[\widetilde{Q}_k(nt) - \widetilde{W}_k(nt)\right]}{\varphi(n)}$$

$$= -\widetilde{Q}_k^n(t) + \frac{\lambda_k^{1/2} c_{a,k} W_{a,k}(nt)}{\varphi(n)} \rightrightarrows \mathcal{G}(\lambda_k^{1/2} c_{a,k}), \quad \text{w.p.1},$$

because $\widetilde{Q}_k^n(t) \rightrightarrows \eta(t) \equiv 0$ w.p.1 and $\lambda_k^{1/2} c_{a,k} W_{a,k}(t)$ is a driftless Brownian motion with variance parameter $\lambda_k^{1/2} c_{a,k}$. Hence we get all the functional LIL limits in this underloaded regime.

**The LIL for the UL Regime.** Now we are ready to find the LIL limits $\mathcal{X}_{k,sup}^*$ and $\mathcal{X}_{k,inf}^*$ in (3.2) based on the functional LIL sets in (3.7). Firstly, we note that if $f^n \rightrightarrows \mathcal{K}_f = \{0\}$, then $\liminf_{n\to\infty} f^n = \limsup_{n\to\infty} f^n = \lim_{n\to\infty} f^n = 0$. Because $\mathcal{K}_{Q_k} = \mathcal{K}_{Z_k} = \{\eta\}$ for $k = 1, 2, \ldots, K$, we have

$$Q_{k,sup}^* = Q_{k,inf}^* = Z_{k,sup}^* = Z_{k,inf}^* = 0, \quad \text{w.p.1}.$$

Secondly, for any $\delta > 0$,

$$\sup_{x \in \mathcal{G}(\delta)} x(1) = \delta, \qquad \inf_{x \in \mathcal{G}(\delta)} x(1) = -\delta, \tag{4.8}$$

where the supremum is actually attained for the functions $x(t) = \delta t$ and the infimum is from $x(t) = -\delta t$. In doing so, we can get the LIL limits $T_{k,sup}^*, T_{k,inf}^*, I_{k,sup}^*, I_{k,inf}^*, D_{k,sup}^*, D_{k,inf}^*$ for $k = 1, 2, \ldots, K$. Hense, (3.8) holds. $\square$

**Proof of Theorem** 3.2. In this critically loaded regime, it is easy to see that the first $K - 1$ classes form a underloaded queueing system $(GI/GI)^{K-1}/1/PPSD$ because $\sum_{k=1}^{K-1} \rho_k < 1$, and the first $K - 1$ classes does not receive the influence from class $K$ under the PPSD policy. So, the functional LIL limits $\mathcal{K}_k^*$ satisfy (3.7), the LIL limits $\mathcal{X}_{k,sup}^*$ and $\mathcal{X}_{k,inf}^*$ satisfy (3.8) for $k = 1, 2, \ldots, K - 1$. So, it remains to find the functional LIL and the LIL limits for class $K$. For applications, we first give the fluid solution to (2.1) in this critically loaded regime: for $t \geq 0$,

$$\bar{\mathbb{X}}_k(t) = (0, 0, \rho_k t, (1 - \sum_{i=1}^k \rho_i)t, \lambda_k t) \quad k = 1, 2, \ldots, K.$$

**The Functional LIL for Class $K$.** We are ready to find the functional LIL for this critically loaded regime. We first find the functional LIL limits for $Q_k$ and $I_K$. Notice that $\bar{X}_k(t) = \bar{Q}_k(t) = 0$ for all $k = 1, 2, \ldots, K$ and $t \geq 0$, by (4.3), for all $t \in [0, 1]$,

$$\frac{\widetilde{X}_K(nt)}{\varphi(n)} = \frac{\widetilde{X}_K(nt) - \bar{X}_K(nt)}{\varphi(n)} = \frac{-\sum_{l=1}^{K-1} \frac{\mu_k}{\mu_l}\left[\widetilde{Q}_l(nt) - \bar{Q}_l(nt)\right]}{\varphi(n)} + \frac{W_K(nt)}{\varphi(n)}$$

$$= -\sum_{l=1}^{K-1} \frac{\mu_k}{\mu_l}\widetilde{Q}_l^n(t) + \frac{W_K(nt)}{\varphi(n)} \Rrightarrow \mathcal{G}(\mu_K\sigma_K), \quad \text{w.p.1}$$

because $\mathcal{K}_{\widetilde{Q}_k} = \{0\}$ for all $k = 1, 2, \ldots, K-1$, and $W_K(t)$ is a driftless Brownian motion with variance parameter $\mu_K\sigma_K$. Since the mappings $\Phi$ and $\Psi$ are continuous in uniform norm, by Strassen's continuous mapping theorem (see Lemma 4.3 in [10]) we have, for all $t \in [0,1]$,

$$\widetilde{Q}_K^n(t) = \frac{\widetilde{Q}_K(nt)}{\varphi(n)} = \frac{\Phi(\widetilde{X}_K)(nt)}{\varphi(n)} \Rrightarrow \Phi(\mathcal{G}(\mu_K\sigma_K)), \quad \text{w.p.1},$$

$$\widetilde{I}_K^n(t) = \frac{\widetilde{I}_K(nt)}{\varphi(n)} = \frac{1}{\mu_K}\frac{\Psi(\widetilde{X}_K)(nt)}{\varphi(n)} = \frac{\Psi(\widetilde{X}_K/\mu_K)(nt)}{\varphi(n)} \Rrightarrow \Psi(\mathcal{G}(\sigma_K)), \quad \text{w.p.1}, \quad (4.9)$$

where the equalities above hold because $\bar{Q}_K(t) = \bar{I}_K(t) = \Phi(\bar{X}_K)(t) = \Psi(\bar{X}_K)(t) = 0$ for $t \geq 0$. Since $\widetilde{Q}_K(t) = \widetilde{Z}_K(t)/\mu_K$ and $\bar{Q}_K(t) = \bar{Z}_K(t)/\mu_K = 0$ for all $t \geq 0$, we have, for all $t \in [0,1]$,

$$\widetilde{Z}_K^n(t) = \frac{\widetilde{Q}_K(nt)}{\mu_K\varphi(n)} = \frac{\Phi(\widetilde{X}_K/\mu_K)(nt)}{\varphi(n)} \Rrightarrow \Phi(\mathcal{G}(\sigma_K)), \quad \text{w.p.1},$$

For $T_K(t)$,

$$\widetilde{T}_K^n(t) = \frac{\widetilde{T}_K(nt) - \bar{T}_K(nt)}{\varphi(n)} = \frac{1}{\mu_K\varphi(n)}\left\{\widetilde{W}_K(nt) - [\widetilde{X}_K(nt) + \widetilde{Y}_K(nt)]\right\}$$

$$= \frac{1}{\varphi(n)}\left\{\sum_{l=1}^{K-1}\frac{1}{\mu_l}\widetilde{Q}_l(nt) - \sum_{l=1}^{K-1}\frac{1}{\mu_l}\widetilde{W}_l(nt) - \frac{1}{\mu_K}\widetilde{Y}_K(nt)\right\}$$

$$= \frac{1}{\varphi(n)}\left\{\sum_{l=1}^{K-1}\frac{1}{\mu_l}\widetilde{Q}_l(nt) - \sum_{l=1}^{K-1}\frac{1}{\mu_l}\widetilde{W}_l(nt) - \frac{1}{\mu_K}\Psi(-\sum_{l=1}^{K-1}\frac{\mu_K}{\mu_l}\widetilde{Q}_l + W_K)(nt)\right\}$$

$$= \sum_{l=1}^{K-1}\frac{1}{\mu_l}\frac{\widetilde{Q}_l(nt)}{\varphi(n)} - \sum_{l=1}^{K-1}\frac{1}{\mu_l}\frac{\widetilde{W}_l(nt)}{\varphi(n)} - \frac{1}{\varphi(n)}\frac{1}{\mu_K}\Psi(-\sum_{l=1}^{K-1}\frac{\mu_K}{\mu_l}\widetilde{Q}_l + \mu_K\sum_{l=1}^{K-1}\frac{\widetilde{W}_l}{\mu_l} + \widetilde{W}_K)(nt)$$

$$= \sum_{l=1}^{K-1}\frac{1}{\mu_l}\frac{\widetilde{Q}_l(nt)}{\varphi(n)} - \sum_{l=1}^{K-1}\frac{1}{\mu_l}\frac{\widetilde{W}_l(nt)}{\varphi(n)}$$

$$- \frac{1}{\varphi(n)}\frac{1}{\mu_K}\sup_{0 \leq s \leq nt}\left\{\sum_{l=1}^{K-1}\frac{\mu_K}{\mu_l}\widetilde{Q}_l(s) - \mu_K\sum_{l=1}^{K-1}\frac{\widetilde{W}_l}{\mu_l}(s) - \widetilde{W}_K(s)\right\}$$

$$= \sum_{l=1}^{K-1}\frac{1}{\mu_l}\frac{\widetilde{Q}_l(nt)}{\varphi(n)} - \sum_{l=1}^{K-1}\frac{1}{\mu_l}\frac{\widetilde{W}_l(nt)}{\varphi(n)} + \inf_{0 \leq s \leq nt}\left\{-\sum_{l=1}^{K-1}\frac{1}{\mu_l}\frac{\widetilde{Q}_l(s)}{\varphi(n)} + \sum_{l=1}^{K-1}\frac{1}{\mu_l}\frac{\widetilde{W}_l(s)}{\varphi(n)} + \frac{\widetilde{W}_K(s)}{\mu_K\varphi(n)}\right\}$$

$$= G_1(\frac{\widetilde{W}_K}{\mu_K\varphi(n)}, \sum_{l=1}^{K-1}\frac{1}{\mu_l}\frac{\widetilde{Q}_l}{\varphi(n)} - \sum_{l=1}^{K-1}\frac{1}{\mu_l}\frac{\widetilde{W}_l}{\varphi(n)})(nt),$$

where the third equality holds because, by (4.3),

$$\widetilde{X}_K(t) = -\sum_{l=1}^{K-1}\frac{\mu_K}{\mu_l}\widetilde{Q}_l(t) + W_K(t) \quad \text{and} \quad W_K(t) = \mu_K\sum_{l=1}^{K-1}\frac{\widetilde{W}_l(t)}{\mu_l} + \widetilde{W}_K(t),$$

and this follows the forth and fifth equalities, the sixth equality comes from (2.2) and

$$\sum_{l=1}^{K-1} \frac{\mu_K}{\mu_l} \widetilde{Q}_l(0) - \mu_K \sum_{l=1}^{K-1} \frac{\widetilde{W}_l}{\mu_l}(0) - \widetilde{W}_K(0) = 0,$$

the last equality comes from (3.10). Since $\widetilde{Q}_l^n(t) = \widetilde{Q}_l(nt)/\varphi(n) \to 0$ as $n \to \infty$ for all $t \in [0, 1]$ and $l = 1, 2, \ldots, K-1$, we have, w.p.1,

$$\sum_{l=1}^{K-1} \frac{1}{\mu_l} \frac{\widetilde{Q}_l(nt)}{\varphi(n)} = \sum_{l=1}^{K-1} \frac{1}{\mu_l} \widetilde{Q}_l^n(t) \to 0 \quad \text{for all} \quad t \in [0, 1].$$

This, together with Lemma 4.4 in [10], implies that

$$(\frac{\widetilde{W}_K}{\mu_K \varphi(n)}, \sum_{l=1}^{K-1} \frac{1}{\mu_l} \frac{\widetilde{Q}_l}{\varphi(n)} - \sum_{l=1}^{K-1} \frac{1}{\mu_l} \frac{\widetilde{W}_l}{\varphi(n)})(nt) \quad \text{and} \quad (\frac{\widetilde{W}_K}{\mu_K \varphi(n)}, - \sum_{l=1}^{K-1} \frac{1}{\mu_l} \frac{\widetilde{W}_l}{\varphi(n)})(nt)$$

have identical relatively compact limits. Since $\widetilde{W}_l(t)$ are mutually independent, $l = 1, 2, \ldots, K$, and the variance $Var(\widetilde{W}_K(t)/\mu_K) + Var(-\sum_{l=1}^{K-1} \widetilde{W}_l)(t)/\mu_l) = \sigma_K^2$. So, by Lemma 4.2, for all $t \in [0, 1]$,

$$(\frac{\widetilde{W}_K(nt)}{\mu_K \varphi(n)}, - \sum_{l=1}^{K-1} \frac{1}{\mu_l} \frac{\widetilde{W}_l(nt)}{\varphi(n)}) \rightrightarrows \mathcal{G}_2(\sigma_K), \quad \text{w.p.1.}$$

This and Strassen's continuous mapping theorem (see Lemma 4.3 in [10]) gives that, for all $t \in [0, 1]$,

$$\widetilde{T}_K^n(t) = G_1(\frac{\widetilde{W}_K}{\mu_K \varphi(n)}, \sum_{l=1}^{K-1} \frac{1}{\mu_l} \frac{\widetilde{Q}_l}{\varphi(n)} - \sum_{l=1}^{K-1} \frac{1}{\mu_l} \frac{\widetilde{W}_l}{\varphi(n)})(nt) \rightrightarrows G_1(\mathcal{G}_2(\sigma_K)), \quad \text{w.p.1.}$$

Now, we go to find the functional LIL of $D_K$. By (4.2) and (4.5),

$$\widetilde{D}_K^n(t) = \frac{\widetilde{D}_K(nt) - \bar{D}_K(nt)}{\varphi(n)} = \frac{-\widetilde{Q}_K(nt) + \lambda_K^{1/2} c_{a,K} W_{a,K}(nt)}{\varphi(n)}$$

$$= \frac{1}{\varphi(n)} \left( -\tilde{X}_K(nt) - \tilde{Y}_K(nt) + \lambda_K^{1/2} c_{a,K} W_{a,K}(t) \right)$$

$$= \frac{1}{\varphi(n)} \left( \sum_{l=1}^{K-1} \frac{\mu_K}{\mu_l} \widetilde{Q}_l(nt) - W_K(nt) - \tilde{Y}_K(nt) + \lambda_K^{1/2} c_{a,K} W_{a,K}(nt) \right)$$

$$= \frac{1}{\varphi(n)} \left( \sum_{l=1}^{K-1} \frac{\mu_K}{\mu_l} \widetilde{Q}_l(nt) - \sum_{l=1}^{K-1} \frac{\mu_K \widetilde{W}_l(nt)}{\mu_l} - \tilde{Y}_K(nt) + \mu_K^{1/2} c_{s,K} W_{s,K}(\bar{T}(nt)) \right)$$

$$= \sum_{l=1}^{K-1} \frac{\mu_K \widetilde{Q}_l(nt)}{\varphi(n)\mu_l} - \sum_{l=1}^{K-1} \frac{\mu_K \widetilde{W}_l(nt)}{\varphi(n)\mu_l} + \frac{\mu_K^{1/2} c_{s,K} W_{s,K}(\bar{T}(nt))}{\varphi(n)}$$

$$+ \inf_{0 \le s \le nt} \left\{ -\sum_{l=1}^{K-1} \frac{\mu_K \widetilde{Q}_l(s)}{\varphi(n)\mu_l} + \sum_{l=1}^{K-1} \frac{\mu_K \widetilde{W}_l(s)}{\varphi(n)\mu_l} - \frac{\mu_K^{1/2} c_{s,K} W_{s,K}(\bar{T}(s))}{\varphi(n)} + \frac{\lambda_K^{1/2} c_{a,K} W_{a,K}(s)}{\varphi(n)} \right\}$$

$$= G_1(\frac{\lambda_K^{1/2} c_{a,K} W_{a,K}}{\varphi(n)}, \sum_{l=1}^{K-1} \frac{\mu_K \widetilde{Q}_l}{\varphi(n)\mu_l} - \sum_{l=1}^{K-1} \frac{\mu_K \widetilde{W}_l}{\varphi(n)\mu_l} + \frac{\mu_K^{1/2} c_{s,K} W_{s,K}(\bar{T}(\cdot))}{\varphi(n)})(nt),$$

where the third, the fifth and the sixth equalities follow from (4.2), the forth follows from (4.3), the seventh equality follows from (3.10). Since $\widetilde{Q}_l^n(t) = \widetilde{Q}_l(nt)/\varphi(n) \to 0$ as $n \to \infty$ for all $t \in [0,1]$ and $l = 1, 2, \ldots, K-1$, the sequence

$$\left( \frac{\lambda_K^{1/2} c_{a,K} W_{a,K}(nt)}{\varphi(n)}, \sum_{l=1}^{K-1} \frac{\mu_K \widetilde{Q}_l(nt)}{\varphi(n)\mu_l} - \sum_{l=1}^{K-1} \frac{\mu_K \widetilde{W}_l(nt)}{\varphi(n)\mu_l} + \frac{\mu_K^{1/2} c_{s,K} W_{s,K}(\bar{T}((nt)))}{\varphi(n)} \right)$$

has the same relatively compact limits as

$$\left( \frac{\lambda_K^{1/2} c_{a,K} W_{a,K}(nt)}{\varphi(n)}, -\sum_{l=1}^{K-1} \frac{\mu_K \widetilde{W}_l(nt)}{\varphi(n)\mu_l} + \frac{\mu_K^{1/2} c_{s,K} W_{s,K}(\bar{T}((nt)))}{\varphi(n)} \right),$$

which is relatively compact, by Lemma 4.2, the corresponding convergence limit set is $\mathcal{G}_2(\mu_K \sigma_K)$ because the sum of variances of $\lambda_K^{1/2} c_{a,K} W_{a,K}(t)$ and $-\sum_{l=1}^{K-1} \mu_K \widetilde{W}_l(t)/\mu_l + \mu_K^{1/2} c_{s,K} W_{s,K}(\bar{T}((t)))$ is $\mu_K^2 \sigma_K^2 t$. Notice that $G_1$ is a continuous binary function, together with Strassen's continuous mapping theorem (see Lemma 4.3 in [10]), $\widetilde{D}_K^n(t) \Rightarrow G_1(\mathcal{G}_2(\mu_K \sigma_K))$ w.p.1 for all $t \in [0,1]$. So far, we finish the functional LIL's proof for the critically loaded regime.

**The LIL for Class $K$.** By Corollaries 3.1 and 3.2 in [15], for any $\delta > 0$,

$$\sup_{x \in \Phi(\mathcal{G}(\delta))} x(1) = \delta \quad \text{and} \quad \inf_{x \in \Phi(\mathcal{G}(\delta))} x(1) = 0, \tag{4.10}$$

where the supremum is actually attained for the functions $x(t) = \delta t$ and the infimum is from $x(t) = 0$. In dong so, we can get the LIL:

$$Q_{K,sup}^* = \mu_K Z_{K,sup}^* = \sigma_K, \quad Q_{K,inf}^* = Z_{K,inf}^* = 0, \quad \text{w.p.1.}$$

For the LIL of $I_K$, for $\delta > 0$ given above, by (2.1) on page 169 in [15], $|y(b)| \le \delta\sqrt{b} \le \delta$ for any $y \in \mathcal{G}(\delta)$ and $0 \le b \le 1$, then, for any $\delta > 0$,

$$\sup_{x \in \Psi(\mathcal{G}(\delta))} x(1) = \sup_{y \in \mathcal{G}(\delta)} \sup_{0 \le s \le 1} \{-y(s)\} = \delta,$$

$$\inf_{x \in \Psi(\mathcal{G}(\delta))} x(1) = \inf_{y \in \mathcal{K}(\delta)} \sup_{0 \le s \le 1} \{-y(s)\} = 0, \tag{4.11}$$

where, as for (4.10), the supremum is attained for the functions $y(s) = -s$ and the infimum is from $y(s) = 0$. This gives the LIL: $I_{K,sup}^* = \sigma_K/\mu_K$ and $I_{K,inf}^* = 0$ w.p.1. That is, (3.11) holds.

Hence, we finished the proof. $\qquad\square$

## 5   Conclusion

We introduced a multi-class single server queue $(GI/GI)^K/1/PPSD$ queueing system, which consists of $K$ classes of customers and one server. Specially, all customers come to the system from outside and leave the system after service completion. A preemptive priority service discipline is in force: class 1 taking the highest priority, class 2 the second highest priority, and so on, class $K$ the lowest priority.

This paper can be thought of as a companion of [10], which focus on the functional LIL and LIL for the overloaded regime for the $(GI/GI)^K/1/PPSD$ queueing system, different from the underloaded and critically loaded regimes. In this paper, we develop the functional LIL and the LIL in Lévy's type for the five key performance measures: queue length, workload, busy time, idle time, and departure processes.

The functional LIL and LIL limits help us to understand the asymptotic performance of the queueing system well. From the definition, the functional LIL and the LIL limits refine the fluid limits, which characterizes the deterministic dynamics of performance for queueing systems. They show us the maximal asymptotic deviation around the fluid limit, and are identified respectively to be the functional compact set and functions by the first and second moments of primitive data: the arrival and service.

The main results: Theorems 3.1 and 3.2, are proved by an approach based on the strong approximation, which follows four steps: First, to establish the fluid and strong approximations for the performance functions of interest (e.g., the queue length and workload processes); Second, to associate the functional LILs of these performance functions with the functional LILs of their corresponding strong approximations; Third, to treat the Brownian motion related processes to obtain closed-form functional LIL limits, which is generally a compact set of continuous functions; Finally, to obtain the LIL limits through analyzing the compact set of their corresponding functional LIL limits.

Finally, we note that it is also interesting to find the functional LIL and LIL limits for more queueing systems. They can give us some more engineering insight on the asymptotic fluctuation through some numerical value and simple functions based on the primitive data, which is not be given by other approximations, such fluid, diffusion and strong approximations.

**Acknowledgements.** This work is supported by NSFC grants 11871116 and 11971074.

# References

1. Chen, H.: Fluid approximations and stability of multiclass queueing networks I: Wrok-conserving disciplines. Ann. Appl. Prob. **5**(3), 637–665 (1995)
2. Chen, H., Mandelbaum, A.: Hierarchical modeling of stochastic networks, Part I: fluid models. In: Yao, D.D. (ed.) Stochastic Modeling and Analysis of Manufacturing Systems, pp. 47–105. Springer, New York (1994). https://doi.org/10.1007/978-1-4612-2670-3_2
3. Chen, H., Shen, X.: Strong approximations for multiclass feedforward queueing networks. Annal. Appl. Probab. **10**(3), 828–876 (2000)
4. Chen, H., Yao, D.D.: Fundamentals of Queueing Networks. Springer, New York (2001)

5. Csörgő, M., Révész, P.: Strong Approximations in Probability and Statistics. Academic Press, New York (1981)

6. Dai, J.G.: On the positive Harris recurrence for multiclass queueing networks: a unified approach via fluid limit models. Ann. Appl. Probab. **5**(1), 49–77 (1995)

7. Ethier, S.N., Kurtz, T.G.: Markov Processes: Characterization and Convergence. Wiley, New York (1986)

8. Glynn, P.W., Whitt, W.: A new view of the heavy-traffic limit for infinite-server queues. Adv. Appl. Probab. **23**(1), 188–209 (1991)

9. Glynn, P.W., Whitt, W.: Departures from many queues in series. Ann. Appl. Probab. **1**(4), 546–572 (1991)

10. Guo, Y., Hou X.: Functional law of the iterated logarithm for multiclass queues with preemptive priority service discipline: the overloaded case. Submitted (2018)

11. Guo, Y., Li, Z.: Asymptotic variability analysis for a two-stage tandem queue, part I: the functional law of the iterated logarithm. J. Math. Anal. Appl. **450**, 1479–1509 (2017)

12. Guo, Y., Li, Z.: Asymptotic variability analysis for a two-stage tandem queue, part II: the law of the iterated logarithm. J. Math. Anal. Appl. **450**, 1510–1534 (2017)

13. Guo, Y., Liu, Y.: A law of iterated logarithm for multiclass queues with preemptive priority service discipline. Queueing Syst. **79**(3), 251–291 (2015)

14. Guo, Y., Liu, Y., Pei, R.: Functional law of iterated logarithm for multi-server queues with batch arrivals and customer feedback. Ann. Oper. Res. **264**, 157–191 (2018)

15. Iglehart, G.L.: Multiple channel queues in heavy traffic: IV. Law of the iterated logarithm. Z.Wahrscheinlichkeitstheorie verw. Geb. **17**, 168–180 (1971)

16. Lévy, P.: Théorie de l'addition des variables aléatories. Gauthier-Villars, Paris (1937)

17. Lévy, P.: Procesus stochastique et mouvement Brownien. Gauthier-Villars, Paris (1948)

18. Minkevičius, S., Steišūnas, S.: A law of the iterated logarithm for global values of waiting time in multiphase queues. Stat. Probab. Lett. **61**(4), 359–371 (2003)

19. Minkevičius, S.: On the law of the iterated logarithm in multiserver open queueing networks. Stochastics **86**(1), 46–59 (2014)

20. Sakalauskas, L.L., Minkevičius, S.: On the law of the iterated logarithm in open queueing networks. European Journal of Operational Research. 20(3), 632–640 (2000)

21. Strassen, V.: An invariance principle for the law of the iterated lorgarith. Z. Wahrscheinlichkeitstheorie verw. Geb. **3**(3), 211–226 (1964)

22. Zhang, H.: Strong approximations of irreducible closed queueing networks. Adv. Appl. Probab. **29**(2), 498–522 (1997)

23. Zhang, H., Hsu, G.X., Wang, R.X.: Strong approximations for multiple channels in heavy traffic. J. Appl. Probab. **27**(3), 658–670 (1990)

24. Zhang, H., Hsu, G.X.: Strong approximations for priority queues: head-of-the-line-first discipline. Queueing Syst. **10**(3), 213–234 (1992)

# Tail Asymptotics for the Waiting Time in an M/G/1/ROS Vacation Queue with Regularly-Varying Service

Bin Liu[(⊠)]

School of Mathematics and Physics, Anhui Jianzhu University, Hefei 230601, China
bliu@amt.ac.cn

**Abstract.** We study the asymptotic behavior of tail probability for the waiting time in the steady-state $M/G/1/ROS$ multiple-vacation queue with regularly-varying service time and vacation time distributions. Conditioning on the server being busy or on vacation, the asymptotic conditional tail probabilities are obtained explicitly. We also verify that the waiting-time tail for $M/G/1/ROS$ queue with multiple-vacation is asymptotically equivalent to that for the standard $M/G/1/ROS$ queue (without vacation), as long as the vacation time has a tail probability lighter than the service time.

**Keywords:** $M/G/1$ queue · Random order of service · Vacation · Waiting time · Tail asymptotics

## 1 Introduction

Triggered by the desire for measuring the quality of service (QoS) in modern communication networks (see, e.g., [14] and [15]), there has been much interest in studying the asymptotic behaviors for queues with heavy-tailed service time distributions. The tail asymptotics for queueing quantities, such as queue length and waiting time, is of fundamental importance due to the stringent QoS requirements often requiring these tail probabilities to be significantly small.

Queueing systems with vacations are a type of very important queueing systems, which find many applications in abroad range of areas, e.g., production, computer, and communication systems. A variety of queues with vacations have been extensively studied for more than 40 years. Literature reviews on vacation queues can be found in, e.g., the survey [8] and the book [17].

In this paper, we are interested in the asymptotic behavior of tail probability for the stationary waiting time in the $M/G/1$ queue with multiple-vacation and random order service (ROS) discipline. The customers are assumed to arrive

Supported in part by the National Science Foundation of China (No. 71571002), the Natural Science Foundation of the Anhui Higher Education Institutions of China (No. KJ2017A340), and the Research Project of Anhui Jianzhu University.

Q.-L. Li et al. (Eds.): Cao Festschrift 2019, CCIS 1102, pp. 361–373, 2019.
https://doi.org/10.1007/978-981-15-0864-6_18

according to a Poisson process with rate $\lambda$. The service time $T_\beta$ is assumed to be i.i.d. r.v.'s having the distribution $F_\beta(t)$ with $F_\beta(0) = 0$ and mean $\beta_1 < \infty$. Each time a service is completed and the system is not empty, the next customer to be served is selected at random from all the customers waiting in the queue. Each time a busy period ends and system becomes empty, the server undergoes a vacation of random length of time $T_\alpha$. Whenever the server returns from a vacation and finds one or more customers waiting, the server goes on serving a customer immediately, otherwise, on return from a vacation, the server finds no customer waiting, the server takes on a vacation again. The generic vacation time $T_\alpha$ is assumed to have the distribution $F_\alpha(t)$ with $F_\alpha(0) = 0$ and mean $\alpha_1 < \infty$. Besides, we use the notations $\alpha(s)$ and $\beta(s)$ to represent the Laplace-Stieltjes (LS) transforms of $F_\alpha(t)$ and $F_\beta(t)$, respectively. It is well known that the system is stable if and only if (iff) $\rho = \lambda\beta_1 < 1$, which is assumed to hold throughout this paper.

There are many references on asymptotic analysis for queueing systems with heavy-tailed distributions, e.g., Asmussen, Klüppelberg and Sigman [1], Boxma and Denisov [4], and more references can be found in two excellent surveys: Borst et al. [3], and Boxma and Zwart [6]. As far as the ROS discipline concerned, we refer readers to [5] and [12]. Under the assumption of regularly-varying service time distribution, Borst, et al. [5] and Kim, et al. [12] obtained asymptotic expressions for the waiting time distributions in the ordinary $M/G/1$ queue (without vacation) and the $M/G/1$ queue with retrials, respectively.

Our focus in this paper is to study the asymptoic behavior for the tail probability of the waiting time in the $M/G/1/ROS$ vacation queue with regularly-varying service time and vacation time distributions, which is one of typical and commonly used heavy-tailed distributions. Conditioning on the server being busy or on vacation, the asymptotic conditional tail probabilities are obtained explicitly. As a side product (Remark 2) of main results obtained in this paper, we verify that the waiting-time tail for $M/G/1/ROS$ queue with multiple vacation is asymptotically equivalent to that for the standard $M/G/1/ROS$ queue (without vacation), as long as the vacation time has a tail probability lighter than the service time.

The rest of the paper is organized as follows: Sect. 2 provides preliminaries to facilitate our analysis. In Sects. 3 and 4, we study the asymptotic behaviors for the conditional tail probabilities of waiting time conditioning on the server being busy and on vacation, respectively.

## 2   Preliminary

In this section, we present some definitions, notations and useful literature results, which will be used in later sections

**Definition 1 (Bingham, Goldie and Teugels [2]).** *A measurable function $U : (0, \infty) \to (0, \infty)$ is regularly varying at $\infty$ with index $\sigma \in (-\infty, \infty)$ (written $U \in \mathcal{R}_\sigma$) iff $\lim_{t \to \infty} U(xt)/U(t) = x^\sigma$ for all $x > 0$. If $\sigma = 0$ we call $U$ slowly varying, i.e., $\lim_{t \to \infty} U(xt)/U(t) = 1$ for all $x > 0$.*

We will use $L(t)$ to represent a slowly varying function at $\infty$ (see Definition 1) and make the following basic assumptions on the service time $T_\beta$ and the vacation time $T_\alpha$:

**A1.** *The service time $T_\beta$ has tail probability $P\{T_\beta > t\} \sim t^{-b}L(t)$ as $t \to \infty$, where $1 < b < 2$.*

**A2.** *The vacation time $T_\alpha$ has tail probability $P\{T_\alpha > t\} \sim \gamma P\{T_\beta > t\}$ as $t \to \infty$, where $\gamma \geq 0$.*

*Remark 1.* When $\gamma = 0$, Assumption A2 is to be interpreted as $P\{T_\alpha > t\} = o(P\{T_\beta > t\})$, which means that the vacation time has a tail probability lighter than the service time. When $\gamma > 0$, two tail probabilities are asymptotically equivalent up to a prefactor $\gamma$.

Let $T_\pi$ be the busy period of the standard $M/G/1$ queue with arrival rate $\lambda$ and service time $T_\beta$. It is well known that $\pi_1 \overset{\text{def}}{=} E(T_\pi) = \beta_1/(1-\rho)$. By $\pi(s)$, we denote the LS transform of the probability distribution function of $T_\pi$. Under Assumption A1, the tail probability $P\{T_\pi > t\}$ is regularly varying according to de Meyer and Teugels [7]:

$$P\{T_\pi > t\} \sim \frac{1}{(1-\rho)^{b+1}} \cdot t^{-b}L(t) \quad \text{as } t \to \infty. \tag{1}$$

Let $F_\beta^{(e)}(t)$ be the so-called equilibrium distribution of $F_\beta(t)$, which is defined as $F_\beta^{(e)}(t) = \beta_1^{-1}\int_0^t(1 - F_\beta(x))dx$. Similarly, we define $F_\alpha^{(e)}(t) = \alpha_1^{-1}\int_0^t(1 - F_\alpha(x))dx$ and $F_\pi^{(e)}(t) = \pi_1^{-1}\int_0^t(1 - F_\pi(x))dx$. Denote by $\beta^{(e)}(s)$, $\alpha^{(e)}(s)$ and $\pi^{(e)}(s)$ the LS transforms of $F_\beta^{(e)}(t)$, $F_\alpha^{(e)}(t)$ and $F_\pi^{(e)}(t)$, respectively.

By Karamata's theorem (e.g., p. 28 in Bingham, Goldie and Teugels [2]) and Assumptions A1 and A2, we know that $1 - F_\beta^{(e)}(t) \sim c_\beta t^{-b+1}L(t)$, $1 - F_\alpha^{(e)}(t) \sim c_\alpha t^{-b+1}L(t)$ and $1 - F_\pi^{(e)}(t) \sim c_\pi t^{-b+1}L(t)$ as $t \to \infty$, where

$$c_\beta = \frac{1}{(b-1)\beta_1}, \tag{2}$$

$$c_\alpha = \frac{\gamma}{(b-1)\alpha_1}, \tag{3}$$

$$c_\pi = \frac{1}{(b-1)\pi_1} \cdot \frac{1}{(1-\rho)^{b+1}} = \frac{c_\beta}{(1-\rho)^b}. \tag{4}$$

Applying Theorem 8.1.6, p. 333 in Bingham et al. [2], we further obtain the asymptotic properties for LS transforms $\beta^{(e)}(s)$, $\alpha^{(e)}(s)$ and $\pi_0^{(e)}(s)$:

$$1 - \beta^{(e)}(s) = c_\beta c(b)s^{b-1}L(1/s)(1 + o(1)) \quad s \downarrow 0, \tag{5}$$

$$1 - \alpha^{(e)}(s) = c_\alpha c(b)s^{b-1}L(1/s)(1 + o(1)) \quad s \downarrow 0, \tag{6}$$

$$1 - \pi^{(e)}(s) = c_\pi c(b)s^{b-1}L(1/s)(1 + o(1)) \quad s \downarrow 0, \tag{7}$$

where $c(b) = \Gamma(b-1)\Gamma(2-b)/\Gamma(b-1)$.

Let $W$ be the waiting time of a generic customer, $W_b$ and $W_v$ be two r.v.s whose probability distributions coincide with the conditional probability distributions of $W$ given that the generic customer finds the server busy and on vacation upon its arrival, respectively. Precisely, $P\{W_b \leq t\} = P\{W \leq t|\text{busy}\}$ and $P\{W_v \leq t\} = P\{W \leq t|\text{vacation}\}$. Therefore, the probability distributions $P\{W_b \leq t\}$ and $P\{W_v \leq t\}$ have the LS transforms $W_b(s) \stackrel{\text{def}}{=} E(e^{-sW}|\text{busy})$ and $W_v(s) \stackrel{\text{def}}{=} E(e^{-sW}|\text{vacation})$, respectively. Our starting point for tail asymptotic analysis on $P\{W_b > t\}$ and $P\{W_v > t\}$ is based on the expressions for $W_b(s)$ and $W_v(s)$, which can be found in [17].

$$W_b(s) = \frac{1-\rho}{\alpha_1 \rho s} \int_{\pi(s)}^1 \frac{[1 - \alpha(\lambda - \lambda u)][\beta(\lambda - \lambda u) - \beta(s + \lambda - \lambda u)]}{[\beta(\lambda - \lambda u) - u][u - \beta(s + \lambda - \lambda u)]}$$
$$\cdot \exp\{-G(s, u)\} \, du, \tag{8}$$

$$W_v(s) = \frac{1}{\alpha_1 s} \int_{\pi(s)}^1 \frac{\alpha(\lambda - \lambda u) - \alpha(s + \lambda - \lambda u)}{u - \beta(s + \lambda - \lambda u)} \cdot \exp\{-G(s, u)\} \, du, \tag{9}$$

where

$$G(s, u) = \int_u^1 \frac{1}{v - \beta(s + \lambda - \lambda v)} dv. \tag{10}$$

## 3    Asymptotic Tail Probability of $W_b$

In this section, we are going to derive the asymptotic expression of $P\{W_b > t\}$ as $t \to \infty$ based on $W_b(s)$, the LS transform of distribution function $P\{W_b \leq t\}$. Let us rewrite (8) as follows

$$W_b(s) = \frac{1-\rho}{\rho s} \int_{\pi(s)}^1 \alpha^{(e)}(\lambda - \lambda u) \cdot \left[ \frac{\lambda - \lambda u}{\beta(\lambda - \lambda u) - u} + \frac{\lambda - \lambda u}{u - \beta(s + \lambda - \lambda u)} \right]$$
$$\cdot \exp\{-G(s, u)\} \, du. \tag{11}$$

Setting $u = u(t) = 1 - st/\lambda$ in (11), and noting that

$$\frac{\lambda - \lambda\pi(s)}{s} = \frac{\rho}{1-\rho} \cdot \frac{1 - \pi(s)}{sE(T_\pi)} = \frac{\rho}{1-\rho}\pi^{(e)}(s), \tag{12}$$

we get that

$$W_b(s) = \frac{1-\rho}{\rho} \int_0^{\frac{\rho}{1-\rho}\pi^{(e)}(s)} \alpha^{(e)}(st) \left[ \frac{st}{st - \lambda + \lambda\beta(st)} + \frac{st}{-st + \lambda - \lambda\beta(s + st)} \right]$$
$$\cdot \exp\{-G(s, 1 - st/\lambda)\} dt. \tag{13}$$

It follows from (10) that

$$G(s, 1 - st/\lambda) = \int_{1-(st/\lambda)}^{1} \frac{1}{v - \beta(s + \lambda - \lambda v)} dv$$

$$= \int_{0}^{1} \frac{st}{-stw + \lambda - \lambda\beta(s + stw)} dw$$

$$= \int_{0}^{1} \frac{1}{D(s, t, w)} dw, \tag{14}$$

where

$$D(s, t, w) = \frac{-stw + \lambda - \lambda\beta(s + stw)}{st} = -w + \rho(w + 1/t)\beta^{(e)}(s + stw). \tag{15}$$

Let

$$B_1(s, t) = \frac{st}{st - \lambda + \lambda\beta(st)} = \frac{1}{1 - \rho\beta^{(e)}(st)}, \tag{16}$$

$$B_2(s, t) = \frac{st}{-st + \lambda - \lambda\beta(s + st)} = \frac{1}{D(s, t, 1)}, \tag{17}$$

$$C(s, t) = \exp\left\{ -\int_{0}^{1} \frac{1}{D(s, t, w)} dw \right\}, \tag{18}$$

$$g(s, t) = \alpha^{(e)}(st)(B_1(s, t) + B_2(s, t))C(s, t). \tag{19}$$

Then we can further rewrite (13) as

$$W_b(s) = \frac{1 - \rho}{\rho} \int_{0}^{\frac{\rho}{1-\rho}\pi^{(e)}(s)} g(s, t)dt = W_{b_1}(s) - W_{b_2}(s) \tag{20}$$

where

$$W_{b_1}(s) = \frac{1 - \rho}{\rho} \int_{0}^{\frac{\rho}{1-\rho}} g(s, t)dt, \qquad W_{b_2}(s) = \frac{1 - \rho}{\rho} \int_{\frac{\rho}{1-\rho}\pi^{(e)}(s)}^{\frac{\rho}{1-\rho}} g(s, t)dt. \tag{21}$$

In following subsections, we are going to discuss the asymptotic properties for $W_{b_1}(s)$ and $W_{b_2}(s)$ as $s \downarrow 0$, which will be used to obtain the asymptotics of $P\{W_b > t\}$ as $t \to \infty$ later. For this purpose, we start with studying the asymptotic behaviors for $B_1(s, t)$, $B_2(s, t)$ and $C(s, t)$ as $s \downarrow 0$.

### 3.1 Asymptotic Properties for $B_1(s, t)$, $B_2(s, t)$ and $C(s, t)$ as $s \downarrow 0$

It follows from (16) and (5) that

$$B_1(s, t) = \frac{1}{1 - \rho + \rho(1 - \beta^{(e)}(st))} = \frac{1}{1 - \rho} \cdot \frac{1}{1 + \frac{\rho}{1-\rho}\beta_0^{(e)}(st)}$$

$$= \frac{1}{1 - \rho} - \frac{c_\beta c(b)\rho}{(1 - \rho)^2}(st)^{b-1} L(1/s)(1 + o(1)). \tag{22}$$

where we have used the fact that $1/(1 - x) = 1 + x + x^2 + \cdots$ for $|x| < 1$.

It follows from (15) and (5) that

$$D(s,t,w) = -w + \rho w + \rho/t - \rho(w + 1/t)\beta_0^{(e)}(s + stw)$$
$$= (-w + \rho w + \rho/t)\left[1 - \frac{\rho(w + 1/t)\beta_0^{(e)}(s + stw)}{-w + \rho w + \rho/t}\right]$$
$$= (-w + \rho w + \rho/t)$$
$$\cdot \left[1 - c_\beta c(b)\rho\frac{(1/t)(1 + tw)^b}{-w + \rho w + \rho/t} \cdot s^{b-1}L(1/s)(1 + o(1))\right], \quad (23)$$

which implies that

$$\frac{1}{D(s,t,w)} = \frac{1}{-w + \rho w + \rho/t}$$
$$+ c_\beta c(b)\rho\frac{(1/t)(1 + tw)^b}{(-w + \rho w + \rho/t)^2} \cdot s^{b-1}L(1/s)(1 + o(1)). \quad (24)$$

Therefore, by (18),

$$C(s,t) = \varphi(t) \cdot \exp\left\{-c_\beta c(b)\rho H(t)s^{b-1}L(1/s)(1 + o(1))\right\}, \quad (25)$$

where

$$\varphi(t) = \exp\left\{-\int_0^1 \frac{1}{-w + \rho w + \rho/t}dw\right\} = \left(\frac{-t + \rho t + \rho}{\rho}\right)^{\frac{1}{1-\rho}}, \quad (26)$$
$$H(t) = \int_0^1 \frac{(1/t)(1 + tw)^b}{(-w + \rho w + \rho/t)^2}dw. \quad (27)$$

Note that the fact that $e^{-x} = 1 - x + (-x)^2/2! + \cdots$. Then (25) yields

$$C(s,t) = \varphi(t)\left[1 - c_\beta c(b)\rho H(t)s^{b-1}L(1/s)(1 + o(1))\right]. \quad (28)$$

In addition, by (17) and (24), we get

$$B_2(s,t) = \frac{1}{-1 + \rho + \rho/t} + c_\beta c(b)\rho\frac{(1/t)(1 + t)^b}{(-1 + \rho + \rho/t)^2}s^{b-1}L(1/s)(1 + o(1)). (29)$$

## 3.2   Asymptotic Property for $W_{b_1}(s)$ as $s \downarrow 0$

It follows from (22) and (29) that

$$B_1(s,t) + B_2(s,t) = \psi(t) - c_\beta c(b)\rho K(t)s^{b-1}L(1/s)(1 + o(1)), \quad (30)$$

where

$$\psi(t) = \frac{1}{1-\rho} \cdot \frac{\rho}{-t + \rho t + \rho}. \tag{31}$$

$$K(t) = K_0(t) - K_1(t) \tag{32}$$

$$K_0(t) = \frac{t^{b-1}}{(1-\rho)^2}, \tag{33}$$

$$K_1(t) = \frac{(1/t)(1+t)^b}{(-1+\rho+\rho/t)^2}. \tag{34}$$

By (19), (6), (28) and (30), we know

$$g(s,t) = \varphi(t)\psi(t) - \Big[c_\alpha c(b) t^{b-1}\varphi(t)\psi(t) + c_\beta c(b)\rho\varphi(t)\big(K(t) + \psi(t)H(t)\big)\Big]$$
$$\cdot s^{b-1}L(1/s)(1+o(1)). \tag{35}$$

Recalling the expression of $W_{b_1}(s)$ in (21), along with (35), we can write

$$W_{b_1}(s) = \frac{1-\rho}{\rho}\int_0^{\frac{\rho}{1-\rho}} \varphi(t)\psi(t)dt - \Big[c_\alpha c(b)\frac{1-\rho}{\rho}\int_0^{\frac{\rho}{1-\rho}} t^{b-1}\varphi(t)\psi(t)dt +$$
$$c_\beta c(b)(1-\rho)\int_0^{\frac{\rho}{1-\rho}} \varphi(t)\Big(K(t) + \psi(t)H(t)\Big)dt\Big]$$
$$\cdot s^{b-1}L(1/s)(1+o(1)). \tag{36}$$

In the following, we are going to calculate the integrals in (36). By (26) and (31),·

$$\varphi(t)\psi(t) = \frac{1}{1-\rho}\Big(\frac{-t+\rho t + \rho}{\rho}\Big)^{\frac{1}{1-\rho}-1}, \tag{37}$$

hence

$$\frac{1-\rho}{\rho}\int_0^{\frac{\rho}{1-\rho}} \varphi(t)\psi(t)dt = \frac{1}{1-\rho}\int_0^1 (1-x)^{\frac{1}{1-\rho}-1}dx = 1, \tag{38}$$

$$\frac{1-\rho}{\rho}\int_0^{\frac{\rho}{1-\rho}} t^{b-1}\varphi(t)\psi(t)dt = \frac{1}{\rho}\Big(\frac{\rho}{1-\rho}\Big)^b \int_0^1 (1-x)^{\frac{1}{1-\rho}-1}x^{b-1}dx. \tag{39}$$

By (26) and (33),

$$\int_0^{\frac{\rho}{1-\rho}} \varphi(t)K_0(t)dt = \frac{1}{(1-\rho)^2}\int_0^{\frac{\rho}{1-\rho}} \Big(\frac{-t+\rho t+\rho}{\rho}\Big)^{\frac{1}{1-\rho}} \cdot t^{b-1}dt$$
$$= \frac{1}{(1-\rho)^2}\Big(\frac{\rho}{1-\rho}\Big)^b \int_0^1 (1-x)^{\frac{1}{1-\rho}} \cdot x^{b-1}dx. \tag{40}$$

By (26) and (34),

$$\int_0^{\frac{\rho}{1-\rho}} \varphi(t)K_1(t)dt = \int_0^{\frac{\rho}{1-\rho}} \Big(\frac{-t+\rho t+\rho}{\rho}\Big)^{\frac{1}{1-\rho}} \cdot \frac{t(1+t)^b}{(-t+\rho t+\rho)^2}dt$$
$$= \frac{1}{(1-\rho)^2}\int_0^1 (1-x)^{\frac{1}{1-\rho}-2} \cdot x\Big(1 + \frac{\rho}{1-\rho}x\Big)^b dx. \tag{41}$$

By (37) and (27),

$$\int_0^{\frac{\rho}{1-\rho}} \varphi(t)\psi(t)H(t)dt$$

$$= \int_0^{\frac{\rho}{1-\rho}} \varphi(t)\psi(t) \cdot \left( \int_0^t \frac{(1+y)^b}{(-y+\rho y+\rho)^2} dy \right) dt$$

$$= \frac{1}{(1-\rho)} \int_0^{\frac{\rho}{1-\rho}} \frac{(1+y)^b}{(-y+\rho y+\rho)^2} \left[ \int_y^{\frac{\rho}{1-\rho}} \left( \frac{-t+\rho t+\rho}{\rho} \right)^{\frac{\rho}{1-\rho}} dt \right] dy$$

$$= \frac{1}{(1-\rho)\rho} \int_0^{\frac{\rho}{1-\rho}} \left( \frac{-y+\rho y+\rho}{\rho} \right)^{\frac{\rho}{1-\rho}-1} (1+y)^b dy$$

$$= \frac{1}{(1-\rho)^2} \int_0^1 (1-x)^{\frac{1}{1-\rho}-2} \left( 1+\frac{\rho}{1-\rho}x \right)^b dx. \qquad (42)$$

Noting that $K(t)$ is given in (32) and substituting (38)–(42) into (36), we obtain

$$W_{b_1}(s) = 1 - d_b c(b) \cdot s^{b-1} L(1/s)(1+o(1)), \qquad (43)$$

where

$$d_b = \left( \frac{c_\alpha}{\rho} + \frac{c_\beta}{1-\rho} \right) \left( \frac{\rho}{1-\rho} \right)^b \int_0^1 (1-x)^{\frac{1}{1-\rho}} \cdot x^{b-1} dx$$

$$+ \frac{c_\beta}{1-\rho} \int_0^1 (1-x)^{\frac{1}{1-\rho}-1} \left( 1+\frac{\rho}{1-\rho}x \right)^b dx. \qquad (44)$$

### 3.3    Asymptotic Property for $W_{b_2}(s)$ as $s \downarrow 0$

Recall (21). By the integration middle value theorem, there exists $\xi(s) \in (0,1)$ such that

$$W_{b_2}(s) = g\big(s, h(s)\big)\Big(1 - \pi^{(e)}(s)\Big), \qquad (45)$$

where $h(s) = \frac{\rho}{1-\rho}\Big[\pi^{(e)}(s) + \xi(s)(1-\pi^{(e)}(s))\Big]$. It follows from (7) that

$$h(s) = \frac{\rho}{1-\rho}\Big(1+O(1) \cdot s^{b-1}L(1/s)\Big). \qquad (46)$$

Next, we will prove that $g\big(s, h(s)\big) = o(1)$. By (35), we have

$$g\big(s,h(s)\big) = \varphi(h(s))\psi(h(s)) - \Big[c_\alpha c(b)(h(s))^{b-1}\varphi(h(s))\psi(h(s))$$

$$+ c_\beta c(b)\rho\varphi(h(s))\Big(K(h(s)) + \psi(h(s))H(h(s))\Big)\Big]$$

$$\cdot s^{b-1}L(1/s)(1+o(1)). \qquad (47)$$

By (46), we know that $-h(s) + \rho h(s) + \rho = O(1) \cdot s^{b-1}L(1/s)$, which together with (26) and (37) leads to

$$\varphi(h(s)) = O(1) \cdot s^{(b-1)/(1-\rho)} (L(1/s))^{\frac{1}{1-\rho}}, \qquad (48)$$

$$\varphi(h(s))\psi(h(s)) = O(1) \cdot s^{(b-1)(\frac{1}{1-\rho}-1)} (L(1/s))^{\frac{1}{1-\rho}-1}. \qquad (49)$$

Because $\lim_{s\to 0} h(s) = \rho/(1-\rho)$, it follows from (33), (34) and (27) that

$$K_0(h(s)) = \frac{(h(s))^{b-1}}{(1-\rho)^2} = O(1), \qquad (50)$$

$$K_1(h(s)) = \frac{(1/h(s))(1+h(s))^b}{(-1+\rho+\rho/h(s))^2} = O(1) \cdot \left(\frac{1}{s^{b-1}L(1/s)}\right)^2, \qquad (51)$$

$$H(h(s)) = \int_0^1 \frac{(1/h(s))(1+h(s)w)^b}{(-w+\rho w+\rho/h(s))^2}dw = O(1). \qquad (52)$$

Further, by (48)–(52),

$$\varphi(h(s))K(h(s))s^{b-1}L(1/s) = o(1), \qquad (53)$$

$$\varphi(h(s))\psi(h(s))H(h(s))s^{b-1}L(1/s) = o(1). \qquad (54)$$

It follows from (47), (49), (53) and (54) that $g(s, h(s)) = o(1)$, which, together with (45) and (7), results in

$$W_{b_2}(s) = o(1) \cdot s^{b-1}L(1/s). \qquad (55)$$

## 3.4   Tail Probability Asymptotics for $W_b$

By (20), (43) and (55),

$$W_b(s) = 1 - c(b)d_b s^{b-1}L(1/s)(1+o(1)). \qquad (56)$$

Applying Theorem 8.1.6, p. 333 in Bingham et al. [2], we obtain

$$P\{W_b > t\} = P\{W > t|busy\} \sim d_b t^{-b+1}L(t) \qquad \text{as } t \to \infty. \qquad (57)$$

where $d_b$ is given in (44).

## 4   Asymptotic Tail Probability of $W_v$

In this section, we are going to derive the asymptotic expression of $P\{W_v > t\}$ as $t \to \infty$ based on $W_v(s)$ given in (9). Let

$$A(s,t) = \frac{\alpha(st) - \alpha(s+st)}{\alpha_1 st} = (1+1/t)\alpha^{(e)}(s+st) - \alpha^{(e)}(st), \qquad (58)$$

$$g_v(s,t) = A(s,t)B_2(s,t)C(s,t). \qquad (59)$$

Similar to derivation of (20), we get that

$$W_v(s) = \int_0^{\frac{\rho}{1-\rho}\pi^{(e)}(s)} g_v(s,t)dt = W_{v1}(s) - W_{v2}(s), \qquad (60)$$

where

$$W_{v1}(s) = \int_0^{\frac{\rho}{1-\rho}} g_v(s,t)dt, \qquad W_{v2}(s) = \int_{\frac{\rho}{1-\rho}\pi^{(e)}(s)}^{\frac{\rho}{1-\rho}} g_v(s,t)dt. \qquad (61)$$

In following subsections, we will discuss the asymptotic behavior of $W_{v_1}(s)$ and $W_{v_2}(s)$ as $s \downarrow 0$, which will be used later to obtain the asymptotics of $P\{W_v > t\}$ as $t \to \infty$.

## 4.1   Asymptotic Property for $W_{v_1}(s)$ as $s \downarrow 0$

It follows from (6) and (58) that

$$A(s,t) = \frac{1}{t}\left[1 - c_\alpha c(b)\big((1+t)^b - t^b\big)s^{b-1}L(1/s)(1+o(1))\right], \qquad (62)$$

which, together with (29), implies that

$$\begin{aligned}
A(s,t)B_2(s,t) &= \frac{1}{-t+\rho t+\rho}\left[1 - \Big(c_\alpha c(b)\big((1+t)^b - t^b\big) - c_\beta c(b)\rho\frac{(1+t)^b}{-t+\rho t+\rho}\Big)\right.\\
&\qquad\qquad \left.\cdot s^{b-1}L(1/s)(1+o(1))\right]\\
&= \frac{1-\rho}{\rho}\psi(t)\left[1 - c(b)\rho\Big(c_\alpha R_1(t) - c_\beta R_2(t)\Big)\right.\\
&\qquad\qquad\qquad\qquad \left.\cdot s^{b-1}L(1/s)(1+o(1))\right].
\end{aligned} \qquad (63)$$

where

$$R_1(t) = \frac{(1+t)^b - t^b}{\rho}, \qquad (64)$$

$$R_2(t) = \frac{(1+t)^b}{-t+\rho t+\rho}. \qquad (65)$$

By (63) and (28), we obtain

$$\begin{aligned}
g_v(s,t) &= \varphi(t)\psi(t)\left[\frac{1-\rho}{\rho} - c(b)(1-\rho)\Big(c_\beta H(t) + c_\alpha R_1(t) - c_\beta R_2(t)\Big)\right.\\
&\qquad\qquad \left.\cdot s^{b-1}L(1/s)(1+o(1))\right].
\end{aligned} \qquad (66)$$

Recalling the expression of $W_{v_1}(s)$ given in (61), along with (38) and (66), we can write

$$\begin{aligned}
W_{v_1}(s) &= 1 - c(b)(1-\rho)\left[\int_0^{\frac{\rho}{1-\rho}} \varphi(t)\psi(t)\Big(c_\beta H(t) + c_\alpha R_1(t) - c_\beta R_2(t)\Big)dt\right]\\
&\qquad\qquad \cdot s^{b-1}L(1/s)(1+o(1)).
\end{aligned} \qquad (67)$$

Next, let us calculate the integrals in (67). By (37) and (64),

$$\int_0^{\frac{\rho}{1-\rho}} \varphi(t)\psi(t)R_1(t)dt$$

$$= \frac{1}{\rho(1-\rho)} \int_0^{\frac{\rho}{1-\rho}} \left(\frac{-t+\rho t+\rho}{\rho}\right)^{\frac{1}{1-\rho}-1} \cdot \left[(1+t)^b - t^b\right]dt$$

$$= \frac{1}{(1-\rho)^2} \int_0^1 (1-x)^{\frac{1}{1-\rho}-1} \left[\left(1+\frac{\rho x}{1-\rho}\right)^b - \left(\frac{\rho x}{1-\rho}\right)^b\right]dx. \qquad (68)$$

By (37) and (65),

$$\int_0^{\frac{\rho}{1-\rho}} \varphi(t)\psi(t)R_2(t)dt = \frac{1}{1-\rho} \int_0^{\frac{\rho}{1-\rho}} \left(\frac{-t+\rho t+\rho}{\rho}\right)^{\frac{1}{1-\rho}-1} \cdot \frac{(1+t)^b}{-t+\rho t+\rho}dt$$

$$= \frac{1}{(1-\rho)^2} \int_0^1 (1-x)^{\frac{1}{1-\rho}-2} \cdot \left(1+\frac{\rho}{1-\rho}x\right)^b dx. \qquad (69)$$

Substituting (68), (69) and (42) into (67), we obtain

$$W_{v_1}(s) = 1 - c(b)d_v \cdot s^{b-1}L(1/s)(1+o(1)), \qquad (70)$$

where

$$d_v = \frac{c_\alpha}{1-\rho} \int_0^1 (1-x)^{\frac{1}{1-\rho}-1} \left[\left(1+\frac{\rho x}{1-\rho}\right)^b - \left(\frac{\rho x}{1-\rho}\right)^b\right]dx. \qquad (71)$$

## 4.2   Asymptotic Property for $W_{v_2}(s)$ as $s \downarrow 0$

Recall (61). By the integration middle value theorem, there exists $\xi_v(s) \in (0,1)$ such that

$$W_{v_2}(s) = g_v\big(s, h_v(s)\big)\big(1 - \pi^{(e)}(s)\big), \qquad (72)$$

where $h_v(s) = \frac{\rho}{1-\rho}\left[\pi^{(e)}(s) + \xi_v(s)(1-\pi^{(e)}(s))\right]$. Further, by (7),

$$h_v(s) = \frac{\rho}{1-\rho}\left[1 + O(1)\cdot s^{b-1}L(1/s)\right]. \qquad (73)$$

Next, we will prove that $g_v\big(s, h_v(s)\big) = o(1)$. By (66), we have

$$g_v\big(s, h_v(s)\big)$$

$$= \varphi(h_v(s))\psi(h_v(s))\left[\frac{1-\rho}{\rho} - c(b)(1-\rho)\right.$$

$$\left.\cdot\big(c_\beta H(h_v(s)) + c_\alpha R_1(h_v(s)) - c_\beta R_2(h_v(s))\big)s^{b-1}L(1/s)(1+o(1))\right]. \qquad (74)$$

Immediately, by (73), we know that $-h_v(s) + \rho h_v(s) + \rho = O(1)\cdot s^{b-1}L(1/s)$, which together with (37) leads to

$$\varphi\big(h_v(s)\big)\psi\big(h_v(s)\big) = O(1)\cdot s^{(b-1)(\frac{1}{1-\rho}-1)}\big(L(1/s)\big)^{\frac{1}{1-\rho}-1}. \qquad (75)$$

Because $\lim_{s \to 0} h_v(s) = \rho/(1-\rho)$, it follows from (64) and (65) that

$$R_1(h_v(s)) = \frac{(1+h_v(s))^b - (h_v(s))^b}{\rho} = O(1), \tag{76}$$

$$R_2(h_v(s)) = \frac{(1+h_v(s))^b}{-h_v(s) + \rho h_v(s) + \rho} = O(1) \cdot \left(\frac{1}{s^{b-1}L(1/s)}\right), \tag{77}$$

$$H(h_v(s)) = \int_0^1 \frac{(1/h_v(s))(1+h_v(s)w)^b}{(-w+\rho w + \rho/h_v(s))^2} dw = O(1). \tag{78}$$

Further, by (75) and (77),

$$\varphi(h_v(s))\psi(h_v(s))R_2(h_v(s))s^{b-1}L(1/s) = o(1). \tag{79}$$

It follows from (74)–(76) and (78)–(79) that $g_v(s, h_v(s)) = o(1)$, which, together with (72) and (7), results in

$$W_{v_2}(s) = o(1) \cdot s^{b-1}L(1/s). \tag{80}$$

## 4.3 Tail Probability Asymptotics for $W_v$ and $W$

By (60), (70) and (80)

$$W_v(s) = 1 - c(b)d_v \cdot s^{b-1}L(1/s)(1+o(1)). \tag{81}$$

Applying Theorem 8.1.6, p. 333 in Bingham et al. [2], we obtain

$$P\{W > t | \text{vacation}\} = P\{W_v > t\} \sim d_v \cdot t^{-b+1}L(t) \qquad \text{as } t \to \infty. \tag{82}$$

where $d_v$ is given in (71).

Note that $P\{W > t\} = \rho P\{W_b > t\} + (1-\rho)P\{W_v > t\}$. By (57) and (82), we have

$$P\{W > t\} \sim \left(\rho d_b + (1-\rho)d_v\right) \cdot t^{-b+1}L(t) \qquad \text{as } t \to \infty. \tag{83}$$

**A special case: $\gamma = 0$.**
This is the case when the vacation time $T_\alpha$ has a tail lighter than the service time $T_\beta$, in which $c_\alpha = 0$. Thus, by (83), (44) and (71),

$$P\{W > t\} \sim \frac{\rho}{1-\rho}c_\beta c_W \cdot t^{-b+1}L(t) \qquad \text{as } t \to \infty. \tag{84}$$

where

$$c_W = \left(\frac{\rho}{1-\rho}\right)^b \int_0^1 (1-x)^{\frac{1}{1-\rho}} \cdot x^{b-1}dx$$
$$+ \int_0^1 (1-x)^{\frac{1}{1-\rho}-1}\left(1+\frac{\rho}{1-\rho}x\right)^b dx. \tag{85}$$

*Remark 2.* In [5], Boxma et al. (2004) have shown that the asymptotic result (84), along with (85) is true for the standard $M/G/1/ROS$ queue (without vacation). As one of side products in this paper, we have verified that such a result is still valid even for the $M/G/1/ROS$ queue with multiple vacations, as long as the vacation time has a tail probability lighter than the service time.

# References

1. Asmussen, S., Klüppelberg, C., Sigman, K.: Sampling at subexponential times, with queueing applications. Stochast. Processes Appl. **79**(2), 265–286 (1999)
2. Bingham, N.H., Goldie, C.M., Teugels, J.L.: Regular Variation. Cambridge University Press, Cambridge (1989)
3. Borst, S.C., Boxma, O.J., Nunez-Queija, R., Zwart, A.P.: The impact of the service discipline on delay asymptotics. Perform. Eval. **54**, 175–206 (2003)
4. Boxma, O., Denisov, D.: Sojourn time tails in the single server queue with heavy-tailed service times. Queueing Syst. **69**, 101–119 (2011)
5. Boxma, O.J., Foss, S.G., Lasgouttes, J.M., Queija, R.: Waiting time and asymptotics in the single server queue with service in random order. Queueing Syst. **46**, 35–73 (2004)
6. Boxma, O., Zwart, B.: Tails in scheduling. ACM Sigmetrics Perform. Eval. Rev. **34**, 13–20 (2007)
7. de Meyer, A., Teugels, J.L.: On the asymptotic behavior of the distributions of the busy period and service time in $M/G/1$. J. Appl. Probab. **17**, 802–813 (1980)
8. Doshi, B.T.: Queueing systems with vacations - a survey. Queueing Syst. **1**, 29–66 (1986)
9. Embrechts, P., Klüppelberg, C., Mikosch, T.: Modelling Extremal Events. AM, vol. 33. Springer, Heidelberg (1997). https://doi.org/10.1007/978-3-642-33483-2
10. Feller, W.: An Introduction to Probability Theorey and its Applications, vol. II. Wiley, London (1971)
11. Foss, S., Korshunov, D., Zachary, S.: An Introduction to Heavy-Tailed and Subexponential Distributions, 2nd edn. Springer, New York (2011). https://doi.org/10.1007/978-1-4614-7101-1
12. Kim, J., Kim, J., Kim, B.: Regularly varying tail of the waiting time distribution in $M/G/1$ retrial queue. Queueing Syst. **65**(4), 365–383 (2010)
13. Liu, B., Wang, J., Zhao, Y.Q.: Tail asymptotics of the waiting time and the busy period for the $M/G/1/K$ queues with subexponential service times. Queueing Syst. **76**(1), 1–19 (2014)
14. Park, K., Willinger, W.: Self-Similar Network Traffic and Performance Evaluation. Wiley, New York (2000)
15. Resnick, S.I.: Heavy-Tail Phenomena: Probabilistic and Statistical Modeling. Springer Series in Operations Research and Financial Engineering. Springer, New York (2007). https://doi.org/10.1007/978-0-387-45024-7
16. Shang, W., Liu, L., Li, Q.-L.: Tail asymptotics for the queue length in an $M/G/1$ retrial queue. Queueing Syst. **52**(3), 193–198 (2006)
17. Takagi, H.: Queueing Analysis: A Foundation of Performance Evaluation. Volume 1: Vacation and Priority Systems, Part 1. North-Holland (1991)

# Strategic Joining in an M/M/1 Constant Retrial Queue with Reserved Idle Time Under N-Policy

Jinting Wang[1(✉)] and Lulu Li[2]

[1] School of Management Science and Engineering,
Central University of Finance and Economics, Beijing 100081,
People's Republic of China
jtwang@cufe.edu.cn
[2] Department of Mathematics, Beijing Jiaotong University,
Beijing 100044, People's Republic of China

**Abstract.** This paper considers an M/M/1 constant retrial queueing model with reserved idle time under $N$-policy. A customer can occupy the server instantaneously when the server is turned on and idle upon arrival. After the service of the last customer, the server stays idle for some random time. During this period, if a customer arrives, he obtains service immediately. Otherwise, the sever will shut down for saving energy and be reactivated if the number of waiting customers in retrial orbit reaches a given threshold $N(N > 1)$. The probabilities of the server in different states are derived through generating function method. Moreover, based on the reward-cost function and the expected payoff, all customers will decide whether to join or balk the system upon arrival. Given these strategic behaviors we establish the net profit of the service provider per unit of time. Finally, some numerical examples are presented to illustrate the necessity of the reserved idle time's existence from the perspective of the service provider. It is found the longer the reserved idle time, the greater the server's profit.

**Keywords:** Retrial queue · Reserved idle time · N-policy ·
Equilibrium strategies

## 1 Introduction

Since the work of Naor [11] and Edelson and Hilderbrand [5], the game-theoretic analysis of queueing systems has been paid considerable attention. In such an issue, given a reward-cost structure, information on the queue length and other performance measures is a crucial factor for customers who make decision on whether to join or to balk the system upon arrival. In principle, there are two

This work was supported in part by the National Natural Science Foundation of China under grant nos. 71871008 and 71571014.

© Springer Nature Singapore Pte Ltd. 2019
Q.-L. Li et al. (Eds.): Cao Festschrift 2019, CCIS 1102, pp. 374–388, 2019.
https://doi.org/10.1007/978-981-15-0864-6_19

groups regarding the information levels in rational queueing: observable and unobservable queues. Interested readers are referred to Hassin and Haviv [10] and Hassin [9] for more comprehensive reviews.

It is very significant to study how the information level influences customers' decision and how to control queueing systems under different information levels. For example, Burnetas and Economou [1] considered a Markovian single-server queueing system with setup times, and they derived equilibrium strategies of customers under various information levels and analyzed the stationary behavior of system under these strategies. Economou and Kanta [3] studied the Markovian single-server queue that alternates between on and off periods. The equilibrium threshold balking strategies in fully observable and almost observable queues were derived.

These issues are studied in the scope of queueing game with vacations. Due to the versatile applicability of vacation queueing systems, considerable interests have been focused on the study of performance evaluation and optimal control of these systems. For a detailed review of the main results and methodologies on this topic, the interested readers are referred to Doshi [2], Takagi [12], Tian and Zhang [13]. In reality, server vacations often occur in practical problems. For example, a server may deactivate for economic reasons, suffer random failures, go under preventive maintenance or attend a secondary service system. During last decades, the economic analysis of vacation queueing systems with strategic customer behavior has gained a great amount of interest. Guo and Hassin [6] considered a single-server vacation queue with $N$-policy and exhaustive service, where the server reactivates once there are $N$ customers arrived in the system and shuts down when the system becomes empty. It was shown that the equilibrium and socially optimal strategies are multiple for unobservable queueing situation and unique for the observable case. The work was extended by Guo and Hassin [7] to the case of heterogeneous customers. They studied both unobservable and observable queues and considered two situations regarding customers' delay sensitivity. Later on, Guo and Li [8] studied strategic behavior and social optimization in partially observable Markovian vacation queues. Recently, Wang et al. [14] considered customers strategic behavior and the corresponding social maximization problem in an M/M/1 constant retrial queue with the $N$-policy. They examined customers strategic responses to different information levels, compared the individual equilibrium with the socially optimal behavior and studied the Price of Anarchy of the system numerically.

Note that all the work mentioned above are conducted without the reserved idle time. So in this paper, we try to model an M/M/1 constant retrial queueing model with reserved idle time under $N$-policy. After completing the service of the last customer, the server stays idle for some random time waiting for a new customer. If there is a new arrival during the idle period, the server continues serving the new arrival, otherwise it shuts down. And it's resumed only when the queue length reaches a given critical length $N(N > 1)$. We consider the partially observable queues where a customer can only observe the state of the server at his arrival instant and does not observe the number of customers in the system.

We study the customers' equilibrium arrival rate and the net profit of the service provider. What is more, we give a proof of the necessity of the reserved idle time by numerical experiments.

Thus, the main contributions of this paper are two-fold. Firstly, among all the works with $N$-policy, we are the first to introduce the reserved idle time and setup time to it. From the perspective of service provider, it is shown that the longer reserved idle time can bring more profit. That is, the existence of reserved idle time is necessary. Then for such a model we investigate customers' equilibrium behavior, where both the phenomena of "follow-the-crowd" and "avoid-the-crowd" can arise.

The outline of this paper is given as follows. In Sect. 2, we give a detailed model description and the reward-cost structure. In Sect. 3, we obtain an exact solution for the probabilities of the server in different states in terms of partial generating functions. The equilibrium arrival rate for joining the orbit when only the server's state is observable is given in Sect. 4. And in Sect. 5, we establish the net profit function and the cost of the service provider, and some numerical examples are provided in order to verify the rationality of the reserved idle time. Then the conclusion is summarized in Sect. 6.

## 2    Model Description

We consider an M/M/1 constant retrial queueing model with reserved idle time under $N$-policy in which customers arrive according to a Poisson process with rate $\lambda$. The service times are assumed to be exponentially distributed with rate $\mu$. A customer will be served instantaneously when finding the server turned on and idle (called "normal idle state") upon arrival. After the service completion of the last customer, the server stays idle for some random time, and the reserved idle time is assumed to be exponentially distributed with rate $\alpha$. During this period, if a customer arrives, he obtains service immediately. Otherwise, the sever will shut down (called "sleep mode state") and becomes active (called "normal working state") again if the number of waiting customers in retrial orbit reaches a given threshold $N$. The setup time is assumed to be exponentially distributed with rate $\beta$. An arriving customer who finds the server busy (serving a customer) joins a "virtual" retrial orbit in accordance with a first-come-first-served (FCFS) discipline. After finishing service, a customer leaves the system and the server seeks to serve a customer from the retrial orbit. The time required to seek a customer from the orbit is assumed to be exponentially distributed with rate $\theta$. However, a new customer maybe arrive during the seeking process. In this case, the server interrupts the seeking process and serves the new arriving customer. We assume that the interarrival times, service times and seeking times are mutually independent.

We are interested in the behavior of customers that whether to join or to balk upon their arrival. After service, we assume that every customer receives a reward of $R$ units and each customer needs to pay a waiting cost $C$ units per time unit until the time he leaves after being served. Moreover, we assume that the server charges all the customers entering the system an admission fee $P(< R)$. Customers are risk neutral and wish to maximize their expected benefit. We assume that

$$R - P > \frac{C}{\mu}. \tag{1}$$

If this condition fails to hold, which means that the customers who find the server idle don't want to enter the system because their reward is equal to or smaller than their expected total waiting cost. We further stress that the decisions are irrevocable and the queueing system does not permit retrials of balking customers or reneging of entering customers.

In our model, we assume that the customers are informed upon their arrival only about the state of the server. In principle, the customers' joining probabilities upon arrival are conditional on the state of the server $C(t)$. Therefore, we denote the equilibrium joining probabilities of customers by $q_i$ $(i = 0, 1, 2)$, and consequently the effective equilibrium arrival rates are $\lambda_i \equiv \lambda q_i$ which are equal to or less than $\lambda$. From the condition (1), we know that the customer's reward exceeds the expected waiting cost when observing the server is in the normal idle state. Thus, he or she will definitely enter the system, which means $\lambda_0 = \lambda$. So joining such a system is a dominant strategy. Therefore, we need only study the behavior of the customers when $C(t) = 1, 2$, that is, the server is in sleep mode state and normal working state. Instead of working on the joining probabilities, we can alternatively consider the corresponding arrival rates.

## 3    Performance Measures

In this section, we have the fact that $\{X(t) = (C(t), N(t)); t \geq 0\}$ forms a Markov chain on the state space $S = \{(0, i), i \geq 0; (1, j), j \geq 0; (2, k), 0 \leq k \leq N\}$ where $C(t)$ denotes the state of the server (0: normal idle; 1: normal working; 2: sleep mode) and $N(t)$ denotes the number of customers in the orbit at time $t$, respectively. We assume that the system is stable, i.e., the stationary distribution exists if and only if (see [4])

$$\frac{\lambda_1(\lambda + \theta)}{(\mu\theta)} < 1.$$

The transition rate diagram is shown in Fig. 1. Letting $\pi_{i,j} = \lim_{t\to\infty} P(C(t) = i, N(t) = j)$ denote the joint stationary distribution of $X(t)$. We define the corresponding partial generating functions as follows.

$$\Pi_0(z) = \sum_{n=0}^{\infty} \pi_{0,n} z^n, \Pi_1(z) = \sum_{n=0}^{\infty} \pi_{1,n} z^n, \Pi_2(z) = \sum_{n=0}^{N} \pi_{2,n} z^n. \, |z| \leq 1$$

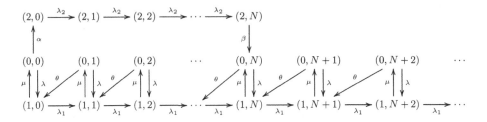

**Fig. 1.** Transition rate diagram of the original model

**Lemma 1.** *For the M/M/1 constant retrial queueing model with reserved idle time under N-policy, given the arrival rates $(\lambda, \lambda_1, \lambda_2)$, the stationary probabilities of the states 0, 1 and 2 are given, respectively, by*

$$\Pi_0(1) = \left[ \frac{\mu}{\lambda + \theta} \frac{\lambda\theta + N\alpha(\lambda + \theta)}{(\lambda + \theta)(\mu - \lambda_1) - \lambda\mu} + \frac{\theta}{\lambda + \theta} \right] \pi_{0,0}, \tag{2}$$

$$\Pi_1(1) = \frac{\lambda\theta + N\alpha(\lambda + \theta)}{(\lambda + \theta)(\mu - \lambda_1) - \lambda\mu} \pi_{0,0}, \tag{3}$$

$$\Pi_2(1) = (\frac{N\alpha}{\lambda_2} + \frac{\alpha}{\beta})\pi_{0,0}. \tag{4}$$

*Moreover, we have the following equations*

$$\Pi'_0(z)|_{z=1} = \left\{ \frac{\mu}{\lambda + \theta} \left\{ \frac{2N\alpha\lambda + N(N-1)\alpha(\lambda + \theta)}{2[(\lambda + \theta)(\mu - \lambda_1) - \lambda\mu]} \right. \right.$$
$$\left. \left. + \frac{\lambda_1(\lambda + \theta)[\lambda\theta + N\alpha(\lambda + \theta)]}{[(\lambda + \theta)(\mu - \lambda_1) - \lambda\mu]^2} \right\} + \frac{N\alpha}{\lambda + \theta} \right\} \pi_{0,0}, \tag{5}$$

$$\Pi'_1(z)|_{z=1} = \left\{ \frac{2N\alpha\lambda + N(N-1)\alpha(\lambda + \theta)}{2[(\lambda + \theta)(\mu - \lambda_1) - \lambda\mu]} + \frac{\lambda_1(\lambda + \theta)[\lambda\theta + N\alpha(\lambda + \theta)]}{[(\lambda + \theta)(\mu - \lambda_1) - \lambda\mu]^2} \right\} \pi_{0,0}, \tag{6}$$

$$\Pi'_2(z)|_{z=1} = \left[ \frac{N(N-1)}{2} \frac{\alpha}{\lambda_2} + \frac{N\alpha}{\beta} \right] \pi_{0,0}, \tag{7}$$

*where*

$$\pi_{0,0} = \left[ \frac{\mu}{\lambda + \theta} \frac{\lambda\theta + N\alpha(\lambda + \theta)}{(\lambda + \theta)(\mu - \lambda_1) - \lambda\mu} + \frac{\theta}{\lambda + \theta} + \frac{\lambda\theta + N\alpha(\lambda + \theta)}{(\lambda + \theta)(\mu - \lambda_1) - \lambda\mu} + \frac{N\alpha}{\lambda_2} + \frac{\alpha}{\beta} \right]^{-1}.$$

*Proof.* The balance equations for states $(i, j)$ are given as follows.

$$(\lambda + \alpha)\pi_{0,0} = \mu\pi_{1,0}, \tag{8}$$

$$(\lambda + \theta)\pi_{0,n} = \mu\pi_{1,n}, 1 \leq n \leq N - 1, n \geq N + 1, \tag{9}$$

$$(\lambda + \theta)\pi_{0,N} = \mu\pi_{1,N} + \beta\pi_{2,N}, \tag{10}$$

$$(\lambda_1 + \mu)\pi_{1,n} = \lambda_1\pi_{1,n-1} + \lambda\pi_{0,n} + \theta\pi_{0,n+1}, n \geq 1, \tag{11}$$

$$(\lambda_1 + \mu)\pi_{1,0} = \lambda\pi_{0,0} + \theta\pi_{0,1}, \tag{12}$$

$$\lambda_2\pi_{2,0} = \alpha\pi_{0,0}, \tag{13}$$

$$\lambda_2\pi_{2,n} = \lambda_2\pi_{2,n-1}, 1 \leq n \leq N - 1, \tag{14}$$

$$\beta\pi_{2,N} = \lambda_2\pi_{2,N-1}. \tag{15}$$

Using equations (13)–(15), we obtain

$$\Pi_2(z) = \frac{1-z^N}{1-z}\frac{\alpha}{\lambda_2}\pi_{0,0} + \frac{\alpha}{\beta}z^N\pi_{0,0}. \tag{16}$$

Multiplying equations (8)–(12) by the corresponding powers of $z$ and summing up, we have the following system of equations for generating functions

$$(\lambda + \theta)\Pi_0(z) + \alpha\pi_{0,0} = \mu\Pi_1(z) + \alpha z^N\pi_{0,0} + \theta\pi_{0,0}, \tag{17}$$

$$(\lambda_1 + \mu)\Pi_1(z) = \lambda\Pi_0(z) + \frac{\theta}{z}\Pi_0(z) + \lambda_1 z\Pi_1(z) - \frac{\theta}{z}\pi_{0,0}, \tag{18}$$

which yields

$$\Pi_0(z) = \frac{\mu}{\lambda+\theta}\Pi_1(z) + \frac{\alpha(z^N - 1) + \theta}{\lambda+\theta}\pi_{0,0}, \tag{19}$$

$$\Pi_1(z) = \frac{(\lambda z + \theta)[\alpha(z^N - 1) + \theta] - \theta(\lambda + \theta)}{(\lambda+\theta)z(\lambda_1 + \mu - \lambda_1 z) - (\lambda z + \theta)\mu}\pi_{0,0}. \tag{20}$$

Letting $z = 1$ in equations (16), (19) and (20), combining the normalization condition and using L'Hôspital's rule we can derive $\pi_{0,0}$. Then we can obtain $\Pi_0(1), \Pi_1(1), \Pi_2(1)$ as given by (2)–(4). Taking the first derivative of the equations (16), (19) and (20) with respect to $z$ and letting $z = 1$, we have $\Pi_0'(z)|_{z=1}, \Pi_1'(z)|_{z=1}, \Pi_2'(z)|_{z=1}$. which are given by (5)–(7). This completes the proof. □

## 4    Equilibrium Arrival Rates

We define $T(i,j)(i = 0, j \geq 1; i = 1, j \geq 0; i = 2, 1 \leq j \leq N)$ as the mean waiting time of a tagged customer that he is at the $j$th position in the orbit, given that the server is in state $i$.

**Lemma 2.** *For the M/M/1 constant retrial queueing model with reserved idle time under N-policy, the expected waiting time of a tagged customer given that he is at the $j$th position in the orbit and the servers state is $i$ $(i = 0,1,2)$, are given by, respectively,*

$$T(0,j) = \frac{\lambda+\theta+\mu}{\mu\theta}j, j \geq 1, \tag{21}$$

$$T(1,j) = \frac{\lambda+\theta+\mu}{\mu\theta}j + \frac{1}{\mu}, j \geq 0, \tag{22}$$

$$T(2,j) = \frac{N-j}{\lambda_2} + \frac{1}{\beta} + \frac{\lambda+\theta+\mu}{\mu\theta}j, 1 \leq j \leq N. \tag{23}$$

*Proof.* By a first step argument, we know $T(i, j)$ satisfies the linear system of equations

$$T(1, 0) = \frac{1}{\mu}, \tag{24}$$

$$T(1, j) = \frac{1}{\lambda_1 + \mu} + \frac{\lambda_1}{\lambda_1 + \mu} T(1, j) + \frac{\mu}{\lambda_1 + \mu} T(0, j), j \geq 1, \tag{25}$$

$$T(0, j) = \frac{1}{\lambda + \theta} + \frac{\lambda}{\lambda + \theta} T(1, j) + \frac{\theta}{\lambda + \theta} T(1, j - 1), j \geq 1, \tag{26}$$

$$T(2, j) = \frac{N - j}{\lambda_2} + \frac{1}{\beta} + T(0, j), 1 \leq j \leq N. \tag{27}$$

Solving (25) with respect to $T(1, j)$ and substituting in (26), we obtain that

$$T(1, j) - T(1, j - 1) = \frac{\lambda + \theta + \mu}{\mu \theta}, j \geq 1.$$

Taking into account (24), we obtain (22). Substituting (22) in (26) we can obtain (21) and then substituting (21) into (27), we obtain (23). This completes the proof. □

**Theorem 1.** *For the M/M/1 constant retrial queueing model with reserved idle time under N-policy, given the arrival rates $(\lambda, \lambda_1, \lambda_2)$, the expected waiting time of an arrival upon seeing a busy sever and the expected waiting time upon seeing a sleep server are*

$$W_1 = \frac{\lambda + \theta + \mu}{\mu \theta} + \frac{1}{\mu} + \frac{\lambda + \theta + \mu}{\mu \theta} \left( \frac{2N\alpha\lambda + N(N-1)\alpha(\lambda + \theta)}{2[\lambda\theta + N\alpha(\lambda + \theta)]} \right.$$

$$\left. + \frac{\lambda_1(\lambda + \theta)}{[(\lambda + \theta)(\mu - \lambda_1) - \lambda\mu]} \right), \tag{28}$$

$$W_2 = \frac{1}{\beta} + \frac{N - 1}{\lambda_2} + \frac{\lambda + \theta + \mu}{\mu \theta} + \left( \frac{\lambda + \theta + \mu}{\mu \theta} - \frac{1}{\lambda_2} \right) \frac{N(N-1)\beta + 2N\lambda_2}{2N\beta + 2\lambda_2}. \tag{29}$$

*Proof.* Let $P(k|s)$ be the probability that the number of customers in the orbit is $k$ conditional on observing the server's status $s$, where $s = 1, 2$. Note that, we define $W_s(k)$ be the expected waiting time if the number of customers in the orbit is $k$ and the server's status is $s$ when a customer arrives.

If the arriving customer observes a busy server, the conditional probability is $P(k|1) = \frac{\pi(1, k)}{\Pi_1(1)}$, and according to (22), we have $W_1(k) = T(1, k + 1) = \frac{\lambda + \theta + \mu}{\mu \theta}(k + 1) + \frac{1}{\mu}$. Thus, the expected waiting time upon seeing a busy server is

$$W_1 = \sum_{k=0}^{\infty} W_1(k) P(k|1) = \frac{\sum_{k=0}^{\infty} \pi(1, k) T(1, k + 1)}{\Pi_1(1)} = \frac{\sum_{k=0}^{\infty} \pi(1, k)[\frac{\lambda + \theta + \mu}{\mu \theta}(k + 1) + \frac{1}{\mu}]}{\Pi_1(1)}$$

$$= \frac{\lambda + \theta + \mu}{\mu \theta} + \frac{1}{\mu} + \frac{\lambda + \theta + \mu}{\mu \theta} \frac{\Pi_1'(z)|_{z=1}}{\Pi_1(1)}.$$

From Lemma 1, we obtain (28).

Similarly, if the arriving customer observes the server in sleep mode, the corresponding conditional probability is $P(k|2) = \frac{\pi(2,k)}{\Pi_2(1)}, 0 \le k \le N - 1$ and $W_2(k) = T(2, k + 1) = \frac{N-k-1}{\lambda_2} + \frac{1}{\beta} + \frac{\lambda+\theta+\mu}{\mu\theta}(k + 1)$, are derived by (23). Thus, the expected waiting time upon seeing a server in sleep mode is

$$
\begin{aligned}
W_2 &= \sum_{k=0}^{N-1} W_2(k)P(k|2) = \frac{\sum_{k=0}^{N-1} \pi(2,k)T(2,k+1)}{\Pi_2(1)} \\
&= \frac{\sum_{k=0}^{N-1} \pi(2,k)[\frac{N-k-1}{\lambda_2} + \frac{1}{\beta} + \frac{\lambda+\theta+\mu}{\mu\theta}(k+1)]}{\Pi_2(1)} \\
&= \frac{1}{\beta} + \frac{N-1}{\lambda_2} + \frac{\lambda+\theta+\mu}{\mu\theta} + (\frac{\lambda+\theta+\mu}{\mu\theta} - \frac{1}{\lambda_2})\frac{\Pi_2'(z)|_{z=1}}{\Pi_2(1)}.
\end{aligned}
$$

From Lemma 1, we obtain (29). This completes the proof. $\qquad\square$

From (28) and (29), we can see that the expected waiting time $W_1(W_2)$ is independent of the arrive rate $\lambda_2(\lambda_1)$. Thus, we can determine the equilibrium arrival rates separately. So in the remainder of this work, we can regard $W_1$ (or $W_2$) as $W_1(\lambda_1)$ (or $W_2(\lambda_2)$).

**Theorem 2.** *For the M/M/1 constant retrial queueing model with reserved idle time under N-policy, the equilibrium arrival rate when the server is busy is given by*

$$
\lambda_1^e = \begin{cases}
0, & \text{if } R - P \le C\left\{ \frac{\lambda+\mu+\theta}{\mu\theta} + \frac{1}{\mu} + \frac{\lambda+\mu+\theta}{\mu\theta} \frac{2N\alpha\lambda+N(N-1)\alpha(\lambda+\theta)}{2[\lambda\theta+N\alpha(\lambda+\theta)]} \right\}; \\[2ex]
\frac{A\mu\theta}{(1+A)(\lambda+\theta)}, & \text{if } C\left\{ \frac{\lambda+\mu+\theta}{\mu\theta} + \frac{1}{\mu} + \frac{\lambda+\mu+\theta}{\mu\theta} \frac{2N\alpha\lambda+N(N-1)\alpha(\lambda+\theta)}{2[\lambda\theta+N\alpha(\lambda+\theta)]} \right\} < R - P \\[2ex]
& \quad < C\left\{ \frac{\lambda+\theta+\mu}{\mu\theta} + \frac{1}{\mu} + \frac{\lambda+\theta+\mu}{\mu\theta} \right. \\[2ex]
& \quad \left. \times (\frac{2N\alpha\lambda+N(N-1)\alpha(\lambda+\theta)}{2[\lambda\theta+N\alpha(\lambda+\theta)]} + \frac{\lambda(\lambda+\theta)}{[(\lambda+\theta)(\mu-\lambda)-\lambda\mu]}) \right\}; \\[2ex]
\lambda, & \text{if } R - P \ge C\left\{ \frac{\lambda+\theta+\mu}{\mu\theta} + \frac{1}{\mu} \right. \\[2ex]
& \quad \left. + \frac{\lambda+\theta+\mu}{\mu\theta} (\frac{2N\alpha\lambda+N(N-1)\alpha(\lambda+\theta)}{2[\lambda\theta+N\alpha(\lambda+\theta)]} + \frac{\lambda(\lambda+\theta)}{[(\lambda+\theta)(\mu-\lambda)-\lambda\mu]}) \right\}.
\end{cases} \tag{30}
$$

*where*

$$
A = \frac{(R-P)\mu\theta}{C(\lambda+\theta+\mu)} - 1 - \frac{\theta}{\lambda+\theta+\mu} - \frac{2N\alpha\lambda+N(N-1)\alpha(\lambda+\theta)}{2[\lambda\theta+N\alpha(\lambda+\theta)]}.
$$

*And the equilibrium arrival rate when the server is in sleep mode is given as follows: If $\lambda_{22} \ge \lambda$,*

$$\lambda_2^e = \begin{cases} 0, & \text{if } R - P < C\{\frac{1}{\beta} + \frac{N-1}{\lambda} + \frac{\lambda+\theta+\mu}{\mu\theta} + (\frac{\lambda+\theta+\mu}{\mu\theta} - \frac{1}{\lambda})\frac{N(N-1)\beta+2N\lambda}{2N\beta+2\lambda}\}; \\ \{0, \lambda\}, & \text{if } R - P = C\{\frac{1}{\beta} + \frac{N-1}{\lambda} + \frac{\lambda+\theta+\mu}{\mu\theta} + (\frac{\lambda+\theta+\mu}{\mu\theta} - \frac{1}{\lambda})\frac{N(N-1)\beta+2N\lambda}{2N\beta+2\lambda}\}; \\ \{0, \lambda_2', \lambda\}, & \text{if } R - P > C\{\frac{1}{\beta} + \frac{N-1}{\lambda} + \frac{\lambda+\theta+\mu}{\mu\theta} + (\frac{\lambda+\theta+\mu}{\mu\theta} - \frac{1}{\lambda})\frac{N(N-1)\beta+2N\lambda}{2N\beta+2\lambda}\}. \end{cases}$$

(31)

where $\lambda_{22}$ is the positive root of $B = 0$, $B = [2\mu\theta + N(N+1)\beta(\lambda+\theta+\mu)]\lambda_2^2 + 2\mu\theta\beta N(1-N)\lambda_2 - (N-1)\mu\theta\beta^2 N^2$ and $\lambda_2'$ is the solution of equation $R - P - CW_2 = 0$.
If $\lambda_{22} < \lambda$,

$$\lambda_2^e = \begin{cases} 0, & \text{if } R - P < CW_2(\lambda_{22}); \\ \{0, \lambda_{21}', \lambda_{22}'\}, & \text{if } CW_2(\lambda_{22}) < R - P < CW_2(\lambda); \\ \{0, \lambda_{23}', \lambda\}, & \text{if } R - P > CW_2(\lambda). \end{cases}$$

(32)

where $\lambda_{21}'$ and $\lambda_{22}'$ satisfies the equation $R - P - CW_2 = 0$, and $\lambda_{23}'$ is the smaller root of the equation $R - P - CW_2 = 0$.

*Proof.* First, we consider the equilibrium arrival rate upon seeing the server in normal working state, denoted by $\lambda_1^e$. Taking the first derivative about $\lambda_1$ of the equation (28), we have

$$W_1'(\lambda_1) = \frac{(\lambda+\theta+\mu)(\lambda+\theta)}{[(\lambda+\theta)(\mu-\lambda_1) - \lambda\mu]^2} > 0.$$

So the mean waiting time $W_1$ increases with $\lambda_1$. This demonstrates "avoid-the-crowd" (ATC) customer behavior, and consequently, at most one equilibrium arrival rate exists. According to the reward-cost structure, the net payoff of the customer joining the system equals the difference between his reward $R$ and his waiting cost. So, from Theorem 1, his expected net payoff is

$$S_1(\lambda_1) = R - P - CW_1(\lambda_1)$$
$$= R - P - C\Big[\frac{\lambda+\theta+\mu}{\mu\theta} + \frac{1}{\mu} + \frac{\lambda+\theta+\mu}{\mu\theta}\Big(\frac{2N\alpha\lambda + N(N-1)\alpha(\lambda+\theta)}{2[\lambda\theta + N\alpha(\lambda+\theta)]}\Big)$$
$$+ \frac{\lambda_1(\lambda+\theta)}{[(\lambda+\theta)(\mu-\lambda_1) - \lambda\mu]}\Big)\Big].$$

(33)

We observe that $S_1(\lambda_1)$ is strictly decreasing for $\lambda_1 \in [0, \lambda]$ and thus have the following results:

1. If $R - P \leq C\{\frac{\lambda+\mu+\theta}{\mu\theta} + \frac{1}{\mu} + \frac{\lambda+\mu+\theta}{\mu\theta}\frac{2N\alpha\lambda+N(N-1)\alpha(\lambda+\theta)}{2[\lambda\theta+N\alpha(\lambda+\theta)]}\}$, i.e., $S_1(0) \leq 0$, then $S_1(\lambda_1)$ in (33) is non-positive for every $\lambda_1$, so the best response is balking and the unique equilibrium point is $\lambda_1^e = 0$, which gives the first part of (30);

2. If $C\{\frac{\lambda+\mu+\theta}{\mu\theta} + \frac{1}{\mu} + \frac{\lambda+\mu+\theta}{\mu\theta}\frac{2N\alpha\lambda+N(N-1)\alpha(\lambda+\theta)}{2[\lambda\theta+N\alpha(\lambda+\theta)]}\} < R - P < C\{\frac{\lambda+\theta+\mu}{\mu\theta} + \frac{1}{\mu} + \frac{\lambda+\theta+\mu}{\mu\theta}(\frac{2N\alpha\lambda+N(N-1)\alpha(\lambda+\theta)}{2[\lambda\theta+N\alpha(\lambda+\theta)]} + \frac{\lambda(\lambda+\theta)}{[(\lambda+\theta)(\mu-\lambda)-\lambda\mu]})\}$, there exists an arrival rate that satisfies $R - P = CW_1(\lambda_1)$, which derives the unique equilibrium rate $\lambda_1 = \frac{A\mu\theta}{(1+A)(\lambda+\theta)}$, which gives the second part of (30);

3. If $R - P \geq C\{\frac{\lambda+\theta+\mu}{\mu\theta} + \frac{1}{\mu} + \frac{\lambda+\theta+\mu}{\mu\theta}(\frac{2N\alpha\lambda+N(N-1)\alpha(\lambda+\theta)}{2[\lambda\theta+N\alpha(\lambda+\theta)]} + \frac{\lambda(\lambda+\theta)}{[(\lambda+\theta)(\mu-\lambda)-\lambda\mu]})\}$,
i.e., $S_1(\lambda) \geq 0$, then $S_1(\lambda_1)$ in (33) is non-negative for every $\lambda_1$, so the unique
equilibrium point is $\lambda_1^e = \lambda$. Thus, we have the third part of (30).

Similarly, we consider the equilibrium arrival rate upon seeing the server is
in sleep mode state, denoted by $\lambda_2^e$. If all other customers choose to balk when
seeing the server in sleep mode, the system will never be active and the best
response for the tagged customer is also to balk. Obviously, all balking is always
an equilibrium strategy. So we now only consider a positive equilibrium arrival
rate upon seeing a sleep mode server. Similarly, we take the first derivative about
$\lambda_2$ of the equation (29), we have

$$W_2'(\lambda_2) = \frac{B}{2\lambda_2^2\mu\theta(N\beta + \lambda_2)^2},$$

where $B = [2\mu\theta + N(N+1)\beta(\lambda+\theta+\mu)]\lambda_2^2 + 2\mu\theta\beta N(1-N)\lambda_2 - (N-1)\mu\theta\beta^2 N^2$.
Supposing that the solutions of $B = 0$ are $\lambda_{21}$ and $\lambda_{22}$, by calculation we can
see that $\lambda_{21} < 0$ and $\lambda_{22} > 0$. Thus, we have
**Case 1:** if $\lambda_{22} \geq \lambda$, $W_2$ decreases with $\lambda_2$ in $[0, \lambda]$. This demonstrates "follow-
the-crowd" (FTC) customer behavior, and therefore, multiple equilibrium arrival
rates could exist.
**Case 2:** if $\lambda_{22} < \lambda$, $W_2$ decreases with $\lambda_2$ in $[0, \lambda_{22}]$, while increases with $\lambda_2$
in $[\lambda_{22}, \lambda]$.
So, from Theorem 1, if an arriving customer finds the server in sleep mode
and decides to enter, his expected net benefit is

$$S_2(\lambda_2) = R - P - CW_2(\lambda_2)$$
$$= R - P - C\left[\frac{1}{\beta} + \frac{N-1}{\lambda_2} + \frac{\lambda+\theta+\mu}{\mu\theta}\right.$$
$$\left. + (\frac{\lambda+\theta+\mu}{\mu\theta} - \frac{1}{\lambda_2})\frac{N(N-1)\beta + 2N\lambda_2}{2N\beta + 2\lambda_2}\right]. \tag{34}$$

In **Case 1**, we observe that $S_2(\lambda_2)$ is strictly increasing for $\lambda_2 \in [0, \lambda]$ and
thus have the following results:

1. if $R - P < C\{\frac{1}{\beta} + \frac{N-1}{\lambda} + \frac{\lambda+\theta+\mu}{\mu\theta} + (\frac{\lambda+\theta+\mu}{\mu\theta} - \frac{1}{\lambda})\frac{N(N-1)\beta+2N\lambda}{2N\beta+2\lambda}\}$, i.e., $S_2(\lambda) < 0$,
   then $S_2(\lambda_2)$ in (34) is negative for every $\lambda_2$, so the best response is balking
   and the unique equilibrium point is $\lambda_2^e = 0$, which gives the first branch of
   (31);
2. if $R - P = C\{\frac{1}{\beta} + \frac{N-1}{\lambda} + \frac{\lambda+\theta+\mu}{\mu\theta} + (\frac{\lambda+\theta+\mu}{\mu\theta} - \frac{1}{\lambda})\frac{N(N-1)\beta+2N\lambda}{2N\beta+2\lambda}\}$, i.e., $S_2(\lambda) = 0$,
   then $S_2(\lambda_2)$ in (34) is negative for every $\lambda_2 < \lambda$, so all the customers decide
   to join the system and the unique equilibrium point is $\lambda_2^e = \lambda$;
3. if $R - P > C\{\frac{1}{\beta} + \frac{N-1}{\lambda} + \frac{\lambda+\theta+\mu}{\mu\theta} + (\frac{\lambda+\theta+\mu}{\mu\theta} - \frac{1}{\lambda})\frac{N(N-1)\beta+2N\lambda}{2N\beta+2\lambda}\}$, i.e., $S_2(\lambda) > 0$,
   then there exist two equilibrium arrival rates $\lambda_2^e = \lambda_2' \in [0, \lambda]$ and $\lambda_2^e = \lambda$,
   where $\lambda_2'$ is the unique root of $S_2(\lambda_2) = 0$. Thus we have the last part of (31).

In **Case 2**, $S_2(\lambda_2)$ achieves the maximum in $\lambda_{22}$. Thus, we have the following results:

1. if $R - P < CW_2(\lambda_{22})$, i.e., $S_2(\lambda_{22}) < 0$, then $S_2(\lambda_2)$ in (34) is negative for every $\lambda_2 \in [0, \lambda]$, so the best response is balking and the unique equilibrium point is $\lambda_2^e = 0$, which gives the first branch of (32);
2. if $CW_2(\lambda_{22}) \leq R - P < CW_2(\lambda)$, then there exist two equilibrium arrival rates $\lambda_2^e = \lambda_{21}'$ and $\lambda_2^e = \lambda_{22}'$, where $\lambda_{21}'$ and $\lambda_{22}'$ satisfy the equation $R - P - CW_2 = 0$;
3. if $R - P \geq CW_2(\lambda)$, then there exist two equilibrium arrival rates $\lambda_2^e = \lambda_{23}'$ and $\lambda_2^e = \lambda$, where $\lambda_{23}'$ is the smaller root of the equation $R - P - CW_2 = 0$. Thus we have the last part of (32). This completes the proof. $\qquad\square$

*Remark 1.* From Eqs. (28) and (29), it is easily seen that upon finding the server is busy, the expected waiting time is increasing on $\lambda_1$. So we have an ATC situation, and the individual equilibrium arrival rate $\lambda_1^e$ is given in (30). Upon finding the server is in sleep mode, in Case 1, the expected waiting time decreases in $\lambda_2$. So we have an FTC situation, there are multiple equilibrium arrival rates $\lambda_2^e$ given in (31). However, the value $\lambda_2'$ is unstable. That is because, with any small increase of the arrival rate, the expected waiting time decreases and more customers will enter the system. This will further increase the arrival rate.

## 5    The Optimal Reserved Idle Time

Given the customers' equilibrium behavior, in this section, we study the optimal reserved idle time to maximize the server's profit. Considering the server incurs some costs when it is on, assume the consumption of the server per time unit when it is in setup, idle and busy states is $c_s, c_i, c_b$ respectively, and there is no consumption when the server is off. Moreover, the consumption of the space per customer per time unit is $c_h$.

According to Lemma 1, we obtain the average number of customers in the system is $LQ = \Pi_0'(z)|_{z=1} + \Pi_1'(z)|_{z=1} + \Pi_2'(z)|_{z=1} + \Pi_1(1)$. Accordingly, the cost of the service provider per unit of time is

$$S_{cost} = c_h LQ + c_s \pi_{2,N} + c_i \Pi_0(1) + c_b \Pi_1(1). \tag{35}$$

On the other hand, the effective arrival rate of the system in equilibrium is

$$\lambda_{eff} = \lambda_2^e(\Pi_2(1) - \pi_{2,N}) + \lambda_1^e \Pi_1(1) + \lambda \Pi_0(1), \tag{36}$$

where $\Pi_2(1), \Pi_1(1), \Pi_0(1)$ are given in Lemma 1 and $\lambda_2^e, \lambda_1^e$ are illustrated in Theorem 2. Thus, the net profit of the service provider per unit of time can be expressed as

$$S_{pro} = \lambda_{eff} P - S_{cost}. \tag{37}$$

Next, we try to analyze the influences of the parameter $\alpha$ on the net profit of the service provider per unit of time, i.e., $S_{pro}$. Considering the complexity of the monotonic proof of our objective function, we tend to adopt numerical experiments to investigate the optimal reserved idle time and verify the necessity of $\alpha's$ existence.

From Figs. 2, 3, 4 and 5, we observed that the net profit is decreasing with $\alpha$, because the server shuts down faster when no one is in the system as $\alpha$ gets larger and reduces the customers' willingness to enter when the system is idle, which leads to the smaller effective arrival rate and the net profit of the service provider per unit of time. Thus, the smaller $\alpha$, the larger net profit of the service provider is. In other words, from the perspective of the server, the setting of reserved idle time can bring more profit.

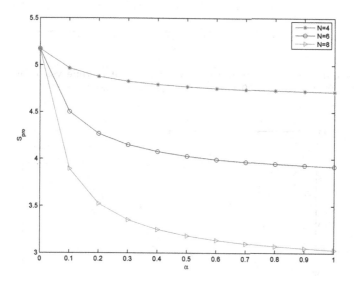

**Fig. 2.** The profit $S_{pro}$ of the service provider per unit of time vs. $\alpha$ for $\lambda = 1, \mu = 5, \theta = 2, \beta = 2, R = 20, P = 10, C = 1, ch = 1, c_s = 1, c_i = 4, c_b = 6$.

*Remark 2. Note that as $\alpha \to \infty$ and $\beta \to \infty$, i.e., the reserved idle time and the setup time tend to zero. That is, the system degenerates into the model of Wang et al. [14].*

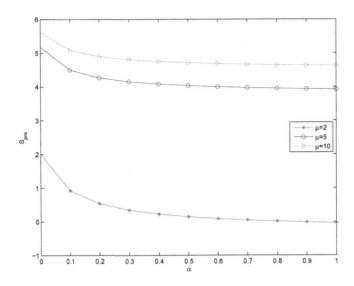

**Fig. 3.** The profit $S_{pro}$ of the service provider per unit of time vs. $\alpha$ for $\lambda = 1, \theta = 2, N = 6, \beta = 2, R = 20, P = 10, C = 1, ch = 1, c_s = 1, c_i = 4, c_b = 6$.

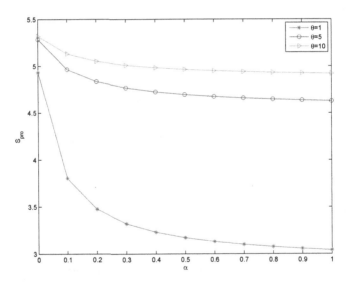

**Fig. 4.** The profit $S_{pro}$ of the service provider per unit of time vs. $\alpha$ for $\lambda = 1, \mu = 5, N = 6, \beta = 2, R = 20, P = 10, C = 1, ch = 1, c_s = 1, c_i = 4, c_b = 6$.

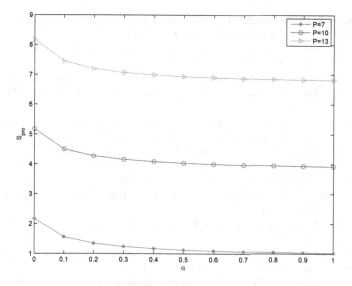

**Fig. 5.** The profit $S_{pro}$ of the service provider per unit of time vs. $\alpha$ for $\lambda = 1, \mu = 5, \theta = 2, N = 6, \beta = 2, R = 20, C = 1, ch = 1, c_s = 1, c_i = 4, c_b = 6$.

## 6   Conclusions

In this paper we studied an M/M/1 constant retrial queueing model with reserved idle time under $N$-policy. The model was explicitly analyzed in terms of generating function and the analysis yielded an exact solution for the probabilities of the server in different states. Furthermore, we studied the equilibrium arrival rates upon finding the server is busy and in sleep mode, which is conditioned on the state of the server is observable to the potential customers. Furthermore, we studied the net profit of the service provider with respect to $\alpha$ numerically and verified the necessity of $\alpha's$ existence. In addition to this, there is still some work to do in the future. For example, the retrial rate is not constant but linear in many practical cases, which means the retrial rate is proportionable to the number of customers in the orbit.

## References

1. Burnetas, A., Economou, A.: Equilibrium customer strategies in a single server Markovian queue with setup times. Queueing Syst. **56**(3), 213–228 (2007). https://doi.org/10.1007/s11134-007-9036-7
2. Doshi, B.T.: Queueing systems with vacations—a survey. Queueing Syst. **1**(1), 29–66 (1986). https://doi.org/10.1007/BF01149327
3. Economou, A., Kanta, S.: Equilibrium balking strategies in the observable single-server queue with breakdowns and repairs. Oper. Res. Lett. **36**(6), 696–699 (2008)
4. Economou, A., Kanta, S.: Equilibrium customer strategies and social-profit maximization in the single-server constant retrial queue. Naval Res. Logistics **58**, 107–122 (2011)

5. Edelson, N.M., Hilderbrand, D.K.: Congestion tolls for Poisson queuing processes. Econometrica J. Econometric Soc. **43**, 81–92 (1975)
6. Guo, P., Hassin, R.: Strategic behavior and social optimization in Markovian vacation queues. Oper. Res. **59**(4), 986–997 (2011)
7. Guo, P., Hassin, R.: Strategic behavior and social optimization in Markovian vacation queues: the case of heterogeneous customers. Eur. J. Oper. Res. **222**(2), 278–286 (2012)
8. Guo, P., Li, Q.: Strategic behavior and social optimization in partially-observable Markovian vacation queues. Oper. Res. Lett. **41**(3), 277–284 (2013)
9. Hassin, R.: Rational Queueing. CRC Press, Boca Raton (2016)
10. Hassin, R., Haviv, M.: To Queue or Not to Queue: Equilibrium Behavior in Queueing Systems. Springer, Boston (2003). doi: 10.1007/978-1-4615-0359-0
11. Naor, P.: The regulation of queue size by levying tolls. Econometrica J. Econometric Soc. **37**, 15–24 (1969)
12. Takagi, H.: Queueing analysis: a foundation of performance analysis. Vacation Priority Syst. **1** (1991)
13. Tian, N., Zhang, Z.G.: Vacation Queueing Models: Theory and Applications. Springer, Boston (2006). doi: 10.1007/978-0-387-33723-4
14. Wang, J., Zhang, X., Huang, P.: Strategic behavior and social optimization in a constant retrial queue with the N-policy. Eur. J. Oper. Res. **256**(3), 841–849 (2017)

# System Safety, Manufacturing Systems and Others

# Sequential Common Change Detection and Isolation of Changed Panels in Panel Data

Yanhong Wu$^{(\boxtimes)}$

California State University Stanislaus, Turlock, CA 95382, USA
ywu1@csustan.edu

**Abstract.** Quick detection of common changes is critical in sequential monitoring of multi-stream data where a common change is referred as a change that only occurs in a portion of panels. After briefly reviewing the CUSUM and Shiryayev-Roberts (SR) procedures for a single sequence under an exponential family model, we propose a combined CUSUM-SR procedure that is locally optimal in terms of the delay detection time. The design based on the Average In-control Run Length and comparisons with other procedures are discussed. After a common change is detected, a classifier formed by the post-change parameter estimations is used to isolate the possible candidates for the changed panels, that are then used to estimate the common change-point. To reduce the false discovery rate (FDR), supplementary runs are proposed. Dow Jones 30 Industrial Stock Prices are used for demonstration.

**Keywords:** Common change · Sequential detection · CUSUM and SR procedure · False Discovery Rate

## 1 Introduction

The traditional change-point detection is focused on quick detection of a change in a single sequence (panel) where the change is typically caused from internal sources. Two most well-known procedures are the CUSUM and Shiryayev-Roberts procedures that are briefly introduced in Sect. 2. Recent research on sequential change-point detection problem has focused on detecting common changes in multi-stream data or panel data. Here, a common change is referred as a change that occurs only in a portion of the N panels; usually caused by external sources. Several typical detection procedures have been discussed and extended; see Xie and Siegmund [23], Mei [5], and Tartakovsky and Veeravalli [12]. Chan [3] discussed the optimality of detection procedures.

The estimation of common change-point and isolation of changed panels after sequential detection is also important. For example, when a common change is detected in 30 Dow Jones Industrial Indexed Stocks, we shall be interested in finding out in which sectors of the 30 stocks the change occurred. In multi-sensor detection, the estimation of the location of the signal is also important after a

© Springer Nature Singapore Pte Ltd. 2019
Q.-L. Li et al. (Eds.): Cao Festschrift 2019, CCIS 1102, pp. 391–409, 2019.
https://doi.org/10.1007/978-981-15-0864-6_20

signal is detected. Wu [21] proposed a combined SR-CUSUM procedure that uses the sum of N Shiryayev-Roberts processes to detect the common change and N individual CUSUM processes to isolate the changed panels and estimate the change point. The alarming limit is chosen such that the average in-control run length is equal to a designated value. The procedure is introduced in Sect. 3 along with asymptotic results for the average run lengths. Those panels with large scores formed by the post-mean estimates along with the delay detection time will be considered as the candidates of changed panels. As the simulation results in Sect. 4 show, when the signal strength is weak, i.e. the post-change parameter is close to the reference value, it is difficult to discriminate the true alarms from those false alarms. To reduce the false discovery rate (FDR), supplementary runs are necessary for those candidates of true changed panels. In Sect. 5, we propose to run supplementary Sequential Probability Ratio Tests (SPRTs) with two sided boundary $[0, d)$ starting at initial value $c$ $(0 < c < d)$ on each of the candidates for changed panels. Those panels with runs that down-cross the boundary 0 will be classified as false alarms; while those with runs that up-cross the boundary $d$ are classified as true alarms. The value of $d$ and $c$ will be chosen depending on the estimated false alarm rate or the estimated proportion of false alarms. The simulation results show that the FDR is reduced significantly with the supplementary runs. Theoretical results for the distribution of estimated delay detection time and normalized post-change parameter estimation are presented in Sect. 6 for both unchanged and changed panels. Dow Jones 30 stock daily closing prices are used for illustration.

## 2    Sequential Change Detection in a Single Sequence

### 2.1    The CUSUM Procedure

Consider a single sequence of independent observations $\{X_k\}$ with change point $\nu$ which follow a standard exponential family distribution with the baseline density $f_0(x)$ for $k < \nu$ and

$$f_\theta(x) = exp(\theta x - c(\theta))f_0(x)$$

for $k \geq \nu$ and $\theta > 0$, where $c(0) = c'(0) = 0$ and $c''(0) = 1$.

For testing $H_0 : 0 \leq \nu < \infty$ vs $H_1 : \nu = \infty$, the sequential CUSUM procedure based on likelihood ratio process makes an alarm at

$$\tau_{CUSUM} = \inf\{n > 0 : T_n > A\},$$

where $T_n = max(0, T_{n-1} + \delta X_n - c(\delta))$ for a predesigned alarming boundary $A$ and a reference value $\delta > 0$ for $\theta$. When an alarm is made, the change-point $\nu$ is estimated as the last zero epoch of the CUSUM process $T_n$:

$$\hat{\nu} = \max\{n < \tau_{CUSUM} : T_n = 0\},$$

and the post-change parameter $\theta$ is estimated by the mean equation:

$$c'(\hat{\theta}) = \frac{T_{\tau_{CUSUM}}}{\tau_{CUSUM} - \hat{\nu}} + \frac{c(\delta)}{\delta}.$$

The conditional bias and absolute bias of $\hat{\nu}$ and $\hat{\theta}$ given $\tau_{CUSUM} > \nu$ are studied in Wu [15] along with related inference problems. To design the CUSUM procedure, the following accurate approximation (Siegmund [10]) for Average In-control Run Length ($ARL_0$) can be used to calculate $A$ in the normal case:

$$ARL_0 = E_\infty[\tau_{CUSUM}] \approx \frac{e^{A+\delta\rho} - 1 - (A + \delta\rho)}{\delta^2/2},$$

where $E_\infty[.]$ denote the expectation when there is no change and $\rho \approx 0.5826$.

## 2.2   The Shiryayev-Roberts Procedure

Alternatively, we can use the Bayesian approach by first assuming the change point $\nu$ follows a geometric prior distribution

$$P[\nu = k] = (1 - \lambda)^{k-1}\lambda,$$

for $k = 1, 2, \dots$. For observations up to time $t$, the posterior distribution of $\nu$ given $X_1, \dots, X_t$ equals to

$$\pi_{k,t} = P[\nu = k | x_1, \dots, x_t]$$
$$= \frac{e^{\sum_k^t (\delta x_j - c(\delta))} \Pi_1^t f_0(x_j) p (1 - p)^{k-1}}{\sum_{k=1}^t e^{\sum_k^t (\delta x_j - c(\delta))} \Pi_1^t f_0(x_j) \lambda (1 - \lambda)^{k-1} + (1 - \lambda)^t \Pi_1^t f_0(x_j)},$$

for the reference value $\delta$. Define

$$\pi_t = \sum_{k=1}^t \pi_{k,t} = P[\nu \le t | x_1, .., x_t]$$

as the posterior probability that the change occurred before time $t$ and the conditional posterior mean as

$$E[\nu | \nu \le t; x_1, \dots, x_t] = \frac{\sum_{k=1}^t k \pi_{k,t}}{\pi_t}.$$

An alarm should be raised when the posterior odd process $\pi_t/(1 - \pi_t)$ crosses a threshold.

As $\lambda \to 0$ (change occurs far away from beginning), the standardized posterior odd process approaches the Shiryayev-Roberts (SR) process (Pollak and Siegmund [7]).

$$R_t = R_t(\delta) = \lim_{\lambda \to 0} \frac{\pi_t}{\lambda(1 - \pi_t)}$$
$$= \sum_{k=1}^t e^{\sum_k^t (\delta x_j - c(\delta))}$$
$$= (1 + R_{t-1}(\delta)) e^{\delta x_t - c(\delta)}.$$

where $R_0(\delta) = 0$ and the conditional mean approaches

$$E[\nu|\nu \leq t; x_1, ..., x_t] \to \frac{Q_t}{R_t},$$

where

$$Q_t = \sum_{k=1}^{t} k e^{\sum_{k}^{t}(\delta x_j - c(\delta))} = (t + Q_{t-1})e^{\delta x_t - c(\delta)},$$

for $Q_0 = 0$. An alarm is raised at

$$\tau_{SR} = \inf\{t > 0 : R_t(\delta) > B\}.$$

After an alarm, the conditional posterior mean given $\nu \leq \tau_{SR}$ is

$$\tilde{\nu} = \lfloor \frac{Q_{\tau_{SR}}}{R_{\tau_{SR}}} \rfloor,$$

since the change-point is assumed to take the integer values only. The true post-change parameter $\theta$ is estimated by the equation.

$$c'(\tilde{\theta}) = \frac{\sum_{[\tilde{\nu}]+1}^{T_0} x_i}{T_0 - [\tilde{\nu}]}.$$

In the normal case, the following accurate approximation for $ARL_0 = E_\infty[\tau_0]$ (Pollak [6]) can be used to design $B$:

$$ARL_0 \approx Be^{\delta\rho},$$

By using the recursive form of $R_t$, the following lemma gives the monotone property of the change-point estimation and its variance:

**Lemma 1.** (i) The recursive change-point estimation $\{\nu_n\} = \frac{Q_n}{R_n}$ is monotone non-decreasing.
(ii) Under $P_\theta(.)$,

$$Var(R_n) = e^{c(2\theta)-2c(\theta)}Var(R_{n-1}) + n^2(e^{c(2\theta)-2c(\theta)} - 1).$$

**Remark 1.** When the post-change parameter $\theta = \delta$, the CUSUM procedure minimizes the maximum conditional delay detection time; while the Shiryayev-Roberts procedure minimizes the stationary conditional delay detection time.

**Remark 2.** A numerical comparison in terms of the conditional bias (absolute bias) of $\tilde{\nu}$ and $\hat{\nu}$ by using the two procedures is studied in Tables 2.1 and 10.1 of Wu [15] for $\delta = 0.5, 1.0$ in the normal case. It showed that there are no significant differences between the two procedures. The CUSUM procedure performs better for $\delta = 0.5$ and the Shiryayev-Roberts procedure performs slightly better for $\delta = 1.0$. However, when $\theta$ is unknown, a smaller reference value $\delta$, say, 0.5, is chosen as the minimum change magnitude subject to detect and in addition, it is inconvenient to calculate the post-change parameter estimation by using the Shiryayev-Roberts procedure. Thus, the CUSUM procedure is preferred in terms of change-point and post-change parameter estimation.

# 3   A Combined SR-CUSUM Procedure

## 3.1   Definition

Assume there are $N$ independent panels and the observations $\{X_j^{(i)}\}$ in each panel $i$ follow the same change point model defined in Sect. 2. Also, a change may only occur in a $p = \frac{K}{N}$ proportion of the $N$ panels, called common change. The $N$ panels can be assumed following a mixture model with probability $p$ of change in each panel. For convenience of discussion, we select the same reference parameter $\delta$ for all panels.

For observations $\{X_j^{(i)}\}$ for $i = 1, ..., N$ and $j = 1, ..., t$ with possible change points $0 \le \nu = k \le t$, the log-likelihood ratio for testing $H_0 : \nu = \infty$ vs $H_1 : 0 \le \nu < \infty$ can be written as

$$l_t(k, p) = \sum_{i=1}^{N} \ln(1 + p(e^{\sum_{k+1}^{t}(\delta X_j^{(i)} - c(\delta))} - 1)).$$

By taking the local approach as discussed in Wu [13], we let $p \to 0$ and the local score is obtained as

$$l_t(k) = \frac{\partial l_t(k, p)}{\partial p}\Big|_{p=0} = \sum_{i=1}^{N}(e^{\sum_{k+1}^{t}(\delta X_j^{(i)} - c(\delta))} - 1).$$

By using the same Bayesian approach as in Sect. 2, the limiting local posterior-odd process will be equivalent to

$$R_t(\delta) - Nt = \sum_{k=1}^{t}\sum_{i=1}^{N}(e^{\sum_{k+1}^{t}(\delta X_j^{(i)} - c(\delta))} - 1)$$

$$= \sum_{i=1}^{N}\sum_{k=1}^{t}(e^{\sum_{k+1}^{t}(\delta X_j^{(i)} - c(\delta))} - 1)$$

$$= \sum_{i=1}^{N}(R_t(i) - t).$$

where $R_t = R_t(\delta) = \sum_{i=1}^{N} R_t(i)$ is the sum of the N Shiryayev-Roberts processes and

$$R_t(i) = \sum_{k=1}^{t} e^{\sum_{k+1}^{t}(\delta X_j^{(i)} - c(\delta))}$$

is the Shiryayev-Roberts process formed from the observations in panel $i$.

An alarm will be raised at the stopping time

$$\tau = \inf\{t > 0 : R_t > B\},$$

where $B$ is chosen such that the $ARL_0$ is equal to the designated value.

## 3.2    Average Run Lengths

The design for $B$ is based on the following approximation for the average in-control run length and its proof is given in Wu [21], which extends the results of Pollak [6].

**Theorem 1.** *As $B \to \infty$,*

$$ARL_0 \approx \frac{B}{N}\gamma(\delta),$$

*where*

$$1/\gamma(\delta) = \lim_{x \to \infty} E_\delta e^{-(S_{\tau_x} - x)}$$

*with*

$$\tau_x = \inf\{t > 0 : S_t^{(1)} = \sum_{i=1}^{t}(\delta X_i^{(1)} - c(\delta)) > x\}.$$

*In the normal case, when $\delta$ is small,*

$$ARL_0 = E_\infty \tau \approx \frac{B}{N}e^{\rho\delta},$$

*where $\rho \approx 0.583$.*

Further results on the exponential property of false alarm time can be seen in Pollak and Tartakovsky [8] and Polunchenko [9]. The following theorem also gives an upper bound for the Average Out-of-Control Run Length $ARL_1$ when the number of changed panels is $K \leq N$.

**Theorem 2.** *For $\theta$ satisfying $c'(\theta) > c(\delta)/\delta$,*

$$ARL_1 = E_0[\tau] \leq O\left(\frac{\ln(B/K)}{\delta c'(\theta) - c(\delta)}\left(1 - \frac{N-K}{B(\delta c'(\theta) - c(\delta))} + o\left(\frac{1}{B}\right)\right)\right).$$

*In the normal case for $\theta > \delta/2$,*

$$ARL_1 \leq O\left(\frac{\ln(B/K)}{\delta(\theta - \delta/2)}\left(1 - \frac{N-K}{B\delta(\theta - \delta/2)} + o\left(\frac{1}{B}\right)\right)\right).$$

In the normal case, Table 1 gives the simulated average conditional delay detection times (CDDT) $E[\tau - \nu|\tau > \nu]$ for $ARL_0 = 1000$ with number of panels $N = 10, 20, 60, 100$ and change portion $p = 0.05, 0.10, 0.20, 0.30, 0.50$ and $1.00$ in the case $\delta = 0.5$ with the change point $\nu = 100$. The value of $B$ is calculated from the approximation given in Theorem 1. Numbers in the brackets are the corresponding false alarm rates (FAR) with 5000 replications. Table 2 gives the corresponding results for $\delta = 1.0$. Comparison with other procedures as shown in Wu [21] demonstrated that the proposed procedure is very competitive when the proportion is small.

**Table 1.** Average delay detection time and false alarm rate with $\delta = 0.5$ and $ARL_0 = 1000$

| Panel number | | $N = 10$ | $N = 20$ | $N = 60$ | $N = 100$ |
|---|---|---|---|---|---|
| p | $\mu$ | $B = 7472.92$ | $B = 14945.83$ | $B = 44837.5$ | $B = 74729.15$ |
| 0.05 | 0.5 | | 49.38 (0.052) | 36.22 (0.047) | 32.85 (0.036) |
| 0.05 | 1.0 | | 18.55 (0.058) | 15.35 (0.059) | 14.57 (0.042) |
| 0.05 | 1.5 | | 11.47 (0.061) | 9.84 (0.036) | 9.56 (0.034) |
| 0.05 | 2.0 | | 8.42 (0.067) | 7.47 (0.044) | 7.12 (0.047) |
| 0.10 | 0.5 | 43.51 (0.054) | 34.70 (0.051) | 27.74 (0.049) | 26.19 (0.040) |
| 0.10 | 1.0 | 16.66 (0.048) | 14.61 (0.043) | 12.73 (0.049) | 12.27 (0.053) |
| 0.10 | 1.5 | 10.37 (0.056) | 9.33 (0.053) | 8.26 (0.047) | 8.15 (0.043) |
| 0.10 | 2.0 | 7.69 (0.053) | 6.99 (0.051) | 6.26 (0.047) | 6.07 (0.041) |
| 0.20 | 0.5 | 30.35 (0.049) | 26.50 (0.047) | 22.81 (0.039) | 21.18 (0.044) |
| 0.20 | 1.0 | 13.00 (0.054) | 11.68 (0.062) | 10.56 (0.045) | 10.06 (0.048) |
| 0.20 | 1.5 | 8.33 (0.076) | 7.72 (0.057) | 7.10 (0.045) | 6.78 (0.046) |
| 0.20 | 2.0 | 6.17 (0.054) | 5.87 (0.044) | 5.34 (0.043) | 5.14 (0.045) |
| 0.30 | 0.5 | 26.33 (0.069) | 22.94 (0.060) | 19.80 (0.053) | 18.50 (0.043) |
| 0.30 | 1.0 | 11.25 (0.072) | 10.33 (0.053) | 9.39 (0.055) | 8.98 (0.036) |
| 0.30 | 1.5 | 7.41 (0.063) | 6.89 (0.065) | 6.28 (0.046) | 6.14 (0.048) |
| 0.30 | 2.0 | 5.55 (0.062) | 5.18 (0.062) | 4.76 (0.051) | 4.65 (0.043) |
| 0.50 | 0.5 | 20.28 (0.063) | 18.41 (0.052) | 16.54 (0.049) | 15.41 (0.036) |
| 0.50 | 1.0 | 9.41 (0.056) | 8.75 (0.064) | 7.90 (0.041) | 7.68 (0.041) |
| 0.50 | 1.5 | 6.29 (0.066) | 5.92 (0.047) | 5.39 (0.033) | 5.23 (0.037) |
| 0.50 | 2.0 | 4.76 (0.055) | 4.47 (0.042) | 4.16 (0.041) | 4.06 (0.043) |
| 1.0 | 0.5 | 15.17 (0.061) | 14.25 (0.045) | 12.52 (0.053) | 12.05 (0.044) |
| 1.0 | 1.0 | 7.41 (0.065) | 6.94 (0.058) | 6.29 (0.042) | 6.22 (0.043) |
| 1.0 | 1.5 | 4.97 (0.065) | 4.64 (0.055) | 4.31 (0.035) | 4.26 (0.038) |
| 1.0 | 2.0 | 3.81 (0.058) | 3.62 (0.056) | 3.38 (0.048) | 3.28 (0.036) |

# 4    Isolation and Estimation After Detection

## 4.1    Isolation of Changed Panels

As mentioned in Wu [17], for the change-point estimation, we are mainly interested in the absolute bias as the change-point is a location parameter where for a single sequence, it was proved that the MLE from the CUSUM test has smaller absolute bias than the Bayesian estimation. The combined SR-CUSUM procedure calculates the $N$ CUSUM processes recursively as

$$T_t(i) = \max(0, T_{t-1}(i) + \delta X_t^{(i)} - c(\delta))$$

**Table 2.** Average delay detection time and false alarm rate with $\delta = 1.0$ and $ARL_0 = 1000$

| Panel number | | $N = 10$ | $N = 20$ | $N = 60$ | $N = 100$ |
|---|---|---|---|---|---|
| p | $\mu$ | $B = 5582.21$ | $B = 11164.42$ | $B = 33493.26$ | $B = 55822.1$ |
| 0.05 | 0.5 | | 72.08 (0.074) | 40.05 (0.075) | 34.58 (0.075) |
| 0.05 | 1.0 | | 15.34 (0.088) | 11.62 (0.076) | 10.50 (0.076) |
| 0.05 | 1.5 | | 8.46 (0.082) | 6.87 (0.073) | 6.36 (0.076) |
| 0.05 | 2.0 | | 5.86 (0.083) | 4.94 (0.079) | 4.70 (0.070) |
| 0.10 | 0.5 | 63.54 (0.085) | 43.36 (0.084) | 28.56 (0.074) | 24.30 (0.069) |
| 0.10 | 1.0 | 14.03 (0.078) | 11.35 (0.099) | 9.22 (0.077) | 8.65 (0.070) |
| 0.10 | 1.5 | 7.65 (0.096) | 6.73 (0.093) | 5.72 (0.084) | 5.43 (0.077) |
| 0.10 | 2.0 | 5.39 (0.083) | 4.73 (0.084) | 4.23 (0.069) | 4.05 (0.069) |
| 0.20 | 0.5 | 38.34 (0.079) | 28.50 (0.107) | 20.50 (0.092) | 18.27 (0.069) |
| 0.20 | 1.0 | 10.23 (0.091) | 8.85 (0.088) | 7.66 (0.092) | 7.20 (0.087) |
| 0.20 | 1.5 | 6.06 (0.077) | 5.43 (0.095) | 4.84 (0.072) | 4.67 (0.087) |
| 0.20 | 2.0 | 4.41 (0.082) | 3.98 (0.085) | 3.62 (0.088) | 3.53 (0.069) |
| 0.30 | 0.5 | 28.50 (0.085) | 22.18 (0.099) | 16.98 (0.095) | 15.25 (0.084) |
| 0.30 | 1.0 | 8.86 (0.091) | 7.72 (0.088) | 6.80 (0.083) | 6.46 (0.078) |
| 0.30 | 1.5 | 5.32 (0.088) | 4.86 (0.102) | 4.35 (0.082) | 4.19 (0.093) |
| 0.30 | 2.0 | 3.81 (0.095) | 3.60 (0.094) | 3.31 (0.086) | 3.18 (0.097) |
| 0.50 | 0.5 | 19.84 (0.087) | 16.65 (0.077) | 13.65 (0.104) | 12.56 (0.080) |
| 0.50 | 1.0 | 7.21 (0.077) | 6.62 (0.078) | 5.77 (0.085) | 5.65 (0.077) |
| 0.50 | 1.5 | 4.57 (0.099) | 4.25 (0.088) | 3.86 (0.082) | 3.74 (0.079) |
| 0.50 | 2.0 | 3.36 (0.084) | 3.16 (0.079) | 2.95 (0.072) | 2.86 (0.076) |
| 1.0 | 0.5 | 13.84 (0.082) | 12.24 (0.084) | 10.47 (0.077) | 9.96 (0.074) |
| 1.0 | 1.0 | 5.75 (0.068) | 5.17 (0.090) | 4.76 (0.106) | 4.58 (0.077) |
| 1.0 | 1.5 | 3.70 (0.096) | 3.47 (0.083) | 3.21 (0.079) | 3.07 (0.081) |
| 1.0 | 2.0 | 2.82 (0.082) | 2.68 (0.072) | 2.56 (0.087) | 2.50 (0.090) |

in each individual panel and at the alarm time $\tau$ the change-point for the $i^{th}$ panel is estimated as the last zero point of $T_t(i)$,

$$\hat{\nu}_i = \max\{t < \tau : T_t(i) = 0\},$$

for $i = 1, ..., N$, which is indeed the MLE when $\mu = \delta$. The post-change parameter estimate in panel $i$ is given by the moment estimation equation

$$c'(\hat{\theta}_i) = \frac{\sum_{j=\hat{\nu}_i+1}^{\tau} X_j^{(i)}}{\tau - \hat{\nu}_i} = \frac{1}{\delta}\left[\frac{T_\tau(i)}{\tau - \hat{\nu}_i} + c(\delta)\right].$$

After an alarm is raised for a common change-point, we want to isolate the changed panels and estimate the common change-point. A common criterion is to use the false discovery rate (FDR) defined as the ratio of falsely claimed "changed" panels among all claimed changed panels.

Apparently, to isolate the "true" changed panels, both the change-point estimation (or estimated delay detection time $\tau - \hat{\nu}_i$ after the change-point estimation) and estimated strength of signals (the post-change parameter estimations) provide related information.

First, we note that the induced distribution formed by all the change-point estimations $\hat{\nu}_i$ for $i = 1, 2, ..., N$ is a mixture distribution from those "true" changed panels and other panels. However, note that the change-point estimation for those "unchanged" panels is highly related to the alarm limit and the delay detection time $\tau - \hat{\nu}_i$ for $i = 1, 2, ..., N$ are relatively short. We further note that the change-point estimations for unchanged panels are highly dependent on the alarming limit that makes us difficult to use it as a classifier. Instead, the isolation of changed panels should rely mainly on the post-change parameter estimation by comparing with reference value.

Here we use a modified score defined as

$$Z_i = \frac{\sqrt{\tau - \hat{\nu}_i}(c'(\hat{\theta}_i))}{\sqrt{c''(\hat{\theta})}},$$

as the classifier, which is the multiplication of the square root of the delay detection time and the standardized post-mean estimation. In the normal case, this reduces to

$$Z_i = \sqrt{\tau - \hat{\nu}_i}\frac{T_\tau(i)/(\tau - \hat{\nu}_i) + \delta^2/2}{\delta}.$$

On the other hand, the classifier can also be thought as the sum of normalized CUSUM process value and the square root of the delay detection time.

A panel with score $Z_i > z^*$ will be considered as a changed pane where $z^*$ is selected as any common critical value. Second order corrections of this score in a large scale of sequential probability ratio tests and one-sided truncated sequential tests are discussed in Wu ([15,21]).

To show how the distribution of the change point estimation and z-score behaves for unchanged and changed panels after detection, we conduct a simulation study. For $ARL_0 = 1000$ and $\nu = 100$, we take $\delta = 0.5$, $\mu = 0.5, 1.0, 1.5, 2.0$, $N = 100$ and $K = 10$. $B$ is calculated from the approximation given in Theorem 1. The simulated distributions show that although the delay detection time distributions are distinguishable between the changed and unchanged panels when $\mu = \delta = 0.5$, the difference becomes less obvious as $\mu$ is larger than $\delta$. However, the difference between z-score distributions becomes obvious when $\mu$ is larger than $\delta$.

For $N = 100$ and $p = 0.05, 0.10, 0.20, 0.50$, Table 3 reported the median conditional delay detection time (CDDT), the median change-point estimation $\tilde{\nu}$ based on isolated changed panels, the false non-discovery rate(FNR)defined as the rate of non-discovered changed panels among all true changed panels, the false discovery rate (FDR) defined as the rate of non-changed panels among all isolated change panels. The simulation is replicated for 5000 times conditioning on those trials with $\tau > \nu$.

Note that the estimate that minimizes the absolute bias from the induced distribution formed from the change-point estimators is known as the median estimate as stated in the following standard lemma.

**Lemma 2.** *For any distribution function $F(x)$, the value $c$ which minimizes $\int |x - c| dF(x)$ is the median of $F(x)$ defined as any number $M$ such that $F(M) \geq 1/2$ and $1 - F(M-) \geq 1/2$.*

The simulation procedure is summarized into the following several steps:

*Step 1:* The sum of the N Shiryayev-Roberts processes is used to detect the common change and the 100 individual CUSUM processes for 100 panels are calculated in parallel, so are the corresponding change-point estimations and post-change mean estimations.

*Step 2:* When an alarm is made, conditioning on $\tau > \nu$, those panels with scores $Z_i > z^* = 1.645$ are isolated as the changed panels and used to estimate the common change point and changed panels. Meanwhile the simulated false alarm rate (FAR) ($P(\tau \leq \nu)$) is reported.

*Step 3:* The common change-point is calculated as the median $\tilde{\nu}$ of the corresponding estimations from the isolated changed panels. The median conditional delay detection time (CDDT) is also reported.

*Step 4:* FNR and FDR are reported.

The results show that the bias of the change-point estimation becomes more negative as $\mu$ gets larger. By treating $\delta$ as the smallest change-magnitude we want to detect, we found that $\delta = 0.5$ is preferred in terms of the FNR and FAR. However, the FDRs are large.

To show how the value of $z^*$ affects the error rates, we select several common values $z^* = 1.645, 1.96, 2.326$ and $2.576$. For $\delta = 0.5$, $ARL_0 = 5000$, $\nu = 1000$, $N = 100$, and $K = 10$ (B $= 373647.5$), with 5000 simulations, Table 4 reported the FAR (False Alarm Rate), median conditional delay detection time (CDDT), median change-point estimation $\tilde{\nu}$, FNR, and FDR. Although the FDR is smaller for larger values of $z^*$, it increases the FNR when $\mu = \delta$.

## 4.2    Supplementary Sequential Tests

Tables 3 and 4 show that no matter what the post-change mean is, the false discovery rate is very high as the false alarms are not ignorable for large number of panels. To overcome this difficulty, we propose to conduct supplementary SPRTs for those panels with large values of $z^*$ as discussed in Wu [18]. The procedure in the normal case is defined in the following several steps.

(i) *We select all the panels with $Z_i = \sqrt{\tau - \hat{\nu}_i} \frac{T_\tau(i)/(\tau - \hat{\nu}_i) + \delta^2/2}{\delta} > z^*$ as potential candidates for changed panels where $z^* = 1.645$ for supplementary sequential tests; (as we use $[0, \delta]$ as the indifference zone).*

(ii) *The supplementary sequential tests are run on these panels starting at $d/2$ with boundary $[0, d]$, where $d$ is selected such that $d = o(\ln(B/K))$ (without increasing average detection length theoretically as stated in Theorem 2). If the run ends crossing the lower boundary 0, then the panel is classified without change; otherwise it is classified as a true changed panel.*

**Table 3.** Estimation of common change point and isolation error rates with $ARL_0 = 1000$ and $\delta = 0.5$.

| p | $\mu$ | FAR | CDDT | $\tilde{\nu}$ | FNR | FDR |
|---|---|---|---|---|---|---|
| $\delta = 0.5$ | | | | | | |
| 0.05 | 0.5 | 0.045 | 32 | 114.5 | 0.0545 | 0.836 |
| | 1.0 | 0.044 | 15 | 100 | 0.0082 | 0.829 |
| | 1.5 | 0.043 | 10 | 96.5 | 0.0021 | 0.828 |
| | 2.0 | 0.040 | 7 | 95 | 0.0007 | 0.827 |
| 0.10 | 0.5 | 0.046 | 26 | 106 | 0.0814 | 0.714 |
| | 1.0 | 0.0428 | 12 | 98 | 0.017 | 0.700 |
| | 1.5 | 0.0454 | 8 | 96 | 0.0046 | 0.696 |
| | 2.0 | 0.037 | 6 | 95 | 0.0011 | 0.695 |
| 0.20 | 0.5 | 0.037 | 21 | 101 | 0.118 | 0.536 |
| | 1.0 | 0.035 | 10 | 97.5 | 0.0295 | 0.513 |
| | 1.5 | 0.040 | 7 | 96 | 0.0081 | 0.507 |
| | 2.0 | 0.0442 | 5 | 96 | 0.0022 | 0.504 |
| 0.50 | 0.5 | 0.040 | 16 | 99 | 0.179 | 0.240 |
| | 1.0 | 0.045 | 8 | 97 | 0.0622 | 0.216 |
| | 1.5 | 0.0408 | 5 | 97 | 0.022 | 0.208 |
| | 2.0 | 0.0414 | 4 | 97 | 0.0083 | 0.206 |
| $\delta = 1.0$ | | | | | | |
| 0.05 | 1.0 | 0.0934 | 10 | 104 | 0.036 | 0.776 |
| | 1.5 | 0.0864 | 6 | 100 | 0.0103 | 0.771 |
| | 2.0 | 0.0786 | 5 | 100 | 0.0029 | 0.771 |
| 0.10 | 1.0 | 0.0764 | 9 | 101 | 0.0622 | 0.630 |
| | 1.5 | 0.0774 | 5 | 100 | 0.0201 | 0.620 |
| | 2.0 | 0.0846 | 4 | 99.5 | 0.0066 | 0.615 |
| 0.20 | 1.0 | 0.0814 | 7 | 100 | 0.0900 | 0.440 |
| | 1.5 | 0.081 | 5 | 100 | 0.0345 | 0.426 |
| | 2.0 | 0.0784 | 4 | 99 | 0.0125 | 0.420 |
| 0.50 | 1.0 | 0.0868 | 6 | 100 | 0.1411 | 0.175 |
| | 1.5 | 0.0864 | 4 | 100 | 0.0629 | 0.162 |
| | 2.0 | 0.081 | 3 | 100 | 0.0285 | 0.156 |

(iii) *For the final isolated panels, the common change-point will be estimated as the mean or median of the corresponding change-point estimations.*

We notice that the supplementary runs will increase the total average delay detection time in lower order, but should reduce the false alarm rate significantly.

**Table 4.** Effect of $z^*$ on isolation error rates

| $z^*$ | $\mu$ | FAR | $CDTT$ | $\tilde{\nu}$ | FNR | FDR |
|---|---|---|---|---|---|---|
| 1.645 | 0.5 | 0.168 | 35 | 1013 | 0.0433 | 0.705 |
| | 1.0 | 0.165 | 16 | 999.5 | 0.0047 | 0.697 |
| | 1.5 | 0.170 | 11 | 997 | 0.00009 | 0.695 |
| | 2.0 | 0.173 | 8 | 996 | 0.00001 | 0.695 |
| 1.96 | 0.5 | 0.177 | 35 | 1006 | 0.0787 | 0.607 |
| | 1.0 | 0.164 | 16 | 998 | 0.011 | 0.586 |
| | 1.5 | 0.176 | 11 | 996.5 | 0.002 | 0.586 |
| | 2.0 | 0.160 | 8 | 995 | 0.00008 | 0.586 |
| 2.326 | 0.5 | 0.170 | 35 | 1001 | 0.146 | 0.454 |
| | 1.0 | 0.173 | 16 | 997.5 | 0.029 | 0.419 |
| | 1.5 | 0.173 | 11 | 996 | 0.006 | 0.413 |
| | 2.0 | 0.173 | 8 | 995.5 | 0.001 | 0.413 |
| 2.576 | 0.5 | 0.170 | 35 | 1000 | 0.212 | 0.342 |
| | 1.0 | 0.159 | 16 | 997 | 0.049 | 0.299 |
| | 1.5 | 0.176 | 11 | 996.5 | 0.012 | 0.289 |
| | 2.0 | 0.178 | 8 | 996 | 0.003 | 0.289 |

**Table 5.** FDR after supplementary runs for $ARL_0 = 1000$ and $\delta = 0.5(1.0)$ with $N = 100$

| p | $\mu$ | FAR | CDDT | FNR | FDR | $\hat{\nu}$ | $\tilde{\nu}$ | SRL |
|---|---|---|---|---|---|---|---|---|
| 0.05 | 0.5 | 0.036 | 32 | 0.142 | 0.316 | 102.6 | 101 | 12 |
| | 1.0 | 0.043 | 15 (10) | 0.0096 (0.1071) | 0.294 (0.1922) | 96.8 | 98 (100) | 10 (4) |
| | 1.5 | 0.040 | 10 (6) | 0.0024 (0.0171) | 0.282 (0.1723) | 95.6 | 97 (100) | 10 (3) |
| | 2.0 | 0.044 | 7 (5) | 0.0009 (0.0042) | 0.282 (0.1747) | 95.6 | 97 (100) | 10 (3) |
| 0.10 | 0.5 | 0.045 | 26 | 0.164 | 0.195 | 99.9 | 100 | 12 |
| | 1.0 | 0.044 | 12 (9) | 0.0183 (0.1259) | 0.168 (0.110) | 97.1 | 98 (100) | 9 (4) |
| | 1.5 | 0.044 | 8 (5) | 0.0041 (0.0257) | 0.166 (0.097) | 96.7 | 97.5 (100) | 8 (3) |
| | 2.0 | 0.044 | 6 (4) | 0.0017 (0.0074) | 0.162 (0.096) | 96.5 | 97 (100) | 8 (3) |
| 0.20 | 0.5 | 0.042 | 21 | 0.2007 | 0.104 | 98.9 | 99.5 | 12 |
| | 1.0 | 0.043 | 10 (7) | 0.0314 (0.1557) | 0.086 (0.057) | 97.4 | 98 (100) | 8 (4) |
| | 1.5 | 0.047 | 7 (5) | 0.0081 (0.0404) | 0.085 (0.049) | 97.2 | 97.5 (100) | 6 (3) |
| | 2.0 | 0.043 | 5 (4) | 0.0030 (0.0140) | 0.085 (0.047) | 97.0 | 97 (100) | 4.5 (2) |
| 0.50 | 0.5 | 0.044 | 16 | 0.2546 | 0.031 | 98.4 | 99 | 12 |
| | 1.0 | 0.038 | 8 (6) | 0.0634 (0.2018) | 0.024 (0.0156) | 97.5 | 98 (100) | 6 (4) |
| | 1.5 | 0.046 | 5 (4) | 0.0210 (0.0699) | 0.024 (0.0136) | 97.4 | 98 (100) | 4 (3) |
| | 2.0 | 0.042 | 4 (3) | 0.0094 (0.0287) | 0.023 (0.0131) | 97.4 | 97.5 (100) | 3 (2) |

Table 5 gives the simulation results with 5000 runs for $ARL_0 = 1000$, $N = 100$, and $\delta = 0.5$ and $1.0$ as in Table 1. After a common change is detected, sequential tests for those panels with Z-score larger than $Z^* = 1.645$ are run with two sided boundaries $[0,4]$ with starting values 2.0. The selection of $d = 4$ makes both the first type and second of error equal to $e^{-(d+\rho)} = e^{-2.583} \approx 0.0755$ (as discussed in Wu [18]). For change point $\nu = 100$, the proportion of changed panels is taken as 0.05, 0.10, 0.20, and 0.50. Reported are the False Alarm Rate (FAR), 1st kind err rate, False Discovery Rate (FDR) calculated conditioning on those true alarms, the estimations of the change point calculated as the mean ($\hat{\nu}$) and median ($\tilde{\nu}$) based on the change-point estimations from those final isolated "changed" panels, and the median average supplementary run length (SRL).

**Remark 3.** As stated in (iii), to eliminate the possible large runs, we can truncate all tests at time $C$ such that $d = o(C)$ but $C = o(\ln(B))$. An alternative procedure by using the sequential truncated one-sided tests was discussed in Wu [20].

**Remark 4.** Control of the two types of errors for sequential multiple hypotheses testing procedure has been a topic of recent focuses. We refer to De and Baron [4] and Song and Fellouris [11] for some recent results. Further investigation in our case seems necessary as we already have preliminary information available.

Figure 1 plotted the histograms for the FDR and change-point estimations based on 5000 simulations.

Based on the simulation results, we can see that the supplementary runs can reduce the FDR dramatically with slight increase on the first kind of error rate. Also the median estimation for the change-point has smaller bias than the mean estimation.

# 5    Distribution of Change Point Estimation and Standardized Z-Score

In this section, we present some theoretical results for the estimated delay detection time and the z-score. Wu [14] studied second order approximations for the bias of change-point estimation by using the CUSUM procedure for a single sequence. Second order corrections of the z-scores for a single sequence with true alarm and false alarms are further studied in Wu ([16,19]). In the single sequence, the distributions of the CUSUM process and Shiryayev-Roberts process have been studied in Pollak and Siegmund [7], Pollak and Tartakovsky [8], and Polenchenko [9] in both the discrete and continuous time cases. However, multiple sequence tests raise quite different scenario as the alarm will be mostly raised by the true changed panels that, in turn, do not affect the probability behaviours for the unchanged panels. Therefore, we split our discussions for those unchanged panels from changed panels.

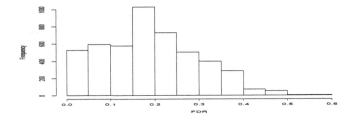

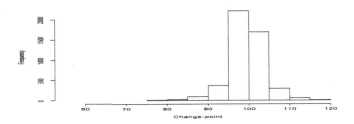

**Fig. 1.** Histograms of FDR and change-point estimation

## 5.1   Null Distribution for Change-Point Estimation

By conditioning on a common change being detected, the alarm will mostly be raised from those changed panels. Those unchanged panels can be studied by treating the alarm time as a random stopping time. Here, we used the approach as in Wu [15].

Consider a random walk $\{S_n = S_0 + \sum_{i=1}^{n}(X_i - \delta/2)\}$ with $S_0 = 0$ where $X_i$ are i.i.d $N(0,1)$ random variables. By looking backward at the alarm time $\tau$, the paths of observations from those unchanged panels are approximately i.i.d copies of $\{S_n\}$ and $\tau - \hat{\nu}_i$ is the maximum point $\sigma_M$ with maximum value $M$:

$$M = \sup_{0 \le k < \infty} S_k.$$

In the following we give some results for $(\sigma_M, M)$. Let $\tau_+^{(0)} = 0$. Define

$$\tau_+ = \tau_+^{(1)} = \inf\{n > 0 : S_n > 0\}$$

and for $k \ge 2$

$$\tau_+^{(k)} = \inf\{n > \tau_+^{(k-1)} : S_n > S_{\tau_+^{(k-1)}}\},$$

and

$$L = \sup\{k > 0 : \tau_+^{(k)} < \infty\},$$

and $L = 0$ if $\tau_+ = \infty$. It can be easily seen that

$$P[L = k] = p^k(1 - p),$$

for $k = 0, 1, 2, \ldots$ where $p = P[\tau_+ < \infty]$.

Now we note that
$$(\sigma_M, M) =_d (\tau_+^{(L)}, S_{\tau_+^{(L)}}).$$

For given $L = k > 0$, $(\tau_+^{(k)}, S_{\tau_+^{(k)}})$ is in distribution equivalent to the sum of k i.i.d. copies of $(\tau_+, S_{\tau_+}) | \tau_+ < \infty$. This leads to the following Laplace transform for $(\sigma_M, M)$ in the normal case.

**Theorem 3.**
$$E[s^{\sigma_M} e^{\lambda M}] = \frac{1 - G_+(1,0)}{1 - G_+(s,\lambda)},$$

where $G_+(s,\lambda) = 1 - E[s^{\tau_+} e^{\lambda S_{\tau_+}}; \tau_+ < \infty]$.

*Proof.* By conditioning on the value of $L$, we have

$$E[s^{\sigma_M} e^{\lambda M}] = \sum_{k=0}^{\infty} E[s^{\tau_+^{(L)}} e^{\lambda S_{\tau_+^{(L)}}}; L = k]$$

$$= \sum_{k=0}^{\infty} (E[s^{\tau_+} e^{\lambda S_{\tau_+}}; \tau_+ < \infty])^k P(\tau_+ = \infty)$$

$$= \frac{1 - P(\tau_+ < \infty)}{1 - E[s^{\tau_+} e^{\lambda S_{\tau_+}}; \tau_+ < \infty]}.$$

**Remark 5.** From Theorem 3, the exact forms for moments of $\sigma_M$ and $M$ can be obtained. For example,

$$E[\sigma_M] = \frac{E[\tau_+; \tau_+ < \infty]}{P(\tau_+ = \infty)}; \quad E[M] = \frac{E[S_{\tau_+}; \tau_+ < \infty]}{P(\tau_+ = \infty)}.$$

Since the random walk $\{S_n\}$ has negative drift, the Wiener-Hopf factorization (e.g. Theorem 8.41 of Siegmund [10]) gives

$$(1 - G_+(s,\lambda))(1 - G_-(s,\lambda)) = 1 - se^{\lambda^2/2 - \delta\lambda/2},$$

where $G_-(s,\lambda) = E[s^{\tau_-} e^{\lambda S_{\tau_-}}]$. By letting $s \to 1$ and $\lambda = 0$, we have

$$1 - G_+(1,0) = \frac{1}{E\tau_-}.$$

Thus, we have an alternative form

$$E[s^{\sigma_M} e^{\lambda M}] = \frac{1}{E\tau_-} \frac{1 - E[s^{\tau_-} e^{\lambda S_{\tau_-}}]}{1 - se^{\lambda^2/2 - \delta\lambda/2}}.$$

The following corollary gives the approximate exponential property for $M$ as $\delta \to 0$.

**Corollary 1.** *As $\delta \to 0$,*

$$\lim_{n \to \infty} P(T_n^{(1)} < x) = P(\delta M < x) \to 1 - e^{-x}.$$

*Proof.* We take $s = 1$ and note that $E[S_{\tau_-}] = -(\delta/2)E[\tau_-]$. Using the alternative form, as $\delta \to 0$, the moment generating function of $\delta M$ equals to

$$
\begin{aligned}
E[e^{\lambda \delta M}] &= \frac{1}{E[\tau_-]} \frac{1 - E[e^{\lambda \delta S_{\tau_-}}]}{1 - e^{\lambda^2 \delta^2/2 - \lambda \delta^2/2}} \\
&= \frac{-\delta/2}{E[S_{\tau_-}]} \frac{\lambda \delta E[S_{\tau_-}] + o(\delta)}{\lambda^2 \delta^2/2 - \lambda \delta^2/2 + o(\delta^2)} \\
&= \frac{1 + o(\delta)}{1 - \lambda + o(\delta)} \\
&= \frac{1}{1 - \lambda} + o(\delta),
\end{aligned}
$$

the moment generating function of a standard exponential distribution.

To study the distribution of $\sigma_M$, denote by $P^*(.)$ the probability measure with mean $\delta/2$ and $\tau_c^* = \inf\{n : S_n^* > c\}$ for an independent copy $\{S_n^*\}$ of $\{S_n\}$ for $c > 0$.

**Corollary 2.** $P[\sigma_M = 0] = P(\tau_+ = \infty]$ and for any $n > 0$,

$$
P[\sigma_M \leq n] = E[P^*[\tau_M^* \leq n|M]].
$$

*Proof.*

$$
\begin{aligned}
P[\sigma_M \leq n] &= P[\sup_{n \leq k < \infty} S_k \leq \sup_{0 \leq k \leq n} S_k] \\
&= P[\sup_{n \leq k < \infty} (S_k - S_n) \leq - \sup_{0 \leq k \leq n} (S_n - S_k)] \\
&= P[M \leq \sup_{0 \leq k \leq n} S_k^*] \\
&= E[P^*(\tau_M^* \leq n|M)].
\end{aligned}
$$

When $\delta \to 0$, $P(\tau_+ = \infty) \to 0$. From Corollary 1, $\delta M$ is asymptotically an exponential variable. For a given large value of $M$, Equation (3.30) of Siegmund [10] gives the following inverse Gaussian approximation:

$$
P^*[\tau_M^* \leq n|M] \approx 1 - \Phi(\frac{M + \rho}{\sqrt{n}} - (\delta/2)\sqrt{n}) + e^{\delta(M + \rho)} \Phi(-\frac{M + \rho}{\sqrt{n}} - (\delta/2)\sqrt{n}).
$$

Therefore the unconditional distribution of $\sigma_M$ is approximately a mixture of inverse Gaussian distribution.

**Remark 6.** More recently, Wu [22] gives the explicit joint distribution for $(\sigma_M, M)$ under the continuous Brownian motion model. Also, a second order approximation for the distribution of $M$ is used to propose a FDR-controlled isolation rule. Comparisons with the proposed procedure will be conducted in future communications.

## 5.2   A Real Example

Here we use the adjusted closing prices of Dow Jones 30 stocks from Sept 5, 2018 to Sept 4, 2018 (total 252 days) as illustration. We use the differences of logarithm as the original data. The Auto-Correlation Function (ACF) plots show that the autocorrelation can be ignored at the first step. The first 100 data for each stock are used to standardize with mean 0 and standard deviation 1. Also, the observations with absolute values larger than 3 are truncated to 3 to avoid any potential effect of outliers. To detect a common change of decrease in mean, we use the negatives of the observations.

For $N = 30$ and $ARL_0 = 10,000$, the approximate value for $B$ is 224142.6. The S-R process detected the common change at $\tau = 107$. The histograms of change-points and z-scores in Fig. 2 showed that all the panels are isolated as changed panels.

Supplementary SPRTs are run as described in Sect. 5 which can also identify whether the change in panels are temporary. The histogram of the run lengths showed that the supplementary runs have median run length 2 and the 20 stocks that their means showed decreases are "AAPL" "AXP" "CAT" "CVX" "CSCO" "KO" "DWDP" "XOM" "GS" "HD" "JNJ" "JPM" "MCD" "MSFT" "PFE" "TRV" "UTX" "VZ" "V" "WMT".

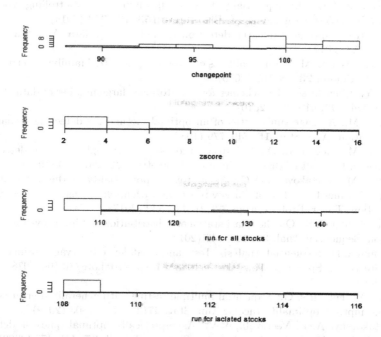

**Fig. 2.** Plots of change-point estimation and Z-scores

## 6    Conclusion

We proposed a combined SR-CUSUM procedure to detect common changes in multi-panel data. To reduce the FDR for isolating changed panels and estimating the common change point, supplementary SPRTs are proposed to run on possible candidates after detection. An alternative method is to run one-sided truncated sequential tests by just finding the true changed panels as discussed in Wu [20]. Further discussions on sequential multiple tests on the control of FDR can also be used in the supplementary run; see Bartroff [1] and Bartroff and Song [2]. As discussed in Wu [21], we may also use the adaptive combined SR-CUSUM procedure which can eliminate high biases when the true change magnitudes are unknown. The results will be presented in future communications.

**Acknowledgement.** This research is partially supported by a RSCA grant from California State University Stanislaus.

## References

1. Bartroff, J.: Multiple hypothesis tests controlling generalized error rates for sequential data. Statistica Sinica **28**, 363–398 (2018)
2. Bartroff, J., Song, J.: Sequential tests of multiple hypotheses controlling type I and II familywise error rates. J. Stat. Plan. Infer. **153**, 100–114 (2014)
3. Chan, H.P.: Optimal sequential detection in multi-stream data. Ann. Stat. **45**(6), 2636–2763 (2017)
4. De, S.K., Baron, M.: Sequential tests controlling generalized familywise error rates. Stat. Methodol. **23**, 88–102 (2015)
5. Mei, Y.: Efficient scalable schemes for monitoring a large number of data streams. Biometrika **97**, 419–433 (2010)
6. Pollak, M.: Average run lengths of an optimal method for detecting a change in distribution. Ann. Stat. **15**, 749–779 (1987)
7. Pollak, M., Siegmund, D.: A diffusion process and its applications to detecting a change in the drift of Brownian motion. Biometrika **72**, 267–280 (1985)
8. Pollak, M., Tartakovsky, A.G.: Asymptotic exponentiality of the distribution of first exit times for a class of Markov processes with applications to quickest change detection. Theor. Probab. Appl. **53**(2), 430–442 (2017)
9. Polunchenko, A.S.: On the quasi-stationary distribution of Shiryayev-Roberts diffusion. Sequential Anal. **36**, 129–149 (2017)
10. Siegmund, D.: Sequential Analysis: Tests and Confidence Intervals. Springer Series in Statistics. Springer, New York (1985). https://doi.org/10.1007/978-1-4757-1862-1
11. Song, Y., Fellouris, G.: Sequential multiple testing with generalized error control: an asymptotic optimality theory. Ann. Stat. **47**(3), 1776–1803 (2019)
12. Tartakovsky, A.G., Veeravalli, V.V.: Asymptotically optimal quickest detection change detection in distributed sensor. Sequential Anal. **27**, 441–475 (2008)
13. Wu, Y.: Supplementary score tests in mixture models. Commun. Stat.: Theor. Methods **31**(5), 753–780 (2002)
14. Wu, Y.: Bias of estimator of change point detected by a CUSUM procedure. Ann. Inst. Stat. Math. **56**, 142–157 (2004)

15. Wu, Y.: Inference for Change-Point and Post-Change Means After a CUSUM Test. Lecture Notes in Statistics, vol. 180. Springer, New York (2005). https://doi.org/10.1007/b100107

16. Wu, Y.: Inference for post-change mean detected by a CUSUM procedure. J. Stat. Plan. Infer. **136**, 3625–3646 (2006)

17. Wu, Y.: Discussion on "sequential design and estimation in heteroscedastic nonparametric regression" by Sam Efromovich. Sequential Anal. **26**, 33–36 (2007a)

18. Wu, Y.: Detecting sparse signals with a large scale of sequential probability ratio tests. In: Proceedings of 2006 Joint Statistical Meeting, ASA, Seattle, pp. 1253–1259 (2007b)

19. Wu, Y.: False alarms and sparse change segment detection by using a CUSUM procedure. Sequential Anal. **26**(4), 321–334 (2007c)

20. Wu, Y.: Supplementary score test for sparse signals in large-scale truncated sequential tests. J. Stat. Theor. Pract. **12**(4), 744–756 (2018)

21. Wu, Y.: A combined SR-CUSUM procedure for detecting common changes in panel data. Commun. Stat.: Theor. Methods **48**(17), 4302–4319 (2019a)

22. Wu, Y.: Estimation of common change point and isolation of changed panels after sequential detection. arXiv:1907.02097 (2019b)

23. Xie, Y., Siegmund, D.: Sequential multi-sensor change-point detection. Ann. Stat. **41**, 670–692 (2013)

# A Production-Inventory System with a Service Facility and Production Interruptions for Perishable Items

Dequan Yue[1(✉)], Sai Wang[1], and Yuying Zhang[2]

[1] School of Science, Yanshan University, Qinhuangdao 066004, China
ydq@ysu.edu.cn, wangsai.ng@qq.com
[2] School of Economics and Management, Yanshan University,
Qinhuangdao 066004, China
756552686@qq.com

**Abstract.** In this paper, we consider a production-inventory system with a service facility and production interruptions. Customers arrive in the system according to a Poisson process and require a random time of the service from a single service facility. The service time is assumed to be exponentially distributed. The items are produced according to an $(s, S)$ policy. Each customer leaves the system with one item from the inventory at his service completion epoch if the inventory is available. The production is interrupted for a vacation of random time once the inventory level becomes $S$. The vacations are exponentially distributed. On return from a vacation, if the inventory level depletes to $s$, then the production is immediately switched on. It then starts production and is kept in the on mode until the inventory level becomes $S$. The items in stock are perishable and have exponential life times. It is assumed that no customers is allowed to join the queue when the inventory level is zero. We first derive the stability condition of the system. Then, We obtain the product form solution for the stationary joint distribution of the number of customers and the on-hand inventory level. Based on this stationary distribution, we compute explicitly some performance measures and develop a cost function. Finally, some numerical results are presented.

**Keywords:** Production-inventory system · Service time · Lost sales · Perishable items · Cost function

## 1 Introduction

Queueing-inventory systems are queueing systems with attached inventory control. In such a system, satisfying each demand needs not only an on-hand inventory item but also some service time. Over the last decades, queueing-inventory systems have attracted much attention due to their close connection with the research on integrated supply chain management. The first work on queuing-inventory system appears in Sigman and Simchi-Levi [1]. The authors investigate an M/G/1 queueing-inventory system and develop a light traffic heuristic for

© Springer Nature Singapore Pte Ltd. 2019
Q.-L. Li et al. (Eds.): Cao Festschrift 2019, CCIS 1102, pp. 410–428, 2019.
https://doi.org/10.1007/978-981-15-0864-6_21

finding performance descriptions for their model. This is followed by a sequence of papers (see Berman and Kim [2,3], Berman and Sapna [4–6], Schwarz et al. [7], Krishnamoorthy et al. [8], Viswanath et al. [9] and a survey paper Krishnamoorthy et al. [10] and reference therein).

Krishnamoorthy and Viswanath [11] extend a queueing-inventory system with $(s, S)$ policy to a production-inventory system with a positive service time and server's vacation. They investigate the system stability and the system state distribution. Several performance measures are also computed. Krishnamoorthy et al. [12] consider an $(s, S)$ production-inventory system with a positive service time under the assumptions of a Poisson arrival process and exponentially distributed service time and production time. They obtain product form solution for the steady state distribution under assumption that customers do not join when the inventory level is zero.

Baek and Moon [13] study a production-inventory system with a positive service time in which the stocks are delivered both by an outside supplier and an international production. The stocks are replenished either by an external order under a $(r, Q)$ policy, or by an internal production. They obtain a product form solution for the stationary joint distribution of the queue length and the on-hand inventory level. Baek and Moon [14] consider an $(s, S)$ production-inventory system with an attached $M/M/c/\infty$ queue. The production process and arrival process are assumed to be Poisson processes. The service times are assumed to be exponentially distributed. They prove the independence of the inventory level process and the queue length process, and they derive the explicit stationary joint probability in product form. Recently, Yue and Qin [15] extend the model studied by Krishnamoorthy et al. [12] by considering the production vacation. They assume that the production facility takes multiple vacation when the on-hand inventory level reaches $S$. They obtain the product form solution for the stationary joint distribution of the queue length and the on-hand inventory level under the assumption that all arriving customers are lost during the stock out period. Besides, they compute explicitly some performance measures and develop a cost function based on these performance measures.

In all work quoted above, items in inventory are stored indefinitely to meet demands. However, when dealing with perishables inventory, the product life time must be taken into account in inventory models. Examples of perishable products include fresh food, seasonal products, blood products, chemicals, medicines and so on. Since the last few decades, analysis of the models for perishable products has drawn much attention of researchers because of its important applications in inventory and production systems. A comprehensive survey on perishable inventory systems is referred to Karaesmen et al. [16] and Bakker et al. [17]. For an early review of work, we refer the reader to Nahmias [18], Raafat [19], and Goyal and Giri [20]. Queueing-inventory systems with perishable inventory have been studied by Sivakumar and Arivarignan [21], Manuel et al. [22], [23] and several others, just mention a few of them.

In this paper, we consider a production-inventory system with perishable items and a service facility who serves customers. The items are produced according to an $(s, S)$ policy. Each customer leaves the system with one item from the inventory at his service completion epoch if the inventory is available. The production is interrupted for a vacation of random time once the inventory level becomes $S$. This model extends the system model in Yue and Qin [15] to the model with perishable inventory items and interruption of the production.

The main contributions of this paper are as follows: (i) We extend the model considered by Yue and Qin [15] by considering perishable inventory items. (ii) We derive an explicit stationary condition which is independent on the parameters of the production and the vacation by using quasi-birth-and-death (QBD) process theory. (iii) We obtain product form solution for the stationary joint distribution of the queue length and the on-hand inventory level by using the matrix analytic approach. (iv) We compute explicitly some performance measures and develop a cost function based on these performance measures.

The rest of the paper is organized as follows. The system model is described in Sect. 2. In Sect. 3, we perform the steady state analysis. Firstly, the stability condition of the system is derived by using QBD process theory. Then, the joint distribution of the number of customers and the inventory level in product form is obtained. Finally, some performance measures are computed. A mean cost function is developed in Sect. 4, and numerical results are also presented in this section. Conclusions are given in Sect. 5.

## 2    System Model

The production-inventory system considered in this paper is described as follows (see Fig. 1):

Customers arrive in the system according to a Poisson process with rate $\lambda$. There is a single service facility (server) who serves the customers one by one under a First-Come, First-Served (FCFS) discipline. The service times are exponentially distributed with rate $\mu$. Each customer leaves the system with one item from the inventory at his service completion epoch. It is assumed that all arriving customers are lost when the inventory level is zero.

The system has a single production facility that produces one type of product, and the production times are exponentially distributed with rate $\eta$. The production is interrupted for a vacation of random time once the inventory level becomes $S$. The vacations are exponentially distributed with rate $\theta$. On return from a vacation, if the inventory level depletes to $s$, then the production is immediately switched on and is kept in the on mode until the inventory level becomes $S$. Otherwise, on return from the vacation, if the inventory level is higher than $s$, then the production facility goes on another vacation of random duration at the same distribution rate as earlier. This type of vacation is called multiple vacation in literature of queueing system with vacations (see, e.g., Tian and Zhang [24]).

The items in stock are perishable and their life times are distributed exponentially with rate $\gamma$.

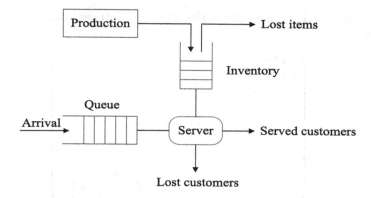

**Fig. 1.** Diagram of the proposed production-inventory model.

# 3   Steady-State Analysis

In this section, we first formulate the system state process as a QBD process and then derive the stability condition of the system. Then, we derive the joint stationary distribution of the number of customers, the inventory level and the status of the production in the steady state. Using this distribution, we compute some performance measures.

## 3.1   Stability Condition

Let $\{X(t), t \geq 0\} = \{(N(t), I(t), J(t)), t \geq 0\}$ be the system state process, where $N(t)$ denotes the number of customers at time $t$, $I(t)$ denotes the inventory level at time $t$, and $J(t)$ denotes the status of the production process at time $t$ which is defined as either 0 or 1 according to whether the production facility is taking on a vacation or in production mode, respectively. Then, the process $\{X(t), t \geq 0\}$ is a continuous-time Markov process (CTMP) which is a QBD process with state space:

$$\Omega = \{(i, j, k), i \geq 0, 0 \leq j \leq S - 1, k = 0, 1\} \cup \{(i, S, 0), i \geq 0\}.$$

The infinitesimal generator of the process $\{X(t), t \geq 0\}$ is given as follows:

$$Q = \begin{pmatrix} A_0 & C & & \\ B & A & C & \\ & B & A & C \\ & & \ddots & \ddots & \ddots \end{pmatrix}$$

where $A_0$, $A$, $B$ and $C$ are all $(2S+1) \times (2S+1)$ matrices, and they are given as follows:

$$A_0 = \begin{pmatrix} F_0 & U_0 & & & & & & \\ L_1 & F_1 & U_0 & & & & & \\ & \ddots & \ddots & \ddots & & & & \\ & & L_s & F_s & U_0 & & & \\ & & & L_{s+1} & F_{s+1} & U_0 & & \\ & & & & \ddots & \ddots & \ddots & \\ & & & & & L_{S-2} & F_{S-2} & U_0 & \\ & & & & & & L_{S-1} & F_{S-1} & U_1 \\ & & & & & & & L_S & F_S \end{pmatrix}$$

with

$$F_0 = \begin{pmatrix} -\theta & \theta \\ 0 & -\eta \end{pmatrix},$$

$$F_i = \begin{pmatrix} -(\lambda+\theta+i\gamma) & \theta \\ 0 & -(\lambda+\eta+i\gamma) \end{pmatrix}, \quad i = 1,2,..,s,$$

$$F_i = \begin{pmatrix} -(\lambda+i\gamma) & 0 \\ 0 & -(\lambda+i\gamma+\eta) \end{pmatrix}, \quad i = s+1, s+2, ..., S-1,$$

$$F_S = -(\lambda+S\gamma), \quad U_0 = \begin{pmatrix} 0 & 0 \\ 0 & \eta \end{pmatrix}, \quad U_1 = \begin{pmatrix} 0 \\ \eta \end{pmatrix},$$

$$L_i = \begin{pmatrix} i\gamma & 0 \\ 0 & i\gamma \end{pmatrix}, \quad i = 1,2,...,S-1,$$

$$L_S = (S\gamma \; 0)$$

and

$$B = \begin{pmatrix} 0 & & & & \\ M_1 & 0 & & & \\ & \ddots & \ddots & & \\ & & M_1 & 0 & \\ & & & M_2 & 0 \end{pmatrix}$$

with

$$M_1 = \begin{pmatrix} \mu & 0 \\ 0 & \mu \end{pmatrix}, M_2 = (\mu \; 0)$$

and $C$ is given by

$$
C = \begin{pmatrix} 0 & & & & \\ & M_3 & & & \\ & & \ddots & & \\ & & & M_3 & \\ & & & & \lambda \end{pmatrix}
$$

where

$$
M_3 = \begin{pmatrix} \lambda & 0 \\ 0 & \lambda \end{pmatrix}
$$

and $A$ is given by

$$
A = A_0 - \frac{\mu}{\lambda} C.
$$

Using QBD process theory, we can derive the stability condition of the system which is given by the following theorem.

**Theorem 1.** *The process $\{X(t), t \geq 0\}$ with the infinitesimal generator $Q$ is positive recurrent if and only if $\rho = \frac{\lambda}{\mu} < 1$.*

*Proof.* To derive the stability condition of the process $\{\Phi(t), t \geq 0\}$, we consider the matrix $H = A + B + C$, which is given by

$$
H = \begin{pmatrix} F_0 & U_0 & & & & & & \\ D_1 & G_1 & U_0 & & & & & \\ & \ddots & \ddots & \ddots & & & & \\ & & D_s & G_s & U_0 & & & \\ & & & D_{s+1} & G_{s+1} & U_0 & & \\ & & & & \ddots & \ddots & \ddots & \\ & & & & & D_{S-2} & G_{S-2} & U_0 \\ & & & & & & D_{S-1} & G_{S-1} & U_1 \\ & & & & & & & D_S & G_S \end{pmatrix}
$$

where

$$
G_i = \begin{pmatrix} -(\mu + \theta + i\gamma) & \theta \\ 0 & -(\mu + \eta + i\gamma) \end{pmatrix}, \quad i = 1, 2, ..., s,
$$

$$
G_i = \begin{pmatrix} -(\mu + i\gamma) & 0 \\ 0 & -(\mu + i\gamma + \eta) \end{pmatrix}, \quad i = s+1, s+2, ..., S-1,
$$

$$
G_S = -(\mu + S\gamma),
$$

$$D_i = \begin{pmatrix} \mu + i\gamma & 0 \\ 0 & \mu + i\gamma \end{pmatrix}, \ i = 1, 2, ..., S - 1,$$

$$D_S = \begin{pmatrix} \mu + S\gamma & 0 \end{pmatrix}$$

and the matrices $U_0$ and $U_1$ are defined previously.

Let $x = (x(0,0), x(0,1), ..., x(S-1,0), x(S-1,1), x(S,0))$ be the steady state probability vector of the generator $H$. Then, $x$ satisfies equations $xH = 0$ and $xe = 1$, where $e$ is a column vector of 1's of appropriate dimension.

From the structure of matrices $B$ and $C$, we have

$$xCe = \lambda \left( \sum_{i=1}^{S-1} [x(i,0) + x(i,1)] + x(S,0) \right) = \lambda \left[ 1 - x(0,0) - x(0,1) \right]$$

and

$$xBe = \mu \left( \sum_{i=1}^{S-1} [x(i,0) + x(i,1)] + x(S,0) \right) = \mu \left[ 1 - x(0,0) - x(0,1) \right].$$

From Neuts [25], the process $\{\Phi(t), t \geq 0\}$ is positive recurrent if and only if

$$xCe < xBe.$$

Thus, we get $\lambda < \mu$ from this equality since it is not difficult to get the steady state probability vector $x$ of the generator $H$, then we have $x(0,0) + x(0,1) \neq 1$. So, the system is stable if and only if $\rho = \frac{\lambda}{\mu} < 1$. □

*Remark 1.* Theorem 1 shows that the stability condition for the present model is the same as that of the classical M/M/1 queueing system, and it is independent to the parameters of production and the vacation. So, the interruption (vacation) of the production dose not influence the stability of the system for our system model.

### 3.2    Steady-State Distribution

In this section, we use matrix analytic approach to derive the steady-state distribution of the system. The idea of the method is as used in Krishnamoorthy and Viswanath [12].

Firstly, we consider a production-inventory system where the service time of customers is zero. For this case, the corresponding Markov process is defined as $\{\hat{X}(t), t \geq 0\} = \{(I(t), J(t)), t \geq 0\}$, where $I(t)$ denotes the inventory level at time $t$, and $J(t)$ denotes the status of the production process at time $t$ which is defined as that in last subsection. The process $\{\hat{X}(t), t \geq 0\}$ is a CTMP with finite state space

$$\hat{\Omega} = \{(j,k), 0 \leq j \leq S - 1, k = 0, 1\} \cup \{S, 0\},$$

and its infinitesimal generator is given by

$$\hat{Q} = \begin{pmatrix} F_0 & U_0 & & & & & & \\ \hat{L}_1 & F_1 & U_0 & & & & & \\ & \ddots & \ddots & \ddots & & & & \\ & & \hat{L}_s & F_s & U_0 & & & \\ & & & \hat{L}_{s+1} & F_{s+1} & U_0 & & \\ & & & & \ddots & \ddots & \ddots & \\ & & & & & \hat{L}_{S-2} & F_{S-2} & U_0 \\ & & & & & & \hat{L}_{S-1} & F_{S-1} & U_1 \\ & & & & & & & \hat{L}_S & F_S \end{pmatrix}$$

with

$$\hat{L}_i = \begin{pmatrix} \lambda + i\gamma & 0 \\ 0 & \lambda + i\gamma \end{pmatrix}, \quad i = 1, 2, ..., S - 1,$$

$$\hat{L}_S = (\lambda + S\gamma \ 0)$$

and the other matrices in $\hat{Q}$ are as defined previously in last subsection.

Let $\pi = (\pi(0,0), \pi(0,1), ..., \pi(S-1,0), \pi(S-1,1), \pi(S,0))$ be the steady-state probability vector of the process $\{\hat{X}(t), t \geq 0\}$. Then $\pi$ satisfies the set of equations

$$\begin{cases} \pi\hat{Q} = 0 \\ \pi e = 1 \end{cases} \tag{1}$$

where $e$ is a column vector of 1's of appropriate dimension. Thus, Eq. (1) reduce to the following set of equations:

$$\theta\pi(0,0) = (\lambda + \gamma)\pi(1,0), \tag{2}$$
$$(\lambda + i\gamma + \theta)\pi(i,0) = (\lambda + (i+1)\gamma)\pi(i+1,0), \quad i = 1, 2, ..., s, \tag{3}$$
$$(\lambda + i\gamma)\pi(i,0) = (\lambda + (i+1)\gamma)\pi(i+1,0), \quad i = s+1, s+2, ..., S-1, \tag{4}$$
$$(\lambda + S\gamma)\pi(S,0) = \eta\pi(S-1,1), \tag{5}$$
$$\eta\pi(0,1) = (\lambda + \gamma)\pi(1,1) + \theta\pi(0,0), \tag{6}$$
$$(\lambda + \eta + i\gamma)\pi(i,1) = (\lambda + (i+1)\gamma)\pi(i+1,1) + \eta\pi(i-1,1) + \theta\pi(i,0),$$
$$i = 1, 2, ..., s, \tag{7}$$
$$(\lambda + \eta + i\gamma)\pi(i,1) = (\lambda + (i+1)\gamma)\pi(i+1,1) + \eta\pi(i-1,1),$$
$$i = s+1, s+2, ..., S-2, \tag{8}$$
$$(\lambda + \eta + (S-1)\gamma)\pi(S-1,1) = \eta\pi(S-2,1), \tag{9}$$
$$\sum_{i=0}^{S-1}[\pi(i,0) + \pi(i,1)] + \pi(S,0) = 1 \tag{10}$$

**Theorem 2.** *The steady-state probability distribution of the process* $\{\hat{X}(t), t \geq 0\}$
*is given by*

$$\pi(i,0) = \alpha_i \pi(S,0), \quad i = 0, 1, ..., S-1, \tag{11}$$
$$\pi(i,1) = \beta_i \pi(S,0), \quad i = 0, 1, ..., S-1 \tag{12}$$

*and*

$$\pi(S,0) = \left[1 + \sum_{i=0}^{S-1} (\alpha_i + \beta_i)\right]^{-1} \tag{13}$$

*where*

$$\alpha_0 = \frac{(\lambda + \gamma)}{\theta} \alpha_1, \tag{14}$$

$$\alpha_i = \alpha_{s+1} \prod_{k=i}^{s} \frac{\lambda + (k+1)\gamma}{\lambda + k\gamma + \theta}, \quad i = 1, 2, ..., s, \tag{15}$$

$$\alpha_i = \prod_{k=i}^{S-1} \frac{\lambda + (k+1)\gamma}{\lambda + k\gamma}, \quad i = s+1, s+2, ..., S-1 \tag{16}$$

*and for* $i = 0, 1, ..., s-1$, $\beta_i$ *is determined by the following iterative formula*

$$\beta_i = \frac{\lambda + (i+1)\gamma}{\eta} \beta_{i+1} + \frac{\lambda + S\gamma}{\eta} - \frac{\theta}{\eta} \sum_{k=i+1}^{s} \alpha_k \tag{17}$$

*and for* $i = s, s+1, ..., S-2$, $\beta_i$ *is given by*

$$\beta_i = \frac{\lambda + S\gamma}{\eta} \sum_{j=1}^{S-i-1} \left\{1 + \frac{1}{\eta^k} \prod_{k=1}^{j} [\lambda + (i+k)\gamma]\right\} \tag{18}$$

*and*

$$\beta_{S-1} = \frac{\lambda + S\gamma}{\eta}. \tag{19}$$

*Proof.* Equations. (3) and (4) can be rewritten as

$$\pi(i,0) = \frac{\lambda + (i+1)\gamma}{\lambda + i\gamma + \theta} \pi(i+1,0), \quad i = 1, 2, ..., s \tag{20}$$

*and*

$$\pi(i,0) = \frac{\lambda + (i+1)\gamma}{\lambda + i\gamma} \pi(i+1,0), \quad i = s+1, s+2, ..., S-1. \tag{21}$$

Using Eq. (21) recursively, we get

$$\pi(i,0) = \alpha_i \pi(S,0), \quad i = s+1, s+2, ..., S-1 \tag{22}$$

where $\alpha_i$ is defined by Eq. (16) for $i = s + 1, s + 2, ..., S - 1$. Using Eq. (20) recursively, we get

$$\pi(i, 0) = \prod_{k=i}^{s} \frac{\lambda + (k + 1)\gamma}{\lambda + k\gamma + \theta} \pi(s + 1, 0), \quad i = 1, 2, ..., s. \tag{23}$$

Substituting Eq. (22) with $i = s + 1$ into Eq. (23), we have

$$\pi(i, 0) = \alpha_{s+1} \prod_{k=i}^{s} \frac{\lambda + (k + 1)\gamma}{\lambda + k\gamma + \theta} \pi(S, 0)$$

$$= \alpha_i \pi(S, 0), \quad i = 1, 2, ..., s \tag{24}$$

where $\alpha_i$ is defined by Eq. (15) for $i = 1, 2, ..., s$. For Eq. (2), using Eq. (24) with $i = 1$, we get

$$\pi(0, 0) = \frac{\lambda + \gamma}{\theta} \pi(1, 0) = \alpha_0 \pi(S, 0)$$

where $\alpha_0$ is defined by Eq. (14).

Equation (8) can be written as

$$(\lambda + i\gamma)\pi(i, 1) - \eta\pi(i - 1, 1) = (\lambda + (i + 1)\gamma)\pi(i + 1, 1) - \eta\pi(i, 1),$$
$$i = s + 1, \ s + 2, ..., S - 2. \tag{25}$$

Repeating Eq. (25) and using Eqs. (5) and (9), we get

$$\pi(i, 1) = \frac{\lambda + (i + 1)\gamma}{\eta} \pi(i + 1, 1) + \frac{\lambda + S\gamma}{\eta} \pi(S, 0), \quad i = s, s + 1, ..., S - 2. \tag{26}$$

From Eq. (5), we have

$$\pi(S - 1, 1) = \beta_{S-1}\pi(S, 0). \tag{27}$$

where $\beta_{S-1}$ is defined by Eq. (19). Let

$$\pi(i, 1) = \beta_i \pi(S, 0), \quad i = s, s + 1, ..., S - 2. \tag{28}$$

then we get from Eq. (26) that

$$\beta_i = \frac{\lambda + (i + 1)\gamma}{\eta} \beta_{i+1} + \beta_{S-1}, \quad i = s, s + 1, ..., S - 2, \tag{29}$$

Repeating Eq. (29), we get $\beta_i$ for $i = s, s + 1, ..., S - 2$ as given by Eq. (18).

Equation (7) can be written as

$$(\lambda + i\gamma)\pi(i, 1) - \eta\pi(i - 1, 1) = (\lambda + (i + 1)\gamma)\pi(i + 1, 1) - \eta\pi(i, 1) + \theta\pi(i, 0),$$
$$i = 1, 2, ..., s. \tag{30}$$

Repeating Eq. (30) and noting Eq. (25), we have

$$\pi(i,1) = \frac{\lambda + (i+1)\gamma}{\eta}\pi(i+1,1) + \frac{\lambda + S\gamma}{\eta}\pi(S,0) - \frac{\theta}{\eta}\sum_{k=i+1}^{s}\pi(k,0),$$
$$i = 0,1,...,s-1. \qquad (31)$$

Let

$$\pi(i,1) = \beta_i\pi(S,0), \ i = 0,1,...,s-1.$$

Using Eq. (11), then we get $\beta_i$ for $i = 0,1,...,s-1$ from Eq. (31) as determined by Eq. (17). Using Eqs. (11) and (12), $\pi(S,0)$ can be derived from the normalizing condition given by Eq. (10) which is as given by Eq. (13).  $\square$

Next, we find the steady-state distribution of the process $\{X(t), t \geq 0\}$. Let $P = (P_0, P_1, ...)$ be the steady-state probability vector of the process $\{X(t), t \geq 0\}$, where

$$P_i = (P(i,0,0), P(i,0,1), ..., P(i, S-1,0), P(i, S-1,1), P(S,0))$$

is a row vector with dimension $2S+1$. Then, $P$ satisfies the following set of equations:

$$\begin{cases} PQ = 0 \\ Pe = 1 \end{cases} \qquad (32)$$

where $e$ is a column vector of 1's of appropriate dimension. The steady-state probability vector $P$ can be obtained by solving Eq. (32).

**Theorem 3.** *If $\rho < 1$, the elements of the steady-state probability vector $P = (P_0, P_1, ...,)$ of the process $\{X(t), t \geq 0\}$ is given by*

$$P_i = (1 - \rho)\rho^i\pi, \ i = 0,1,... \qquad (33)$$

*where $\pi = (\pi(0,0), \pi(0,1), ..., \pi(S-1,0), \pi(S-1,1), \pi(S,0))$ is given by Theorem 2.*

*Proof.* Eq. (32) can be rewritten as follows:

$$P_0A_0 + P_1B = 0, \qquad (34)$$
$$P_iC + P_{i+1}A + P_{i+2}B = 0, \ i = 0,1,..., \qquad (35)$$
$$\sum_{i=0}^{\infty} P_i\, e = 1. \qquad (36)$$

Let

$$P_i = \kappa\rho^i\pi, \ i = 0,1,... \qquad (37)$$

where $\kappa$ is a constant. Now, we verify that Eqs. (34) and (35) are satisfied with the above assumption.

Substituting Eq. (37) into the left sides of Eq. (34) and (35), we have

$$P_0 A_0 + P_1 B = \kappa\pi(A_0 + \rho B) \tag{38}$$

and

$$P_i C + P_{i+1} A + P_{i+2} B = \kappa\rho^i\pi\left(C + \rho A + \rho^2 B\right)$$
$$= \kappa\rho^i\pi\left[C + \rho\left(A_0 - \frac{1}{\rho}C\right) + \rho^2 B\right]$$
$$= \kappa\rho^{i+1}\pi\left(A_0 + \rho B\right), \quad i = 0, 1, \dots. \tag{39}$$

From the structure of the matrices $A_0$, $B$ and $\hat{Q}$, it is easy to verify that

$$A_0 + \rho B = \hat{Q}. \tag{40}$$

From Eq. (1), we have $\pi\hat{Q} = 0$. Hence, the right sides of Eqs. (34) and (35) are verified under the assumption given by Eq. (37). Using Eq. (36) and noting $\pi e = 1$, we have

$$\kappa\sum_{i=0}^{\infty}\rho^i = 1. \tag{41}$$

Hence, if $\rho < 1$, we have $\kappa = 1 - \rho$.    $\square$

*Remark 2.* Theorem 3 shows that the steady-state distribution of the system has a product form of two distributions. One is the distribution of the queue length in the classical M/M/1 system with the same parameters $\lambda$ and $\mu$, and the other one is the distribution of the inventory level in the production-inventory system with interruption of production and zero service time.

## 3.3    Performance Measures

In this subsection, we derive the following performance measures of system by using the steady-state probability distribution obtained in the last section.

(1) The mean number of customers in the system is given by

$$E_{nc} = \sum_{i=0}^{\infty} iP_i e = \frac{\rho}{1-\rho}. \tag{42}$$

(2) The mean inventory level is given by

$$E_{in} = \sum_{i=0}^{\infty}\sum_{j=0}^{S-1} j\left[P(i,j,0) + P(i,j,1)\right] + \sum_{i=0}^{\infty} SP(i,S,0)$$
$$= \left[S + \sum_{j=0}^{S-1} j(\alpha_j + \beta_j)\right]\pi(S,0). \tag{43}$$

(3) The mean production rate is given by

$$E_{pr} = \sum_{i=0}^{\infty} \sum_{j=0}^{S-1} \eta P(i,j,1) = \eta \sum_{j=0}^{S-1} \beta_j \pi(S,0). \tag{44}$$

(4) The mean loss rate of perishable inventory items is given by

$$E_{lp} = \gamma E_{in} \tag{45}$$

where $E_{in}$ is given by Eq. (43).

(5) The mean loss rate of customers is given by

$$E_{lc} = \sum_{i=0}^{\infty} \lambda \left[ P(i,0,0) + P(i,0,1) \right] = \lambda(\alpha_0 + \beta_0)\pi(S,0). \tag{46}$$

*Remark 3.* We observe from Eq. (42) that the mean number of customers in the system is the same as that in the classical M/M/1 queue, and it is independent with the other parameters $\eta$, $\theta$ and $\gamma$. On the other hand, Eqs. (43)–(46) show that the performance measures related to production-inventory are independent with the service rate $\mu$. This is because of the property of product form solution presented in Theorem 3 (see Remark 2).

## 4    Numerical Analysis

In this section, we develop the mean cost function $F(s,S)$ by using these performance measures. Then, we consider effect of some system parameters on these performance measures and the mean cost function by numerical examples.

Firstly, we develop the mean cost function by means of the above performance measures. The mean total cost rate per unit time is given by

$$F(s,S) = C_1 E_{nc} + C_2 E_{in} + C_3 E_{pr} + C_4 E_{lp} + C_5 E_{lc} \tag{47}$$

where $C_1$ is the cost of serving each customer per unit time, $C_2$ is the holding cost per unit time per inventory item, $C_3$ is the cost of production per unit time per inventory item, $C_4$ is the cost incurred due to the loss of perishable inventory items and $C_5$ is the cost incurred due to loss of customers.

In the following, we consider the effect of some parameters of the system on the performance measures related with inventory and the mean cost function by using numerical examples. The numerical results on performance measures are shown in Tables 1, 2, 3 and 4. The numerical results on mean cost function are shown in Figs. 2, 3, 4 and 5. The cost parameters are given as follows: $C_1 = 10$, $C_2 = 10$, $C_3 = 100$, $C_4 = 500$, $C_5 = 2000$. The values of the parameters are considered in the following four cases:

Case 1. $\lambda$ varies from 0.7 to 2.1, and the other parameters are fixed as follows: $\mu = 12$, $\eta = 6$, $\theta = 1.2$, $\gamma = 0.02$, $s = 3$ and $S = 27$.

Case 2. $\eta$ varies from 3.5 to 7.0, and the other parameters are fixed as follows: $\lambda = 1.5$, $\mu = 12$, $\gamma = 0.02$, $s = 3$ and $S = 27$.

Case 3. $\theta$ varies from 0.3 to 2.4, and the other parameters are fixed as follows: $\lambda = 1.5$, $\mu = 12$, $\eta = 6$, $\gamma = 0.02$, $s = 3$ and $S = 27$.

Case 4. $\gamma$ varies from 0.01 to 0.09, and the other parameters are fixed as follows: $\lambda = 1.5$, $\mu = 12$, $\eta = 6$, $\theta = 1.2$, $s = 3$ and $S = 27$.

**Table 1.** The effect of the arrival rate $\lambda$ on some performance measures.

| $\lambda$ | $E_{in}$ | $E_{pr}$ | $E_{lc}$ | $E_{lp}$ |
|---|---|---|---|---|
| 0.7 | 14.1441 | 0.9815 | 0.0014 | 0.2829 |
| 0.9 | 14.1838 | 1.1805 | 0.0032 | 0.2837 |
| 1.1 | 14.1768 | 1.3775 | 0.0060 | 0.2835 |
| 1.3 | 14.1418 | 1.5729 | 0.0099 | 0.2828 |
| 1.5 | 14.0890 | 1.7688 | 0.0150 | 0.2818 |
| 1.7 | 14.0243 | 1.9692 | 0.0212 | 0.2815 |
| 1.9 | 13.9513 | 2.1540 | 0.0286 | 0.2790 |
| 2.1 | 13.8723 | 2.3404 | 0.0370 | 0.2774 |

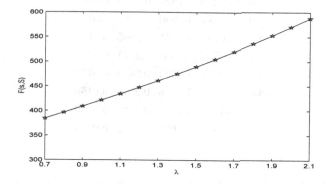

**Fig. 2.** The effect of the arrival rate $\lambda$ on the mean cost function $F(s, S)$.

It is observed from Table 1 that the mean production rate $E_{pr}$ and the mean loss rate of customers $E_{lc}$ increase with $\lambda$. This agrees with our expected observation. As the increasing of the mean arrival rate, more items are taken by customers from the inventory which increases the chance for switching on the production and hence increases the mean production rate $E_{pr}$. Since the more inventory items are needed with the increasing of the arrival rate, thus the probability that on-hand inventory level goes to be zero will increase. Therefore, the mean loss rate of customers $E_{lc}$ increase with the arrival rate. However, the concave property for both the mean inventory level $E_{in}$ and the mean loss rate of perishable inventory items $E_{lp}$ is observed from Table 1. This can be explained as follows: when $\lambda < \theta$, the production is switched on more frequently than arrival of demands. Thus, the addition to the inventory occurs more frequently than arrivals. This explains why the mean inventory level $E_{in}$ and the mean loss rate of perishable inventory items $E_{lp}$ increase with the arrival rate $\lambda$ when $\lambda < \theta$. However, when $\lambda > \theta$, the arrivals of demands occurs more frequently than the chance that the production is switching on. Thus, this leads to the decreasing of both the mean inventory level $E_{in}$ and the mean loss rate of perishable inventory items $E_{lp}$. From Fig. 2, it is observed that the mean cost function $F(s, S)$ increases with the arrival rate $\lambda$.

*Remark 4.* An arrival customer is either receiving one inventory item or lost. Therefore, the relation $\lambda = E_{pr} - E_{lp} + E_{lc}$ is expected. This relation can be observed from Table 1.

**Table 2.** The effect of the production rate $\eta$ on some performance measures.

| $\eta$ | $E_{in}$ | $E_{pr}$ | $E_{lp}$ | $E_{lc}$ |
|--------|----------|----------|----------|----------|
| 3.5 | 13.9273 | 1.7621 | 0.2785 | 0.0165 |
| 4.0 | 13.9877 | 1.7640 | 0.2798 | 0.0158 |
| 4.5 | 14.0266 | 1.7651 | 0.2805 | 0.0154 |
| 5.0 | 14.0537 | 1.7659 | 0.2811 | 0.0152 |
| 5.5 | 14.0737 | 1.7664 | 0.2815 | 0.0151 |
| 6.0 | 14.0890 | 1.7668 | 0.2818 | 0.0150 |
| 6.5 | 14.1011 | 1.7671 | 0.2820 | 0.0150 |
| 7.0 | 14.1109 | 1.7673 | 0.2822 | 0.0150 |

From Table 2, we observe that the mean inventory level $E_{in}$, the mean production rate $E_{pr}$ and the mean loss rate of perishable inventory items $E_{lp}$ increase with parameter $\eta$, and the mean loss rate of customers $E_{lc}$ decreases with parameter $\eta$. Since more items are replenished to the inventory with the parameter $\eta$, then the mean inventory level $E_{in}$ and the mean production rate $E_{pr}$ will increase, which results in the increasing of $E_{lp}$ and the decreasing of $E_{lc}$. This explains our observation for the effect of parameter $\eta$. Moreover, Fig. 3 shows that the mean cost function $F(s, S)$ increases with the parameter $\eta$, but the scale of the increase is very small.

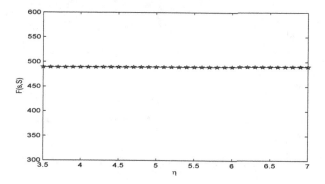

**Fig. 3.** The effect of the production rate $\eta$ on the mean cost function $F(s, S)$.

**Table 3.** The effect of the vacation rate $\theta$ on some performance measures.

| $\theta$ | $E_{in}$ | $E_{pr}$ | $E_{lp}$ | $E_{lc}$ |
|---|---|---|---|---|
| 0.3 | 12.4789 | 1.6118 | 0.2496 | 0.1318 |
| 0.6 | 13.4948 | 1.7188 | 0.2699 | 0.0511 |
| 0.9 | 13.8824 | 1.7520 | 0.2776 | 0.0256 |
| 1.2 | 14.0890 | 1.7668 | 0.2818 | 0.0150 |
| 1.5 | 14.2178 | 1.7747 | 0.2844 | 0.0097 |
| 1.8 | 14.3058 | 1.7794 | 0.2861 | 0.0067 |
| 2.1 | 14.3698 | 1.7825 | 0.2874 | 0.0049 |
| 2.4 | 14.4184 | 1.7846 | 0.2884 | 0.0038 |

Table 3 shows that the same effect of the parameter $\theta$ on the performance measures as the parameter $\eta$. The explanation for this can be observed from the fact that the chance that the production is switching on increases with the parameter $\theta$ and the previous explanation for the effect of the parameter $\eta$ on the performance measures. However, we find that the effect of the parameter $\theta$ on the mean cost function $F(s, S)$ is very different from that of the parameter $\eta$. Figure 4 shows that the mean cost function $F(s, S)$ decreases significantly with the parameter $\theta$. This may be due to the significant decreasing of the mean loss rate of customers $E_{lc}$ which can be seen from the data for $E_{lc}$ in Table 3.

It is observed from Table 4 that the mean inventory level $E_{in}$ decreases with the failure rate $\gamma$, while the other performance measures $E_{pr}$, $E_{lp}$ and $E_{lc}$ increase with the failure rate $\gamma$. This agrees with the expected observation. When the failure rate $\gamma$ increases, the more items in inventory are lost due to the shorter life time. Thus, the mean loss rate of inventory items $E_{lp}$ will increases, and then the mean inventory level $E_{in}$ will decreases. So, this will lead to the increasing of the other two performance measures $E_{pr}$ and $E_{lc}$. Moreover, we observe from Fig. 5 that the mean cost function $F(s, S)$ increases significantly with the failure rate $\gamma$.

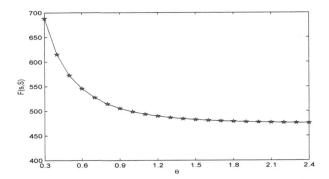

**Fig. 4.** The effect of the vacation rate $\theta$ on the mean cost function $F(s, S)$.

**Table 4.** The effect of the failure rate $\gamma$ on some performance measures.

| $\gamma$ | $E_{in}$ | $E_{pr}$ | $E_{lp}$ | $E_{lc}$ |
|------|---------|---------|---------|---------|
| 0.01 | 14.2875 | 1.6288 | 0.1439 | 0.0140 |
| 0.02 | 14.0890 | 1.7668 | 0.2818 | 0.0150 |
| 0.03 | 13.9342 | 1.9021 | 0.4180 | 0.0159 |
| 0.04 | 13.8148 | 2.0359 | 0.5526 | 0.0167 |
| 0.05 | 13.7252 | 2.1688 | 0.6863 | 0.0174 |
| 0.06 | 13.6614 | 2.2316 | 0.8197 | 0.0181 |
| 0.07 | 13.6206 | 2.4348 | 0.9534 | 0.0187 |
| 0.08 | 13.6006 | 2.7045 | 1.2240 | 0.0195 |

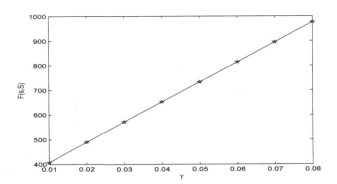

**Fig. 5.** The effect of the failure rate $\gamma$ on the mean cost function $F(s, S)$.

## 5    Conclusions

In this paper, we studied a production-inventory system with a service facility and a $(s, S)$ production policy where the production was interrupted for a vacation of random time. The steady-state joint distribution of the queue length and the on-hand inventory level of the system in product form was obtained by using the matrix technique. Some performance measures and the mean cost function were derived based on the performance measures. The effect of system parameters on some performance measures and the mean cost function were investigated using numerical examples. Numerical investigations illustrated the monotonic effect of some parameters. Although we obtain the explicit cost function given by Eq. (47), it is not immediate to prove structural property (e.g., convexity) of the function $F(s, S)$. So, it seems hard to perform an analytical sensitivity analysis for the cost function. However, some numerical optimization methods such as genetic algorithm may used to obtain the optimal policy $(s^*, S^*)$ to minimize the cost function. Future work could consider various generalizations to make the system more realistic. For example, one extension would be to consider more general distributions of the service times, the production times and the vacation times.

**Acknowledgments.** This work was supported by the Natural Science Foundation of Hebei Province, China (No. A2017203078).

# References

1. Sigman, K., Simchi-Levi, D.: Light traffc heuristic for an M/G/1 queue with limited inventory. Ann. Oper. Res. **40**(1), 371–380 (1992)
2. Berman, O., Kim, E.: Stochastic models for inventory management at service facilities. Stoch. Models **15**(4), 695–718 (1999)
3. Berman, O., Kim, E.: Dynamic inventory strategies for profit maximization in a service facility with stochastic service, demand and lead time. Math. Methods Oper. Res. **60**(3), 497–521 (2004)
4. Berman, O., Sapna, K.P.: Inventory management at service facilities for systems with arbitrary distributed service times. Stoch. Models **16**(3–4), 343–360 (2000)
5. Berman, O., Sapna, K.P.: Optical control of service for facilities holding inventory. Comput. Oper. Res. **28**, 429–441 (2001)
6. Berman, O., Sapna, K.P.: Optimal service rates of a service facility with perishable inventory items. Naval Res. Logistics **49**, 464–482 (2002)
7. Schwarz, M., Sauer, C., Daduna, H., Kulik, R., Szekli, R.: M/M/1 queueing systems with inventory. Queueing Syst. **54**(1), 55–78 (2006)
8. Krishnamoorthy, A., Deepak, T.G., Narayanan, V.C., Vineetha, K.: Control policies for inventory with service time. Stochastic Anal. Appl. **24**(4), 889–899 (2006)
9. Viswanath, C.N., Deepak, T.G., Krishnamoorthy, A., Krishkumar, B.: On $(s, S)$ inventory policy with service time, vacation to server and correlated lead time. Qual. Technol. Quanti. Manage. **5**(2), 129–144 (2008)
10. Krishnamoorthy, A., Lakshmy, B., Manikandan, R.: A survey on inventory models with positive service time. Opsearch **48**(2), 153–169 (2011)
11. Krishnamoorthy, A., Narayanan, V.C.: Production inventory with service time and vacation to the server. IMA J. Manage. Math. **22**, 33–45 (2011)
12. Krishnamoorthy, A., Narayanan, V.C.: Stochastic decomposition in production inventory with service time. Eur. J. Oper. Res. **228**, 358–366 (2013)

13. Baek, J.W., Moon, S.K.: The M/M/1 queue with a production-inventory system and lost sales. Appl. Math. Comput. **233**, 534–544 (2014)

14. Baek, J.W., Moon, S.K.: A production-inventory system with Markovian service queue and lost sales. J. Korean Stat. Soc. **45**(1), 14–24 (2016)

15. Yue, D., Qin, Y.: A production inventory system with service time and production vacations. J. Syst. Sci. Syst. Eng. **28**(2), 168–180 (2019)

16. Karaesmen, I., Scheller-Wolf, A., Deniz, B.: Managing perishable and aging inventories: review and future research directions. In: Kempf, K., Keskinocak, P., Uzsoy, P. (eds.) International Series in Operations Research and Management Science: Planning Production and Inventories in the Extended Enterprise, vol. 1, pp. 393–438. Springer, Heidelberg (2011). https://doi.org/10.1007/978-1-4419-6485-4_15

17. Bakker, M., Riezebos, J., Teunter, R.H.: Review of inventory systems with deterioration since 2001. Eur. J. Oper. Res. **221**(2), 275–284 (2012)

18. Nahmias, S.: Perishable inventory theory, a review. Oper. Res. **30**, 680–708 (1982)

19. Raafat, F.: Survey of literature on continuously deteriorating inventory models. J. Oper. Res. Soc. **42**(1), 27–37 (1991)

20. Goyal, S., Giri, B.: Recent trends in modeling of deteriorating inventory. Eur. J. Oper. Res. **134**, 1–16 (2001)

21. Sivakumar, B., Arivariganan, G.: A perishable inventory system at service facilities with negative customers. Int. J. Inf. Manage. Sci. **17**(2), 1–18 (2006)

22. Manuel, P., Sivakumar, B., Arivarignan, G.: A perishable inventory system with service facilities, MAP arrivals and PH-service times. J. Syst. Sci. Syst. Eng. **16**(1), 62–73 (2007)

23. Manuel, P., Sivakumar, B., Arivarignan, G.: A perishable inventory system with service facilities and retrial demands. Comput. Ind. Engineering **54**, 484–501 (2007)

24. Tian, N., Zhang, Z.G.: Vacation Queuing Models: Theory and Applications. Springer-Verlag, New York (2006). https://doi.org/10.1007/978-0-387-33723-4

25. Neuts, M.F.: Matrix-Geometric Solutions in Stochastic Models: an Algorithmic Approach. John Hopkins Press, Baltimore (1981)

# Deposit Design for a Production System with Impatient Customers

Na Li[1(✉)], Wei Wang[1], and Rui-Na Fan[2]

[1] Department of Industrial Engineering, Shanghai JiaoTong University,
Shanghai 200240, China
na-li03@sjtu.edu.cn
[2] School of Economics and Management, Yanshan University,
Qinhuangdao 066004, China

**Abstract.** Most researchers who study make-to-stock production control systems either assume that all orders are eventually met (complete backordering) or that no customers are willing to wait (lost sales). In an actual manufacturing system, however, customers may queue in line but renege after some time due to impatience. To alleviate the loss in sales caused by the reneging behavior of impatient customers, deposits from the customers may be required by a production system. In this paper, an axial turbine blade production system with impatient customers and a deposit policy is modeled as an assembly queue network. We derive a mathematical model of system performance based on Markov chain methods, and we discuss the optimal deposit amount from the point of view of net profit optimization.

**Keywords:** Impatient customer · Deposit · Risk · Optimization

## 1 Introduction

We consider a production system with impatient customers. Customers impatience may influence the demand of a production system. Most researchers who study production system analysis assume that customers either wait until their orders are met or leave when their orders are not fulfilled immediately. In an actual manufacturing system, however, customers may leave the system (called balking) when they find that the waiting list is too long. In other cases, customers may place orders knowing that the orders cannot be met immediately. Subsequently, some of the customers may become impatient and leave the system after deciding not to wait longer (called reneging).

In this paper, an axial turbine blade factory is studied. The turbine blade is one of the key parts of gas turbine. The studied axial turbine blade is high-technology product and very few local factories can provide it. Nowadays, a lot of new power stations are emerging and many old power stations are updating their gas turbines. They both need to order the turbine blade based on their requirements, which lead the order list of the studied axial turbine blade increase.

© Springer Nature Singapore Pte Ltd. 2019
Q.-L. Li et al. (Eds.): Cao Festschrift 2019, CCIS 1102, pp. 429–445, 2019.
https://doi.org/10.1007/978-981-15-0864-6_22

Some customers will balk and search for product from oversea market. Others may join the order list. While waiting for the product, some may renege by finding other alternative providers. Since the production process is very long and the specified resource for each order is prepared in advance, the renege of customers will damage the profit. Therefore, the deposit policy is proposed to reduce the renege loss of the system. Every customer who places an order is required to pay a deposit. To some extent, a deposit policy guarantees that customers, despite being impatient, will renege less often.

A key challenge confronting managers is determining the optimal deposit amount. To make the decision, we build an optimization model with the objective of maximizing net profit. To solve the optimization problem, we first investigate a mathematical model of the customers balking and reneging behavior with respect to deposit. Then, we evaluate the performance of the system with regard to the behavior model and propose a straightforward solution based on the performance measures. Finally, we examine, through numerical experiments, the optimal decisions with respect to different system characteristics and propose some insights for system management.

This paper has three main contributions to the literature. First, we provide a modeling method for representing the effect of paying a deposit on customer behavior. Second, we construct the system using an assembly queue line and assess its performance based on the proposed customer behavior model. Third, we offer insight into making decisions regarding the deposit value.

The remainder of this paper is organized as follows: In Sect. 2, we present literature relevant to the study. In Sect. 3, we describe the details of the proposed optimization problem and the methodology for solving the problem. In Sect. 4, we demonstrate, through numerical studies, the sensitivity of the optimal deposit decision with respect to customers deposit risk perception, customers impatience level and the system arrival rate. In Sect. 5, we state our conclusion and propose future research opportunities.

## 2    Literature Review

Many researchers have investigated customer impatience modeling (e.g., Kawanishi [8], Ou and Rao [14], Wang et al. [23]). Both balking and reneging have been considered. Various balking rules have been presented in the literature. Waiting time is the main factor that affects one's decision to either join or balk (e.g., Liu and Kulkarni [9,10]). However, customers always decide on the basis of the queue length given that the queue length can be easily obtained by customer. Some work-concerned customers balk if the number of customers before them is beyond a value (e.g., Christ and Avi-Itzhak [2], Economou and Kanta [4]). Several other papers have examined queues with balking probabilities depending on the number of waiting customers in the system upon arrival (e.g., Singh [17], Lozano and Moreno [12]). In modeling reneging behaviors, most studies assume that customers renege after a waiting time period if the queue time exceeds some tolerable value. The simplest case of the tolerable value is a fixed constant (see for example Haight

[6], Boots and Tijms [1] and Xiong and Altiok [20], and their references). In other studies, researchers assume that the maximal waiting time has an exponential distribution (see for example Rao [16], Whitt [25], Wang and Chang [22]). In addition, customers' impatience may result from a slow service rate (Perel and Yechiali [15]) or a disastrous breakdown (Yechiali [21]).

Alternatively, no research has been conducted yet on a deposit policy and how it affects customer behaviors. Nevertheless, we can draw lessons from the literature on price-dependent demand rate. Some studies consider that price will influence demand, such as Wee [24], Datta and Paul [3]), Mondal et al. [13], Teng and Chang [19]. Most of them assume that the demand rate is a decreasing function with respect to price.

System design and optimization problems regarding customer impatience have been investigated extensively. For example, Sumita, Masuda and Yamakawa [18] studied optimal pricing and capacity planning for a service facility with impatient customers and limited buffer; Lodree, Jang & Klein [11] proposed an optimal scheduling method for a two-stage supply chain system dependent on customer impatience; Jouini et al. [7] studied call center performance optimization problems by considering impatience and real-time anticipated delays; and Economopoulos, Kouikoglou & Grigoroudis [5] investigated a production control problem with impatient customers.

To the best of our knowledge, no studies have been conducted to determine an optimal deposit level for a system with impatient customers. We attempt to provide a modeling method for understanding how a deposit influences customer behavior and propose a method for making the deposit decision.

## 3   Optimization Model

The production system we discuss is a basic make-to-stock system with a single manufacturing facility that meets the demand for a single item. When the predetermined maximum stock level is reached the production stops. If the system is out of stock, arriving customers may balk or enter a queue. If they enter the queue they are required to pay a non-refundable deposit. Some impatience customers may renege while waiting in the queue. We assume that the production time has an exponential distribution with parameter $\lambda_p$. The customers are assumed to arrive at the system from the outside according to a Poisson process at rate $\lambda_d$. Each customer desires a unit of the product, and each must pay a sum of money $S$ as a deposit if he or she decides to place an order.

The customers are impatient and sensitive to the risk of losing their deposit. When a customer arrives at the system, their first choice is to either place an order or balk from the system. If the customers decide to place an order, they still have the option of waiting until the product arrives or reneging if the waiting time is too long.

By customer interview we found that imposing on the system will reduce the customers' reneging probability while simultaneously increasing their balking probability. The reason that they balk is that they fear to loss the deposit since they may have other better purchase choice while waiting. Consequently, it has a multifaceted impact on system performance.

We are interested in the design of $S$. The challenge is to find the optimal $S$ that will be most likely to help us achieve our objectives. For example, if our goal is to maximize the net profit of the system, the optimization problem can be expressed as follows:

$$Max \ obj = NP = R \times TH + S \times R_R - h \times E_p - b \times R_L - c \times E_o \quad (1)$$
$$s.t. \ [TH, R_R, R_L, E_o] = Performance\,(S)\,, 0 \le S \le LS$$

In the formulas, $NP$ is the unit time net profit. $TH$ is the throughput rate, and $R$ is the profit achieved from every order satisfied by the production factory. The profit from the satisfied orders per time unit is $TH \times R$. $R_R$ is the renege rate of the customers, and the average obtained deposit from default orders is $S \times R_R$. If $h$ is the average holding cost per unit time per product, and $E_p$ is the average number of products, then the holding cost is $h \times E_p$. If $b$ is the cost for per lost order each time unit, and $R_L$ is the customer loss rate, then the cost caused by lost order is $b \times R_L$. If $c$ is the cost per unit time when a customer is waiting to be accommodated, and $E_o$ is the average number of waiting customers, then the customer waiting cost is $c \times E_o$. Furthermore, $Performance(S)$ is a function to derive $TH, R_R, E_p, R_L$ and $E_o$ for a given system with specified $S$.

Formulation (1) is a complex nonlinear optimization problem. To solve this problem, we must appreciate the impact of customer behavior on the system and understand the system's performance. In Subsect. 3.1, we will discuss how the customer numbers and deposit value affect the system. In Subsect. 3.2, we develop a method based on a queuing model to derive $Performance(S)$. A simple exhaustive search algorithm is proposed in Subsect. 3.3.

## 3.1   Customer Behavior Modeling

The mathematical model of the customers' balking and reneging behavior with respect to $S$ is essential for the development of the performance of the system. In this section, we study a method for the mathematical modeling of customer behavior.

Both the waiting time and the deposit amount will affect the balking behavior of a customer. Let "$Pr_BCD$" be the probability of balking caused by the deposit requirement. We interviewed 24 customers and got some general ideas. Firstly, the reason that they balk is that they fear to loss the deposit since they may have other better purchase choice. Therefore, the risk attitude is one factor that will influence their balk probability. Secondly, the balking preference is related with the expected product delivery waiting time. If the waiting time is short, they would like to pay more deposit, whereas, they will consider the risk that they may switch to other supplier and loss the deposit. The effect is similar as that the deposit has a perception value related with the waiting time. For simplification, we define it as $S\,(1 + \delta)^n$, $\delta$ in which is a parameter reflecting the customers' attitude to the influence of waiting time. Besides, we found that most of the customers will consider the value of the product, if the product is more valuable, they would like to pay more deposit. The proportion of the perception deposit value to the value of the product is one of the critical factors.

Therefore, we assume that "$Pr_BCD$" is the ratio of the perceived value of $S$ to the largest acceptable deposit $LS$, and we define it as $f(S, n)$

$$Pr\_BCD = f(S, n) = \begin{cases} 0 & n = 0 \\ (S(1+\delta)^n /LS)^{(1-\alpha)} & S(1+\delta)^n < LS \\ 1 & S(1+\delta)^n \geq LS \end{cases} \qquad (2)$$

where $S(1+\delta)^n$ indicates the customer's perceived value of the deposit relative to $n$. $LS$ is the value of the order. $\alpha$ is a parameter which represents the risk attitude to the deposit for a customer and $0 < \alpha < 1$. Figure 1 displays how $\alpha$ and $\delta$ influence the $Pr\_BCD$. The curves confirm that higher $\alpha$ and bigger $\delta$ imply more preference for balking. Note that this formulation is only a rough and specific modeling method which derived from the case we studied. By choosing suitable parameters $\alpha$ and $\delta$, it can model the deposit caused balking in the specified case. However, more general studies should be done in future to justify its suitable range.

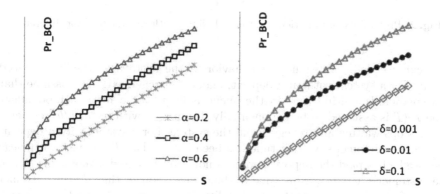

**Fig. 1.** Probability curves for deposit risk balking with respect to $\alpha$ and $\delta$

Considering the impatience-caused balking, a customer who arrives at the system will balk due to the long waiting time. Lozano and Moreno [8] pointed that the probability a customer balks due to impatience. $Pr\_ICB$, can be expressed as a nonlinear increasing function of the number of the customers waiting in the system, $n$, as

$$Pr\_ICB = 1 - r^n, \quad 0 < r < 1 \qquad (3)$$

in which $r$ represents the level of the customer's patience. Figure 2 indicates that a higher $r$ means more impatient customer type and a higher $n$ implies higher probability of impatient customer balking for a given type of customer.

Therefore, the rate that customers enter the system for a given waiting number of orders $n$ and deposit $S$ will be

$$\lambda_d^{'}(n) = \lambda_d \left(1 - \mathrm{Pr}\_BCD\right)\left(1 - \mathrm{Pr}\_ICB\right) \tag{4}$$

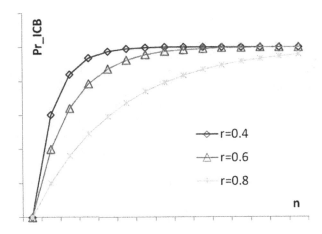

**Fig. 2.** Probability of impatient customer balking with respect to $n$ for different $r$

Next, we define the reneging behavior of the customers who have placed orders. For a system without a deposit, many researchers ([12–14]) assume that if a customer's waiting time in the queue is longer than his upper boundary time $T$ ($T$ is assumed to be exponentially distributed with mean $1/\theta$), the customer stops waiting and reneges from the system. For a system with a deposit, even if the customer's waiting time has been longer than $T$, he may not renege because he has paid the deposit and does not want to lose it. We assume that if the customer's actual waiting time in the queue is longer than $T$, the customer reneges from the system with a probability function $g(S)$. Given that the customer would not like to lose a larger deposit, the probability function is a convex decreasing function (Fig. 3), which we define as

$$g(S) = \beta^S, \tag{5}$$

where $0 < \beta < 1$ is a parameter that indicates the perspective of the customer to lose the deposit he has paid. A larger $\beta$ signifies a customer who is more willing to give up the deposit. Assuming that customers are independent, the average reneging rate $\mu(n)$ for customers waiting in the system for a given $n$ is

$$\mu(n) = n\beta^S\theta \tag{6}$$

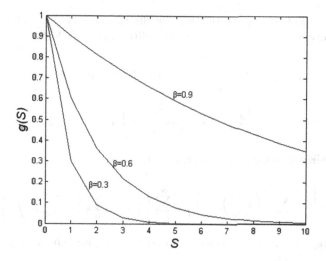

**Fig. 3.** Probability function $g(S)$ with respect to $S$ for different $\beta$

## 3.2 Performance Analysis of the Production System

We model the system as a virtual assembly production line (Fig. 4), in which the product and the order are seen as two parts to be assembled by a virtual machine. $B_o$ is the buffer of the orders. Once an order and a product are simultaneously available, the virtual assembly machine matches them immediately.

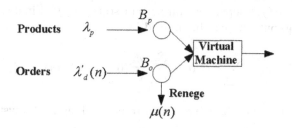

**Fig. 4.** The virtual assembly line model

Let $N_p(t)$ be the number of products in buffer $B_p$ and $N_o(t)$ be the number of orders in buffer $B_o$ at time $t$. The products in $B_p$ and the orders in $B_o$ will not exist at the same time. Then, $X = \{N_p(t), N_o(t), t \geq 0\}$ forms a Markov chain with state space $\Theta = \{0, N_o\} \cup \{N_p, 0\}$, where $N_o = \{0, 1, 2, 3, ...\}$, $N_p = \{0, 1, 2, 3, ..., C\}$ in which the buffer capacity limitation is $C$. The transition rates are

$$q^* = \begin{cases} \lambda_d & (i,0) \to (i-1,0) \\ \lambda_d'(j) & (0,j) \to (0,j+1) \\ \lambda_p & (i,0) \to (i+1,0) \\ \lambda_p + \mu(j) & (0,j) \to (0,j-1) \end{cases} \tag{7}$$

We derive explicit expressions for the distribution of the system using the balance equations; the steady-state probabilities are as follows:

$$\pi(i,0) = \left(\frac{\lambda_d}{\lambda_p}\right)^{C-i} \pi(C,0), \text{ for } i = 0, ..., C-1, \tag{8}$$

$$\pi(0,j) = \left(\frac{\lambda_d}{\lambda_p}\right)^C \prod_{i=1}^{j} \left(\frac{\lambda_d'(i-1)}{\lambda_p + \mu(i)}\right) \pi(C,0), \text{ for } j \geq 1 \tag{9}$$

Using normalizing conditions $\sum_\Theta \pi(i,j) = 1$, we obtain

$$\pi(C,0) = \left[1 + \sum_{i=0}^{C-1}\left(\left(\frac{\lambda_d}{\lambda_p}\right)^{C-i}\right) + \sum_{j=1}^{\infty}\left(\left(\frac{\lambda_d}{\lambda_p}\right)^C \prod_{i=1}^{j}\left(\frac{\lambda_d'(i-1)}{\lambda_p+\mu(i)}\right)\right)\right]^{-1} \tag{10}$$

The average number of products in inventory $E_p$ and the average number of waiting orders $E_o$ are given by

$$E_p = E[N_p] = \sum_{i=1}^{C} i \times \pi(i,0) \tag{11}$$

$$E_o = E[N_o] = \sum_{j=1}^{\infty} i \times \pi(0,j) \tag{12}$$

If product inventory is not zero, the newly arrived order will be satisfied at the rate of $\lambda_d$. If the order inventory is not zero, the system satisfies its order at the rate of $\lambda_d'(j) - \mu(j)$ (dependent on system state $(0,j)$). Finally, the assembly system's throughput rate can be derived by

$$TH = \lambda_d \times \sum_{i=0}^{C} \pi(i,0) + \sum_{j=1}^{\infty}\left(\pi(0,j) \times \left(\lambda_d'(j) - \mu(j)\right)\right) \tag{13}$$

Defining Pr_B as the probability of a customer balking in the system, the probability will be

$$Pr\_B = 1 - (1 - Pr\_BCD)(1 - Pr\_ICB). \tag{14}$$

Then, the average balking rate is given by

$$R_B = E[BalkingRate] = \sum_{j=0}^{\infty}(\pi(0,j)\,Pr\_B(S,j))\lambda_d \tag{15}$$

The average reneging rate is

$$R_R = E[RenegingRate] = \sum_{j=0}^{\infty}(\pi(0,j)\mu(j)). \tag{16}$$

The average rate of customer loss is the sum of the average balking rate and the average reneging rate, which yields

$$R_L = R_B + R_R. \tag{17}$$

The average utilization of the system is

$$U = 1 - \sum_{i=0}^{C} \pi(i, 0). \tag{18}$$

### 3.3  Solution Procedure

Because the feasible deposit space is finite in reality, we propose the following exhaustive search algorithm for obtaining the optimal policy by iteratively evaluating and comparing each policy:

**Step (1)**: Set $obj^*=0$; initialize the deposit S and searching step $m$;

**Step (2)**: Compute $Performance(S)$;

**Step (3)**: Derive $obj(S)$. If $obj(S) > obj^*$, then set $obj^* = obj(S)$ and $S^* = S$;

**Step (4)**: If $S + m \leq LS$, then set $S = S + m$ and go to step 2; Otherwise, stop.

## 4  Numerical Experiments

An axial turbine blade product case is studied. The system allows at most 15 waiting orders ($C = 15$). The arrival rate from the marketing is $\lambda_d = 1$. The production rate is $\lambda_p = 0.8$. The value of the product is normalized as $LS = 20$. The customers' behavior is investigated by interviews and data analysis. 24 customers are interviewed; They are required to answer the questions such as what will make them balk, how many orders will make them refuse to join into the order and balk, how the deposit influences their balking decision and how the deposit affects their decision of renege. The information are applied to construct the customer behavior models proposed in Sect. 3.1 and the customer was consulted frequently to estimate values of parameters used in the model. Finally, the parameters are set as $\delta = 0.0001, \alpha = 0.1, \beta = 0.1, r = 0.97, \theta = 1/50$. The economic parameters are $R = 20, b = 0.25, h = 0.5$ and $c = 0.01$. We change the deposit value and obtain the system performances by the performance analysis method. Table 1 shows some of the results.

In addition to the basic performance measures, we also compute the $DCB$, $DAR$ and $GAP$. $DCB$ is the increased balking rate for a system with a deposit $S$ compared with a no-deposit system. $DAR$ is the reduced reneging rate for a system with a deposit $S$ compared with a no-deposit system. $GAP$ is the difference between the two. These variables can be expressed by the following formulas:

$$DAR = R_R(0) - R_R(S), \tag{19}$$

$$DCB = R_R(S) - R_B(0), \qquad (20)$$

$$GAP = DAR - DCB. \qquad (21)$$

In the system, $DAR$ can be seen as the decrease in the system loss rate resulting from imposing a deposit, and $DCB$ is the increase in $R_L$ brought about by the deposit. $GAP$, therefore, indicates whether the positive effect outweighs the negative effect and by how much. We illustrate in Fig. 5 how $GAP$ changes along with $S$. As indicated, $GAP$ increases first and then decreases with respect to $S$. The maximum value of $GAP$ exists at $S = 0.9$, which means that imposing a deposit of 0.9 offers the most benefit to $R_L$. Similarly, the minimum $R_L$ is located when $S = 0.9$ (Table 1). We also find that $GAP$ is positive only when $S$ is less than 2.3. This result suggests that when imposing a deposit of more than 2.3, the positive impact of decreased reneging behavior is outweighed by the negative consequence of increased balking.

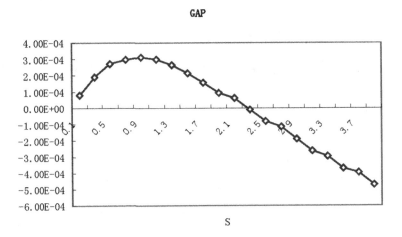

**Fig. 5.** The gap between $DAR$ and $DCB$

When considering the optimal $S^*$ with regard to the net profit, we find from Table 1 that the optimal deposit is 0.7, not 0.9. This is because that imposing a deposit affects not only the performance of $R_L$ but also other factors, such as the inventory level and the number of customers waiting.

In the next few subsections, we discuss how the variables of the problem impact the optimal decision. In each experiment, unless otherwise stated, we modify one parameter, fixing the others to the standard values and computing the optimal deposit $S^*$ and the corresponding $NP^*$. We also compute the gap between $NP^*$ and $NP(0)$, which indicates the difference in the benefit when comparing systems with and without a deposit requirement:

$$Benefit = NP^* - NP(0). \qquad (22)$$

## 4.1   Effect of Customer Arrival Rate

In this subsection, we investigate how the customer arrival rate influences the optimal decision. Because there is complex interplay between arrival rate and performance, this result is difficult to predict. However, we do find some insights, as shown by the numerical results in Table 2.

**Table 1.** Performance results for system with the standard parameter values

| S | TH | $E_p$ | $E_o$ | $R_B$ | $R_R$ | $R_L$ | U | DCB | DAR | GAP | NP |
|---|---|---|---|---|---|---|---|---|---|---|---|
| 0.1 | 0.79693 | 0.85227 | 4.5859 | 0.12847 | 0.072855 | 0.20132 | 0.75901 | 0.012415 | 0.012494 | 7.97E−05 | 15.424 |
| 0.3 | 0.79721 | 7.69E−01 | 5.1367 | 0.14972 | 0.051489 | 0.20121 | 0.78247 | 0.033669 | 0.03386 | 0.00019056 | 15.473 |
| 0.5 | 0.79739 | 7.22E−01 | 5.5215 | 0.16621 | 0.034921 | 0.20113 | 0.79592 | 0.050156 | 0.050428 | 0.00027175 | 15.499 |
| 0.7 | 0.79746 | 7.01E−01 | 5.7392 | 0.17821 | 0.022902 | 0.20111 | 0.80173 | 0.06215 | 0.062446 | 0.00029661 | 15.507 |
| 0.9 | 0.79747 | 7.00E−01 | 5.8143 | 0.18645 | 0.01464 | 0.20109 | 0.80204 | 0.070397 | 0.070709 | 0.00031214 | 15.504 |
| 1.1 | 0.79744 | 7.12E−01 | 5.7841 | 0.19192 | 9.19E−03 | 0.20111 | 0.79854 | 0.075861 | 0.07616 | 0.0002985 | 15.494 |
| 1.3 | 0.79736 | 7.34E−01 | 5.6827 | 0.19544 | 5.70E−03 | 0.20114 | 0.79248 | 0.079389 | 0.079653 | 0.0002633 | 15.48 |
| 1.5 | 0.79725 | 7.61E−01 | 5.5368 | 0.19769 | 3.50E−03 | 0.20119 | 0.78472 | 0.081633 | 0.081847 | 0.00021379 | 15.464 |
| 1.7 | 0.79713 | 7.93E−01 | 5.3654 | 0.19911 | 2.14E−03 | 0.20125 | 0.77585 | 0.083052 | 0.083208 | 0.00015568 | 15.446 |
| 1.9 | 0.797 | 8.27E−01 | 5.1811 | 0.20001 | 1.30E−03 | 0.20131 | 0.76627 | 0.083952 | 0.084044 | 9.28E−05 | 15.427 |
| 2.1 | 0.79689 | 8.62E−01 | 4.99 | 0.20055 | 7.93E−04 | 0.20134 | 0.75621 | 0.084496 | 0.084556 | 6.06E−05 | 15.408 |
| 2.3 | 0.79675 | 8.99E−01 | 4.801 | 0.20093 | 4.81E−04 | 0.20142 | 0.74591 | 0.084878 | 0.084868 | −1.08E−05 | 15.388 |
| 2.5 | 0.7966 | 9.36E−01 | 4.6148 | 0.20119 | 2.92E−04 | 0.20149 | 0.73545 | 0.085139 | 0.085057 | −8.15E−05 | 15.368 |
| 2.7 | 0.79649 | 9.73E−01 | 4.4313 | 0.20134 | 1.77E−04 | 0.20152 | 0.72488 | 0.085288 | 0.085172 | −0.0001161 | 15.349 |
| 2.9 | 0.79634 | 1.01E+00 | 4.2556 | 0.20149 | 1.07E−04 | 0.20159 | 0.71433 | 0.085432 | 0.085242 | −0.0001905 | 15.329 |
| 3.1 | 0.79619 | 1.05E+00 | 4.0858 | 0.2016 | 6.49E−05 | 0.20167 | 0.70378 | 0.085547 | 0.085284 | −0.0002626 | 15.309 |
| 3.3 | 0.79608 | 1.08E+00 | 3.9204 | 0.20166 | 3.93E−05 | 0.2017 | 0.69325 | 0.085605 | 0.08531 | −0.0002958 | 15.29 |
| 3.5 | 0.79593 | 1.12E+00 | 3.7635 | 0.20175 | 2.38E−05 | 0.20177 | 0.68282 | 0.085695 | 0.085325 | −0.0003701 | 15.27 |
| 3.7 | 0.79583 | 1.16E+00 | 3.6105 | 0.20178 | 1.44E−05 | 0.2018 | 0.67243 | 0.08573 | 0.085334 | −3.95E−04 | 15.251 |
| 3.9 | 0.79568 | 1.19E+00 | 3.4662 | 0.20187 | 8.73E−06 | 0.20188 | 0.66219 | 0.085812 | 0.08534 | −4.72E−04 | 15.231 |

To discuss our findings, we illustrate two figures–Figs. 6 and 7–to explicitly show how $NP^*$ and $S^*$, respectively, change in relation to $\lambda_d$.

Figure 6 illustrates that $NP^*$ first increases quickly and then decreases slowly. It is clear that an increase in production rate will enhance the throughput, resulting in higher net profit. This outcome demonstrates that revenue is the dominant factor contributing to net profit at the beginning of the process. However, when the production rate is increased to a level that the system cannot afford (more than its service capacity), loss due to congestion ($R_L$ and $E_o$) will increase and

**Table 2.** Optimal decisions for systems with different $\lambda_d$

| $\lambda_d$ | 0.6 | 0.8 | 1 | 1.2 | 1.4 | 1.6 | 1.8 | 2 |
|---|---|---|---|---|---|---|---|---|
| $S^*$ | 0.3 | 0.7 | 0.7 | 0.4 | 0.3 | 0.3 | 0.3 | 0.3 |
| $NP^*$ | 2.9094 | 11.754 | 15.507 | 15.798 | 15.755 | 15.685 | 15.615 | 15.547 |
| $Benefit$ | 0.00021 | 0.14049 | 0.11899 | 0.0336 | 0.02005 | 0.02118 | 0.02463 | 0.0283 |

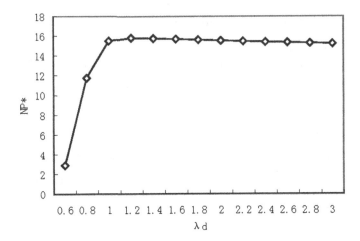

**Fig. 6.** The optimal net profits for systems with different customer arrival rates

dominate the trend of net profit. Therefore, net profit begins to decrease. Because $R$ is greater than both $b$ and $c$, the increase rate of $NP^*$ is much higher than its decrease rate. As a result, we recommend that customer development policies should be linked with system capacity. If measures such as extensive promotion bring in more customers than the system can afford, the system's net profit may suffer. We separate the total benefit of imposing an optimal deposit from the detailed compositions of net profit. These results are shown in Fig. 7. As evidenced, we find that the later increase of net profit occurs due to unreturned deposits; however, it is not our intention to impose this type of deposit strategy. In the long run, a system of achieving net profit through unreturned deposits will hurt the reputation of the firm. Therefore, we consider only that the benefit of the deposit policy increases first and then decreases with the arrival rate. This conclusion signifies that a system with a small arrival rate is not crowded and therefore has little need for the deposit to reduce the reneging effect. While a system with an extremely large arrival rate (such as $\lambda_d = 1.4$, in this case), is too overcrowded to be helped by the deposit. In this instance, systems with arrival rates $\lambda_d = 0.8$ and $\lambda_d = 1$ have significantly more benefit than systems with other arrival rates. This conclusion signifies that a deposit strategy is more suitable for a system with an arrival rate near the production rate or slightly higher.

### 4.2　Effect of Deposit Risk Perception

The risk perception of customers is affected by parameters $\alpha$, $\delta$, $LS$ and $\beta$; we will study the influence of them in turn.

A higher value of $\alpha$ indicates a higher level of customer risk sensitivity, signifying that this customer is less willing to pay a deposit. Based on the standard parameter setting, we change $\alpha$ from 0.1 to 0.6; Table 3 shows the results. We

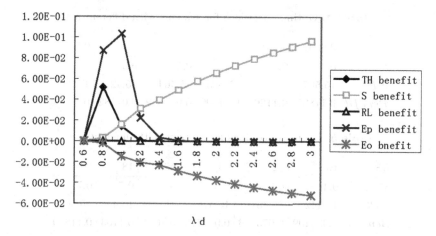

**Fig. 7.** The benefits from each component for different customer arrival rates

find, as $\alpha$ increases, that the optimal deposit is decreased, and $NP^*$ and $Benefit$ are likewise decreased. This result suggests that for customers who are more risk sensitivity, we obtain less benefit from imposing a deposit. In this case, when $\alpha = 0.6$, requiring a deposit derives no benefit, and we should not apply the deposit strategy.

**Table 3.** Optimal decisions for systems with different $\alpha$

| $\alpha$ | 0.1 | 0.2 | 0.3 | 0.4 | 0.5 | 0.6 |
|---|---|---|---|---|---|---|
| $S^*$ | 0.7 | 0.7 | 0.7 | 0.7 | 0.6 | 0 |
| $NP^*$ | 15.507 | 15.494 | 15.477 | 15.454 | 15.421 | 15.388 |
| $Benefit$ | 0.11899 | 0.10599 | 0.089474 | 0.065949 | 0.033107 | 0 |

$\delta$ reflects the customer's perception value to $S$. A larger $\delta$ implies that the customer is more concerned about the time value of the deposit. For a fixed waiting time, a larger $\delta$ indicates that the customer values the deposit more and is therefore less willing to pay it. Table 4 shows the decision results at different $\delta$ settings. The impact of $\delta$ on system decisions is similar to that of in that a larger $\delta$ results in less benefit from imposing a deposit.

Table 5 demonstrates how the decision changes with $LS$. We conclude from the results that for a more valuable order, the deposit can be larger and the system benefit will be greater.

The impact that a deposit policy has on reneging behavior is affected primarily by $\beta$. A larger $\beta$ indicates that customers are more likely to forego the deposit and renege from the system. As shown in Table 6, we conclude that imposing a deposit is more suitable for a system with a small because t $\beta$he benefit decreases

**Table 4.** Optimal decisions for systems with different $\delta$

| $\delta$ | 0.0001 | 0.001 | 0.01 | 0.1 | 0.2 | 0.5 |
|---|---|---|---|---|---|---|
| $S^*$ | 0.7 | 0.7 | 0.7 | 0.5 | 0.3 | 0 |
| $NP^*$ | 15.507 | 15.507 | 15.503 | 15.464 | 15.42 | 15.388 |
| $Benefit$ | 0.11899 | 0.11865 | 0.11509 | 0.075849 | 0.03182 | 0 |

**Table 5.** Optimal decisions for systems with different $LS$

| $LS$ | 10 | 12 | 14 | 16 | 18 | 20 |
|---|---|---|---|---|---|---|
| $S^*$ | 0.4 | 0.5 | 0.6 | 0.6 | 0.7 | 0.7 |
| $NP^*$ | 15.442 | 15.461 | 15.476 | 15.488 | 15.498 | 15.507 |
| $Benefit$ | 0.054409 | 0.073263 | 0.087773 | 0.1001 | 0.1103 | 0.11899 |

with respect to $\beta$. Interestingly, we find that the optimal deposit increases first and then decreases with respect to $\beta$. This outcome is because the deposit is not returned and therefore is not included in the net profit function. Table 7 displays the results when we delete the unreturned deposit value from the $NP$ function. As shown in the table, if we do not consider the unreturned deposit, we should impose a larger deposit for customers who are less concerned about losing the deposit. When the customers are more concerned about losing the deposit, at a level of $\beta = 0.55$ or higher, the system is not suitable for imposing a deposit.

**Table 6.** Optimal decisions for systems with different $\beta$

| $\beta$ | 0.1 | 0.25 | 0.4 | 0.55 | 0.7 | 0.85 |
|---|---|---|---|---|---|---|
| $S^*$ | 0.7 | 0.8 | 1 | 1 | 0.9 | 0.4 |
| $NP^*$ | 15.507 | 15.477 | 15.452 | 15.429 | 15.406 | 15.388 |
| $Benefit$ | 0.11899 | 0.088886 | 0.064257 | 0.040655 | 0.018451 | 9.11E-05 |

### 4.3    Effect of Impatience

In the model, $r$ indicates the impatience level when customers arrive, and $\theta$ denotes their impatience level while waiting in queue. A larger $r$ suggests that customers are more often balk. We conclude from Table 8 that a deposit strategy is more suitable for a system with customers who are less willing to balk because the benefit increases with $r$. However, the optimal deposit decreases first and then increases. This result signifies that, for a system with a very small $r$, one can expect less deposit-caused balking and apply a larger deposit because the customers are more impatient, yet balking little. Therefore, the increase in deposit-caused balking is also small. For a system with a very high $r$, one can

**Table 7.** Optimal decisions for systems with different $\beta$ (the unreturned S is not included in $NP$)

| $\beta$ | 0.1 | 0.25 | 0.4 | 0.55 | 0.7 | 0.85 |
|---|---|---|---|---|---|---|
| $S^*$ | 0.8 | 0.9 | 0.9 | 0 | 0 | 0 |
| $NP^*$ | 15.492 | 15.45 | 15.415 | 15.388 | 15.388 | 15.388 |
| $Benefit$ | 0.10406 | 0.061803 | 0.026632 | 0 | 0 | 0 |

**Table 8.** Optimal decisions for systems with different $r$

| $r$ | 0.2 | 0.35 | 0.5 | 0.65 | 0.8 | 0.95 |
|---|---|---|---|---|---|---|
| $S^*$ | 0.8 | 0.7 | 0.6 | 0.6 | 0.5 | 0.7 |
| $NP^*$ | 14.409 | 14.474 | 14.556 | 14.67 | 14.86 | 15.351 |
| $Benefit$ | 0.008024 | 0.008944 | 0.010713 | 0.01421 | 0.023963 | 0.0852 |

also expect less deposit-caused balking because the customers are balking a great deal already, and any increase in balking makes little difference.

Regarding $\theta$, a larger $\theta$ denotes customers who are more impatient and also more willing to renege. From Table 9, we can conclude that we should impose a larger deposit for systems with a larger $\theta$. Though the system $NP^*$ decreases with $\theta$, the benefit from imposing a deposit increases with $\theta$, suggesting that imposing a deposit is more suitable for customers with higher impatience levels while waiting in queue. The decrease in net profit is caused by elevated reneging behavior when $\theta$ increases.

**Table 9.** Optimal decisions for system with different $\theta$

| $\theta$ | 0.014 | 0.017 | 0.02 | 0.025 | 0.033 | 0.05 |
|---|---|---|---|---|---|---|
| $S^*$ | 0.6 | 0.7 | 0.7 | 0.8 | 0.9 | 1.1 |
| $NP^*$ | 15.519 | 15.513 | 15.507 | 15.499 | 15.49 | 15.476 |
| $Benefit$ | 0.074061 | 0.092875 | 0.11899 | 0.15742 | 0.21758 | 0.31664 |

## 5 Conclusions and Future Work

In this paper, we discuss the deposit design problem in a system with impatient customers. We propose models for behavior when customers are affected by different levels of deposit and impatience. To this end, we determine how to construct a mathematical analysis of this type of problem. On the basis of behavior models, production system performance measures are obtained by analyzing a virtual assembly line. Optimal deposit imposition in different situations is discussed via numerical experiments.

Our suggestions for future studies are as follows: (1) design and validate additional accurate customer behavior models concerning deposit and impatience and (2) explore systems with other production features, such as batch arrival customers or order treatment service time.

**Acknowledgments.** The work described in this paper was supported by a Research Grant from the National Natural Science Foundation of China (70932004, 71002037, 71090404, 71090400), the Specialized Research Fund for the Doctoral Program of Higher Education (20090073110035, 20100073120080), and the Research Fund for the Key Scientific and Innovation Project of Shanghai Municipal Education Commission (09ZZ19).

# References

1. Boots, N., Tijms, H.: An M/M/c queue with impatient customers. Top **7**, 213–220 (1999). https://doi.org/10.1007/BF02564722
2. Christ, D., Avi-Itzhak, B.: Strategic equilibrium for a pair of competing servers with convex cost and balking. Manag. Sci. **48**, 813–820 (2002)
3. Datta, T.K., Paul, K.: An inventory system with stock-dependent, price-sensitive demand rate. Prod. Planning Control **12**(1), 13–20 (2001)
4. Economou, A., Kanta, S.: Equilibrium balking strategies in the observable single-server queue with breakdowns and repairs. Oper. Res. Lett. **36**, 696–699 (2008)
5. Economopoulos, A.A., Kouikoglou, V.S., Grigoroudis, E.: The base stock/base backlog control policy for a make-to-stock system with impatient customers. IEEE Trans. Autom. Sci. Eng. **8**(1), 243–249 (2010)
6. Haight, F.: Queueing with reneging. Metrika **2**, 186–197 (1959). https://doi.org/10.1007/BF02613734
7. Jouini, O., Dallery, Y., Akşin, Z.: Queueing models for full-flexible multi-class call centers with real-time anticipated delays. Int. J. Prod. Econ. **120**(2), 389–399 (2009)
8. Kawanishi, K.: QBD approximations of a call center queueing model with general patience distribution. Comput. Oper. Res. **35**(8), 2463–2481 (2008)
9. Liu, L., Kulkarni, V.: Explicit solutions for the steady state distributions in M/PH/1 queues with workload dependent balking. Queueing Syst. **52**, 251–260 (2006)
10. Liu, L., Kulkarni, V.: Busy period analysis for M/PH/1 queues with workload dependent balking. Queueing Syst. **59**, 37–51 (2008). https://doi.org/10.1007/s11134-008-9074-9
11. Lodree, E., Jang, W., Klein, C.M.: Minimizing response time in a two-stage supply chain system with variable lead time and stochastic demand. Int. J. Prod. Res. **42**, 2263–2278 (2004)
12. Lozano, M., Moreno, P.: A discrete time single-server queue with balking: economic applications. Appl. Econ. **40**, 735–748 (2008)
13. Mondal, B., Bhunia, A.K., Maiti, M.: An inventory system of ameliorating items for price dependent demand rate. Comput. Industr. Eng. **45**(3), 443–456 (2003)
14. Ou, J., Rao, B.M.: Benefits of providing amenities to impatient waiting customers. Comput. Oper. Res. **30**, 2211–2225 (2003)
15. Perel, N., Yechiali, U.: Queues with slow servers and impatient customers. Eur. J. Oper. Res. **201**, 247–258 (2010)

16. Rao, S.S.: Queuing models with balking, reneging, and interruptions. Oper. Res. **13**, 596 (1965)
17. Singh, V.P.: Two-server Markovian queues with balking: heterogeneous vs. homogeneous servers. Oper. Res. **18**, 145–159 (1970)
18. Sumita, U., Masuda, Y., Yamakawa, S.: Optimal internal pricing and capacity planning for service facility with finite buffer. Eur. J. Oper. Res. **128**(1), 192–205 (2001)
19. Teng, J.T., Chang, C.T.: Economic production quantity models for deteriorating items with price- and stock-dependent demand. Comput. Oper. Res. **32**(2), 297–308 (2005)
20. Xiong, W., Altiok, T.: An approximation for multi-server queues with deterministic reneging times. Ann. Oper. Res. **172**, 143–151 (2009). https://doi.org/10.1007/s10479-009-0534-3
21. Yechiali, U.: Queues with system disasters and impatient customers when system is down. Queueing Syst. **56**, 195–202 (2007). https://doi.org/10.1007/s11134-007-9031-z
22. Wang, K.H., Chang, Y.C.: Cost analysis of a finite M/M/R queueing system with balking, reneging, and server breakdowns. Math. Methods Oper. Res. **56**, 169 (2002). https://doi.org/10.1007/s001860200206
23. Wang, K., Li, N., Jiang, Z.: Queueing system with impatient customers: a review. In: Proceedings of 2010 IEEE International Conference on Service Operations and Logistics, and Informatics, SOLI, pp. 82–87 (2010)
24. Wee, H.M.: A replenishment policy for items with a price-dependent demand and a varying rate of deterioration. Prod. Planning Control **8**(5), 494–499 (1997)
25. Whitt, W.: Improving service by informing customers about anticipated delays. Manag. Sci. **45**, 192–207 (1999)

# A Route Option for Different Commodity Groups in International Trade: China Pakistan Economic Corridor

Usman Akbar[1], Rui-Na Fan[1(✉)], and Quan-Lin Li[2]

[1] School of Economics and Management, Yanshan University, Qinhuangdao, China
{usman.akbar,fanruina}@stumail.ysu.edu.cn
[2] School of Economics and Management,
Beijing University of Technology, Beijing, China
liquanlin@tsinghua.edu.cn

**Abstract.** The China Pakistan Economic Corridor (CPEC), an extension of the silk road, is highly expected to bring the regional stability among south Asian countries by acting as a link to the silk road economic belt and the 21st-century maritime silk road. Hence, the increasingly large scale international trade makes the role of CPEC road shipments significant between supply and demand. By considering various risks involved in the CPEC's road transportation routes, a trade-off function based on the factors: Cost $C_s$, profit $B_s$, and safety $P_s$ is set up. We establish a probability density function taking cost, profit and safety as the continuous random variables, and use their weight to capture a better trade-off policy. Thus such a trade-off function is to maximize the total satisfaction of road trading by analyzing the best appropriateness between the commodity group and the CPEC route. The methodology and results given in this paper provide an effective method for several reliable route paths among commodity groups.

**Keywords:** International trade · The China Pakistan Economic Corridor · Trade-off function · Commodity group

## 1 Introduction

The China Pakistan Economic Corridor (CPEC) is an umbrella project of Belt and Road Initiative (BRI). It connects Kashgar and the western region of China with the deep sea port of Gwadar, Pakistan. By viewing from the perspective of China, Pakistan serves as an assurance policy to her energy risks, as it provides diversity, security, and enhancement of energy supplies. China, being the largest consumer of oil in the world, has a long term plan of oil from Gwadar, which is likely to save a distance of over 7580 miles. Not to mention 80% of China's oil, that is currently imported bypassing hostile choke-points through the Straits of Malacca, is likely to increase [11]. The completion of the China Pakistan Economic Corridor (CPEC) will reduce the distance by 50–85% for China's present

© Springer Nature Singapore Pte Ltd. 2019
Q.-L. Li et al. (Eds.): Cao Festschrift 2019, CCIS 1102, pp. 446–464, 2019.
https://doi.org/10.1007/978-981-15-0864-6_23

container traffic. The oil tankers that take 20 to 30 days to reach Shanghai from Gulf are likely to take less than a week time through Gwadar (see Table 1). Strategically, Pakistan lies at the fulcrum of Middle East, South Asia, and Central Asia, which provides the most economical link between Central Asia or Persian Gulf and the energy scant broader South Asia [3]. It exhibits that the trading activity through Pakistan is going to increase.

**Table 1.** Trading distance comparison (via Shanghai vs via Gwadar)

| Sr. | Origin | Destination | Via Shanghai (miles) | Via Pakistan (miles) | Saved (miles) | Saved (%) |
|---|---|---|---|---|---|---|
| 1 | Central China | Middle East | 11206 | 3626 | 7580 | 68 |
| 2 | Central China | Europe | 17801 | 10928 | 6873 | 39 |
| 3 | Central China | Pakistan (Gwadar) | 10601 | 3081 | 7520 | 71 |
| 4 | Western China | Middle East | 12537 | 2295 | 10242 | 82 |
| 5 | Western China | Europe | 19132 | 9597 | 9539 | 50 |
| 6 | Western China | Pakistan (Gwadar) | 11932 | 1750 | 10182 | 85 |

The road infrastructure construction is taking longer than planned, but from short term, medium term and long term projects, the emphasis is on developing a road infrastructure first [19]. Hence, the primary and readily bulk trading of CPEC will start from roads. To see up to what extent the highways and motorways under CPEC are developed in Pakistan, follow the report on Pakistan Economic Survey 2017–2018 [14]. Partially functional China Pakistan Economic Corridor has already started serving two purposes; internationally it is becoming world's most significant trade route scaling down global transport distance between continents, while it is helping in invigorating Pakistan's stagnant economy domestically. Before the CPEC, Pakistan has relatively advanced road transportation infrastructure (than railways or airways) in the form of intercity highways and motorways managed by National Highway and Motorway Authority (NHMA). Now, the massive Chinese investment and her direct civil engineering expertise on the road infrastructure, which includes widening and upgrading existing highways and building new highway routes, has enhanced the CPEC's capacity to meet the challenges of potential transit trade.

Keeping aside the motivational debate on China's massive investment, Pakistan's smooth trade, including 10–15% Chinese trade, is not only dependent on the road capacities to handle the trade volume but also on the risk involved while trading through highways. Pakistan's internal challenges are a major impediment in the smooth execution of trade transportation. The common problems that still exist in Pakistan's transportation are traveling time, long waiting hours, high cost, increased cost of doing business, and above all the poor performance of transport sector which cost around 5% of GDP each year [6]. On the other hand, there exist geopolitical and security risks causing uncertainty, due to the perceived world power's adjustment towards this region. The mix of international, national, and extremist factor is creating disruptive activities, which is the primary threat to CPEC [10].

There are four routes under the CPEC: The western, the central, and the eastern route, also known as three main corridors; and the fourth, the northern route. The northern route connects to Kashgar in western China. The western route is comparatively shorter and runs through the underdeveloped area of Khyber Pakhtunkhwa and Balochistan province. The central route crisscrosses the country, and the eastern route runs through Sindh and Punjab mainly [6]. For some details of all the key node cities under the CPEC routes, please refer to Table 2.

The transit trade from immediate and distant neighbors is likely to increase the stress on transport demand (supply chain). To cope with potential transit challenges, this paper proposes a trade-off function with the help of probability density functions to examine the risk of cost, profit and safety for each CPEC highway route. The traveling cost, delay time, and the unsafe factors of road transportation are considered to drive a better trade-off policy to make road transportation under the CPEC smooth and competitive in the international market. The results allow the policymakers to evaluate the best appropriate route for different commodities groups. The remainder of this paper is organized as follows. Section 2 provides the literature review. Section 3 shows the probability design and analysis. Section 4 contains the performance analysis. And the conclusion is in Sect. 5.

## 2  Literature Review

Several researchers got attracted towards the mega CPEC corridor's infrastructure development. The world's critics and project's challenges have brought a gap for researchers. Another reason is the increasing trade growth of countries on Belt and Road Initiative (BRI). Since the inauguration of CPEC, Pakistan's total export is growing every month i.e. 27.4% increase in September 2018 [5]. Recently, some researchers also have tabled high quality research on both intermodel fright and means of fright. Qi and Wang [21] compared the time and cost on Maritime route and Eurasian land bridge to transport goods from China to Europe. They concluded that it took 10 days less on land bridge to transport goods as compared to the sea. Khalid [1] analyzed the effect of CPEC in terms of transport cost and distance. He compared the existing route of Chinese trade with the new CPEC route. It resulted in decrease of cost and time, i.e. cost from China to Middle East decreased by $1450 and for Europe by $1350, while the travel time decreased by 21 days. The distance through the CPEC to Middle East and Europe is decreased by 11,000 km to 13,000 km.

An admirable research, on structural reliability of highway bridges on the CPEC by using the probability distributions and the live load analysis, showed the maximum probability for the failure of the highway bridges against the world's standard of bridges. Rasheed [23] suggested that the legal load limit to be strictly imposed by National Highway Authority of Pakistan and that the actual data needed to record for the better output results. Celikbilek [4] showed how maximizing the profit while to minimize the total risk associated with supply chain network design. He took the case of Chicago to discuss the risk in supply chain efficiency. A harsh winter and tough transportation conditions were considered, which prevented the companies in establishing a facility. There were different risk levels which influence on the supply chain network efficiency. Kanings [8], a senior economist, analyzed the international trade on the Belt and Road Initiative (BRI) and discussed how the connections between BRI countries were strengthened and the trade cost between them was lowered. She further explained that the transportation infrastructure projects, across Asia and Europe, are the key parts of BRI projects, and transportation cost is consistently found to be an important part of trade cost. The halving transport cost by WTO is 33% on the assumption that the transport cost is two-third of trade cost. When the trade cost falls between BRI connected countries, the trade between both of them increases. Masood et al. [11] explained the unsatisfied transportation system of Pakistan that hardly satisfied the local demand of the country with regional and international transit made from Karachi port. But the trade through deep sea Gwadar port is likely to reduce the travel distance by 50 to 85% towards Europe, Middle East and Africa, and likely to increase the trade volume through the CPEC corridors. Hence it was important to analyze the direct and indirect risks involved on road trading through Pakistan. The challenges of poor transport service, low condition highway network, poor safety record, and poor truck fleet, were likely to overcome with the high investment on road infrastructure. But the common challenges of transport sector that needed to be addressed was long waiting time and traveling time, as well as high cost of travel caused reduction in export. They explained that the transport sector is a back bone of business economy. In Pakistan it costed around 5% of GDP every year.

The major security concerns in Pakistan are religious extremism, sectarian, and ethno-political violence in Sindh and Baluchistan supported by anti-state forces. The number of terror attempts has already been observed that targeted the Chinese interest. A single attack on the CPEC road infrastructure construction can have negative impact on international trade through Pakistan. Recent operations by arm forces has already weakened these groups and the country witnessed about 70% decline in security issues. See Hussain [6] for more details. On the other hand the natural disaster is a main factor that may effect the CPEC road transportation. To this end, an energy efficient transportation plan for the CPEC is read, in which probable impact on carbon emission of Pakistan is analyzed, and the reduction of $CO_2$ emission is suggested by Zubedi et al. [26]. They suggested strategies to the climate researchers and policy makers for adoption and mitigation of greenhouse gases (GHG).

## 3   Probability Design and Analysis

With the help of primary and secondary data, we calculate the normal distributions for the travel cost, travel time with delay time, safety and profit between major city nodes on the four routes of CPEC, and proposes a *trade-off* function against each route to facilitate the sensitivity of different commodity groups in road shipments. The trade-off function can help us to adopt the most suitable route for the commodity groups and can allow us to switch among the routes on the basis of the priority factors of commodities' road shipments. The secondary data is the official data obtained by Ministries of Pakistan, whereas the primary source is the questionnaire survey that are filled by Pakistan's provincial traders and logistics' operators. A total of 322 fillings are achieved and their weights are concluded, where $\alpha_1$, $\alpha_2$, and $\alpha_3$ are the weights of cost, profit and safety respectively. The trade-off function is defined as:

$$\Im = \alpha_1 E[C_S] + \alpha_2 E(B_S) + \alpha_3 E(P_S). \tag{1}$$

The cost $C_S$, profit $B_S$, and safety $P_S$ are calculated by means of the exponential probability density function

$$f(x) = \lambda e^{-\lambda t}. \tag{2}$$

The four routes are named as the Northern, Western, Central and Eastern routes [15]. The distance of the routes, the names of national highways and the names of major city nodes are all real and official [18] (See Table 2). It is worthwhile to note that the northern route is connected with all the other three routes of the CPEC (also called the three corridors of the CPEC). The distance of the northern route adds up to the distance of any other route (Western, Eastern and Central), in order to calculate the total distance between two edges, i.e. the deep sea Gwadar Port in Baluchistan and the border on Karakorum highway "Khunjerab Pass" (See Fig. 1).

**Table 2.** Major city nodes on CPEC routes [15]

| Northern Nodes | Distance (km) | Highways | Western Nodes | Distance (km) | Highways | Central Nodes | Distance(km) | Highways | Eastern Nodes | Distance (km) | Highways |
|---|---|---|---|---|---|---|---|---|---|---|---|
| Khunjrab-Raikot | 335 | N35 | Burhan-D.I.Khan | 285 | N5, N80 | Burhan-D.I.Khan | 288 | N5, N80 | Peshawar-Islamabad | 155 | M1 |
| Raikot-Thakot | 270 | N35 | D.I.Khan-Zhob | 205 | N50 | D.I.Khan-Jampur | 250 | N55 | Islamabad-Pindi Bhattian | 235 | M2 |
| Thakot-Havelian | 120 | N35 | Zhob-Quetta | 331 | N50 | Jampur-Wangu Hills | 363 | M55, N8 | Pindi Bhattian-Multan | 298 | M4 |
| Havelian-Burhan | 59 | E35, M1 | Quetta-Surab | 214 | N25 | Wangu Hills-Khuzdar | 108 | M8, N25 | Multan-Sukkur | 392 | N70, N65 |
| | | | Surab-Hoshab | 449 | N85 | Khuzdar-Basima | 110 | N30 | Sukkur-Hyderabad | 296 | N5 |
| | | | Hoshab-Gwadar | 193 | M8, N10 | Basima-Gwadar | 642 | N10, N85 | Hyderabad-Karachi | 136 | M9 |
| | | | | | | | | | Karachi-Gwadar | 630 | N10 |

[a]The city nodes "Burhan-DI Khan" and "Bisma-Gwadar" on central route are the partially shared highway segment of western route.

## 3.1   CPEC Route Road Travel Time

The travel time is calculated by dividing the total distance of road transport by average speed of truck. The highest allowed speed of the truck is 110 km/h on a smooth highway, and dropped to 30 km/h or below on hilly and mountainous areas [24]. Therefore, an average speed of 50 km/h is considered to fulfill the requirements, see Table 3. Total distance is retrieved from the National Highway Authority (NHA) [15].

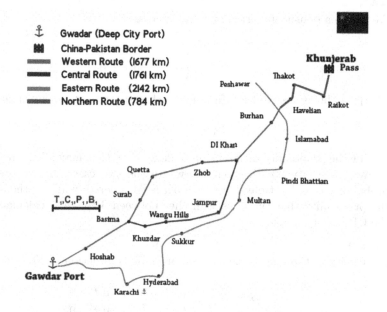

⚓ Gwadar (Deep City Port)
🏛 China-Pakistan Border
▬ Western Route  (1677 km)
▬ Central Route   (1761 km)
▬ Eastern Route   (2142 km)
▬ Northern Route (784 km)

NOTE: Gwadar and Khunherab are considered the two extreme ends for the calculation of route's trade-off functions.
B, C, P, T are the profit, cost, safety, and time between two city nodes.
Chart: Imran Akbar - Source: National Highway Authority, Pakistan - Created with Photoshop CC & Corel Draw

**Fig. 1.** Routes of China Pakistan economic corridor

## 3.2  Delay in Travel Time

There are many different factors that may cause delay in traveling like bad
weather, strikes, traffic jam, driver's rest time, and police inspections etcetera.
Moreover, the administered law and order situation in Pakistan is not at a point
where surprise delays can completely be avoided. Hence, the probability of delay
time is calculated between two city nodes by using the normal probability density
function. Here the Gwadar is treated as origin with delay time (i.e. $T_1$) and city
Hoshab as the next city on the western route with delay time (i.e. $T_2$) and so
on (see Figure 1). Since the travel time increases on continuous behavior, we can
write the transportation time and its mean between two city nodes as follows:
The transportation time is

$$T_R = T_1 + T_2 + T_3 + ... + T_{n+1}. \tag{3}$$

The mean of transportation time is

$$E[T_R] = E[T_1] + E[T_2] + E[T_3] + ... + E[T_{n+1}], \tag{4}$$

and

$$E[T_i] = \frac{1}{\lambda_i}. \tag{5}$$

We define a density function of delay time as

$$f(t) = \begin{cases} e^{-\lambda t}, & \text{if } t \geq 0 \\ 0, & \text{if } t < 0. \end{cases} \tag{6}$$

The probability distribution function with respect to the delay time is

$$F(x) = 1 - e^{-\int_0^x \lambda(t)dt} \tag{7}$$

For the probability distribution of delay time that may affect the travel time between two city nodes, given below are the two cases. In the first case, the probability of delay time more than 5 h is considered; while in the second case, the probability that the delay time lies between 3 to 5 h is calculated between the CPEC city nodes.

**Case 1**
For the delay time more the 5 h, the probability function of Node $i$ is define as

$$P(T_i \leq t) = \begin{cases} 1 - e^{-\lambda_i t}, & \text{if } t \geq 0, \\ 0, & \text{if } t < 0, \end{cases} \tag{8}$$

and the mean of the delay time is given by

$$\mu_{T_i} = E[T_i] = \frac{1}{\lambda_i}. \tag{9}$$

As we know from the real data calculations, the ideal travel time between city nodes is 3.5 h (if there is no travel congestion and smooth driving is continuous). By considering that the maximum tolerance delay is 7 h between two city nodes, we assume that the delay time of 40 ft container vehicles is exponentially distributed with the mean of 4 h.

The delay time mean is given by,

$$\mu_1 = E[T_1] = 4,$$

and the delay rate is given by

$$\lambda_1 = \frac{1}{\mu_1} = \frac{1}{4}.$$

The probability of delay time that may occur for more than 5 h to reach Hoshab city is given by

$$P(T_1 \geq 5) = e^{\frac{-5}{4}} = 0.2865$$

Hence, the probability is about 25% for 40 ft container vehicles to reach Hoshab city with a delay time of 5 h (Fig. 2).

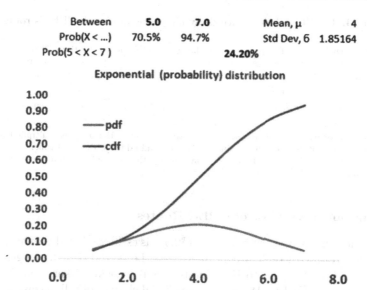

| Between | 5.0 | 7.0 | | Mean, μ | 4 |
| Prob(X < ...) | 70.5% | 94.7% | | Std Dev, δ | 1.85164 |
| Prob(5 < X < 7) | | | 24.20% | | |

**Fig. 2.** Probability of delay time for more than 5 h from Gwadar Port to city Hoshab

## Case 2

The probability of the delay time lies between an interval of 3 and 5 h. When the average delay time is 4 h, the probability function is defined as

$$P\{3 \leq T < 5\} = \int_3^5 \lambda e^{-\lambda t} dt.$$

Hence the probability of delay time which lies between 3 to 5 h is 18%. While from case 1, the probability of delay time more than 5 h is 25%. Because there are more chances of delay with more than 5 h for a 40 ft container vehicle, so the 5 h delay has been added in the traveling time between two city nodes (see Table 3).

### Road Lanes, Toll Plazas, and Intersections Under CPEC Routes

The interchanges and toll plazas on the highways also cause delays in travel time. For this purpose, some official data for numbers of tolls and interchanges on the CPEC highways are collected [16]. Further, the way that they are most likely to encounter on each journey is also observed. The interchanges and toll plazas are most likely to encounter when changing the highway routes. The toll plazas get on the way only when an entry or exit is made from the city [13]. It must be acknowledged that there are a lot of interchanges and toll plazas on each route, only the interchanges and the toll plazas of the city nodes are being considered. Hence the delay of 15 min is added in Table 3.

**Table 3.** Total travel time with delay factors against each CPEC's route

| Route | Distance kilometers | Toll plazas average | Interchanges average | Travel time hours | Delay time hours | Total time hours |
|-------|--------------------|--------------------|---------------------|------------------|-----------------|-----------------|
| Northern | 784 | 2 | 2 | 16 | 25.68 | 26 |
| Western | 1677 | 1 | 1 | 49 | 55.89 | 56 |
| Central | 1761 | 2 | 2 | 51 | 60.90 | 62 |
| Eastern | 2142 | 7 | 10 | 59 | 109.52 | 113 |

[a]Delay time of 5 h were found between first to city nodes. Hence, it is multiplied by average number of interchanges against each route and added with travel time.
[b]Total time is obtained by adding the delay time with the result of multiplication of interchanges by 0.3 h.

### 3.3   Transportation Cost of CPEC Routes

The domestic average cost per kilometer ($0.43) is estimated by dividing the total distance with total cost of 40 ft container truck, i.e. $1260 [1]. The longest route distance from Gwadar port to Khunjerab pass is 2926 km taken from National Highway Authority (NHA), Pakistan. The official charges of toll plazas, as shown in Table 4, are also added in to the total cost [17].

Total Transportation Cost = Road Travel Cost + Cost of toll Plazas

**Table 4.** Route cost of CPEC road transport

| Route | Distance kilometers | Toll plazas | Max toll tax dollars | Total cost dollars |
|-------|--------------------|-------------|---------------------|-------------------|
| Northern | 784 | 2 | 4 | 345 |
| Western | 1677 | 1 | 1 | 723 |
| Central | 1761 | 2 | 1 | 760 |
| Eastern | 2142 | 7 | 4 | 948 |

[a]$0.43 per kilometer cost is applied.
[b]Maximum toll tax has been considered for the calculation of cost.

The individual cost $C_i$ between two city nodes, as well as the total cost $C_S$ of each route is calculated. Moreover, the probability of cost occurrence is also derived by using a cumulative distribution function, which will be mentioned in the next section of this paper. For the cost, we have

$$C_S = C_1 T_1 + C_2 T_2 + C_3 T_3 + \ldots + C_{n+1} T_{n+1}.$$

Taking its mean as

$$E[C_S] = C_1 E[T_1] + C_2 E[T_2] + C_3 E[T_3] + \ldots + C_{n+1} E[T_{n+1}].$$

## 3.4   Transportation Prices by Logistic Companies

Price of standard 40 ft container with full load is being considered to fulfill the purpose of getting the profit against the CPEC route trip. The road freight price of 40 ft container truck is taken by five cargo companies in Pakistan, named Costa Logistics, MG Sky Cargo, Akurate Services, Agility, and Silk Logistics (Table 5). The charges remain valid for short period and also fluctuate a lot, hence the average price $0.81 per kilometers is calculated among the five expensive and cheap logistics companies [9]. The transportation price is calculated by multiplying the average price of a kilometer with total distance of the route.

**Table 5.** Price of the CPEC road transportation

| Company | Price (dollar) | Distance from Gwadar to Khunjerab (kilometers) | Truck load (cubic meter) | Price (per km) | Mean (dollar) |
|---|---|---|---|---|---|
| Costa Logistics | 1807 | | | 0.71 | |
| MG Sky Cargo | 1920 | | | 0.75 | |
| Akurate Services | 2033 | 2545 | 100 | 0.80 | 0.81 |
| Agility | 2259 | | | 0.89 | |
| Silk Logistics | 2250 | | | 0.88 | |

[a]Price of 40 ft container's service is obtained by five companies operating nationwide.

## 3.5   Unsafe Factors of the CPEC Routes

Considering Pakistan's current security issues, traffic situations and geographical challenges and the major factors, that may cause uncertainty in an international trade through CPEC, are natural disasters, terrorism, and accidents. With the help of these factors, we are able to find out how unsafe the route is, and the route's unsafe probability is denominated by $P_R$. Further, natural disasters with sub-factors of temperature, floods, earth quakes, Glacier-Lake Outburst Flood (GLOF), and land sliding are calculated for the routes individually. As the official data were found in weighted form, the normalized values are obtained with the help of absolute function [7,12]. Accidental data is not available city wise, only provincial accidents and highway routes accidents are available. Therefore highway route wise accidents are considered more appropriate for this research [2,22]. On the other hand, for terrorism eight years (2010–2018) of data is considered to get the trend [20]. Unlike accidental data, the data on terrorist attacks is considered as city wise, it will only effect the road shipment if there is a stop at one of the city nodes. Special Economic Zones (SEZ) allocated under CPEC are the points where the trucks most likely transit. The city nodes around the SEZ are considered of high importance. Hence, the average number of city terrorism attacks over the years are taken for all the CPEC routes. The probability between two city nodes to be unsafe is given in two different cases: the first one is that a city terrorism attacks make the roads unusable.

$$P_S = P_1 P_2 P_3 ... P_{n+1}.$$

The other one is for only a node influence

$$P_S = (1 - P_1)(1 - P_2)(1 - P_3)...(P_{n+1}).$$

### 3.6   Profit on Different Routes of CPEC

Profit $B_S$ is simply determined by subtracting the cost from the price. Since the delay time and unsafe factors are different for each route, they impact the direct and indirect cost. Its mean and variance are also calculated for routes and for city nodes. This will help in differentiating the routes from each other. We can write the profit between two city nodes as

$$B_S = B_1 + B_2 + B_3 + ... + B_{n+1},$$

and its mean is given by

$$E[B_S] = E[B_1] + E[B_2] + E[B_3] + ... + E[B_{n+1}].$$

## 4   Performance Analysis

This section is to find an optional trade-route for smooth and safe road shipping of different commodity groups in an increased international trade of CPEC. To calculate the trade-off function for each route, the elements calculated for the trade-off function are the cost $C_S$, the profit $B_S$, and the safety $P_S$ (see Table 6). In between the city nodes of each route, there are many small cities so that the district impact has been taken for unsafe factors, for example, accidents and natural disasters etcetera. The results are conducted using Vertex42 on Microsoft Excel [25]. See Table 6 for the means and variances of the factors given in the trade-off function.

### 4.1   Probabilities of Traveling Times

The cumulative distribution function is used to get the travel time probabilities between city nodes along the northern, western, central, and eastern routes. The result shows that the cumulative probability of a travel time less then 7 h (including delay time), that a random 40 ft container truck takes to cover the distance between each city nodes, is 0.2862. And the cumulative probability that a randomly chosen 40 ft container truck will travel in more then 106 h to cover the distance between each city nodes is almost zero i.e. 0.0003 (Fig. 3). Table 6 shows us the probability of travel time against each city nodes' interval.

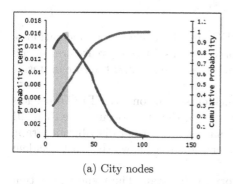

(a) City nodes

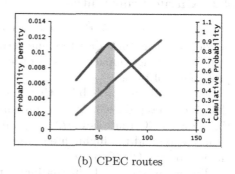

(b) CPEC routes

**Fig. 3.** Probability graph of travel time between city nodes and routes of CPEC. 1. Blue line is indicating probability density function (pdf) and red is the indication of cumulative distribution function (cdf). 2. Highlighted area under the curve shows the area of least travel time. (Color figure online)

**Table 6.** Results of the trade-off factors in China Pakistan economic corridor

| Sr. | City Nodes Under CPEC Routes | Time (Hours) Time Length | Time Probability | Time Mean | Time Variance | Unsafe Factors Probability | Unsafe Mean | Unsafe Variance | Cost($) Total Cost | Cost Probability | Cost Mean | Cost Variance | Profit ($) Total Profit | Profit Probability | Profit Mean | Profit Variance |
|---|---|---|---|---|---|---|---|---|---|---|---|---|---|---|---|---|
| | NORTHERN ROUTE | 26.28 | 0.15 | | | 0.47 | | | 345 | 0.08 | | | $287.33 | 0.08 | | |
| 1 | Khunjrab-Raikot | 12.00 | 0.35 | 0.33 | 0.00 | 0.70 | 0.55 | 0.01 | 145 | 0.61 | 0.33 | 0.06 | $125.05 | 0.68 | $0.37 | 0.06 |
| 2 | Raikot-Thakot | 10.70 | 0.33 | | | 0.51 | | | 117 | 0.53 | | | $100.55 | 0.53 | | |
| 3 | Thakot-Havelian | 7.70 | 0.29 | | | 0.43 | | | 53 | 0.12 | | | $44.00 | 0.19 | | |
| 4 | Havelian-Burhan | 11.78 | 0.35 | | | 0.57 | | | 34 | 0.07 | | | $14.01 | 0.09 | | |
| | WESTERN ROUTE | 56.29 | 0.41 | | | 0.32 | | | 723 | 0.55 | | | $630.62 | 0.58 | | |
| 1 | Burhan-D.I.Khan | 11.00 | 0.34 | 0.34 | 0.00 | 0.46 | 0.41 | 0.01 | 125 | 0.49 | 0.46 | 0.05 | $104.96 | 0.56 | $0.53 | 0.04 |
| 2 | D.I.Khan-Zhob | 9.40 | 0.31 | | | 0.48 | | | 89 | 0.28 | | | $76.04 | 0.37 | | |
| 3 | Zhob-Quetta | 11.92 | 0.35 | | | 0.42 | | | 144 | 0.60 | | | $123.54 | 0.67 | | |
| 4 | Quetta-Surab | 9.58 | 0.32 | | | 0.41 | | | 93 | 0.30 | | | $79.43 | 0.39 | | |
| 5 | Surab-Hoshab | 14.28 | 0.39 | | | 0.24 | | | 194 | 0.86 | | | $168.03 | 0.88 | | |
| 6 | Hoshab-Gwadar | 9.16 | 0.31 | | | 0.44 | | | 84 | 0.25 | | | $71.52 | 0.34 | | |
| | CENTRAL ROUTE | 61.50 | 0.91 | | | 0.11 | | | 760 | 0.60 | | | $661 | 0.63 | | |
| 1 | Burhan-D.I.Khan (Partial Western Route) | 11.06 | 0.34 | 0.34 | 0.00 | 0.40 | 0.36 | 0.01 | 125 | 0.49 | 0.46 | 0.10 | $107.33 | 0.57 | $0.52 | 0.09 |
| 2 | D.I.Khan-Jampur | 10.30 | 0.33 | | | 0.47 | | | 109 | 0.39 | | | $93.01 | 0.48 | | |
| 3 | Jampur-Wangu Hills | 12.56 | 0.36 | | | 0.49 | | | 157 | 0.68 | | | $135.61 | 0.74 | | |
| 4 | Wangu Hills-Khuzdar | 7.46 | 0.29 | | | 0.22 | | | 48 | 0.11 | | | $39.47 | 0.17 | | |
| 5 | Khuzdar-Basima | 7.50 | 0.29 | | | 0.22 | | | 49 | 0.11 | | | $40.23 | 0.18 | | |
| 6 | Basima-Gwadar (Partial Western Route) | 18.14 | 0.44 | | | 0.33 | | | 277 | 0.99 | | | $240.79 | 0.99 | | |
| | EASTERN ROUTE | 112.58 | 0.41 | | | 0.64 | | | 948 | 0.84 | | | $780.75 | 0.82 | | |
| 1 | Peshawar-Islamabad | 50.80 | 0.88 | 0.69 | 0.07 | 0.48 | 0.52 | 0.01 | 112 | 0.41 | 0.61 | 0.06 | $13.14 | 0.18 | $0.51 | 0.08 |
| 2 | Islamabad-Pindi Bhattian | 73.60 | 0.98 | | | 0.30 | | | 150 | 0.64 | | | $39.18 | 0.34 | | |
| 3 | Pindi Bhattian-Multan | 106.66 | 1.00 | | | 0.49 | | | 161 | 0.70 | | | $79.40 | 0.53 | | |
| 4 | Multan-Sukkur | 13.14 | 0.37 | | | 0.55 | | | 176 | 0.78 | | | $140.33 | 0.76 | | |
| 5 | Sukkur-Hyderabad | 11.22 | 0.34 | | | 0.55 | | | 136 | 0.56 | | | $102.89 | 0.57 | | |
| 6 | Hyderabad-Karachi | 45.12 | 0.83 | | | 0.63 | | | 65 | 0.17 | | | $45.06 | 0.19 | | |
| 7 | Karachi-Gwadar | 17.90 | 0.45 | | | 0.63 | | | 272 | 0.99 | | | $236.26 | 0.99 | | |

[a]The probabilities mentioned above are from cumulative distribution function and shows the real probability among them.

[b]Under each route are the city nodes' intervals. The distance against the routes and the distance between city nodes are mentioned in Table 4.

[c]The sum of city nodes' distance will differ from the total distance calculated for routes. This is because if we go in and out city wise, we have to pay toll tax at each city interchange, but if the vehicle enters the highway and does not exit then it will have to pay as per the highways it switches. The highways has been mentioned in Table 2.

## 4.2   Unsafe Probabilities

To find out how unsafe the routes of CPEC are, the background calculation of accident, past natural disasters and terrorist attacks for last 8 years are considered. The average impact of each of the factor is added to find the rate of unsafe probability of routes.

The cumulative probability that the accidents occur on any CPEC route less than 126 times is 0.2263. And the probability that they may occur more then 591 times on CPEC route is 0.0795. Figure 4 shows the normal distribution graph of accidents, and the highlighted area under the curve shows the least accident chances on the CPEC routes.

Terrorism has a high fluctuating history in Pakistan. The trend shows that the cumulative probability of terrorist incident equals to 0 is 0.2863, and the probability that they may occur more than 91 times is 0.0036.

Lastly, Fig. 5 shows the probability graphs of each route for natural disaster. The probability of less then 20 disastrous incidents to happen in western route is 0.12057, whereas the probability of incident for more than 37 is 0.15853. Likewise, all the other probabilities can be followed in Table 7.

Moreover, the combine effect of unsafe factors i.e. accidents, terrorism, and natural disasters is calculated by taking average of the three factors (Please see Table 8).

**Table 7.** Cumulative probabilities of natural disasters on each route

| Natural disasters: Temperature, Earth quakes, Floods, Land sliding, GLOF | | | | |
|---|---|---|---|---|
| Probabilities | Western | Central | Eastern | Northern |
| | $x_{min} = 20,$ $x_{max} = 37$ | $x_{min} = 30,$ $x_{max} = 42$ | $x_{min} = 26,$ $x_{max} = 41$ | $x_{min} = 11,$ $x_{max} = 44$ |
| $\Pr(x < x_{min})$ | 0.12057 | 0.12949 | 0.04727 | 0.08247 |
| $\Pr(x > x_{max})$ | 0.15853 | 0.05501 | 0.147 | 0.12129 |
| $\Pr(x_{min} < x < x_{max})$ | 0.7209 | 0.81551 | 0.80572 | 0.7962411 |

## 4.3   Traveling Cost on CPEC Routes

Even though the CPEC route is cost efficient for China, as it reduces the distance manifold. But for Pakistan it is still a challenge to save more due to the poorly managed past transportation system of the country. Figure 6 shows probability graph of cost against routes as well as against city nodes. The cost of travel on new CPEC routes including the direct cost (toll taxes) and the indirect cost (delay time) are considered to get the valid probabilities. The cumulative probability of 40 ft container truck to travel among city nodes with cost less than $33.6 is 0.072 and probability of cost greater than $277 is 0.0091. However, as discussed earlier that using routes do not require to pay the toll tax at each

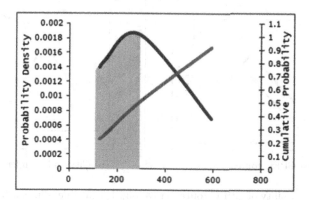

**Fig. 4.** CPEC routes accidental probability graph. 1. Blue line is indicating probability density function (pdf) and red is the indication of cumulative distribution function (cdf). 2. Highlighted area under the curve shows the area of the least accident occurrence. (Color figure online)

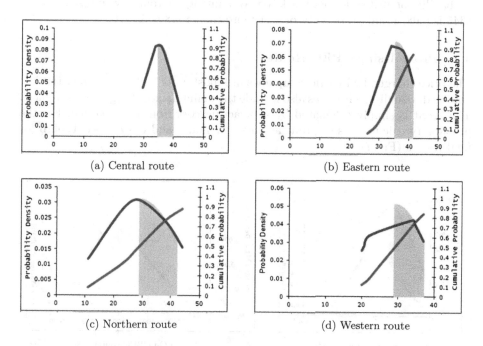

(a) Central route

(b) Eastern route

(c) Northern route

(d) Western route

**Fig. 5.** Probability of natural disasters. 1. Blue line indicates probability density function (pdf) and red indicates cumulative distribution function (cdf). 2. Highlighted area under the curve shows the area with probability of having more chances of natural disasters. (Color figure online)

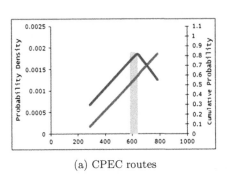
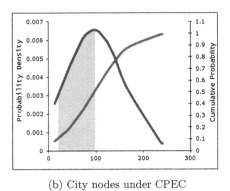

(a) CPEC routes                    (b) City nodes under CPEC

**Fig. 6.** Cost probability of CPEC routes' traveling. 1. Blue line is indicating probability density function (pdf) and red is indication cumulative distribution function (cdf). 2. Highlighted area under the curve shows the area with probability of having more chances of minimum cost. (Color figure online)

interchange, it only needs to pay once exit is required. Hence, the cumulative probability of 40 ft container truck to travel among the routes with cost less than $345 is 0.0834 and probability of cost more than $948 is 0.1570.

### 4.4 Profit Using CPEC Routes

The price charged by logistic companies in Pakistan for a 40 ft container truck is obtained. Usually their prices also include the maintenance charges of the vehicle. The profit is simply calculated by subtracting cost from price earned. For the probability calculations of profit against city nodes and routes. See Table 6 for final results (Fig. 7).

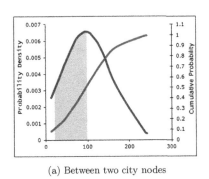
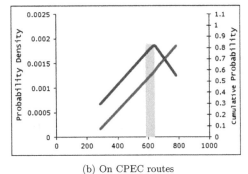

(a) Between two city nodes                    (b) On CPEC routes

**Fig. 7.** Profit probability under CPEC. 1. Blue line is indicating probability density function (pdf) and red is the indication of cumulative distribution function (cdf). 2. Highlighted area under the curve shows the area with probability of minimum profit that is only 8.5% considering city nodes and 7.6% considering CPEC routes. (Color figure online)

## 4.5 Weights of Cost, Profit and Safety

The fully functional CPEC route is the eastern route whereas other routes are partially in use due to some parts of them under construction and they will get completed in the year 2020. With the help of questionnaire, we are able to find out the current scenario of what people think of CPEC routes and how their trading is getting on with these routes. The traders and logistic business officials from all provinces (Punjab, Baluchistan, Sindh, KPK) are picked to fill the questionnaire. For weights $\alpha_1$, $\alpha_2$, and $\alpha_3$ are assigned for cost, profit and safety, respectively. In which $\alpha_1, \alpha_2, \alpha_3 \geq 0$ and $\alpha_1 + \alpha_2 + \alpha_3 = 1$, see Table 8 for more details.

**Table 8.** Trade-off functions of CPEC

| Trade-off function | CPEC routes | Cost $\alpha_1 \geq 0$ | Profit $\alpha_2 \geq 0$ | Safety $\alpha_3 \geq 0$ | Cost $E[C_S]$ | Profit $E[B_S]$ | Unsafe $E[P_R]$ |
|---|---|---|---|---|---|---|---|
| $\Im_1$ | Northern route | 0.26 | 0.17 | 0.57 | 0.33 | 0.37 | 0.55 |
| $\Im_2$ | Western route | 0.18 | 0.20 | 0.62 | 0.46 | 0.53 | 0.41 |
| $\Im_3$ | Central route | 0.23 | 0.26 | 0.51 | 0.46 | 0.51 | 0.36 |
| $\Im_4$ | Eastern route | 0.21 | 0.18 | 0.60 | 0.84 | 0.51 | 0.52 |

[a] $\alpha_1$, $\alpha_2$ and $\alpha_3$ are the weights of cost, profit, and safety respectively, summing to 1 (100%).
[b] $E[C_S]$, $E[B_S]$ and $E[P_R]$ are the mean of the probabilities against each route.

## 4.6 Trade-Off Policy Against Major Trade Categories

Our results are to easily decide the trade-off for international trade commodities. The CPEC is not only meant for China-Pakistan, but also to bring regional stability. The trade through CPEC can connect many neighboring countries and the land-locked countries. The trade-off policy helps in deciding the route for trade commodities for their smooth transportation. For example, China, being the biggest consumer of oil in the world, wants the CPEC to be its energy assurance corridor. Because oil is expensive and most demanding product, hence the trade-off policy with maximum route safety must be adopted over route cost or profit. Similarly, the perishable products need a route with less distance and travel time. And the product with less profit margin may need to choose the route with less cost.

Considering our results, the northern route has less traveling distance, less cost and more profit but as this route is unavoidable as it connects China with rest of the three routes of the CPEC. On the other hand, Western Route is upgraded but partially built and expected to be fully functional in the year 2020. This route is found to be the most reliable in every aspect. On contrary, the central route has high cost effecting less profit but is considered as the most safe route. Lastly, the eastern route is the longest one with most industrialized cities connected to it in the country. Infrastructure is strong at the eastern route, and it has been the most developed in CPEC so far. Apart from high cost, less safe and longer travel time, the profit is maximum with respect to the other routes under the CPEC.

**Table 9.** Trade-off between CPEC routes for commodity groups

| China-Pakistan Economic Corridor | Trade-offs considered routes | | | |
|---|---|---|---|---|
| Important trade categories | Northern | Western | Central | Eastern |
| | Unavoidable | Upgraded & partially built | Newly built | Developed & industrialized |
| Processed food groups | T, C, B | | | B |
| Machinery & transport groups | T, C, B | All | C, T, P | B |
| Petroleum groups | T, C, B | | C, T, P | |
| Textile groups | T, C, B | All | | B |
| Chemical groups | T, C, B | All | C, T, P | B |
| Metal groups | T, C, B | All | | |
| Plastic and rubbers groups | T, C, B | | C, T, P | B |
| Paper goods groups | T, C, B | | | B |
| Stone and glass groups | T, C, B | All | | |
| Agriculture group | T, C, B | All | C, T, P | |

[a]B, C, P, T are indicating profitably, less costly, mostly safe and time efficiency, respectively.

Table 9 shows that for processed food groups, the eastern route is the best choice because this route passes through the province Punjab and Sindh, which are most developed and agricultural provinces of Pakistan. Hence trading of food group commodity is likely to have more profit on this route. For machinery and transport group all routes are suitably depending upon the scenario of delivery. For example, for profit the eastern route can be used. For less cost, less time, and less safety the central route can be used. Moreover, for petroleum groups the safety matters more, hence the central route with less cost, less transportation time and more safety is suitable. Likewise, all the other groups have been assigned a route. Routes can be changed by depending upon the urgency or diplomacy of the trade.

## 5   Conclusion

The execution of increasingly large-scale international the trade through China-Pakistan Economic Corridor (CPEC) is a challenging task for Pakistan more than it is for its core investor China. Especially when the history of Pakistan's trade is not so good. To prove effectiveness and efficiency of road trading which fulfill the supply and demand for regional countries, Pakistan needs to adopt special measures. Even though there are different means of trade planned in the CPEC to avoid any kind of uncertainty, i.e., sea, air, rail, and direct oil pipeline, but the very first linkage is the highways network of Pakistan. Rest of the means (train or air) are included in a long term plan and they are time taking. Hence, the objective of this research is to evaluate the four routes of the CPEC in terms of cost, profit and safety, and to design a trade-off function for each of its route.

Findings of this paper allow us to deeply analyze each route of the CPEC and to suggest an optional trade-off function in term of commodity-groups' safety, urgency and worthy. The sub-factors of the trade-off function is calculated by using probability density function, and the weighted average is included against each factor.

The CPEC, which is already beneficial in connecting Middle East and Europe to the South Asian region through the shortest distance, would bring more benefit to Pakistan itself, if the efficient trading gets executed. Future researches can be but not limited to trade means in the CPEC, cost effective means of the CPEC and policy measures to cope with trading hurdles under the CPEC.

**Acknowledgments.** Quan-Lin Li was supported by the National Natural Science Foundation of China under grants No. 71671158 and 71932002, and by the Natural Science Foundation of Hebei province under grant No. G2017203277.

# References

1. Alam, K.M., Li, X., Baig, S.: Impact of transport cost and travel time on trade under China-Pakistan economic corridor (CPEC). J. Adv. Transp. **2019**, 16 p. (2019). Article no. 7178507. https://doi.org/10.1155/2019/7178507
2. Annual Performance Report: Accidents. http://nhmp.gov.pk/wp-content/uploads/2018/06/Annual-Performance-Report.pdf. Accessed 13 May 2019
3. Ashraf, M.: China Pakitan Economic Corridor (CPEC): analysis of internal and external challenges. Int. J. Bus. Econ. Manag. **4**(5), 106–111 (2017)
4. Calikbilek, C.: A fuzzy approach for a supply chain network design problem. In: 26th Annual Proceedings of Production and Operation Management Society Conference (2010)
5. CEIC Data: Pakistan total exports growth. https://www.ceicdata.com/en/indicator/pakistan/total-exports-growth. Accessed 13 May 2019
6. Hussain, M.: China Pakistan Economic Corridor. Challenges and the way forward, pp. 1–105. The NPS Institute Archive DSpace Repository, Calhoun (2017)
7. Integrated Context Analysis: On vulnerability to food security and natural hazards. https://pdma.gov.pk/. Accessed 13 May 2019
8. Konings, J.: Trade impacts of the belt and road initiative trade impacts of the belt and road initiative. ING Global Economist (1999)
9. Logistic Companies in Pakistan: Silk logistics, mg sky cargo, costa logistics, akurate services, agility. https://www.silklogistics.com.pk/services/, https://www.searates.com/, https://www.costa.com.pk/, www.akurateservices.com/, https://www.agility.com/EN/countries/Pages/Pakistan.aspx. Accessed 13 May 2019
10. Long Term Plan for China Pakistan Economic Corridor: Possible challenges. http://cpec.gov.pk/long-term-plan-cpec. Accessed 13 May 2019
11. Masood, M.T., Farooq, M., Hussain, B.: Pakistan's potential as a transit trade corridor and transportation Challenges. Pakistan Bus. Rev. **18**, 267–289 (2016)

12. Meteoblue, weather close to you. https://www.meteoblue.com/en/weather/forecast. Accessed 13 May 2019

13. Ministry of Communication: NHA's Interactive Map. https://www.google.com/maps. Accessed 13 May 2019

14. Ministry of Finance (MOF): Pakistan Economic Survey. http://www.finance.gov.pk/. Accessed 13 May 2019

15. National Highways Authority (NHA): Map of national highway network. http://nha.gov.pk/wp-content/uploads/2018/01/NHA-Map.jpg. Accessed 13 May 2019

16. National Highway Authority: Notional highway and motorway toll taxes. http://downloads.nha.gov.pk/images/stories/services/toolrate/tr1.pdf. Accessed 13 May 2019

17. National Highway Authority: Toll rates. http://nha.gov.pk/en/. Accessed 13 May 2019

18. National Highways of Pakistan: List of national highways. http://nha.gov.pk/en/. Accessed 13 May 2019

19. Nazir, M.: Analysis of determinants of CPEC's success and failure. Emerging challenges and lessons for Pakistan. Islamabad Policy Res. Inst. (IPRI) 2(1), 51–73 (2017)

20. Pakistan Bureau of Statistics: Compendium on environmental statistics of Pakistan. http://www.pbs.gov.pk/sites/default/files/. Accessed 13 May 2019

21. Qi, Y., Wang, Y.: Analysis of land bridge transportation. Chin. Geogr. Sci. 1(4), 337–346 (1991)

22. Shah, S.A.R., Khattak, A.: Road traffic accidents analysis of motorways in Pakistan. Int. J. Eng. Res. Technol. 2(11), 3340–3354 (2013)

23. Rasheed, A.: Structural reliability analysis of superstructure of highway bridges on CPEC. Dissertation project (2015)

24. Road Safety Pakistan: Road safety. www.roadsafetypakistan.pk/speed-limits. Accessed 13 May 2019

25. Wittwer, J.: Graphing a normal distribution curve in excel. http://www.vertex42.com/ExcelArticles/mc/NormalDistribution-Excel.html. Accessed 13 May 2019

26. Zubedi, A., Jianqiu, Z., Arain, A.Q., et al.: Sustaining low-carbon emission development. An energy efficient transportation plan for CPEC. J. Inf. Process. Syst. 14(2), 322–345 (2018)

# An Improved Strong Connectivity Discriminant Algorithm for Complex Directed Networks

Cheng-Hong Wang, Zhuo Wang[(⊠)], and Zhen-Dong Wu

The Research Institute of Frontier Science, Beihang University, Beijing 100191,
People's Republic of China
zhuowang@buaa.edu.cn

**Abstract.** The existing network connectivity discriminant algorithms have high
time complexity, which could not satisfy the requirement of quick connectivity
discrimination for large-scale networks, such as computer networks, communication networks and energy networks, etc. Aiming at the strong connectivity
discriminant problem of directed networks, this paper extends the Warshall
algorithm from simple directed networks to complex ones and gives an
improved discriminant algorithm, which can reduce the time complexity of the
Warshall algorithm by half. This is of great practical value to the connectivity
discrimination for various large-scale complex directed networks.

**Keywords:** Complex directed network · Strong connectivity · Connectivity
discriminant algorithm · Time complexity

## 1 Introduction

Connectivity discriminant algorithms for complex directed networks, including self-
rings and multiple directed edges, not only play an important role in graph theory
[1–5], but also have wide application in communication [6, 7], computer [8], control
[7, 9, 10], transportation and all kinds of networks [3, 4, 6, 8, 11].

During the last twenty years, thousands of articles have been published in the field
of consistency control of multi-agent systems [7, 11]. In all these articles, it is assumed
that the communication network between the multi-agents is always connected, but
none of them studied the connectivity problem for the communication networks of
multi-agent systems. In the practical application of multi-agent consistency control
(such as the formation flight control of multi-UAVs), the connectivity of communication network between multi-agents must be tested and determined at all time,
otherwise it is difficult to ensure that the predetermined control objectives can be
achieved [3–5]. When the number of agents is huge, or the scale of communication
network is large, due to the high time complexity, the existing network connectivity
discriminant algorithms could not satisfy the requirement of quick discrimination.

This work was supported by the National Natural Science Foundation (NNSF) of China under Grant
61673041.

Q.-L. Li et al. (Eds.): Cao Festschrift 2019, CCIS 1102, pp. 465–474, 2019.
https://doi.org/10.1007/978-981-15-0864-6_24

There are many similar practical applications, which are not described here. The above indicates that we urgently need to develop efficient connectivity discriminant algorithms, in order to satisfy the requirement of connectivity discrimination for various large-scale complex networks.

Networks can be divided into undirected networks, directed networks, undirected and directed hybrid networks (referred to as hybrid networks) [4]. In terms of connectivity, any undirected edge can be replaced by two opposite directed edges, so that both undirected networks and hybrid networks can be represented by directed networks, but not vice versa. For this reason, complex directed networks have greater universality, and the research on efficient discriminant algorithms for strong connectivity has greater theoretical and practical significance.

Strong connectivity discriminant algorithms of directed networks can be mainly classified into two categories: one is the algorithms based on depth-first or breadth-first searching technologies [1], such as Tarjan algorithm and Gabow algorithm; the other is the algorithms based on reachability matrix [2] and the Warshall algorithm based on relation transitive closure [1]. Tarjan algorithm and Gabow algorithm are a kind of network-oriented direct search algorithms. Their advantage is the low algorithm complexity (both are $O(n+m)$, $n$ is the number of nodes, and $m$ is the number of edges), but the disadvantages are that they are only suitable for simple directed networks (no self-rings and multiple directed edges) and are not easy to program (stack and labeling techniques are required). The advantages of the Warshall algorithm is that its structure is very concise and the bit operation method with higher efficiency can be used. The disadvantages are that the algorithm has high complexity ($O(2n^3)$) and can only be used in simple directed networks [1]. Though connectivity discriminant algorithm based on the reachability matrix can be used in complex directed networks, it is generally not applicable to large-scale practical networks due to its high complexity.

In this paper, the Warshall algorithm, which is only suitable for simple directed networks, is extended to complex directed networks, and an improved algorithm is given, which can reduce the time complexity of the Warshall algorithm by half. This provides theoretical basis and support algorithm for various practical applications, such as reliability analysis of network systems [3, 6, 8], consistency control or formation control of large-scale multi-agent systems [7, 9, 10].

## 2    Problem Description and Knowledge Preparation

**Definition 1:** A directed network $D$ is defined as an even pair $D = (V, E)$, where: (1) $V$ is a nonempty set, whose elements are called nodes; (2) $E$ is a subset of ordered product $V \times V$ whose elements are called directed edges.

**Definition 2** (Yin & Wu [2]): Given a directed network $D$, if there exists a directed chain $(u, v)$ so that node $u$ can reach node $v$, then $v$ is said to be reachable from $u$.

(1)  If any two nodes in $D$ are mutually reachable, then $D$ is strongly connected.
(2)  If at least one of the two nodes in $D$ can reach the other node, $D$ is said to be unidirectionally connected.

Definition 2 shows that strong connectivity implies unidirectional connectivity, but not vice versa. Without special declaration, all directed networks discussed in this paper refer to complex directed networks.

**Definition 3** (Yin & Wu [2]): Given a directed network $D = (V, E)$. Suppose that its node set is $V = \{v_1, v_2, \cdots, v_n\}$, and call $P = P(p_{ij}) \in R^{n \times n}$ the reachability matrix of $D$, where

$$p_{ij} = \begin{cases} 1, & \text{there is at least one directed chain from } v_i \text{ to } v_j; \\ 0, & \text{there is no directed chain from } v_i \text{ to } v_j. \end{cases}$$

At least one directed chain (a directed edge is considered as a directed chain of length 1) from $v_i$ to $v_j$ is equivalent to the statement that $v_i$ can reach $v_j$, otherwise $v_i$ is not reachable to $v_j$.

From Definitions 2 and 3, we can conclude that the necessary and sufficient condition for strong connectivity of directed network $D$ is that all elements of reachability matrix $P(P_{ij})$ is 1.

**Example 1** (Yin & Wu [2]): Find the reachability matrix of the directed network shown in Fig. 1, and decide the connectivity of the directed network according to the reachability matrix.

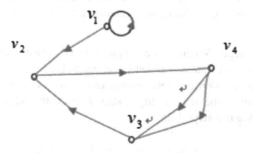

**Fig. 1.** A directed network with 4 nodes

(1) Suppose that the numbering sequence of the nodes is $v_1, v_2, v_3$ and $v_4$, then the reachability matrix of the directed network shown in Fig. 1 is

$$P_4(p_{ij}) = \begin{bmatrix} 1 & 1 & 1 & 1 \\ 0 & 1 & 1 & 1 \\ 0 & 1 & 1 & 1 \\ 0 & 1 & 1 & 1 \end{bmatrix}.$$

The directed network shown in Fig. 1 is not strongly connected, because not all the elements of $P_4(p_{ij})$ are 1.

(2) Consider the remaining directed subnetworks after removing node $v_1$ and its adjacent edges. Suppose that the numbering sequence of nodes is $v_2, v_3$ and $v_4$, then the corresponding reachability matrix is

$$P_3\left(p_{ij}\right) = \begin{bmatrix} 1 & 1 & 1 \\ 1 & 1 & 1 \\ 1 & 1 & 1 \end{bmatrix}.$$

Because all the elements of $P_3\left(p_{ij}\right)$ are 1, the directed subnetworks consisting of nodes $v_2, v_3$ and $v_4$, have all their adjacent edges strongly connected.

The above discussion shows that if we know the reachability matrix of a directed network, we can easily determine the strong connectivity of that network. When the scale of the directed network is small, we can directly obtain the reachability matrix according to Definition 3. However, when the scale of the directed network is very large, it is very difficult to calculate the reachability matrix directly according to Definition 3.

**Definition 4** (Yin & Wu [2]): Given a directed network $D = (V, E)$, suppose that its node set is $V = \{v_1, v_2, \cdots, v_n\}$, and call $A = A\left(a_{ij}\right) \in R^{n \times n}$ the adjacency matrix of $D$, where

$$a_{ij} = \begin{cases} m_{ij}, & v_i \text{ is adjacent to } v_j, \text{ and there are } m_{ij} \text{ directed edges from } v_i \text{ to } v_j; \\ 0, & v_i \text{ is not adjacent to } v_j, \text{ or adjacent to but there are no } edges \text{ from } v_i \text{ to } v_j. \end{cases}$$

Here, $m_{ij} \geq 1$ is an integer. When $D$ is a simple directed network, $m_{ij} = 1$; when $D$ is a complex directed network, there is at least one $m_{ij} > 1$.

There is a definite relationship between adjacency matrix $A$ and reachability matrix $P$. In order to calculate the reachability matrix $P$ from the adjacency matrix $A$, we should take the following steps [2]:

(1) Find $B = A + A^2 + \cdots + A^n$;

(2) Change the nonzero elements in $B$ to 1, while keeping the zero elements unchanged. Suppose that the matrix obtained by this operation is $\overline{B}$, then $\overline{B} = P$. Bit operation cannot be used in the process of solving $B$ and the computational complexity is higher than $O(n^4)$, so this method is not usually used in practical application.

**Definition 5:** Given a directed network $D = (V, E)$, the set of nodes is assumed to be $V = \{v_1, v_2, \cdots, v_n\}$, and $M_R = W_0\left(w_{ij}^{[0]}\right) \in R^{n \times n}$ is called the adjacency relation matrix of $D$, where

$$w_{ij}^{[0]} = \begin{cases} 1, & v_i \text{ is adjacent to } v_j \text{ and there are directed edges from } v_i \text{ to } v_j; \\ 0, & v_i \text{ is not adjacent to } v_j, k \text{ or adjacent but there are no directed edges from } v_i \text{ to } v_j. \end{cases}$$

In the above definition, adjacency relation $R$ between nodes in complex directed network $D$ is a binary relation defined on node set $V$, and the $0 - 1$ matrix representation of $R$ is the adjacency relation matrix $M_R$.

Comparing Definitions 3, 4 and 5, we can conclude that the connectivity of $D$ will not change when the multiple edges between all nodes in complex directed network $D$ are reduced to one. Therefore, when $D$ is a simple directed network, $A = M_R$; when $D$ is a complex directed network, by changing all the elements in $A$ which are greater than 1 to 1 while keeping the others unchanged, we will get $M_R$. In this way, $M_R$ can be directly obtained according to Definition 5, and the connectivity of $D$ remains unchanged.

The meaning of Definition 5 is to extend the adjacency matrix of a simple directed network to the adjacency relation matrix of a complex directed network, such that we can extend the applicability of the Warshall algorithm from simple directed networks to complex directed networks.

**Lemma 1** (Rosen [1]): Given a directed network $D = (V, E)$, suppose the adjacency relation and the adjacency relation matrix is $R$ and $M_R$, respectively. Then

(1) the transitive closure of adjacency relation $R$ is $R^* = \bigcup_{k=1}^{n} R^k$;

(2) $R^*$ equals the connectivity relation of $D$;

(3) the $0 - 1$ matrix of transitive closure $R^*$ is $M_{R^*} = M_R^{[1]} \vee M_R^{[2]} \vee M_R^{[3]} \vee \cdots \vee M_R^{[n]}$.

In Lemma 1(3), "$\vee$" represents the Boolean OR operation, and $M_R^{[k]}(1 \leq k \leq n)$ represents the $k$ th power under Boolean intersection operation. Lemma 1 tells us that transitive closure $R^*$ of the directed network $D$ is the connectivity relation between nodes in $D$, and the transitive closure matrix $M_{R^*}$ is actually the reachability matrix $P$, i.e. $M_{R^*} = P$; if all elements of $M_{R^*}$ are 1, the directed network $D$ is strongly connected, otherwise, $D$ is not strongly connected.

The biggest advantage of solving $M_{R^*}$ is that the bit operation method with higher computational efficiency can be adopted. However, when $M_{R^*}$ is directly calculated according to Lemma 1(3), the computational complexity will be as high as $O(n^4)$. It's obvious that such high computational complexity is not suitable for determining the connectivity of large-scale directed networks, and new efficient algorithms are still need to be developed.

# 3 The Basis of Algorithm Design and the New Algorithm Design

## 3.1 The Basis of Algorithm Design

As aforementioned, the calculation amount for directly solving $M_{R^*}$ according to Lemma 1(3) is huge. For this reason, Warshall made significant improvements to these algorithms in 1960.

The Warshall algorithm is based on the construction of a series of $0 - 1$ matrices. These matrices are $W_0, W_1, \cdots, W_n$, respectively. Among these matrices, $W_0 = M_R$ is the adjacency relation matrix, $W_k = \left[ w_{ij}^{[k]} \right]$ and $W_n = M_{R^*}$ [1].

**Lemma 2** (Rosen [1]): Given a directed network $D$ which has $n(n \geq 2)$ nodes. Suppose that $W_k = \left[ w_{ij}^{[k]} \right]$ is a $0 - 1$ matrix, then $w_{ij}^{[k]} = 1$ if and only if there exists a path from $v_i$ to $v_j$, whose internal nodes are taken from the set $\{v_1, v_2, \cdots, v_k\}$, such that

$$w_{ij}^{[k]} = w_{ij}^{[k-1]} \vee \left( w_{ik}^{[k-1]} \wedge w_{kj}^{[k-1]} \right),$$

where $i, j$ and $k$ are positive integers no more than $n$.

Lemma 2 is the famous Warshall algorithm. We need $2n^2$ bit operations to solve $W_k$ from $W_{k-1}$, but solving $W_n = M_{R^*}$ needs a total number of $n \times 2n^2 = 2n^3$ bit operations. Thus, the computational complexity of the Warshall algorithm is $O(2n^3)$ [1]. This algorithm is most famous for its remarkably simple structure and low computational complexity.

The following analysis shows that the Warshall algorithm can be further improved. Given $k \in \{1, 2, \cdots, n\}$, we can extend the Warshall algorithm as follows

$$w_{11}^{[k]} = w_{11}^{[k-1]} \vee \left( w_{1k}^{[k-1]} \wedge w_{k1}^{[k-1]} \right) \qquad\qquad w_{n1}^{[k]} = w_{n1}^{[k-1]} \vee \left( w_{nk}^{[k-1]} \wedge w_{k1}^{[k-1]} \right)$$

$$w_{12}^{[k]} = w_{12}^{[k-1]} \vee \left( w_{1k}^{[k-1]} \wedge w_{k2}^{[k-1]} \right) \qquad\qquad w_{n2}^{[k]} = w_{n2}^{[k-1]} \vee \left( w_{nk}^{[k-1]} \wedge w_{k2}^{[k-1]} \right)$$

$$w_{13}^{[k]} = w_{13}^{[k-1]} \vee \left( w_{1k}^{[k-1]} \wedge w_{k3}^{[k-1]} \right) \quad \cdots \quad w_{n3}^{[k]} = w_{n3}^{[k-1]} \vee \left( w_{nk}^{[k-1]} \wedge w_{k3}^{[k-1]} \right)$$

$$\vdots \qquad\qquad\qquad\qquad\qquad\qquad\qquad \vdots$$

$$w_{1n}^{[k]} = w_{1n}^{[k-1]} \vee \left( w_{1k}^{[k-1]} \wedge w_{kn}^{[k-1]} \right) \qquad\qquad w_{nn}^{[k]} = w_{nn}^{[k-1]} \vee \left( w_{nk}^{[k-1]} \wedge w_{kn}^{[k-1]} \right)$$

When observing the elements in the first column within brackets on the right side of each equation above, for every $1 \leq j \leq n$, we can find that $w_{1j}^{[k]} = w_{1j}^{[k-1]}$ if $w_{1k}^{[k-1]} = 0$; otherwise, $w_{1j}^{[k]} = w_{1j}^{[k-1]} \vee w_{kj}^{[k-1]}$; $\ldots$; $w_{nj}^{[k]} = w_{nj}^{[k-1]}$ if $w_{nk}^{[k-1]} = 0$, otherwise, $w_{nj}^{[k]} = w_{nj}^{[k-1]} \vee w_{kj}^{[k-1]}$. This shows that every time we add a comparison operation to determine whether $w_{ik}^{[k-1]}$ is 0, we can get $n$ calculation results: $w_{ij}^{[k]} = w_{ij}^{[k-1]} (1 \leq j \leq n)$ or $w_{ij}^{[k]} = w_{ij}^{[k-1]} \vee w_{kj}^{[k-1]} (1 \leq j \leq n)$. The former is an assignment operation and does not increase the amount of calculation (in the algorithm complexity analysis, the assignment operation is not considered), while the latter reduces $2n$ bit operations of $n$ elements needing to be solved, to $n$ bit operations. In this way, the amount of calculation is reduced by half. Based on this observation and analysis, we can improve the Warshall algorithm.

**Corollary 1 (The improved Warshall algorithm):** Given a directed network $D$ with $n(n \geq 2)$ nodes. Suppose that $W_k = \left[ w_{ij}^{[k]} \right]$ is a $0 - 1$ matrix, then $w_{ij}^{[k]} = 1$ if and only if there is a path from $v_i$ to $v_j$, whose internal nodes are taken from set $\{v_1, v_2, \cdots, v_k\}$, such that $w_{ij}^{[k]} = w_{ij}^{[k-1]}$ if $w_{ik}^{[k-1]} = 0$, otherwise $w_{ij}^{[k]} = w_{ij}^{[k-1]} \vee w_{kj}^{[k-1]}$, where $i, j$ and $k$ are positive integers no more than $n$.

According to Corollary 1, we need $(n + n^2)$ bit operations to solve $W_k$ from $W_{k-1}$ (including $n$ comparison operations and $n^2$ bit operations), and solving $W_n = M_{R^*}$

totally needs $n \times (n^2 + n) = n^3 + n^2$ bit operations. In summary, the computational complexity of the algorithm given in Corollary 1 is $O(n^3)$, which reduces the computational complexity of the Warshall algorithm by half. It is not hard to understand that when $D$ is a large-scale sparse network, i.e., $M_R$ is a large-scale sparse matrix, the computational complexity of the algorithm given in Corollary 1 will be lower than $O(n^3)$.

From the previous discussion, we can know that $W_n = M_{R^*} = P$, and $W_n$ can be calculated by the method given in Corollary 1. Therefore, solving the reachability matrix $P$ is converted into solving $W_n$, and we can determine whether $D$ is strongly connected according to whether all the elements of $W_n$ are 1.

### 3.2    The New Algorithm Design

Given a directed network $D$ with $n(n \geq 2)$ nodes, we can design a new strong connectivity discriminant algorithm for $D$ according to Corollary 1. The main steps of the algorithm can be summarized as follows:

(1)  According to Definition 5, input the adjacency relation matrix $W_0 = M_R$;
(2)  According to Corollary 1, calculate the transitive closure matrix $W_n = M_{R^*}$;
(3)  Calculate the number $M(W_n)$ of nonzero elements of $W_n$;
(4)  if $M(W_n) = n^2$, output the result that $D$ is strongly connected; Otherwise, output the result that $D$ is not strongly connected.

**Strong Connectivity Discriminant Algorithm for Directed Networks:**

(1)  When $k = 0$, input the adjacency relation matrix $W_0\left(w_{ij}^{[0]}\right)$ of the directed network $D$ with $n(n \geq 2)$ nodes;
(2)  When $i = 0$, set $k = k + 1$;
(3)  Set $i = i + 1$;
(4)  If $w_{ik}^{[k-1]} = 0$, set $j = 0$ and turn to (5); Otherwise, set $j = 0$ and turn to (9);
(5)  Set $j = j + 1$, $w_{ij}^{[k]} = w_{ij}^{[k-1]}$;
(6)  If $j < n$, turn to (5); Otherwise, turn to (7);
(7)  If $i < n$, turn to (3); Otherwise, turn to (8);
(8)  If $k < n$, turn to (2); Otherwise, turn to (13);
(9)  Set $j = j + 1$ and $w_{ij}^{[k]} = w_{ij}^{[k-1]} \vee w_{kj}^{[k-1]}$;
(10)  If $j < n$, turn to (9); Otherwise, turn to (11);
(11)  If $i < n$, turn to (3); Otherwise, turn to (12);
(12)  If $k < n$, turn to (2); Otherwise, turn to (13);
(13)  Calculate $M(W_n) = \sum_{i=1}^{n} \sum_{j=1}^{n} w_{ij}^{[n]}$;
(14)  If $M(W_n) = n^2$, output the result that $D$ is strongly connected; Otherwise, output the result that $D$ is not strongly connected.

---

**The Improved Warshall Algorithm:**

**Procedure** $(W_0 = M_R \in R^{n\times n})$

$W := M_0$

for $k := 1$ to $n$

    for $i := 1$ to $n$

        if $w_{ik}^{[k-1]} = 0$

            for $j := 1$ to $n$, $w_{ij}^{[k]} = w_{ij}^{[k-1]}$;

    else

            for $j := 1$ to $n$, $w_{ij}^{[k]} = w_{ij}^{[k-1]} \vee w_{kj}^{[k-1]}$.

$M(W_n) = \sum_{i=1}^{n} \sum_{j=1}^{n} w_{ij}^{[n]}$

    if $M(W_n) = n^2$, output: $D$ is strongly connected;

    else, output: $D$ is not strongly connected.

---

A Pseudocode of Strong Connectivity Discriminant Algorithm for Directed Networks:

### 3.3 Example of the Algorithm

**Example 2:** Determine the connectivity of the directed network shown in Fig. 2, with the sequence numbers of the nodes remaining unchanged.

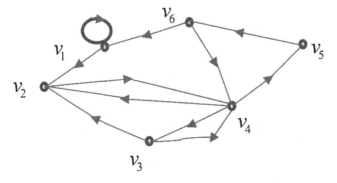

**Fig. 2.** A directed network with 6 nodes

According to the node numbers in Fig. 2 and the improved Warshall algorithm given in this paper, we can obtain

$$W_0 = M_R = \begin{bmatrix} 1 & 1 & 0 & 0 & 0 & 0 \\ 0 & 0 & 0 & 1 & 0 & 0 \\ 0 & 1 & 0 & 0 & 0 & 0 \\ 0 & 1 & 1 & 0 & 1 & 0 \\ 0 & 0 & 0 & 0 & 0 & 1 \\ 1 & 0 & 0 & 1 & 0 & 0 \end{bmatrix}, W_1 = \begin{bmatrix} 1 & 1 & 0 & 0 & 0 & 0 \\ 0 & 0 & 0 & 1 & 0 & 0 \\ 0 & 1 & 0 & 0 & 0 & 0 \\ 0 & 1 & 1 & 0 & 1 & 0 \\ 0 & 0 & 0 & 0 & 0 & 1 \\ 1 & 1 & 0 & 1 & 0 & 0 \end{bmatrix},$$

$$W_2 = W_3 = \begin{bmatrix} 1 & 1 & 0 & 1 & 0 & 0 \\ 0 & 0 & 0 & 1 & 0 & 0 \\ 0 & 1 & 0 & 1 & 0 & 0 \\ 0 & 1 & 1 & 1 & 1 & 0 \\ 0 & 0 & 0 & 0 & 0 & 1 \\ 1 & 1 & 0 & 1 & 0 & 0 \end{bmatrix}, W_4 = \begin{bmatrix} 1 & 1 & 1 & 1 & 1 & 0 \\ 0 & 1 & 1 & 1 & 1 & 0 \\ 0 & 1 & 1 & 1 & 1 & 0 \\ 0 & 1 & 1 & 1 & 1 & 0 \\ 0 & 0 & 0 & 0 & 0 & 1 \\ 1 & 1 & 1 & 1 & 1 & 0 \end{bmatrix},$$

$$W_5 = \begin{bmatrix} 1 & 1 & 1 & 1 & 1 & 1 \\ 0 & 1 & 1 & 1 & 1 & 1 \\ 0 & 1 & 1 & 1 & 1 & 1 \\ 0 & 1 & 1 & 1 & 1 & 1 \\ 0 & 0 & 0 & 0 & 0 & 1 \\ 1 & 1 & 1 & 1 & 1 & 1 \end{bmatrix}, W_6 = M_{R^*} = \begin{bmatrix} 1 & 1 & 1 & 1 & 1 & 1 \\ 1 & 1 & 1 & 1 & 1 & 1 \\ 1 & 1 & 1 & 1 & 1 & 1 \\ 1 & 1 & 1 & 1 & 1 & 1 \\ 1 & 1 & 1 & 1 & 1 & 1 \\ 1 & 1 & 1 & 1 & 1 & 1 \end{bmatrix}.$$

Because $M(W_6) = 6^2 = 36$, we can determine that the directed network shown in Fig. 2 is strongly connected.

Then for this directed network, we will briefly analyze the calculation loads of the connectivity discriminant algorithms based on the reachability matrix, the Warshall algorithm and our improved discriminant algorithm, respectively, in order to compare their time complexities.

(1) The reachability-matrix-based connectivity discriminant algorithm, which needs to calculate the reachability matrix first, requires a total of $5 \times 6^2 + 5 \times 6^2 \times (2 \times 6 - 1) = 2160$ integer operations;
(2) The Warshall connectivity discriminant algorithm totally needs to perform $2 \times 6^3 = 432$ bit operations;
(3) Our connectivity discriminant algorithm performs a total of $6^3 = 216$ bit operations.

Obviously, the algorithm proposed in this paper requires the least number of calculations.

## 4  Conclusion

Connectivity discriminant algorithms for directed networks not only play an important role in graph theory, but also have wide application in communication, computer, control, transportation, energy grids, and various kinds of large-scale networks. The complex directed networks have greater universality than other kinds of networks, and the research on efficient discriminant algorithms for the strong connectivity has greater theoretical and realistic significance.

In Sect. 2, we gave the definition of adjacency relation matrix (Definition 5). The meaning of Definition 5 is to extend the adjacency matrices of simple directed networks to the adjacency relation matrices of complex directed networks, which provides a theoretical basis for extending the Warshall algorithm that is only suitable for simple directed networks to complex directed networks.

The relevant conclusion given in Sect. 3 is the main contribution of this paper. Aiming at the strong connectivity discriminant problem of directed networks, we extended the Warshall algorithm only for simple directed networks to an improved algorithm for complex directed networks, which can reduce the time complexity of the Warshall algorithm by half. This provides theoretical basis and algorithm support for consistency control of multi-agent systems or connectivity discrimination of other large-scale complex directed networks.

Furthermore, after each undirected edge is replaced by two directed edges with opposite directions, the proposed strong connectivity discriminant algorithms for complex directed networks can also be applied to determine the connectivity of complex undirected networks as well as complex hybrid networks.

# References

1. Rosen, K.H., Xu, L.T., Yang, J., Wu, B.: Discrete Mathematics and Its Applications, 1st edn. China Machine Press, Beijing (2016)
2. Yin, J.H., Wu, K.Y.: Graph Theory and Its Algorithm, 1st edn. The University of Science and Technology of China Press, Hefei (2016)
3. Braun, U., et al.: Test-retest reliability of resting-state connectivity network characteristics using fMRI and graph theoretical measures. NeuroImage **59**(2), 1404–1412 (2012)
4. Imae, T., Cai, K.: On algebraic connectivity of directed scale-free networks. J. Franklin Inst. **355**(16), 8065–8078 (2018)
5. Zhen, K., Doucette, E.A., Dixon, W.E.: Distributed connectivity preserving target tracking with random sensing. IEEE Trans. Automat. Contr. **64**(5), 2166–2173 (2019)
6. Min, H., Seo, W., Lee, J., Park, S., Hong, D.: Reliability improvement using receive mode selection in the device-to-device uplink period underlaying cellular networks. IEEE Trans. Wirel. Commun. **10**(2), 413–418 (2011)
7. Yoshioka, C., Namerikawa, T.: Observer-based consensus control strategy for multi-agent system with communication time delay. In: 2008 IEEE International Conference on Control Applications (CCA) Part of the IEEE Multi-Conference on Systems and Control, San Antonio, TX, USA (2008)
8. Mahmood, M.A., Seah, W.K.G., Welch, I.: Reliability in wireless sensor networks: a survey and challenges ahead. Comput. Netw. **79**, 166–187 (2015)
9. Yan, B., Shi, P., Lim, C.C., Wu, C.F., Shi, Z.Y.: Optimally distributed formation control with obstacle avoidance for mixed-order multi-agent systems under switching topologies. IET Control Theor. A. **12**(13), 1853–1863 (2018)
10. Zhao, S.Y.: Affine formation maneuver control of multiagent systems. IEEE Trans. Automat. Contr. **63**(12), 4140–4155 (2018)
11. Su, H.S., Wu, H., Chen, X., Chen, M.Z.Q.: Positive edge consensus of complex networks. IEEE Trans. Syst. Man Cybern. Syst. **48**(12), 2242–2250 (2018)

# Four Extensions of Poisson Distribution: Properties and Different Methods of Estimation

Zhigao Chen[1](✉) and Renyan Jiang[2]

[1] School of Mathematics and Statistics, Changsha University of Science and Technology, Changsha 410114, Hunan, China
chenzhigao@csust.edu.cn
[2] School of Automotive and Mechanical Engineering, Changsha University of Science and Technology, Changsha 410114, Hunan, China

**Abstract.** Generally, when using the Poisson distribution to fit the count data, a basic assumption is that the sample mean and sample variance of the data are roughly equal. If this basic condition is not met, it is not appropriate to model with Poisson distribution. Some extended Poisson distributions have been generalized by researchers to model the Poisson-like distribution data. This paper discusses four new two-parameter extended Poisson distributions: exponentiated Poisson distribution, transmuted Poisson distribution, odd Poisson distribution and Alpha logarithmic transformed Poisson distribution, which provide better fitting effects than traditional Poisson distributions and the traditional Poisson distribution is a special case. The effects of parameters on the distribution function in these distributions are discussed, mainly on the two aspects of mathematical expectation and variance. Compared with the traditional Poisson distribution, the over- or under-Poisson dispersive properties of their distribution are studied. According to the actual three sets of specific data sets, the parameters of these models were estimated by maximum likelihood estimators, least squares estimators, and the suitability of these models was analyzed by comparative analysis.

**Keywords:** Two-parameter extended Poisson distribution · Over- or under-Poisson dispersive · Parameter estimation

## 1 Introduction

The Poisson distribution is a probabilistic model associated with the counting process with a wide range of applications. The Poisson distribution is suitable for describing the probability distribution of the number of random events occurring per unit time. When a random event, such as the calls received by a telephone exchange; passengers coming to a bus stop; particles emitted by a radioactive substance; white blood cells in a certain area under a microscope, etc., occurs at a fixed average instantaneous rate $\lambda$ (When the density or randomness occurs randomly and independently), then the number or number of occurrences of this event in unit time approximately obey the Poisson

© Springer Nature Singapore Pte Ltd. 2019
Q.-L. Li et al. (Eds.): Cao Festschrift 2019, CCIS 1102, pp. 475–488, 2019.
https://doi.org/10.1007/978-981-15-0864-6_25

distribution $P(\lambda)$. Therefore, Poisson distribution plays an important role in some issues of management science, operations research and natural science.

There are many Poisson-like distribution data in real life, and some researchers have developed two-parameter extended Poisson distributions to fit these data. These models include the stuttering Poisson, zero-inflated Poisson (ZIP), hurdle shifted Poisson (HSP), negative binomial, normal and gamma distributions [1–6]. Among these distributions, the stuttering Poisson is computationally intensive; ZIP, HSP and negative binomial distribution are only applicable for over-Poisson dispersive cases.

In order to find some more general models to fit various non-typical count data, and considering the simplicity of the model, we discuss the extension of four new two-parameter Poisson distribution models in this paper. They are exponentiated Poisson distribution, transmuted Poisson distribution, Alpha logarithmic transformed Poisson distribution and odd Poisson distribution. Our purpose is to analyze the characteristics of these models, to understand the performance of their modeling count data, and to enrich the methods of modeling count data.

The paper is organized as follows. In Sect. 2, we summarize two-parameter extensions of the Poisson distribution. The study of four new models is in Sect. 3. The appropriateness and usefulness of the proposed models and approaches are illustrated in Sect. 4. The paper is concluded in Sect. 5.

## 2  Earlier Extensions of the Poisson Distribution

There are several extended Poisson distributions and those extensions with two parameters are outlined as follows.

The first extension is the stuttering Poisson distribution, which is obtained by compounding the Poisson distribution with a geometric distribution [6]. Its probability mass function (pmf) is given by

$$f(x) = \begin{cases} e^{-\lambda}, & x = 0 \\ \sum_{i=1}^{x} C_{x-1}^{i-1}(1-\rho)^{i}\rho^{x-1}\frac{\lambda^{i}e^{-\lambda}}{i!} & x > 0 \end{cases}, \lambda > 0, \rho \in (0,1). \tag{1}$$

Where $x$ represents a non-negative integer and $x$ in the latter model also represents a non-negative integer. The mean and variance are given by

$$\mu = \lambda/(1-\rho), \quad \sigma^2 = \mu(1+\rho)/(1-\rho). \tag{2}$$

From Eq. (2), we have $\sigma^2 > \mu$, implying that it is always over-Poisson dispersive. Clearly, it is computationally intensive and hence its application is limited.

The second extension is the zero-inflated Poisson distribution. It is developed for modeling the excess zeros [2, 4, 7, 8]. The pmf of the ZIP model is given by

$$f(x) = \begin{cases} 1 - \alpha + \alpha e^{-\lambda}, & \text{for } x = 0 \\ \alpha\frac{\lambda^{x}e^{-\lambda}}{x!}, & \text{for } x > 0 \end{cases} \lambda > 0, \alpha \in (0,1). \tag{3}$$

The mean and variance are given by

$$\mu = \alpha\lambda, \quad \sigma^2 = \mu(1 + \lambda - \alpha\lambda). \tag{4}$$

Clearly, we have $\sigma^2 > \mu$, implying that it is always over-Poisson dispersive.

The third extension is the hurdle shifted Poisson distribution (SPD), which is a mixture of a deterministic (or degenerate) distribution and the shifted Poisson distribution [1]. Its pmf is given by

$$f(x) = \begin{cases} q, & x = 0 \\ (1-q)\frac{\lambda^{x-1}e^{-x}}{(x-1)!}, & x > 0 \end{cases}. \tag{5}$$

The mean and variance are given by

$$\mu = (1-q)(\lambda+1), \quad \sigma^2/\mu = q(1+\lambda) + \lambda/(1+\lambda). \tag{6}$$

Clearly, it can be over-Poisson dispersion (when $\lambda$ or $q$ is large) or under-Poisson dispersion (when $\lambda$ and $q$ are small).

## 3 Proposed Extensions of the Poisson Distribution

In this section we discuss four two-parameter Poisson distributions. These four models in this paper are based on the ordinary Poisson distribution. By adding a parameter to construct a monotonically rising function of the original distribution function, a new two-parameter Poisson distribution is generated. These distributions are generalizations of the ordinary Poisson distribution, and the ordinary Poisson distribution is a special case of these generalized Poisson distributions. These functions include exponentiated Poisson distribution, transmuted Poisson distribution, odd Poisson distribution and Alpha logarithmic transformed Poisson distribution.

### 3.1 Exponentiated Poisson Distribution

Jiang proposed an exponentiated Poisson distribution (EPD) for modelling bus-motor failure data [9]. The EPD is defined by

$$F(x) = e^{-\lambda\gamma}(\sum_{i=0}^{x} \frac{\lambda^i}{i!})^{\gamma}, \quad \lambda, \gamma > 0. \tag{7}$$

In fact, this distribution function can be expressed as

$$F(x) = (\sum_{i=1}^{x} \frac{\lambda^i}{i!}e^{-\lambda})^{\gamma} = [G(x)]^{\gamma} \tag{8}$$

where $G(x)$ is a Poisson distribution with a parameter of $\lambda$. In this way we can clearly see the regulation of the original Poisson distribution by the parameter $\gamma$.

To examine its Poisson dispersion, we randomly generate a set of combinations of $\lambda$ and $\gamma$. For each combination, we calculate the values of $\mu$ and $\sigma^2$. Figure 1 shows the plot of $\mu/\sigma^2$ vs. $\gamma$, and Fig. 2 shows the plot of $\mu$ vs. $\lambda$. As can be seen from these two figures, the EPD can be either over-Poisson dispersion (when $\gamma < 1$) or under-Poisson dispersion (when $\gamma > 1$); and the mean value ($\mu$) is highly correlated with the value of $\lambda$.

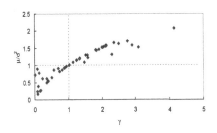

| **Fig. 1.** Plot of $\mu/\sigma^2$ vs. $\gamma$ | **Fig. 2.** Plot of $\mu$ vs. $\lambda$ |
|---|---|

### 3.2   Transmuted Poisson Distribution

For an arbitrary distribution function $G(x)$, Shaw and Buckley [10] define a transmuted distribution as

$$F(x) = (1 + \alpha)G(x) - \alpha G^2(x), \quad |\alpha| \leq 1. \tag{9}$$

When $G(x)$ is the Poisson distribution, we obtain the transmuted Poisson distribution (TPD). In this paper we examine the appropriateness of TPD as an alternative distribution to model the Counting data for Poisson-like distribution.

To examine its Poisson dispersion, we randomly generate a set of combinations of $\lambda$ and $\alpha$. For each combination, we calculate the values of $\mu$ and $\sigma^2$.

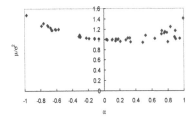
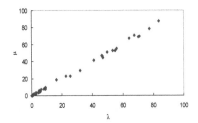

| **Fig. 3.** Plot of $\mu/\sigma^2$ vs. $\alpha$ | **Fig. 4.** Plot of $\mu$ vs. $\lambda$ |
|---|---|

Figure 3 shows the plot of $\mu/\sigma^2$ vs. $\alpha$; and Fig. 4 shows the plot of $\mu$ vs. $\lambda$. As can be seen from these two figures, TPD can be slightly over-Poisson dispersive (when $\alpha > 0$ and $\lambda$ is relatively small) or under-Poisson dispersion; and the mean value ($\mu$) is highly correlated with the value of $\lambda$.

### 3.3 Alpha Logarithmic Transformed Poisson Distribution

Pappas et al. [11] proposed a new method for generating distributions with the following cumulative distribution function (CDF)

$$f(x) = \begin{cases} 1 - \frac{\log[\alpha-(\alpha-1)G(x)]}{\log \alpha}, & \alpha > 0, \alpha \neq 1 \\ G(x), & \alpha = 0 \end{cases}. \tag{10}$$

When $G(x)$ is the Poisson distribution, we obtain the Alpha logarithmic transformed Poisson distribution (ALTPD).

In order to study the characteristics of ALTPD, we used experimental design methods to select different combinations of two parameters $\lambda$ and $\alpha$ ($\lambda$ is the parameter of ordinary Poisson distribution), and discuss their Poisson dispersion by calculating their mean and variance ratio.

ALTPD has a very stable Poisson dispersive. In general, for a given small parameter $\lambda$ (for example, no more than 4), as the parameter $\alpha$ increases, $\mu/\sigma^2$ becomes larger. For the larger parameter $\lambda$, the value of $\mu/\sigma^2$ is very stable close to 1. In addition, when $\alpha$ is less than 1, $\mu/\sigma^2$ is less than 1, and the model is weak over-Poisson dispersive (slightly less than 1). When $\alpha$ is greater than 1, $\mu/\sigma^2$ is generally larger than 1, and the model is weak under-Poisson dispersive (slightly greater than 1). For each given $\alpha$, the mean value ($\mu$) is highly correlated with the value of $\lambda$. Figure 5 shows the plot of $\mu/\sigma^2$ vs. $\alpha$ when $\lambda = 4$; and Fig. 6 shows the plot of $\mu$ vs. $\lambda$.

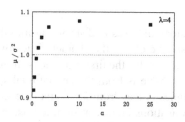

**Fig. 5.** Plot of $\mu/\sigma^2$ vs. $\alpha$

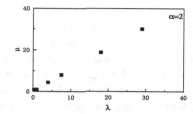

**Fig. 6.** Plot of $\mu$ vs. $\lambda$

### 3.4 Odd Poisson Distribution

Kahadawala [12] proposed a new method for generating distributions with the following cumulative distribution function (CDF):

$$F(x) = \begin{cases} \dfrac{F_\omega^\beta(x)}{F_\omega^\beta(x) + [1-F_\omega(x)]^\beta}, & \lambda > 0, \; \beta > 0 \\[4mm] \dfrac{F_\omega^{-\beta}(x)}{F_\omega^{-\beta}(x) + [1-F_\omega(x)]^{-\beta}}, & \lambda > 0, \; \beta < 0 \end{cases} \quad (11)$$

When $F_w(x)$ is the Poisson distribution, we obtain the Odd Poisson distribution (OPD). OPD has two parameters: $\lambda$ (the rate of occurrence) and $\beta$. In order to study the influence of two parameters on the distribution, this paper uses the different combinations of $\lambda$ and $\beta$. Through calculation and analysis, the following characteristics about the model are obtained.

The parameter $\beta$ has a great influence on the Poisson dispersive of OPD. For a given parameter value $\lambda$, $\mu/\sigma^2$ increases as the parameter $\beta$ increases, and has an exponential function growth pattern. In general, when $\beta$ is less than 1, the value of $\mu/\sigma^2$ is less than 1, the model is over-dispersive; when $\beta$ is greater than 1, the value of $\mu/\sigma^2$ is greater than 1, and the model is over-dispersive. On the other hand, for each given $\alpha$, the mean value ($\mu$) is highly correlated with the value of $\lambda$. Figure 7 shows the plot of $\mu/\sigma^2$ vs. $\alpha$ when $\lambda = 4$; and Fig. 8 shows the plot of $\mu$ vs. $\lambda$ when $\beta = 3.5$.

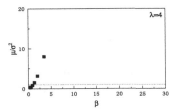

**Fig. 7.** Plot of $\mu/\sigma^2$ vs. $\alpha$

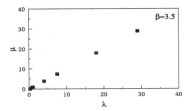

**Fig. 8.** Plot of $\mu$ vs. $\lambda$

## 4   Illustrations

In the current section we will fit the four extensions of Poisson distribution into three different count data sets. By comparative analysis, we obtain the characteristics of the four models in fitting the modeling count data. The data for the first example is from the strike counts data, and the data for the second example is from the annual total spare parts turnover numbers. The data for both examples comes from the count data in the real case. Similar count data in many other applications areas can be fitted using the above four models. In the following case study we take the following steps:

- First we use the maximum likelihood estimation method to fit these count data with five models (including the ordinary Poisson distribution and the four proposed models), and obtain the parameters of the model to determine the specific model. The performance of each model was initially evaluated by comparing the effects of the fit of each model. Since there is no previous information, we must first make a model selection. In this paper we use two common criteria: Akaike information criterion (*AIC*) and the Bayesian information criterion (*BIC*), which are respectively given by

$$AIC = -2\ln(L_{max}) + 2k, \ BIC = -2\ln(L_{max}) + k\ln(n), \tag{12}$$

where $n$ is the sample size, $k$ is the number of the parameters that have been estimated and $L_{max}$ is the maximum value of the likelihood function for the estimated model [13, 14]. Under the two criteria, the method of determining the optimal model is: the smaller the value *AIC*(*BIC*) obtained by calculation, the more appropriate the model is.

- In order to compare the fitting effects of these models, we calculate the error sum of squares (*SSE*, equivalently, residual sum of squares) which is obtained by the maximum likelihood estimation method. *SSE* is denoted by

$$SSE = \sum_i (y_i - \hat{y}_i)^2 \tag{13}$$

where $y_i$ is the actual observed data and $\hat{y}_i$ is the predicted data calculated by the fitted model. A small *SSE* indicates that the model achieved a good fit effect. In addition, the sample mean and sample variance of the original data set and the mathematical expectation and variance of the simulated model are calculated separately. These performances of the five models are comprehensively evaluated by comparing these digital features.
- Fit the five models to three data sets by least squares and compare the fit of each model by calculation.
- Combine the fitted scatter plot, maximum likelihood estimation method and least squares method to make a comprehensive evaluation of the fitting effect of the five models on the three count data sets.

## 4.1   Example 1

The data shown in Fig. 9 come from the number of outbreaks of strikes (in 4-week periods) in the UK coal mining industry for the years 1948–1959. These data can be found in Kendall [15], while they were further analyzed by Athanasios et al. [16]. The abscissa indicates the number of weeks in which the strike continues, and the vertical coordinates indicates the number of times the five strikes occurred. Through simple calculations, we obtain that the sample mean and variance are equal to 0.994 and 0.742, respectively. Therefore, it is obvious that the data set is of under-Poisson dispersion since the sample mean–variance ratio is equal to 0.994/0.742 = 1.3396 > 1.

As can be seen from Fig. 10, except for the ordinary Poisson model and the ALTPD model, the fitting effect is slightly worse, and the other three models have better fitting effects.

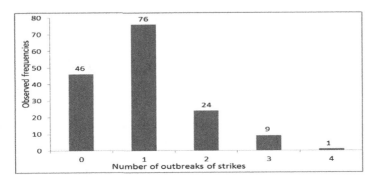

**Fig. 9.**  Bar plot for the strike count data

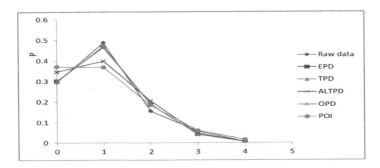

**Fig. 10.**  Scatter plots of raw data and fitted models for Example 1

Table 1 lists each model parameters obtained by the maximum likelihood estimation method (MLEM), in the second and third columns, respectively. The *AIC* and *BIC* values are placed in the fourth and fifth columns, respectively. The last column lists the *SSE* calculated from the model obtained by the maximum likelihood estimation method. From the perspective of the *AIC* and *BIC*, EPD, TPD and OPD are not much different, and they are relatively small. The difference is that the TPD model has the smallest *SSE* value and the best fit. EPD has the smallest *AIC* and *BIC* value and is the most suitable model.

**Table 1.**  Estimated parameters and model performances for Example 1(MLE)

| Model | Parameter1 ($\lambda$) | Parameter2 | AIC | BIC | SSE |
|---|---|---|---|---|---|
| POISSON | 0.9936 | | 387.8724 | 392.9223 | 0.020834 |
| EPD | 0.6199 | 1.9464 | 379.2787 | 385.3784 | 0.001819 |
| TPD | 0.6082 | −0.9986 | 379.2926 | 385.3923 | 0.001661 |
| ALTPD | 0.5231 | 8.5559 | 383.6997 | 389.7994 | 0.013013 |
| OPD | 1.0661 | 1.3106 | 379.3668 | 385.4665 | 0.001898 |

Table 2 lists the Mean, variance, and mean-variance ratios of each model, which are represented by $\mu$, $\sigma^2$, and $\mu/\sigma^2$, respectively, and the calculation results are placed in columns 3 to 7, respectively. In particular, the second column gives the sample mean and sample variance, which are not distinguished by special symbols.

It can be clearly seen from Table 2 that the Mean, variance, and mean-variance ratios of the three models EPD, TPD and OPD are very close to the sample mean and sample variance, and the ratio of the two values are very close, indicating that the three models are relative suitable model.

**Table 2.** Mean, variance, and mean-variance ratio for each model for Example 1

|  | Data | Poisson | EPD | TPD | ALTPD | OPD |
|---|---|---|---|---|---|---|
| $\mu$ | 0.9936 | 0.9936 | 0.9923 | 0.9921 | 0.9732 | 0.9903 |
| $\sigma^2$ | 0.7419 | 0.9936 | 0.7228 | 0.7137 | 0.8044 | 0.7235 |
| $\mu/\sigma^2$ | 1.3393 | 1.0000 | 1.3730 | 1.3900 | 1.2099 | 1.3686 |

Table 3 lists the model parameters obtained by least squares method and *SSE*. By comparing with Table 2, we find that the model parameters obtained by the two parameter estimation methods have little difference except for the ALTPD model. This indicates that the ALTPD model is not robust enough. In addition, the comparison of the SSE values shows that the least squares method improves the maximum likelihood method and the SSE values are all smaller.

**Table 3.** Estimated parameters and model performances for Example 1 (LS)

| Model | Parameter1 ($\lambda$) | Parameter2 | SSE |
|---|---|---|---|
| POISSON | 1.0900 |  | 0.018784 |
| EPD | 0.5239 | 2.2894 | 0.001055 |
| TPD | 0.5925 | −1.0000 | 0.001523 |
| ALTPD | 0.2358 | 69.0121 | 0.011815 |
| OPD | 1.0377 | 1.3963 | 0.001084 |

Through the above analysis and comparison, it is not appropriate to use the simple Poisson distribution and the ALTPD for modeling the data of strike outbreaks, while the other three models EPD, TPD and OPD are appropriate in terms of the *AIC* and *BIC* values, fitting effect and the comparison of digital features. Especially the performance of the EPD model is the best.

## 4.2   Example 2

The data shown in Table 4 come from Guo et al. [17] and deal with the annual total spare parts turnover numbers (which reflect the actual demand of a repairable unit) of

Part 1 and 2. The sample mean and variance for Part 1 [Part 2] are 5.4 and 3.14 [9.3 and 12.70], respectively, implying under-Poisson [over-Poisson] dispersion.

**Table 4.** Demand data for Example 2

| Year | Part 1 | Part 2 | Year | Part 1 | Part 2 | Year | Part 1 | Part 2 | Year | Part 1 | Part 2 |
|---|---|---|---|---|---|---|---|---|---|---|---|
| 1 | 7 | 11 | 14 | 6 | 10 | 27 | 6 | 11 | 40 | 3 | 11 |
| 2 | 6 | 9 | 15 | 8 | 9 | 28 | 5 | 6 | 41 | 4 | 8 |
| 3 | 8 | 10 | 16 | 9 | 16 | 29 | 4 | 4 | 42 | 4 | 7 |
| 4 | 6 | 8 | 17 | 5 | 18 | 30 | 4 | 5 | 43 | 3 | 6 |
| 5 | 5 | 8 | 18 | 8 | 10 | 31 | 5 | 5 | 44 | 4 | 5 |
| 6 | 10 | 9 | 19 | 7 | 18 | 32 | 3 | 12 | 45 | 4 | 7 |
| 7 | 7 | 12 | 20 | 4 | 12 | 33 | 3 | 7 | 46 | 3 | 5 |
| 8 | 6 | 7 | 21 | 6 | 10 | 34 | 5 | 9 | 47 | 5 | 13 |
| 9 | 5 | 5 | 22 | 5 | 8 | 35 | 7 | 8 | 48 | 3 | 8 |
| 10 | 5 | 6 | 23 | 7 | 9 | 36 | 8 | 15 | 49 | 4 | 9 |
| 11 | 6 | 6 | 24 | 5 | 7 | 37 | 4 | 17 | 50 | 4 | 7 |
| 12 | 4 | 13 | 25 | 4 | 7 | 38 | 7 | 9 | | | |
| 13 | 4 | 8 | 26 | 9 | 8 | 39 | 6 | 17 | | | |

Figure 11 shows the scatter plot of two data sets and the models obtained from maximum likelihood estimation. To illustrate the performance of these proposed models in modeling count data, we list some of the modeling information of each model in Tables 5 and 6, respectively, which include the model parameter values (in the second column and Three columns), AIC and BIC values (in the fourth and fifth columns), and SSE (in the last column). By comparing the figure and the table, we have the following observations (Tables 7 and 8):

- According to the results of columns 4–6, the best model for data set 1 (Part 1) is EPD, followed by the OPD; for data set 2 (Part 2), the ordinary Poisson distribution is the most suitable model in terms of the AIC and BIC values, while TPD is the best fit model in terms of SSE.
- It can be seen from Fig. 11 that for Part 1, the fitting effect of all models is slightly better than Part 2. This is partly because the data of Part 2 does not have the typical characteristics of Poisson distribution (unimodal), so the overall fitting effect is not very good.
- In Part 1, the parameters of TPD model are the same as those of the ordinary Poisson distribution, indicating that for larger $\mu/\sigma^2$ (1.72), the TPD model has no advantage over the Poisson distribution. In Part 2, the modeling of ALTPD has similar results.

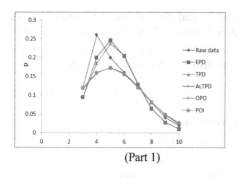 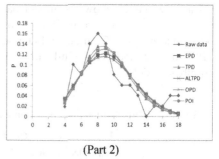

(Part 1)                          (Part 2)

**Fig. 11.** Scatter plot of raw data and fitted models for Example 1

**Table 5.** Estimated parameters and model performances for Example 2 (MLE, Part 1)

| Model | Parameter1 ($\lambda$) | Parameter2 | AIC | BIC | SSE |
|-------|------------|------------|-----|-----|-----|
| POISSON | 5.4000 | | 205.0478 | 208.9599 | 0.010871 |
| EPD | 3.7442 | 2.9630 | 198.0012 | 201.8252 | 0.008825 |
| TPD | 5.4000 | 0.0000 | 207.0478 | 210.8719 | 0.010871 |
| ALTPD | 5.0421 | 1.6602 | 206.9666 | 210.7907 | 0.011118 |
| OPD | 5.4377 | 1.3999 | 200.0431 | 203.8672 | 0.010359 |

**Table 6.** Estimated parameters and model performances for Example 2 (MLE, Part 2)

| Model | Parameter1 ($\lambda$) | Parameter2 | AIC | BIC | SSE |
|-------|------------|------------|-----|-----|-----|
| POISSON | 9.3000 | | 265.9286 | 267.8406 | 0.012575 |
| EPD | 9.8972 | 0.8110 | 267.443 | 271.267 | 0.012773 |
| TPD | 9.9231 | 0.4000 | 266.5151 | 270.3392 | 0.010811 |
| ALTPD | 9.3000 | 0.0000 | 267.9286 | 271.7526 | 0.012575 |
| OPD | 8.9376 | 0.9987 | 266.8197 | 270.6438 | 0.012817 |

**Table 7.** Mean, variance, and mean-variance ratio for each model for Example 1 (Part 1)

| | Data | Poisson | EPD | TPD | ALTPD | OPD |
|---|------|---------|-----|-----|-------|-----|
| $\mu$ | 5.4000 | 5.4000 | 5.3842 | 5.4000 | 5.3628 | 5.3623 |
| $\sigma^2$ | 3.1429 | 5.4000 | 2.8392 | 5.4000 | 5.2209 | 3.0946 |
| $\mu/\sigma^2$ | 1.7182 | 1.0000 | 1.8963 | 1.0000 | 1.0272 | 1.7328 |

**Table 8.** Mean, variance, and mean-variance ratio for each model for Example 1 (Part 2)

| | Data | Poisson | EPD | TPD | ALTPD | OPD |
|---|------|---------|-----|-----|-------|-----|
| $\mu$ | 9.3000 | 9.3000 | 9.3162 | 9.2168 | 9.3000 | 8.9379 |
| $\sigma^2$ | 12.7041 | 9.3000 | 10.7027 | 9.0664 | 9.3000 | 8.9568 |
| $\mu/\sigma^2$ | 0.7320 | 1.0000 | 0.8705 | 1.0166 | 1.0000 | 0.9979 |

Tables 9 and 10 show the model parameters and *SSE* obtained by least squares under the two data sets, respectively. For Part 1, the EPD model is superior to other models in terms of *SSE*, and the OPD model is second. For Part 2, from the perspective of *SSE*, the five models have little difference in terms of *SSE*, and TPD is slightly better than other models.

**Table 9.** Estimated parameters and model performances for Example 2 (LS, Part 1)

| Model | Parameter1 ($\lambda$) | Parameter2 | SSE |
|---|---|---|---|
| POISSON | 4.9961 | | 0.008828 |
| EPD | 3.9498 | 2.0131 | 0.004385 |
| TPD | 4.9961 | 0.0000 | 0.008828 |
| ALTPD | 4.4968 | 1.9795 | 0.008694 |
| OPD | 5.0328 | 1.2477 | 0.005025 |

**Table 10.** Estimated parameters and model performances for Example 2 (LS, Part 2)

| Model | Parameter1 ($\lambda$) | Parameter2 | SSE |
|---|---|---|---|
| POISSON | 8.3339 | | 0.007603 |
| EPD | 7.9735 | 1.1646 | 0.007447 |
| TPD | 9.3235 | 0.5373 | 0.007044 |
| ALTPD | 8.3313 | 0.0000 | 0.007602 |
| OPD | 8.3246 | 1.0604 | 0.007426 |

## 5   Conclusions

As an extension and generalization of a simple Poisson distribution, four two-parameter Poisson distributions are discussed. The article discusses the construction of the model, the relationship between mean and variance and two parameters, and Poisson dispersion. In order to study the performance of these models in terms of counting data, we demonstrate the suitability and performance differences of these proposed models by two parameter estimation methods through two real-world examples. We have the following conclusions:

1. Each of the above distribution is a generalized form of two-parameter Poisson distribution constructed by a distribution function of ordinary Poisson distribution in a certain form. Each distribution is flexible without losing simplicity.
2. The EPD model and the OPD model have relatively good modeling adaptability, which is related to the characteristics of the two models. When the second parameter $\gamma$ of the EPD model is less than 1, the model is over-Poisson dispersive, and while the $\gamma$ is greater than 1, the model is under-Poisson dispersive. Poisson dispersion can be adjusted by parameter $\gamma$. In the OPD model, the influence of the parameter $\beta$ is very large: when $\beta$ is small, the model is over-Poisson dispersive;

when $\beta$ is large, and the model is under-Poisson dispersive, and $\mu/\sigma^2$ increases rapidly with the increase of $\beta$.

3. The parameters of the proposed model are highly correlated with the mean of the distribution, which helps to estimate the initial values of the parameters, and also provides a criterion for the suitability of the model, that is, the estimated value of the parameters should be close to the sample mean.

4. The TPD model is suitable for under-Poisson dispersion, especially for cases that $\mu/\sigma^2$ is slightly greater than 1, and no more than 1.5 is preferred. The ALTPD model is relatively stable, and it's $\mu/\sigma^2$ generally varies in the interval [0.9, 1.1], so the ALTPD model is suitable for modeling data sets with $\mu/\sigma^2$ close to 1. For the case that $\mu/\sigma^2$ is too large or too small, the two models cannot achieve a good fitting effect.

5. For counting data that does not have Poisson distribution characteristics, such as multi-modality, the above model modelling effect is not ideal, so it is necessary for the counting data to initially select the model according to the data characteristics.

These four models enrich the modeling method for modeling Poisson type count data. Combined with the characteristics of specific practical problems, from the generation mechanism of counting data, it is also possible to develop a generalized model of Poisson distribution that is more in line with specific practical problems.

**Acknowledgment.** The research was supported by the National Natural Science Foundation of China (No. 71771029).

# References

1. Snyder, R.D., Ord, J.K., Beaumont, A.: Forecasting the intermittent demand for slow-moving inventories: a modelling approach. Int. J. Forecast. **28**(2), 485–496 (2012)
2. Arab, A., Holan, S.H., Wikle, C.K., Wildhaber, M.L., Wildhaber, M.L.: Semiparametric bivariate zero-inflated Poisson models with application to studies of abundance for multiple species. Environmetrics **23**(2), 183–196 (2012)
3. Begum, A., Mallick, A., Pal, N.: A generalized inflated Poisson distribution with application to modeling fertility data. Thail. Stat. **12**(2), 135–139 (2014)
4. Costantino, F., Gravio, G.D., Patriarca, R., Petrella, L.: Spare parts management for irregular demand items. Omega **81**, 57–66 (2018)
5. Turrini, L., Meissner, J.: Spare parts inventory management: new evidence from distribution fitting. Eur. J. Oper. Res. **273**(1), 118–130 (2019)
6. Chen, C.W.: Using geometric Poisson exponentially weighted moving average control schemes in a compound Poisson production environment. Comput. Ind. Eng. **63**(2), 374–381 (2012)
7. Lambert, D.: Zero-inflated Poisson regression, with an application to defects in manufacturing. Technometrics **3**, 1–14 (1992)
8. Majeske, K.D., Herrin, G.D.: Assessing mixture-model goodness-of-fit with an application to automobile warranty data. In: Proceedings of Annual Reliability and Maintainability Symposium, pp. 378–383 (1995)
9. Jiang, R.: Discrete competing risk model with application to modeling bus-motor failure data. Reliab. Eng. Syst. Saf. **95**(9), 981–988 (2010)

10. Shaw, W.T., Buckley, I.R.: The alchemy of probability distributions: beyond Gram-Charlier expansions, and a skew-kurtotic normal distribution from a rank transmutation map (2009)
11. Pappas, V., Adamidis, K., Loukas, S.: A family of lifetime distributions. Internat. J. Qual. Statist. Reliab. (2012) http://dx.doi.org/10.1155/2012/760687
12. Cooray, K.: A study of moments and likelihood estimators of the odd Weibull distribution. Stat. Methodol. **26**, 72–83 (2015)
13. Akaike, H.T.: A new look at the statistical model identification. IEEE Trans. Autom. Control **19**(6), 716–723 (1974)
14. Acquah, H.D.G.: A bootstrap approach to evaluating the performance of Akaike Information Criterion (AIC) and Bayesian Information Criterion (BIC) in selection of an asymmetric price relationship. J. Agric. Sci. **57**(2), 99–110 (2012)
15. Kendall, M.G.: Natural law in social sciences. The Inaugural Address of the President, OR, vol. 12, no. 4, p. 292 (1961)
16. Rakitzis, A.C., Castagliola, P., Maravelakis, P.E.: A two-parameter general inflated Poisson distribution: properties and applications. Stat. Methodol. **29**, 32–50 (2016)
17. Guo, F., Diao, J., Zhao, Q.H., Wang, D.X., Sun, Q.: A double-level combination approach for demand forecasting of repairable airplane spare parts based on turnover data. Comput. Ind. Eng. **110**, 92–108 (2017)

# The Past Publications by Jinhua Cao

Jing-Yu Ma
School of Economics and Management
Yanshan University, Qinhuangdao 066004, China
mjy0501@126.com

This is a simple summary for The Past Publications by Jinhua Cao according to the order of their published years.

## The Books

1. Cao, J., Cheng, K.: Introduction to Reliability Mathematics. Science Press, Beijing (1986) (in Chinese)
2. Osaki, S., Cao, J.: Reliability Theory and Applications. World Scientific, Singapore (1987)
3. Wu, C.P., Cao, J.: Proceedings of the Second Conference of the Association of Asian-Pacific Operational Research Societies within IFORS. Beijing University Press, Beijing (1992)
4. Wang, Y.Q., Cheng, K., Cao, J.: Operations Research and Decisions. Chengdu University of Science and Technology Press, Chengdu (1992) (in Chinese)
5. Cao, J., Li, W., Liu, B.: Reliability Theory, Methods and Applications. RYORSC'94, Mechanical Industry Press, Beijing (1994) (in Chinese)
6. Cao, J.: Proceedings of the Fifth Symposium on Reliability. RSORSC'95, Mechanical Industry Press, Beijing (1995) (in Chinese)
7. Cao, J.: Proceedings of the Sixth Symposium on Reliability. RSORSC'98, Mechanical Industry Press, Beijing (1998) (in Chinese)
8. Cao, J.: Reliability Theory, Operations Research Handbook. Science Press, Beijing, 517–591 (1999) (in Chinese)

## Papers

9. Cao, J., Yan, J.Y.: Some problems of M/G/1 queueing system in which the probability that a customer joins in queue depends on the queue size. In: Collection of Scientific Papers for the Fifth Anniversary of Chinese University of Science and Technology, pp. 65–75 (1963) (in Chinese)
10. Cheng, K., Cao, J.: Reliability analysis of network systems. Reliability and Environment Tests 1, 7–19 (1977) (in Chinese)
11. Cheng, K., Cao, J.: Reliability analysis of two-unit parallel system. Acta Mathematicae Applagatae Sinica 1(4), 341–352 (1978) (in Chinese)

© Springer Nature Singapore Pte Ltd. 2019
Q.-L. Li et al. (Eds.): Cao Festschrift 2019, CCIS 1102, pp. 489–495, 2019.
https://doi.org/10.1007/978-981-15-0864-6

12. Cao, J., Cheng, K.: Reliability analysis of two-unit warm standby system. Acta Mathematicae Applagatae Sinica 3(2), 147–160 (1980) (in Chinese)

13. Cao, J., Cheng, K.: Reliability analysis of a two-unit cold standby system with priority in use and repair. Journal of Shanghai Institute of Railway Technology 2(4), 33–45 (1981) (in Chinese)

14. Cheng, K., Cao, J.: Reliability analysis of a general repairable system: a Markov renewal model. Acta Mathematicae Applagatae Sinica 4(4), 295–306 (1981) (in Chinese)

15. Cao, J.: Reliability analysis of unrepairable systems. Reliability and Environment Tests (4), 36–51 (1981) (in Chinese)

16. Cao, J., Cheng, K.: Analysis of M/G/1 queueing system with repairable service station. Acta Mathematicae Applagatae Sinica 5(2), 114–127 (1982) (in Chinese)

17. Cao, J.: Reliability analysis of two-unit cold standby repairable system with three states. Journal of Fujian Teachers University (2), 1–12 (1982) (in Chinese)

18. Cao, J.: Reliability analysis of M/G/1 queueing system with repairable service station and finite arrival source. Chinese Journal of Operations Research 2(1), 53–55 (1983) (in Chinese)

19. Cao, J.: Analysis of a machine service model with a repairable service equipment. Journal of Mathematical Research and Exposition 4(4), 93–100 (1985) (in Chinese)

20. Cao, J.: A preventive maintenance policy for a multistate one-unit repairable system with minimal repair at failure. Acta Mathematicae Applicatae Sinica 9(1), 113–123 (1986) (in Chinese)

21. Cao, J., Wu, Y.H.: HDMRL life distribution class. Chinese Journal of Applied Probability and Statistics 2(4), 39–46 (1986) (in Chinese)

22. Xu, G.H., Dong, Z.Q., Cao, J., Cheng, K.: Applied probability: its fields and models. Advances in Mathematics 15(4), 347–366 (1986) (in Chinese)

23. Cao, J.: Availability and failure frequency of a multi-unit parallel system. The Asia-Pacific Journal of Operational Research 4(1), 83–90 (1987)

24. Cao, J.: Availability and failure frequency of two-unit series systems with shut-off rule. In: Reliability Theory and Applications, pp. 1–13. World Scientific, Singapore (1987)

25. Cao, J., Wu, Y.H.: Reliability analysis of a multistate system with a replaceable repair facility. Acta Mathematicae Applicatae Sinica 4(2), 113–121 (1988) (English Series)

26. Cao, J.: Reliability analysis of a repairable system in changing environment subject to a general alternating renewal process. Microelectronics and Reliability 28(6), 889–892 (1988)

27. Cao, J., Wu, Y.H.: Reliability analysis of a two-unit cold standby system with a replaceable repair facility. Microelectronics and Reliability 29(2), 145–150 (1989)

28. Cao, J.: Stochastic behavior of a man-machine system operating under changing environment subject to a Markov process with two states. Microelectronics and Reliability 29(4), 529–531 (1989)

29. Cao, J.: Reliability analysis of Gnedenko system with a repair facility subject to failure and replacement. Chinese Journal of Applied Probability and Statistics 5(3), 65–75 (1989) (in Chinese)

30. Cao, J.: Availability and failure frequency of Gnedenko system. Annals of Operations Research 24, 55–68 (1990)
31. Wang, Y.D., Cao, J.: New kinds of multidimensional IFR distribution. Advances in Applied Probability 22(1), 251–253 (1990)
32. Wang, Y.D., Cao, J.: Optimal allocation for a repairable system. Microelectronics and Reliability 30(6), 1091–1093 (1990)
33. Cao, J., Wang, Y.D.: The NBUC and NWUC classes of life distributions. Journal of Applied Probability 28(2), 473–479 (1991)
34. Cao, J., Wang, Y.D.: Puri birth shock model in changing environment. Mathematical Statistics and Applied Probability 6(1), 123–127 (1991) (in Chinese)
35. Cao, J.: The classes of life distributions defined by convex ordering. In: Proceedings of the 2nd Conference of the Association of Asian-Pacific Operational Research Societies, pp. 357–362, Beijing University Press, Beijing (1992)
36. Cao, J., Wang, Y.D.: Characterization problem of superposition of two renewal processes. Chinese Journal of Applied Probability and Statistics 8(1), 85–90 (1992)
37. Guo, T.D., Cao, J.: Reliability analysis of a multistate one-unit repairable system operating under changing environment. Microelectronics and Reliability 32(3), 439–443 (1992)
38. Guo, T.D., Cao, J.: Reliability analysis of a two-unit paralleled redundant system with a replaceable repair facility. Microelectronics and Reliability 32(9), 1237–1240 (1992)
39. Li, W., Cao, J.: The limiting distribution of the residual lifetime of Markov repairable system. Reliability Engineering and System Safety 41(2), 103–105 (1993)
40. Li, W., Cao, J.: The limiting distributions of the residual lifetimes of several repairable systems. Microelectronics and Reliability 33(8), 1069–1072 (1993)
41. Li, W., Cao, J.: Stochastic scheduling on a single-machine subject to two kinds of breakdowns. In: Proceedings of the First Chinese World Congress on Intelligent Control and Intelligent Automation, pp. 1619–1623. Science Press, Beijing (1993)
42. Cao, J.: Reliability analysis of M/G/1 queueing system with repairable service station of reliability series structure. Microelectronics and Reliability 34(4), 721–725 (1994)
43. Cao, J., Wang, Y.D.: The NBELC and NWELC classes of life distributions. Applied Mathematics, A Journal of Chinese Universities (Series B) 9(3), 237–244 (1994)
44. Wang, Y.D., Cao, J.: Some multivariate DMRL and NBUE distributions based on conditional stochastic order. Acta Mathematicae Applicatae Sinica 10(3), 328–332 (1994) (English Series)
45. Li, W., Cao, J.: Some performance measures of repairable transfer line with general distributions. Systems Science and Mathematical Sciences 7(4), 344–351 (1994) (English Series)
46. Li, W., Cao, J.: The explicit results of a generalized repairable CIMS transfer lines with two stations. Acta Automatica Sinica 20(5), 522–532 (1994) (in Chinese)

47. Li, W., Cao, J.: Stochastic scheduling on an unreliable machine with general uptimes and general set-up times. Journal of Systems Science and Systems Engineering 3(3), 279–288 (1994)

48. Li, W., Cao, J.: Some performance measures of a class of CIMS with reprocessing rule. Chinese Journal of Operations Research 13(1), 69–70 (1994) (in Chinese)

49. Li, W., Cao, J.: Optimal scheduling on an unreliable machine. In: Proc. of RYORSC'94, pp. 200–204. Mechanical Industry Press, Beijing (1994)

50. Liu, B., Cao, J.: Production control of an unreliable manufacturing system with multiple demand states. In: Proc. of RYORSC'94, pp. 205–211. Mechanical Industry Press, Beijing (1994)

51. Guo, T.D., Cao, J.: Some problems of replacement policies in production systems, In: Proc. of RYORSC'94, pp. 169–173. Mechanical Industry Press, Beijing (1994) (in Chinese)

52. Liu, B., Cao, J.: Production control of a class of unreliable manufacturing systems, In: Proc. of CSIAM'94, pp. 236–240. Qinghua University Press, Beijing (1994)

53. Rao, L., Li, P.Q., Cao, J.: A hardware/software reliability growth model. In: Proceedings of the Second International Conference on Reliability, Maintainability and Safety, pp. 577–583. International Academic Publishers, Beijing (1994)

54. Liu, B., Cao, J.: Analysis of a machine service model with a repairable service station of reliability series structure. Microelectronics and Reliability 35(4), 683–690 (1995)

55. Cao, J., Wang, Y.D.: The EBELC and EWELC classes of life distributions. Microelectronics and Reliability 35(6), 969–971 (1995)

56. Cao, J.: The steady-state production rate of semi-Markov manufacturing system. In: Proc. of RSORSC'95, pp. 156–160. Mechanical Industry Press, Beijing (1995) (in Chinese)

57. Li, Q.L., Cao, J.H.: MAR/PH(M/PH)/2 queueing system with repairable service stations. In: Proc. of RSORSC'95, pp. 178–184. Mechanical Industry Press, Beijing (1995) (in Chinese)

58. Liu, B., Cao, J.: A production-inventory system with an unreliable machine and several production rates. In: Proc. of RSORSC'95, pp. 161–165. Mechanical Industry Press, Beijing (1995)

59. Li, W., Cao, J.: T-policy of a manufacturing system with two parallel production facilities. In: Proc. of ISORA'95, pp. 277–284. World Publishing Corporation, Beijing (1995)

60. Liu, B., Cao, J.: A production-inventory system with repairable machines of reliability series structure. In: Proc. of ISORA'95, pp. 336–341. World Publishing Corporation, Beijing (1995)

61. Li, W., Cao, J.: Stochastic scheduling on a single-machine subject to multiple breakdowns. In: Proc. of IFAC YAC'95, pp. 444–449. Beijing (1995)

62. Li, W., Cao, J.: Stochastic scheduling on a single-machine subject to multiple breakdowns according to different probabilities. Operations Research Letters 18, 81–91 (1995)

63. Li, W., Cao, J.: Stochastic scheduling on a repairable manufacturing system. In: Proceedings of International Conference of Intelligent Manufacturing, pp. 167–172. Bellingham, WA, USA (1995)

64. Li, W., Cao, J.: Approximation of two-dissimilar-unit cold standby systems. Acta Mathematicae Applicatae Sinica 12(1), 71–77 (1996) (English Series)

65. Li, Q.L., Cao, J.: Equilibrium behavior of the matched queueing system, In: Proceedings of ISORA'96, pp. 487–499. World Publishing Corporation, Guiling (1996)

66. Yue, D.Q., Cao, J.: Preemptive resume priority queue with batch arrival and a repairable service station. In: Proceedings of ISORA'96, pp. 500–508. World Publishing Corporation, Guiling (1996)

67. Liu, B., Cao, J.: Production control of an unreliable manufacturing system under the assumption of no backlog. Mathematical Methods of Operations Research 46(1), 103–117 (1997)

68. Li, Q.L., Xu, D.J., Cao, J.: Reliability approximation of a Markov queueing system with server breakdown and repair. Microelectronics and Reliability 37(8), 1203–1212 (1997)

69. Yue, D.Q., Cao, J.: Reliability analysis of queueing system $M_1^{x_1}, M_2^{x_2}/G_1, G_2/1$ with repairable service station. Microelectronics and Reliability 37(8), 1225–1231 (1997)

70. Li, W., Cao, J.: Some performance measures of transfer line consisting of two unreliable machines with reprocess rule. Journal of Systems Science and Systems Engineering 6(4), 283–292 (1998)

71. Liu, B., Cao, J.: Inventory analysis of an unreliable production-inventory system. System Science and Mathematical Sciences 11(4), 367–374 (1998) (English Series)

72. Li, Q.L., Cao, J.: Pre-set value policy and optimization of assigning work-pieces in finite CIMS buffer. Operations Research Transactions 2(3), 15–24 (1998) (in Chinese)

73. Cao, J., Wang, C.H.: Monte-Carlo algorithm for the availability indexes of repairable K-terminal network systems. In: RSORSC'98, pp. 176–181. Mechanical Industry Press, Beijing (1998) (in Chinese)

74. Yue, D.Q., Cao, J.: Stochastic comparison properties for a special mixture of distributions. In: RSORSC'98, pp. 112–117. Mechanical Industry Press, Beijing (1998)

75. Li, Q.L., Tan, M., Cao, J., Wang, C.H.: Majorization character analysis of TPH lifetime. In: RSORSC'98, pp. 118–120. Mechanical Industry Press, Beijing (1998) (in Chinese)

76. Wang, C.H., Cao, J.: Reliability analysis of the systems with protect units. In: RSORSC'98, pp. 161–165. Mechanical Industry Press, Beijing (1998) (in Chinese)

77. Wang, J., Cao, J., Li, Q.L.: Limiting distribution of the residual lifetime of Markov repairable system with discrete time. In: RSORSC'98, pp. 199–202. Mechanical Industry Press, Beijing (1998) (in Chinese)

78. Liu, B., Cao, J.: Analysis of a production-inventory system with machine breakdowns and shutdowns. Computers and Operations Research 26, 73–91 (1999)

79. Tian, N.S., Li, Q.L., Cao, J.: Conditional stochastic decompositions in the M/M/C queue with server vacations. Stochastic Models 15(2), 1–10 (1999)

80. Wang, J., Cao, J.: Reliability analysis of the retrial queue with server breakdowns and repairs. In: The 15th Triennial Conference for the International Federation of Operational Research Societies, Beijing, August (1999)

81. Li, Q.L., Cao, J.: Quasi-stationary decomposition of the multi-server retrial queuing system. The 15th Triennial Conference for The International Federation of Operational Research Societies, Beijing, August (1999)

82. Li, Q.L., Cao, J.: The repairable queue MAP/PH(M/PH)/2 with interdependent repairs. Systems Science and Mathematical Sciences 20(1), 78–86 (2000) (in Chinese)

83. Yue, D.Q., Cao, J.: Residual life at random time and its applications in a repairable system. Acta Mathematicae Applicatae Sinica 16(4), 435–443 (2000) (English Series)

84. Liu, B., Yue, D.Q., Cao, J., Wang, H.Q.: Analysis of two-machine CONWIP system: matrix geometric solution. Systems Science and Mathematical Sciences 13(4), 366–375 (2000) (English Series)

85. Yue, D.Q., Cao, J.: The NBUL class of life distribution and replacement polices comparisons. Naval Research Logistics 48(7), 578–591 (2001)

86. Yue, D.Q., Cao, J.: Some results on successive failure times of a system with minimal instantaneous repairs. Operations Research Letters 29(4), 193–198 (2001)

87. Wang, J., Cao, J., Li, Q.L.: Reliability analysis of the retrial queue with server breakdowns and repair. Queueing Systems 38, 363–380 (2001)

88. Wang, J., Cao, J.: Unreliable production-inventory model with a two-phase Erlangian demand arrival process. Computers and Mathematics with Applications 43(1/2), 1–13 (2002)

89. Wang, J., Cao, J.: The limiting distribution of the residual discrete-time Markov repairable systems. Operations Research Transactions 6(2), 27–35 (2002)

90. Wang, J., Cao, J., Liu, B.: Unreliable production-inventory system with superposition of k Poisson demand arrival processes. Acta Mathematicae Applicatae Sinica 26(1), 1–9 (2003) (in Chinese)

91. Wang, J., Cao, J., Liu, B.: Analysis of production-inventory model with Erlangian demand arrival process. Journal of System Science and Complexity 16(2), 184–190 (2003) (English Series)

92. Yun, X., Cao, J.: Reliability analysis of discrete-time series repairable system under variable environment. Systems Science and Mathematical Sciences 23(2), 242–250 (2003) (in Chinese)

93. Li, Q.L., Cao, J.: Two types of RG-factorizations of quasi-birth-and-death processes and their applications to stochastic integral functionals. Stochastic Models 20(3), 299–340 (2004)

94. Yun, X., Cao, J.: Discrete-time repairable system with variable environment. Systems Science and Mathematical Sciences 26(2), 178–186 (2006) (in Chinese)
95. Li, X.H., Liu, K., Cao, J.: Summary and prospect of age characteristics of life distribution. Operations Research and Management Science 23(1), 1–6 (2014) (in Chinese)

# Author Index

Printed in the United States
By Bookmasters